P9-CKW-795

EXPLORING PHYSICS

EXPLORING PHYSICS

CONCEPTS AND APPLICATIONS

Rogers W. Redding
North Texas State University

with
Stuart Kenter

Wadsworth Publishing Company
Belmont, California
A Division of Wadsworth, Inc.

Physics Editor: Thomas P. Nerney
Production: Del Mar Associates
Designer: John Odam
Copy Editor: May Chapman
Illustration Art Director: Tom Gould
Technical Illustrator: Florence Fujimoto
Illustrator: Mark Zingarelli
Cover Designer: Tom Gould
Signing Representative: Bill Hoffman

Printed in the United States of America

1 2 3 4 5 6 7 8 9 10—88 87 86 85 84

ISBN 0-534-02886-1

Library of Congress Cataloging in Publication Data

Redding, Rogers W.
Exploring physics.

Includes index.
1. Physics. I. Kenter, Stuart. II. Title.
QC23.R296 1984 530 83-21663
ISBN 0-534-02886-1

ACKNOWLEDGMENTS

Photos and Illustrations
Cover photo: Chuck O'Rear; galaxy courtesy of NASA.

Chapter 1
Opposite p. 1: Fred Ward/Black Star; 3 (top left): Joyce R. Wilson/Photo Researchers, Inc.; 3 (top right): Federal Bureau of Investigation; 3 (center and bottom left): NASA; 3 (bottom right): Fermilab; 5 (top left): © 1978 Lynwood M. Chace/Photo Researchers, Inc.; 5 (top right): Chuck O'Rear; 5 (bottom left): National Center for Atmospheric Research/National Science Foundation; 5 (bottom right): courtesy of American Museum of Natural History.

Chapter 2
8: Rhodes/Photophile; 10: The Bettmann Archive, Inc.; 19 and 21: cartoons used by permission of Sidney Harris.

Chapter 3
24: Sam Nakamura; 27: cartoon used by permission of Sidney Harris.

Chapter 4
34: NASA; 36: The Bettmann Archive, Inc.; 39: NASA; 40 (left): courtesy of the Atlantic Richfield Co.; 40 (right): Editorial Photocolor Archives; 43: NASA; 45: United Press International Photo.

Chapter 5
50: Jerry Irwin/Black Star; 52: *The Ancient of Days* by William Blake, Whitworth Art Gallery, University of Manchester; 56 and 59: The Bettmann Archive, Inc.; 60: cartoon used by permission of Sidney Harris; 62: NASA; 68 (Figure 5.18): drawn from *The First Three Minutes: A Modern View of the Origin of the Universe,* by Steven Weinberg. Copyright © 1977 by Steven Weinberg. By permission of Basic Books, Inc., Publishers, New York.

Chapter 6
72: © Gerry Cranham/Rapho/Photo Researchers, Inc.; 76: Federal Aviation Administration; 83: Federal Bureau of Investigation.

Chapter 7
86: Tommy Wadelton/Black Star; 90: courtesy of Yamaha.

Chapter 8
102: © Stuart Cohen 1978/Stock, Boston; 105: Einstein, courtesy of the California Institute of Technology Archives.

Chapter 9
116: NASA; 119: *The Persistence of Memory* by Salvadore Dali, Collection, The Museum of Modern Art, New York, given anonymously; 122: *Print Gallery* by M. C. Escher, Collection Haags Gemeentemuseum, The Hague; 126 (Figures 9.6 and 9.7): redrawn from *Physics for Poets* by Robert Marsh. Used with the permission of McGraw-Hill Book Company.

Chapter 10
134: Ginn/Photophile; 141: United Press International Photo; 148: Allen Rokach/The New York Botanical Garden, Bronx, New York; 150: NASA.

Chapter 11
154: The Atlantic Richfield Co.; 156: Tracy/ Photophile; 161: cartoon used by permission of Sidney Harris; 163: courtesy Department of Energy; 165 and 166: Tracy/Photophile; 167 (top): courtesy Department of Energy; 167 (bottom): United Press International Photo.

Chapter 12
170: Menzie/Photophile; 172 and 173: U.S. Department of the Interior, Geological Survey; 176 (Figure 12.5): based on Figure 8.3 in *Physics for the Life Sciences* by Alan Cromer. Used by permission of McGraw-Hill Book Company; 189: NASA; 191: © 1976 Seraillier/Rapho/ Photo Researchers, Inc.; 193: © Tom McHugh, 1979/Rapho/ Photo Researchers, Inc.

Chapter 13
198: Tracy/Photophile; 202 (left): courtesy General Electric Research and Development Center; 202 (right): courtesy American Iron & Steel Institute; 203 and 219: cartoon used by permission of Sidney Harris; 217: courtesy Department of Energy.

Chapter 14
222: © 1981 James Natchwey/Black Star; 227: Stirling engine redrawn from *Popular Science,* January 1982, p. 40. Used by permission. © 1982, Times Mirror Magazines, Inc.; 228: cartoon used by permission of Sidney Harris; 234 (Figures 14.6 and 14.7): William Call.

Chapter 15
238: Menzie/Photophile; 241 (left): Soundcraft Series 2400, BBC Television, Glasgow, Scotland; 241 (right): © 1983 Robert Winner, all rights reserved; 245 (left): The Bettmann Archive, Inc.; 245 (right): National Center for Atmospheric Research; 247 (part b): adapted from "Electrical Location by Fishes" by H. W. Lissman, *Scientific American,* March 1963; 253: courtesy Department of Energy; 258: courtesy Medtronic, Inc., Minneapolis.

Chapter 16
266: United Press International Photo; 271 (top): courtesy Education Development Center, Inc., Newton, MA; 271 (bottom): courtesy General Electric Research Development Center; 272 (Figure 16.9): from Dull, Metcalfe, and Brooks, *Modern Physics.* New York: Henry Holt and Co., 1955. Used by permission of Holt, Rinehart and Winston; 275 (Figure 16.15): redrawn from *The Way Things Work,* copyright © 1967 by George Allen & Unwin. Used by permission of Simon & Schuster, a Division of Gulf & Western Corporation.

Chapter 17
284: Bolster/Photophile; 286: Ben F. Laposky/ Electronic Abstraction; 290 (Figure 17.4) and 293 (Figures 17.8 and 17.9): courtesy Education Development Center, Inc., Newton, MA; 294: courtesy Federal Communications Commission; 296 (Figure 17.11): from *The Project Physics Course,* © 1970 Holt, Rinehart and Winston. Used by permission; 297 (Figure 17.13): from Albert Baez, *The New College Physics.* San Francisco: W. H. Freeman, 1967. Used by permission of the author; 298: United Press International Photo.

Chapter 18
302: © Eric Neurath, all rights reserved/ Stock, Boston; 304: Robert Phillips, *Discover* Magazine, © 1982 Time Inc.; 306: © Flip Schulke/Black Star.

Chapter 19
324: © F. B. Grunzweig/Photo Researchers,

Inc.; 326: Pennsylvania State University, Department of Public Information; inset drawing by Julian D. Maynard, Department of Physics, Pennsylvania State University; 330 (top left): courtesy Peter Marler; 330 (bottom left): courtesy Roger Paine; 330 (right): © 1983 Roger Ressmeyer.

Chapter 20
332: Menzie/Photophile; 337 (Figure 20.3): redrawn from Douglas Giancoli, *The Idea of Physics*, 2nd ed., used by permission of Harcourt Brace Jovanovich; 340 (Figures 20.4 and 20.6) and 341 (Figure 20.7): courtesy Education Development Center, Inc., Newton, MA; 340 (Figure 20.5): from *Atlas of Optical Phenomena* by M. Cagnet, M. Francon, and J. C. Thrierr, plate 34, Springer-Verlag, Berlin, 1962; 344 (Figure 20.13): William Call; 347: Richard Weymouth Brooks/Photo Researchers; 350: courtesy William H. Smith and Dina F. Mandoli, Stanford University Medical Center; 354: courtesy MMT Observatory.

Chapter 21
362: M. Blank/Photophile; 364 and 367: NASA; 370: redrawn from "Dialects in the Language of Bees" by Karl von Frisch, *Scientific American*, August 1962; 374: courtesy Museum of the American Indian, Heye Foundation, New York; 379: © Bruce Roberts/Rapho/Photo Researchers, Inc.

Chapter 22
382: photo by Elizabeth Goldring; image created by Otto Piene and Paul Earls, from "An Earthly Furnace Fueled by Fusion Nears a Crucial Test" by Phillip Boffey, *Smithsonian*, December 1980, pp. 129–137; 397 (top): courtesy Goodyear Tire & Rubber Co.; 397 (bottom): courtesy *California Air Resources Board Bulletin*.

Chapter 23
406: Ben F. Laposky/Electronic Abstraction.

Chapter 24
430: Tracy/Photophile; 439: NASA; 441 (center): courtesy Bell Labs; 441 (bottom): courtesy Britt Corporation.

Chapter 25
444: Fermilab; 456 (Figure 25.6): from George Gamow and John M. Cleveland, *Physics: Foundations and Frontiers*, 3rd ed., p. 483. © 1976. Reprinted by permission of Prentice-Hall, Inc., Englewood Cliffs, NJ; 463 (top): courtesy Radiocarbon Laboratory, University of California, Riverside; 463 (bottom): courtesy Michael L. Goris, Stanford University School of Medicine; 471: courtesy Brookhaven National Laboratory.

Chapter 26
478: Michael Freeman; 472 (Figure 26.1): courtesy Lawrence Radiation Laboratory, University of California, Berkeley; 480: Fermilab; 481: Brookhaven National Laboratory and New York University Medical Center; 482: Fermilab; 483: cartoon used by permission of Sidney Harris.

Text Credits
124–125: quote from pages 46–47 from *The ABC of Relativity* by Bertrand Russell. Copyright 1925 by Harper & Row, Publishers, Inc. Renewed 1953 by Bertrand Russell. Reprinted by permission of Harper & Row, Publishers, Inc.; 482: poem by Harold P. Furth, "Perils of Modern Living," reprinted by permission; © 1956 The New Yorker Magazine, Inc.

For Jennie and for Jeff and Jon

PREFACE

IN THE BEST OF ALL POSSIBLE WORLDS, students would master mathematics and the sciences to a degree necessary for them to use the laws of nature much in the same way as physicists. Worthy as that goal is, it is unrealistic. The concepts and applications of physics, however, are important for all students to comprehend—especially for those students who have chosen the arts, humanities, or social sciences as their majors. To prepare to live in a society that is becoming increasingly and irreversibly technological, it is necessary for them to understand the difference between scientific sense and nonsense and to appreciate the nature of physical laws: what they are and what they are not; what they can tell us and what they cannot; and something about how they are formulated.

This book, intended for a two-semester or three-quarter course sequence, is directed at nonscience majors, who probably will not use physics in their professional careers as would a research physicist or a design engineer. It is not designed to teach the skills of physics—important as they are—but to convey the considerable bearing physical laws have on common experience. We emphasize the *applications* of physics, and due to the excellent work of Stuart Kenter, many examples of those applications in the trades, professions, and industry are provided. In presenting this approach, we have at least two purposes in mind: that students are able to evaluate the evidence of their own experience, through an understanding of physical laws, and that they see the pervasiveness of physics in virtually every aspect of technological society.

We have also tried to include the human side of physics without making its historical development a theme of the book. Historical references and anecdotes about people who have brought physics to its present state are sprinkled throughout the text. Our aim is to show that the doing of physics is an activity carried out by real people who are no less flesh and blood than politicians, rock stars, or the folks next door.

For books of this kind, the question of mathematics—how much and at what level it should be cast—is always an important point. I have tried to keep the mathematics as straightforward as possible: There are essentially no algebraic derivations (except in the supplements at the end of the book); equations, in most cases, are accompanied by a sample calculation; and exercises at the end of each chapter contain simple numerical problems to be solved as aids to understanding the subject matter.

The arrangement of topics in this book is fairly traditional, except that special relativity is treated earlier than in most texts of this genre. There is also a fairly extensive discussion of fundamental particles. Supplementary material is placed at the end of the book, allowing instructors to treat some topics in more depth. Learning Checks, questions that allow the student to pause and reflect on the material just read, appear at the end of chapter sections. Exercises at the end of each chapter are of three kinds: those that simply require recalling the material; those that require some analysis or application; and those that call for numerical calculations.

My several years of teaching Physics 131 and 132 at North Texas State to students with quizzical looks and demands for clearer explanations provided great impetus for this work, and many people have contributed to its completion. I am indebted to my colleagues on the faculty for countless conversations over the years that forced me to examine my own understanding of physics. Marshall Aronson at Wadsworth served as editor during a portion of the book's development and helped give it the final shape. Tom Nerney, who inherited the project midstream at Wadsworth, has been a constant source of encouragement and creativity. May Chapman's careful editing and Nancy Sjoberg's skillful management of the developmental work are significant contributions. The design and art direction talents of John Odam and Tom Gould have created a book that is both visually engaging and a fitting reflection of the text.

The physicists who reviewed the manuscript provided useful criticisms and suggested important changes that I have taken to heart. I am grateful to the following reviewers for their comments and suggestions: Rexford E. Adelberger, Guilford College; Joseph F. Aschner, City University of New York; Donald E. Beaty, College of San Mateo; Ronald A. Brown, State University of New York, Oswego; Alex F. Burr, New Mexico State University; Russell N. Coverdale, Purdue University; Roger B. Creel, University of Akron; Terry T. Crow, Mississippi State University; Michael DeGregorio, College of San Mateo; Robert O. Garrett, Beloit College; Patrick Hamill, San Jose State University; Arthur S. Hobson, University of Arkansas; Larry D. Johnson, Northeast Louisiana University; Roger I. Johnson, North Hennepin Community College; Andrew C. Kowalik, Nassau Community College; James Merkel, University of Wisconsin, Eau Claire; Alan Keith Miller, Pasadena City College;

J. Ronald Mowery, Harrisburg Area Community College; William J. Mullin, University of Massachusetts; Richard T. Obermyer, Pennsylvania State University, McKeesport; Stanley J. Shepherd, Pennsylvania State University; Ronald E. Stoner, Bowling Green State University; Wayne W. Sukow, University of Wisconsin, River Falls; Jack M. Wilson, Sam Houston State University; Francis J. Wunderlich, Villanova University; John S. Zetts, University of Pittsburgh, Johnstown.

Finally, I am greatly indebted to Bill Hoffman at Wadsworth. It was he who first planted the idea. He prodded and cajoled, advised and criticized and, often over cheeseburgers at the cafe in Ponder, Texas, provided the encouragement and enthusiasm that kept me writing. I am forever grateful.

Rogers Redding
Denton, Texas
September 1983

CONTENTS

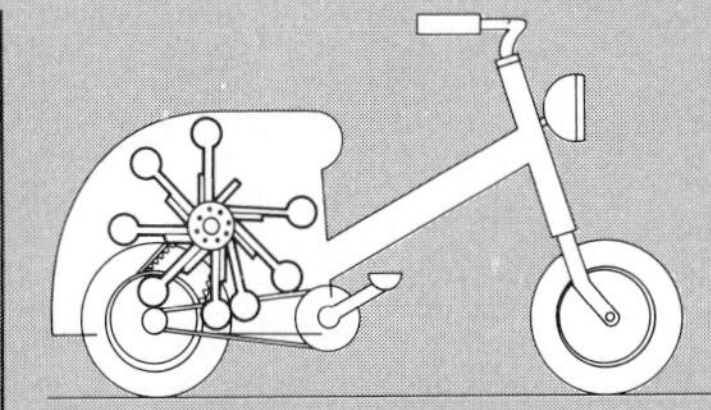

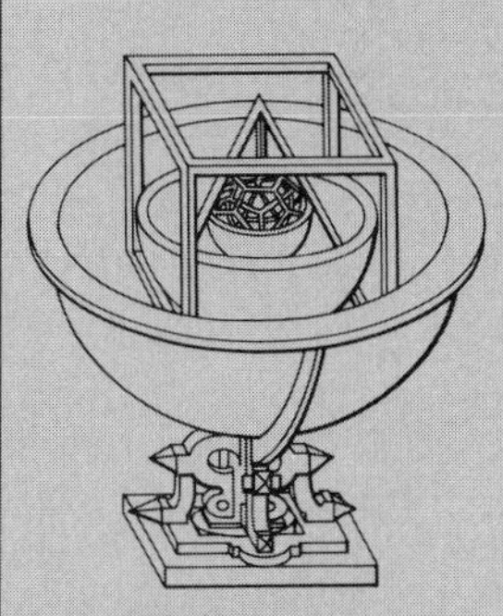

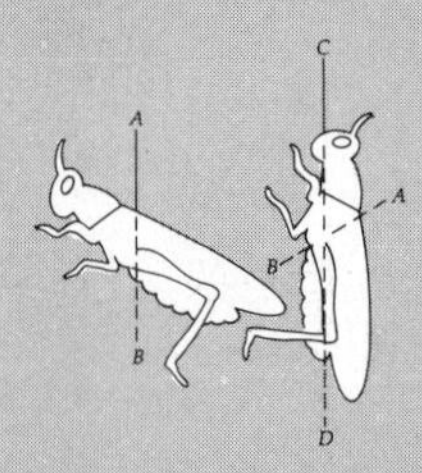

C
A
A
B
B
D

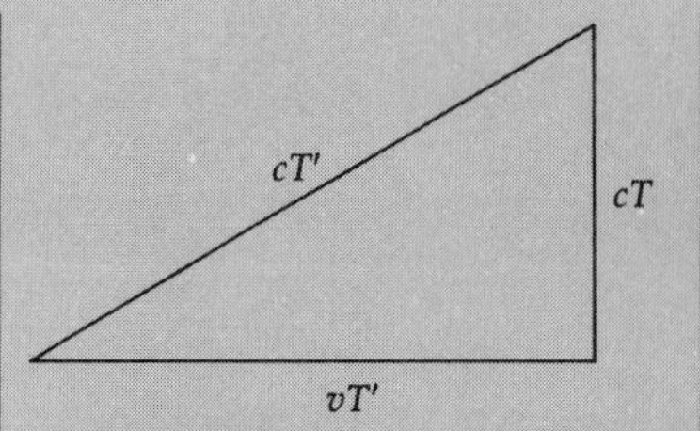

cT'
cT
vT'

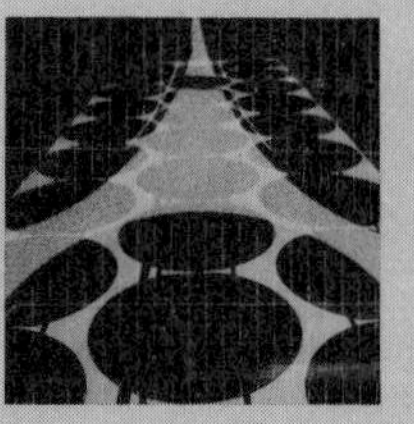

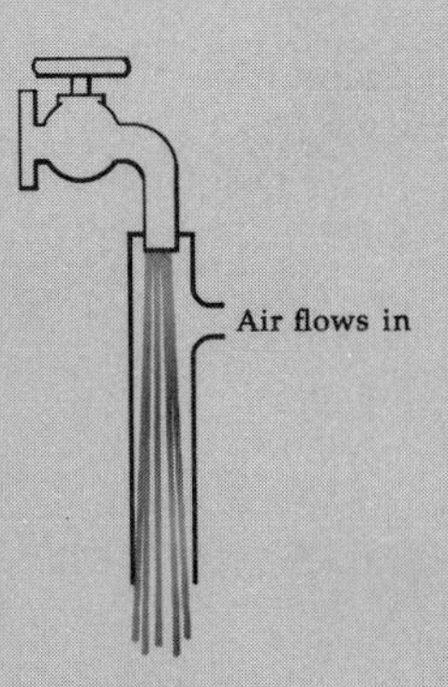

Air flows in

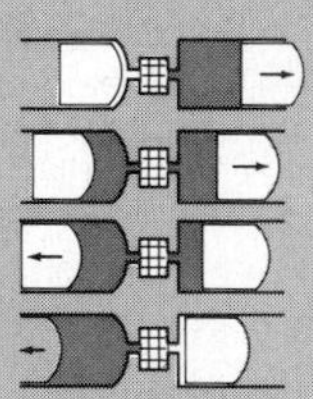

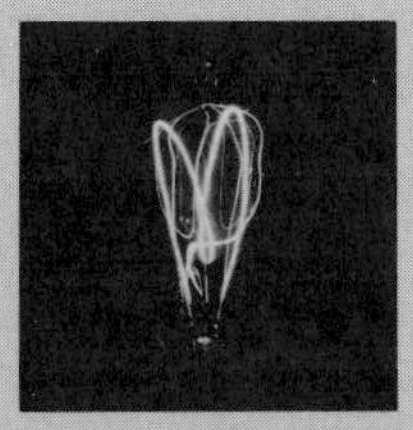

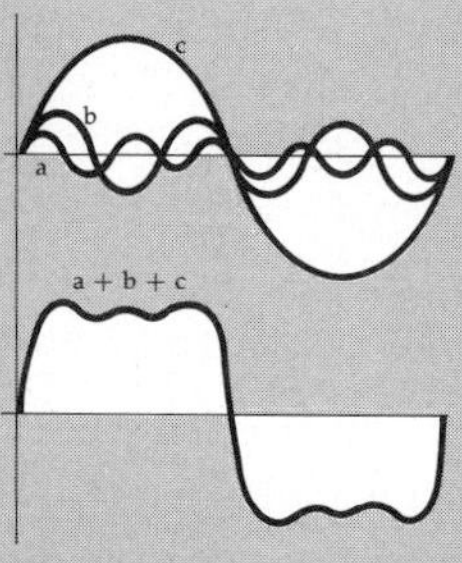
c
b
a
a + b + c

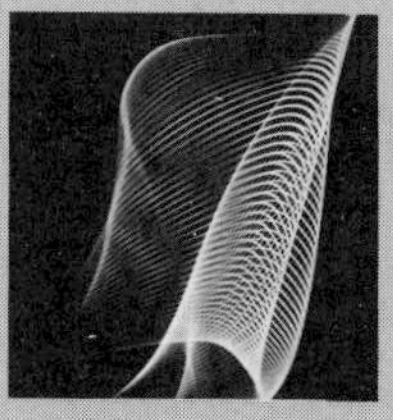

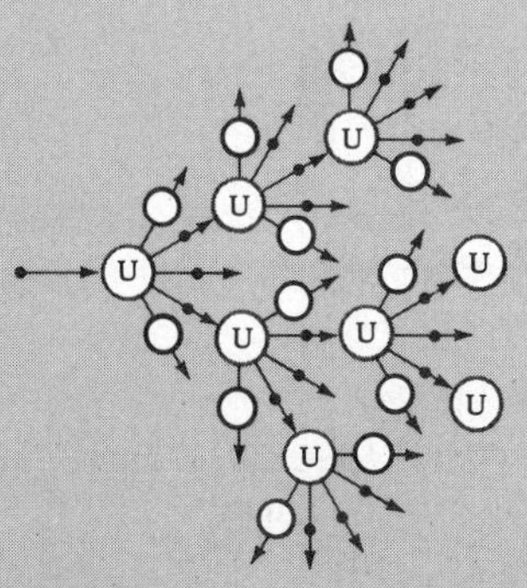
U
U
U
U
U
U
U
U
U

Physical laws give shape to the universe, from the expansive orbits of stars and planets to the intricate formations of shells.

1

THE WORLD OF PHYSICS

The word *physics* comes from the Greek *physikos,* meaning "nature." The world of physics, then, is vast. It includes the environment around us—physical things that we can see and touch and feel, as well as things that our senses don't detect at all. It also covers nature at levels of experience quite different from what we encounter in our daily lives. Physics encompasses the planets, stars, and galaxies of space, as well as the unseen submicroscopic structure of atoms and molecules.

In this course, we will survey physics, discovering a picture of the physical world as modern science now understands it. We will analyze familiar happenings in new ways, exposing you to some new ideas. One of the purposes of our study is to help you look at and appreciate your physical environment in ways you may never have considered before. In so doing, we will show how the concepts and principles of physics appear in many other branches of knowledge and how they apply in the work of many professions and trades.

Put quite simply, the study of physics is the study of nature. Nature appears to us in widely assorted forms and circumstances. Everywhere in the universe, there is a tremendous variety of things going on. Imagine yourself at a seashore, looking about, taking in all you survey. You experience the sea and the sand, the wind and the salt spray, the sun and the rocks, the waves crashing against the shore. Present are motion and stillness, turmoil and quiet. Your curiosity creates questions. Are the waves on the sea related to the wind in your face? Is the blue of the sky like the blue of the ocean? What about the light from the sun and the light reflected from the shore? Are the grains of sand really tiny rocks? Will you understand the sand if you understand the rocks? You begin to look for *similarities* in the events or objects, for *relationships* among the types of motion, for *interactions* among the various bits of

Figure 1-1 (*opposite page*). Physics is at work everywhere — in what you can touch and see, in the far-flung universe, in the atom. Note the similarities of the spiral pattern in these quite different things. Spiraling is a prevalent form of motion, and motion in general is a major concern of physics. (a) Industrial drilling operation. (b) Fingerprints. (c) Satellite picture of a hurricane. (d) A spiral galaxy. (e) An electron track in a bubble chamber.

nature. You try to see beyond the dissimilarities, to find shared likenesses in objects and events that at first glance seem to have nothing in common. (See Figure 1-1.)

The work of physicists involves just such a process of reduction. Physicists examine the world about them and carefully observe what takes place. They speculate about relationships between events and try to draw analogies between phenomena. Designing experiments to test new ideas is a very important part of the process. Eventually, physicists reach conclusions that can lead to a better understanding of nature and, inevitably, to more probing questions.

What do we mean when we say we "understand" something? One answer is this: we understand something if we know the *rules of the game.* In the case of physics and the other sciences, these rules are called the *laws of nature.* Calling them laws can be a bit misleading, however; it gives the impression of commandments carved in stone tablets, waiting to be discovered by some enterprising explorer. This is not the case. The rules of the nature game are more invented than discovered. They arise out of our search for similarities and relationships. They express our own *perception* of nature more than intrinsic truth about nature itself. The rules represent an attempt to simplify the picture of the universe by emphasizing the similarities rather than the differences among events. Be aware of this goal as you begin your study of physics. You will find that a fairly small number of rules account for the infinite variety of phenomena in your world.

An example can help clarify what we mean about the rules of the game. Soccer, the most popular sport in the world, is only recently becoming familiar to the average American. As you watch many soccer games, you gradually begin to figure out some of the rules. The most obvious one is that no player except the goalkeeper is allowed to touch the ball with the hands. A slightly more sophisticated rule, somewhat more difficult to see, concerns "offside." But soon you learn to recognize this one, too. And so it goes. The more you watch and become familiar with the game, the more you gradually come to know the rules. We may say that you have begun to *understand* the game.

So it is with physics. Many of the rules we shall study, such as the laws of gravity and the conservation of energy, were discovered by careful observation followed by deductive reasoning and experiments to check the proposed rule.

Occasionally, something happens in soccer that you don't quite follow. For example, if a defensive player touches the ball with the hand while inside the penalty box, the offensive team is allowed a free shot from a fixed distance in front of the goal, with only the goalkeeper allowed to defend. This rarely happens. You may have to watch many games before you see it

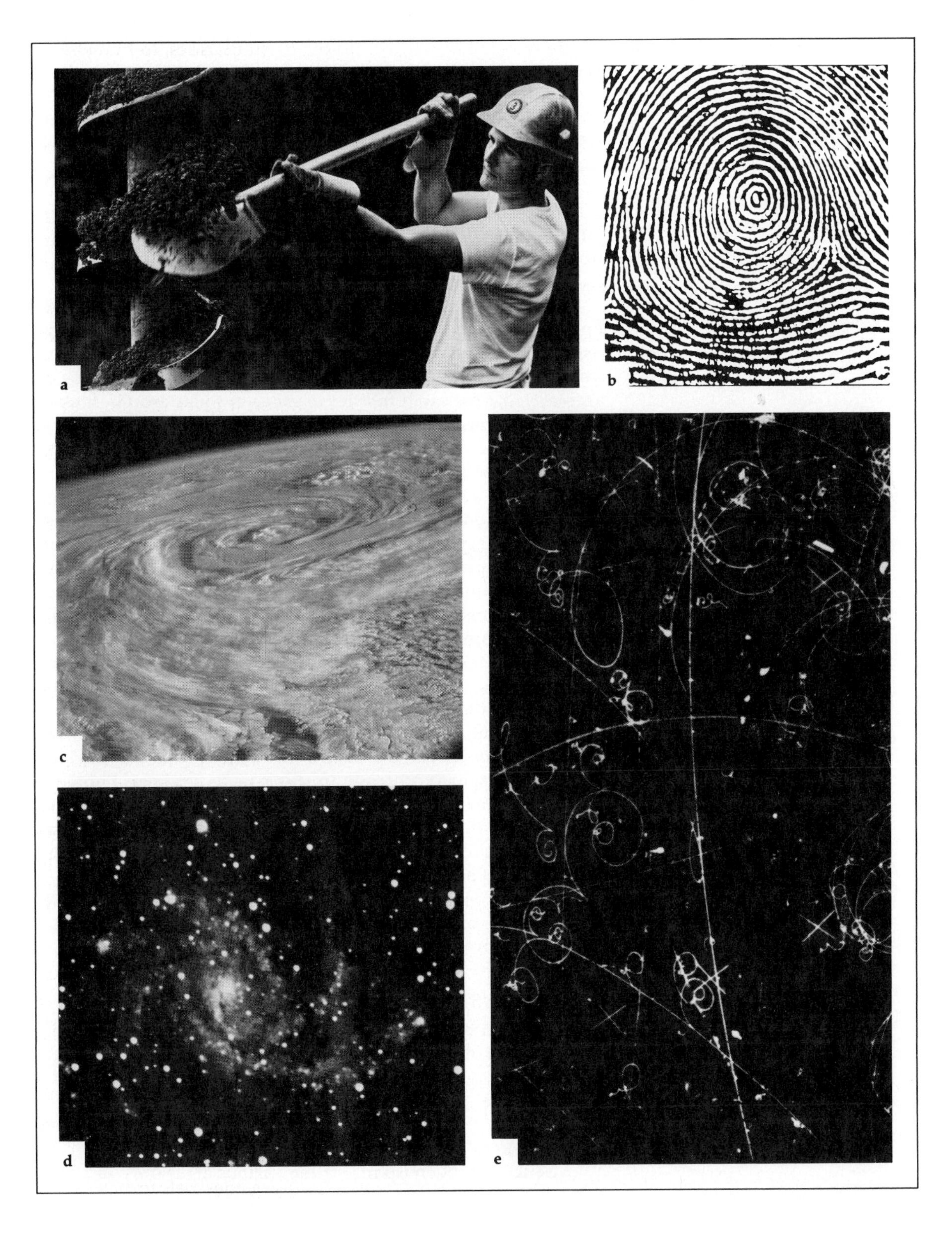
a
b
c
d
e

even once. But when you do see it, you realize that you didn't know all the rules.

The same thing happens in science. For instance, the development of the modern theory called *quantum mechanics* began just over fifty years ago when physicists found that electrons inside atoms and molecules didn't obey the rules of the game known at that time. So a whole new set of rules began to be understood. This process continues today. We still don't know all the rules, especially as they relate to what goes on inside the nucleus of the atom. There is much about nature yet to be learned.

Notice that understanding the rules doesn't mean that you know *why* things happen. In soccer, it isn't always clear why a team uses a particular offensive alignment. Similarly, in physics we can't always determine causes. Everyone knows, for example, that a rock will fall to the ground if you drop it. It obeys the rule of gravity, one that is well understood. We can calculate how long it will take the rock to fall a given distance and how fast it will be traveling when it hits the ground. But nobody knows *why* the rock falls. Physics doesn't answer these "why" questions, the kind that inquire about ultimate causes.

Another aspect of physics to appreciate as you begin your study is this: physics is a human endeavor. Unraveling the mysteries of the universe is not the well-ordered, coldly analytical process it sometimes appears to be. Physicists are people. They have good points, bad points, hang-ups, and concerns, just like everybody else. They make the same false starts, have the same flashes of brilliance, and encounter the same sort of personal problems that you do. They also get excited about gaining some understanding of nature. We hope that you, too, will get this kind of excitement and pleasure in your study of physics.

Much of what you study in this course has been known to science for many years. However, as mentioned earlier, we don't know all the rules of the game yet. Some extremely important discoveries have been made quite recently. Two that come to mind are the structure of DNA (the genetic material of life) and the principles that led to the invention of the laser. This business of understanding nature is very much alive. It is an ongoing process, a drama that continues.

PHYSICS AND THE OTHER SCIENCES

The word *science* covers a broad spectrum of activity that includes such areas as chemistry, physics, psychology, medicine, biology, geology, and anthropology, as well as interdisciplinary areas such as biochemistry and geophysics. (See Figure 1–2.) Physics, though, is a rather special case: not only is it *one* of the sciences, it is the most *basic* of them.

Science is broken up into these various areas only as a prac-

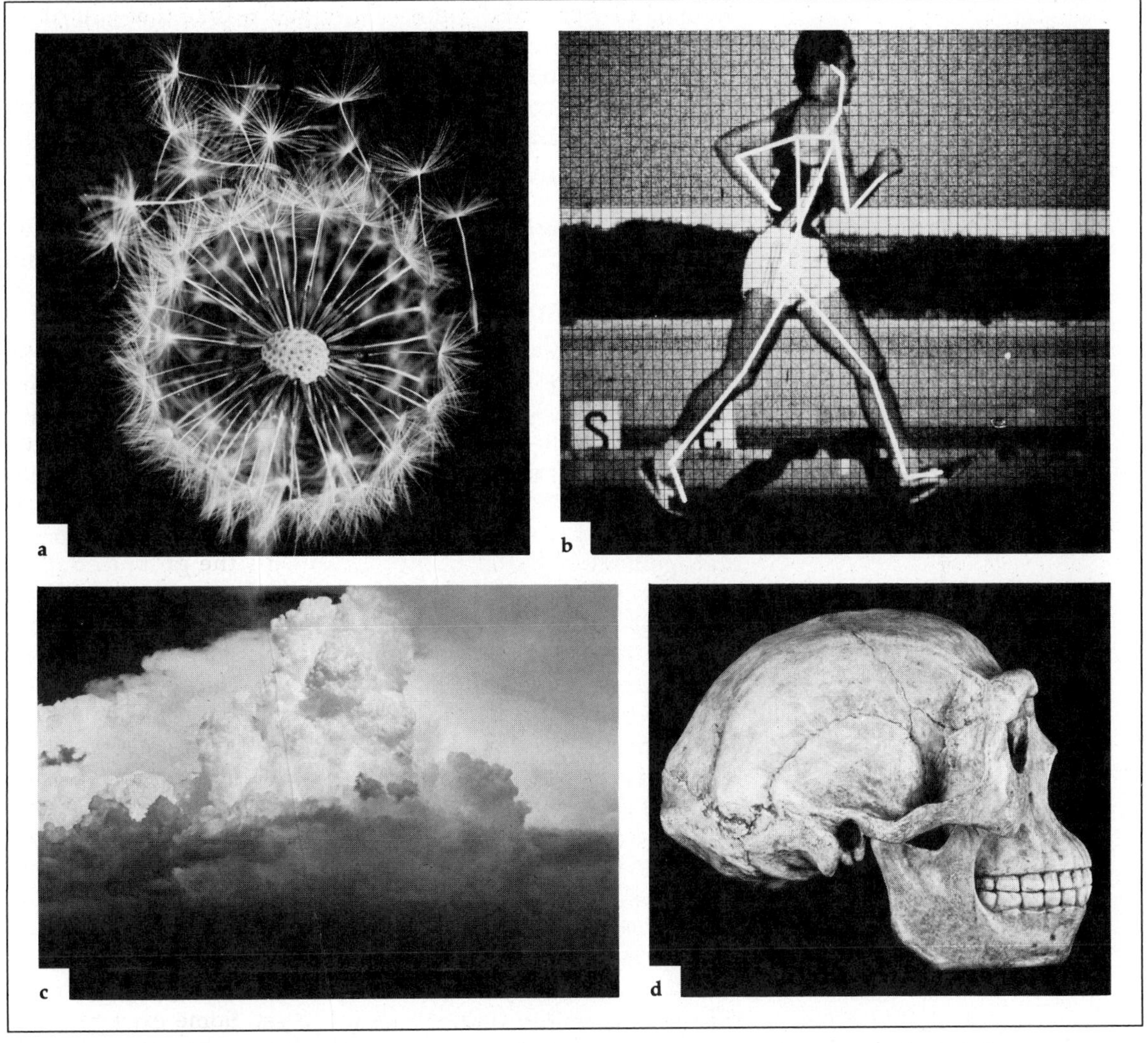

tical matter and a convenience to the interests of different scientists. The division of labor is something superimposed by scientists on nature. Chemistry, for instance, deals mostly with the structure of molecules and their reactions in forming new substances. In principle, all of chemistry is contained in the physics of molecules and their behavior. In practice, however, many molecules and their reactions are too complicated to be readily understood or described in terms of the fundamental laws of nature. Thus, as a practical matter people have needed to specialize in devising theories and experimental techniques for studying chemical systems. It is important, therefore, for chemists to know physics in order to understand nature's rules as they apply to the atoms and molecules they study. Similarly, because biologists are interested in examining processes and

Figure 1-2. Physics plays a part in many scientific disciplines. (a) Botany – plant scientists view the laws of motion at work in the reproductive activities of many plants. (b) Medicine – research physicians study the motion of the human body. (c) Meteorology – weather technicians must know cloud physics to understand the forces that shape clouds. (d) Anthropology – physical anthropologists can infer facts about the diet of prehistoric humans from the forces involved in the act of chewing, which caused some fossil teeth to be shovel-shaped.

systems of plant and animal life, they must also understand basic physics. Similar examples can be drawn from other sciences. Throughout this text, we will point to many illustrations that relate physics to other areas.

It is important to keep in mind that nature does not distinguish between physics and chemistry or between geology and biology. These are artificial distinctions we have devised to make it easier to learn the rules of the game.

PHYSICS AND THE NONPHYSICIST

Most students taking the course for which this textbook is intended probably will never *do* physics in the sense of making direct use of it. So the question sometimes arises: Why should these people take such a course at all? What rationale for understanding physics can be given for those who don't intend to become scientists? Will a knowledge of physics make a journalism major a better journalist or give a business major better management skills?

First, all of us, regardless of vocation, live in a highly technological society, with all its positive and negative traits. As you read these words there is most likely something within your reach that exists because of science or its applications. The material and machinery that science gives us have a direct effect on all our lives—our ability to travel, our standard of living, our health, the way we communicate, the very way we think.

Second, you should understand that, potent as it is, science is not all-powerful. It is not a panacea. It has limitations. We hope that in this course you will gain some appreciation of what science *can* do and what it *cannot* do. Often you will hear it said, "If they can land a man on the moon, they ought to be able to . . . ," when, in fact, whatever it is that fills in the blank may be impossible simply because of the laws of nature. The possibilities and limits of physics, and science in general, should be understood at some level by everyone.

As science and technology become more and more a part of everyday life, it becomes increasingly essential that the populace in general, and educated people in particular, have some basic understanding of things scientific. For example, the American economy is intimately meshed with the economies of other nations primarily because of advances in science. A major portion of our tax dollar goes toward maintaining and developing our technologically based economy. Thus, an electorate informed about science is better able to make wise decisions about the future of society.

Finally, individuals equipped with a fundamental knowledge of science stand a better chance of making sound decisions about matters that affect their work. The concepts and principles of physics have roles to play in a great variety of occupations. We will cite many examples throughout this book. If our journalist and business person gain from exposure to physics nothing more than the ability to ask the right questions, they will be well ahead of the game.

Twentieth-century technology has allowed man to indulge in his age-old fascination with speed and motion — in both progressive and unusual ways.

2

SPEED, VELOCITY, AND ACCELERATION

As we look around, what most attracts our attention are the things that move. Motion enlivens the world. A movie is more exciting than a snapshot. A lone walker in a stationary crowd instantly catches our eye. Movement is an important part of our common experience.

People have wanted to know about all kinds of motion since the dawn of human time. Throwing a spear accurately along a certain path, for example, was a skill that primitive people had to master for hunting and fighting. Early astronomers studying the stars and planets identified *patterns* of motion. Today, we notice patterns in such varied phenomena as biological rhythms and the rise and fall of stock prices.

A large part of physics deals with the study of motion. As we progress, we will see that when we understand various types of motion, we gain important insight into some fundamental features of the universe. We can then tackle such questions as: What makes the earth travel as it does around the sun? Can we predict the path of a tossed football? What determines the motion of a satellite returning from space and heading toward splashdown?

In this chapter, we will briefly look at how ideas of motion have changed over the centuries. Then we will learn the modern way of describing motion.

ARISTOTLE AND GALILEO ON THE STUDY OF MOTION

If we contrast the approaches of two famous intellectual giants, Aristotle and Galileo, we can begin to see how our modern view of motion formed.

The ancient Greek philosopher Aristotle (384–322 B.C.) was a tutor and protegé of Alexander the Great. He joined the phil-

"BUT WE JUST DON'T HAVE THE TECHNOLOGY TO CARRY IT OUT."

osophical school of Plato at the age of seventeen and remained an ardent student until Plato died in 347 B.C. Aristotle later founded his own school in Athens. There, he produced writings on logic, political science, psychology, and various biological problems. His ideas on motion sound strange to modern ears, but they are important because they represent the scientific world view of that time. The sheer weight of Aristotle's great reputation caused people to accept his thoughts on motion as correct for hundreds of years, until Galileo presented a better approach in the seventeenth century. Let's examine the ancient Greek concept of motion.

The thinkers of Aristotle's time considered the universe to be made of four "elements": earth, air, fire, and water, each having its natural place in the scheme of things. (See Figure 2-1.) All objects were thought to be some combination of these four elements, and it was believed that the *natural motion* of the objects would return them to the place where nature intended them to be. Thus, a stone falls to the ground because it is of earth and strives to be there. Rivers flow to the sea, their natural home. A feather, mostly earth but part air, floats gently to the ground, as earth wins the tug-of-war with air. Natural motion, to the ancient Greeks, was nature's way of restoring

Figure 2-1. A later European personification of the Greek view that the universe is composed of four elements: earth (terra), water (aqua), fire (ignis), and air (aer).

things to their place according to the elements of which they were made.

From this natural motion Aristotle distinguished *violent motion*, that is, motion imparted to an object by a push or pull. A stone set rolling on the ground—an initial violent motion—eventually stops by virtue of its natural motion. To keep the stone rolling, an unnatural state, one had to keep on causing violent motion by pushing. Aristotle accounted for the curved path of an object thrown through the air like this: The initial push causes violent motion, which is subsequently overcome by natural motion to return the object to its natural place, earth, and its natural state, rest.

Perhaps the most important feature of Aristotle's system as it applied to falling bodies was the notion that a heavier object falls to earth faster than a lighter one (presumably striving harder to return to its earth home). In his own words: "The downward movement of a mass of gold or lead, or of any other body endowed with weight, is quicker in proportion to its size." To paraphrase a familiar saying, the bigger they are, the faster they fall.

To Galileo, this was an excellent hypothesis, not because it was right but because it was either very right or very wrong. It was a quantitative statement that could easily be tested. Philosophers may debate violent versus natural motion all day, but the question of whether a 10 pound ball falls faster than a 1 pound ball can be resolved quickly by experiments. According to legend, Galileo did such experiments at the famous Leaning Tower in his birthplace of Pisa. He has described them in conversational form in *Dialogues Concerning the Two New Sciences.* Galileo found that, in fact, both the 10 pound ball and the 1 pound ball, when dropped through the air, reach the ground at essentially the same time. By doing similar experiments in water, he found that an extremely important factor is the medium in which the motion takes place. That is, the medium largely determines how long it takes different objects to fall. Galileo reasoned that "in a medium totally devoid of all resistance" (what we now call a *vacuum*), the time would be completely independent of the size, shape, or any other property of the falling bodies.

Galileo's concept of motion forms the basis of our modern view, which this book presents. It continues to be very successful in describing nature. But we should not conclude that Galileo was "right" and Aristotle "wrong." Both men were describing the world as they understood it at the time; "right" or "wrong" has nothing to do with it. What we do know is that Galileo's ideas about motion are more useful—they happen to explain many more observations in a far simpler manner than do Aristotle's ideas. Aristotelian physics collapsed for this reason, not because of any errors or failures on Aristotle's part.

Galileo at the Leaning Tower?

The Leaning Tower of Pisa is frequently cited as the place where Galileo carried out his famous ball-dropping experiment in about 1590. There is plenty of doubt about whether this was actually the case. Scholars still argue the point. So what? you might ask. What difference does it make whether this is truth or legend? Well, this was a pivotal experiment, a turning point in science, really, and we like to be as accurate as possible about these milestone events. After all, it seems that no one else—in 2000 years—had thought to challenge Aristotle by performing this simple test. In a book dated 1605, however, a man named Simon Stevin asserted that he and his friend, one John Grotius, did the same thing as Galileo "long ago." These men claimed they dropped two lead balls, one weighing ten times the other, from a height of 30 feet onto a plank. They claimed the balls landed in such an even manner that there seemed to be just "one thump." Be that as it may, Galileo got the press. But did he perform at Pisa? "Whatever Galileo did, or failed to do at Pisa," scholar Lane Cooper points out, "in all his extant writings, he never once mentions the leaning tower . . ."

Source: Lane Cooper, *Aristotle, Galileo, and the Tower of Pisa.* Ithaca, N.Y.: Cornell University Press, 1935.

Learning Checks

1. How might an Aristotelian explain the following observations?
   a. A helium-filled balloon rises, but an air-filled balloon falls.
   b. A balloon that rises when filled with warm air falls when it cools off.
2. Answer Question 1 from Galileo's viewpoint.

SPEED, VELOCITY, AND ACCELERATION

Galileo's work marked a critical period in the development of *mechanics,* the science of motion. His key contribution was to describe motion in terms of distance and time linked together. Here we will examine three quantities: speed, velocity, and acceleration. Each of these combines distance and time to yield a *rate,* a word meaning "how much in what length of time."

Rates are often used in situations where two quantities change together. For example, the inflation rate tells how prices change as time goes on. When you read in the newspaper that the rate of inflation is 7 percent per year, you know prices are going up so that they will be 7 percent higher after a year's time. Population growth, changes in unemployment, and bank interest are also given as rates. Let us see how the rates describing motion are defined.

Speed

The *speed* at which an object travels simply tells us the distance it goes in some interval of time. We distinguish between average speed and instantaneous speed. *Average speed* is equal to the distance traveled divided by the time required for the trip:

$$\text{average speed} = \frac{\text{distance traveled}}{\text{elapsed time}}$$

A shorthand way of writing this is

$$s = \frac{d}{t}$$

Here, the symbols d and t mean we are considering an interval of distance covered during a certain interval of time. Notice that the distance and the time both change: speed is the rate at which the distance traveled changes as time passes. (See Figure 2–2.)

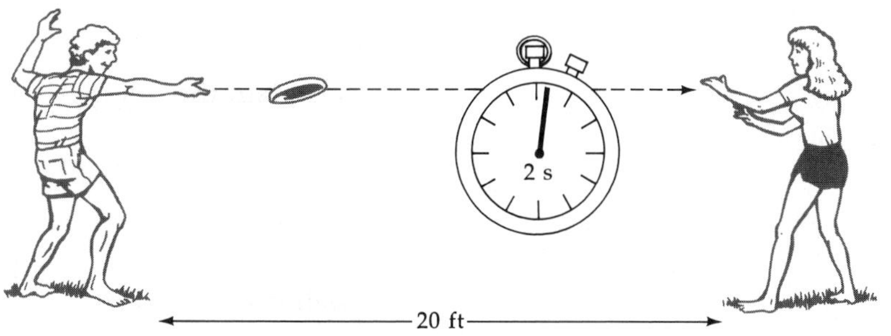

Figure 2-2. The frisbee's speed is 10 ft/s.

Let us look at why we call this the *average* speed. Suppose a sales representative drives the 400 miles from Jackson, Mississippi to Atlanta, Georgia in 8 hours. Then the average speed for the trip is 400 miles divided by 8 hours, or 50 miles per hour (mi/h). Now this does not mean that the speedometer read exactly 50 mi/h during the entire 8 hours. The average speed does not take into account those stretches where the person behind the wheel shot up to 70 mi/h, or the times when, spotting a lurking patrol car, the driver slowed down to 40 mi/h. And it especially does not tell about the rest stops, when the car wasn't moving at all. So the average speed of 50 mi/h simply describes the gross features of the trip and not the details.

To describe this sales route more precisely, we might break up the overall time interval into smaller pieces and use the corresponding distances to compute average speeds for each piece. If we do this for smaller and smaller time intervals, we approach the speed for an instant in time—an infinitely small value of t. We call this the *instantaneous speed.* It would be the number that the speedometer indicates. (The idea of letting t become extremely small but never actually reaching zero is an important part of calculus. For those interested, Supplement 2 discusses this briefly.) For our purposes, however, the average speed will do, since it gets across the idea of speed as the rate at which distance and time change together. Whenever we use the word *speed,* we will always mean average speed.

Velocity

Velocity is a concept closely related to speed. In fact, physicists often slip and use one term for the other. But they are not the same. Velocity is speed with a direction. Suppose that two newspaper carriers on bicycles start their routes at the same place. One rides 10 mi/h to the south and the other 10 mi/h to the west. The two bicycles have the same speed, but obviously something is different because they end up in different places. They have different *velocities* because their directions are different. (See Figure 2-3.) To vary an object's speed, you must make it go either faster or slower. But to alter its velocity, you can change *either* its direction, *or* its speed, or *both.*

Figure 2-3. The speeds are the same, but the velocities are different.

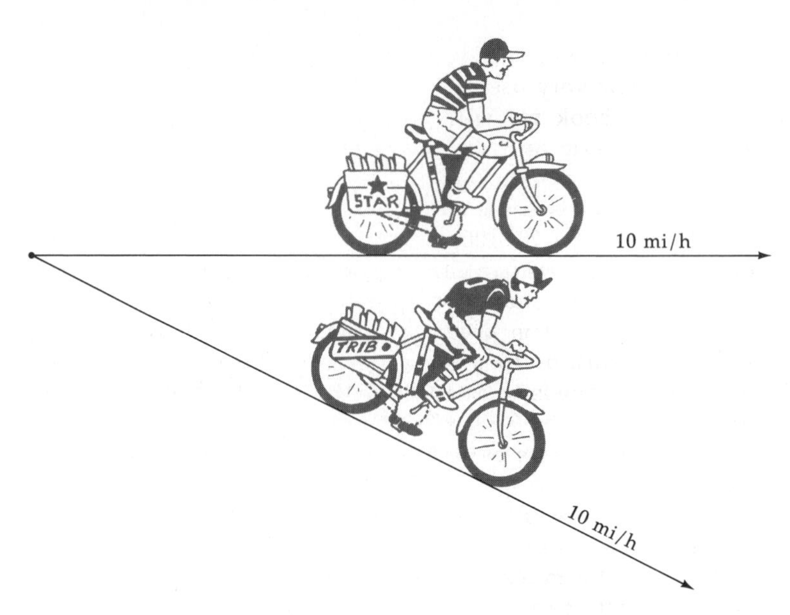

In order to include the direction of motion in our definition of velocity, we must introduce the concept of *displacement,* which means distance traveled in a particular direction. The definition of velocity is then

$$\text{velocity} = \frac{\text{displacement}}{\text{time interval}}$$

or, in physicist's shorthand,

$$\mathbf{v} = \frac{\mathbf{d}}{t}$$

Quantities such as velocity and displacement that have a direction associated with them are called *vectors.* Speed and distance, on the other hand, are independent of direction; we call them *scalars.* It is convenient to write vectors in boldface type, as we have done for **v** and **d**. (Supplement 1 discusses vectors and how to handle them.)

From this definition, we can see that an object moving with constant (or uniform) velocity is moving at constant speed in a straight line, because any deviation from a straight line involves a change in direction. Later on, in studying the laws of motion, we will see a sharp physical distinction between objects moving with constant velocity and those whose velocity is changing.

Note that in discussing speed and velocity we have not specified any particular system of units. That is, should distance be given in meters, miles, inches? Should time be in days, hours, seconds? The choice is mostly a matter of convenience.

We could express speed in furlongs per fortnight, but that would not be very useful.

In this book we will normally use the *metric system*, in which the basic unit of length is the meter (m), about 39.4 inches, or just over a yard. This is a very convenient system, because other distances are obtained by multiplying or dividing the meter by 10, 100, 1000, and so on. Short distances are given in *centimeters* (cm), which is 1/100 of a meter (the prefix *centi-* means one one-hundredth). In this system the *kilometer* (*kilo* means 1000) is used for measuring larger distances. A kilometer (km), or 1000 meters, is about six-tenths of a mile. Automobile speeds in Europe are expressed in kilometers per hour (km/h). We will usually give speeds in meters per second (m/s) or centimeters per second (cm/s). Remember that for velocity we must also specify the direction, such as 30 cm/s west or 25 m/s downward.

When a tennis serve is clocked at 90 mi/h, clearly the ball does not go 90 miles, nor does it travel for a full hour. As a matter of fact, it only goes several feet in a very few seconds. But the 90 mi/h speed expresses the fact that if the ball did continue at this rate for 1 hour, it would go the full 90 miles.

Learning Checks

1. Distinguish between constant speed and constant velocity.
2. If an object is traveling with constant speed, is it necessarily traveling with constant velocity? Explain.
3. If an object is traveling with constant velocity, is it necessarily traveling with constant speed? Explain.
4. Which of the following are proper expressions of velocity: cm/s; ft/h, west; months/s; inches/year; ft/day, upward; m/s, north.

Acceleration

To describe motion, the third concept we will need is *acceleration.* Just as the velocity specifies the rate at which the displacement changes in relation to the time, the acceleration specifies the rate at which the *velocity* changes in time. Suppose we consider some interval of time during which an object's velocity is changing. The *initial velocity* is its velocity at the beginning of the time interval, and the *final velocity* is its velocity at the end. The acceleration is then defined as

$$\text{acceleration} = \frac{\text{final velocity} - \text{initial velocity}}{\text{time interval}}$$

In symbols,

$$\mathbf{a} = \frac{\mathbf{v}_f - \mathbf{v}_i}{t}$$

Note that acceleration is a vector quantity because the direction matters. We indicate this, as before, by writing the symbol for acceleration in boldface type.

Let us first look at the acceleration of an object moving in a straight line. In this case, since there is no change in direction, the acceleration simply reflects a change in how fast the object is going. Imagine that you begin skiing down a slope at 10 km/h and build up speed. After 1 second, your speed is 15 km/h; at the end of the next second it is 20 km/h, and so on. We see that at the end of any second your speed is 5 km/h greater than it was at the end of the previous second. Because your motion is along a straight line, your velocity is also increasing by 5 km/h each second, down the hill. So the rate at which your velocity changes is 5 kilometers per hour per second, which we write as 5 km/h/s. This rate is your acceleration.

Figure 2–4 illustrates a similar situation for a delivery van traveling in a straight line. Here the acceleration is 10 km/h/s, since the van's velocity is increasing by 10 km/h every second.

These two examples illustrate one way in which an object can accelerate—that is, by changing its speed. But acceleration means *any* change in velocity, not just a speed change. That is, since velocity is a vector quantity, any change in the *direction* of motion also is acceleration. For example, suppose you are riding a ferris wheel that goes around at constant speed. Even though you are not speeding up or slowing down, your velocity is changing continuously. At any instant of time, you are traveling in a different direction than before. You are accelerating, even though your speed remains the same. Your velocity changes because your direction of travel is always changing.

This example illustrates the vector properties of acceleration. It isn't just the changing speed that matters—the change in direction is equally important.

An important point about acceleration is that it depends on the *change in velocity,* not the velocity itself. A rocketship that increases its straight-line speed from 25,000 km/h to 25,030 km/h in 10 seconds has the same acceleration as a speedboat that goes from 10 km/h to 40 km/h in 10 seconds. Their velocities are vastly different, but their accelerations are the same. They undergo the same changes in velocity during the same time interval.

We will normally express the numerical value of acceleration in the metric unit of cm/s/s, often written as cm/s^2, and

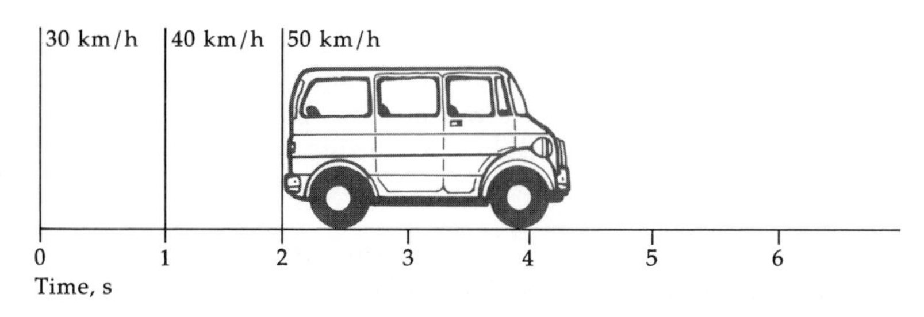

Figure 2-4. The van accelerates at 10 km/h/s.

read "centimeters per second squared." This might sound a little strange, but it is only a shorthand way of saying that an object having an acceleration of 10 cm/s^2 north is changing its velocity by 10 cm/s north every second.

The concept of acceleration can be difficult to grasp because it expresses the rate at which *another* rate—the velocity—changes. For an object to have zero acceleration, its velocity must remain the same, no matter how large or how small it is. Only when there is a change in velocity—either in magnitude, or direction, or both—can there be acceleration. This acceleration tells us how quickly, and in what direction, the velocity is changing.

Negative Acceleration

Often in our study we will run across a term that is used both in physics and in everyday life. Its meaning in physics, however, may be quite different from its meaning in ordinary use. Acceleration is one such term. As we normally use it, the word *acceleration* means speeding up, as in our illustration of skiing down the slopes. But recall that acceleration involves *any change in velocity*. In particular, it can mean slowing down. Let us look at an example.

Suppose that instead of speeding up, the delivery van in Figure 2-4 is gradually coming to a stop. When we first pay attention to it, the van is moving at, say, 50 km/h to the right, and 1 second later it is going 40 km/h to the right. At the end of the next second, its velocity is 30 km/h to the right. This decrease in speed continues until the van stops.

We can see that the van's speed is decreasing by 10 kilometers per hour every second, so it will take 5 seconds for it to come to a halt. Let us calculate the acceleration over this 5 second time interval. The *final* velocity is 0 km/h and the *initial* velocity is 50 km/h to the right. Hence our formula for acceleration gives

$$\mathbf{a} = \frac{(0 \text{ km/h} - 50 \text{ km/h}) \text{ right}}{5 \text{ s}}$$

or

$$\mathbf{a} = -10 \text{ km/h/s to the right}$$

Notice that we get a *negative* number for the acceleration. We can look at the meaning of this minus sign in two ways, both of which say the same thing. First, because the velocity vectors point to the *right*, the negative sign for the acceleration tells us that it is a vector pointing to the *left*. The direction of the acceleration is exactly *opposite* to the direction of the velocity. Second, the minus sign also means that the van is slowing down. You sometimes hear the term *deceleration* used. *Negative acceleration* and *deceleration* are just two different terms for the same concept.

Learning Checks

1. If the motion of an object is described as *uniformly accelerated*, what can you say about the velocity of the object?
2. Distinguish between constant velocity and constant acceleration.
3. Distinguish between constant speed and constant velocity.
4. Why is it incorrect to say, "Acceleration is the velocity divided by the time"?

FREELY FALLING BODIES

Armed with the definitions of velocity and acceleration, let us look again at the motion of freely falling bodies. Using the following familiar example, we can sharpen our understanding of this new term.

By "free fall" we mean that an object falls unhindered by air resistance or any other impediment. Gravity works alone. This is an idealization, of course, unless the object falls in a vacuum or at a place like the moon, for example, where there is no atmosphere. All experiments in physics are like this. We try to eliminate as many factors as possible that might affect the results, and we focus on the particular feature of interest. Free fall is never *exactly* possible on earth—air resistance always interferes. But we can reduce the resistance and get very close to the ideal case.

Suppose that we drop a baseball from the top of the Eiffel Tower and measure its velocity at the end of each second for 5 seconds. We do not give the ball any initial push but just drop it—that is, its initial velocity is zero (we say that the ball is dropped from rest). We would find the velocities very close to those listed in Table 2–1.

Entitled to It

Literature often draws upon fundamental concepts of science and technology to title stories and novels and to make certain points. You can, for example, turn to fiction to discover how writers use the idea of free fall as a metaphor for an emotional experience or a state of being. Two such books are *Free Fall in Crimson*, an adventure by John D. MacDonald, and *Free Fall*, a mainstream novel by William Golding.

Table 2-1. Velocity of the Baseball Dropped from Rest

Elapsed time, s	Velocity, m/s downward
0	0
1	10
2	20
3	30
4	40
5	50

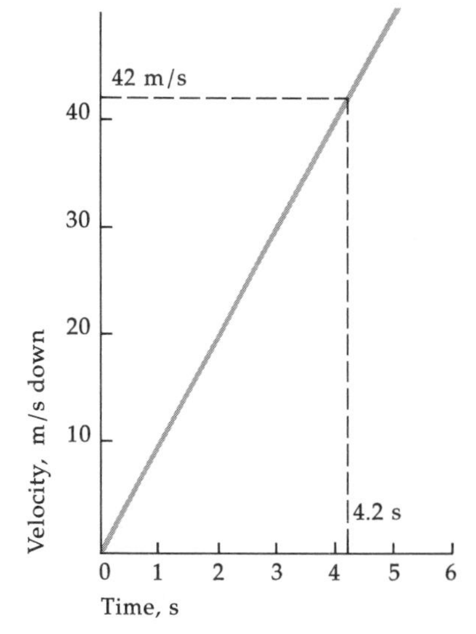

Figure 2-5. Velocity versus time for a freely falling body.

It is helpful to display these numbers in a graph, which we have done in Figure 2-5. We plot the time along the horizontal axis and velocity along the vertical axis. (Supplement 2 discusses graphing techniques in some detail.)

Now let us look at these numbers closely. Note first that the velocity is not constant but is *increasing* in time. Since the velocity is changing, the baseball is accelerating. Furthermore, not only is the velocity increasing, but it is increasing *uniformly.* That is, it is changing in such a way that at the end of each second it is always 10 m/s downward greater than it was the previous second. The velocity is increasing by 10 m/s downward each second, so we may conclude that the falling baseball has a constant acceleration of 10 m/s^2 downward.* Free fall is thus an example of *uniformly accelerated motion.*

Because the acceleration is constant, the points in our graph of velocity versus time lie along a straight line. This line enables us to determine the velocity at some time other than exactly at the end of a second—say, at 4.2 seconds. The dashed lines in Figure 2-5 show how to do it. We find that at the end of 4.2 seconds, the baseball's velocity is 42 m/s downward.

Now suppose that we measure the *distance* that the baseball has fallen at the end of each second. We make another table (Table 2-2) to record the distance (meters) corresponding to the

Table 2-2. Distance of the Baseball for Each Second of Descent

Elapsed time (s)	Distance (m)
0	0
1	5
2	20
3	45
4	80
5	125

*The acceleration of a freely falling body near the earth's surface is closer to 9.8 m/s^2 downward, but our value of 10 m/s^2 downward is good enough for our purposes. It illustrates the point while allowing easy calculations.

(By permission of Johnny Hart and Field Enterprises, Inc.)

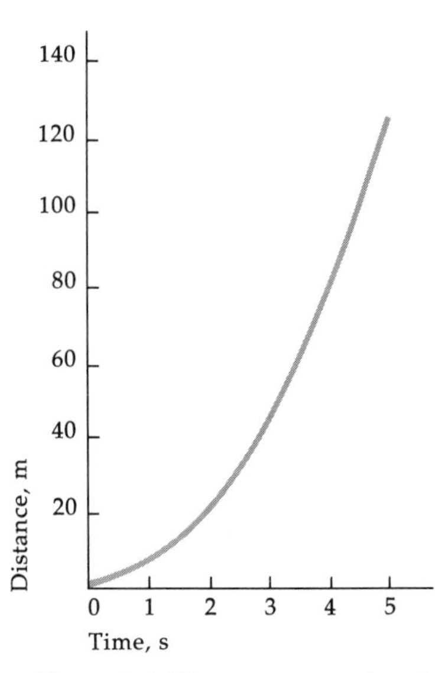

Figure 2-6. Distance versus time for a freely falling body.

elapsed time (seconds). As before, we may display our data in a graph of distance versus time. This is done in Figure 2-6.

Obviously, the distance is changing, but it is not changing *uniformly* with time as did the velocity in our first graph. During the first second, the ball falls 5 meters; during the next second, it falls 15 meters; during the third second, 25 meters, and so on. This simply tells us what we already knew from our analysis of the velocity: namely, that the velocity is not constant but is increasing in time.

Let us summarize our findings. A freely falling body near the earth's surface travels toward the center of the earth with a uniform acceleration of about 10 m/s^2 downward. That is, its velocity is increasing by an amount 10 m/s downward each second, so we may say that the velocity varies *linearly* with time (the velocity-time graph is linear). The distance is related to the time in a different way, since the distance-time graph is *not* a straight line. Our experiment is performed in the absence of air resistance and the ball is dropped from rest.

Suppose we change the circumstances slightly. This time, rather than dropping the ball from rest, we throw it down, giving it an initial velocity of 3 m/s downward. Now our velocity measurements will be those listed in Table 2-3. Notice that even though the velocity values are different from those in Table 2-1, the *changes* in velocity are the same from one second to the next. The baseball is still accelerating at 10 m/s^2 downward. The graph of velocity versus time for these numbers is shown in Figure 2-7.

Figure 2-7. Velocity versus time with initial velocity of 3 m/s down.

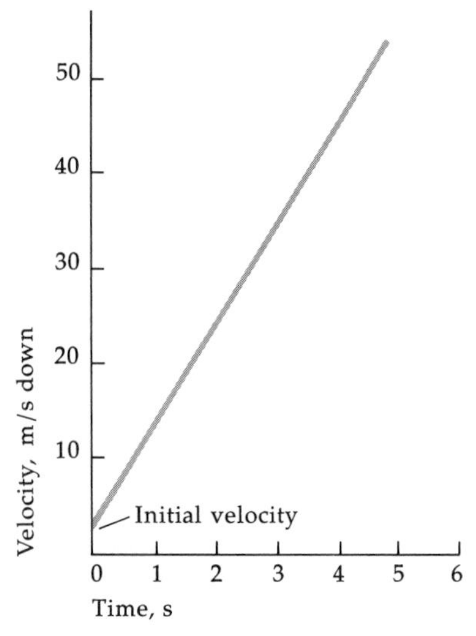

Table 2-3. Velocity of the Baseball When Given an Initial Velocity

Elapsed time, s	Velocity, m/s downward
0	3
1	13
2	23
3	33
4	43
5	53

Notice that nowhere in our discussion have we referred to how heavy the falling object is. We have said it is a baseball, but we could just as easily have been discussing a cannonball, a piece of buckshot, or (again in the absence of air resistance) a feather. A major point of disagreement between Aristotle and Galileo was whether the motion of a freely falling body depended on its size or weight. We shall return to this topic in the next chapter.

Learning Checks

1. What is the acceleration of a freely falling body one second after it is released from rest? Two seconds? Ten seconds?

2. What is the distance a freely falling body released from rest has fallen after five seconds? Six seconds?

3. What is the speed of a freely falling body released from rest after five seconds? Six seconds? Two and one-half seconds?

SUMMARY

Aristotle's early view of motion distinguished between *natural motion,* the striving of an object to return to its natural place, and *violent motion,* which resulted from a push or a pull.

Galileo's studies involved three concepts: *speed, velocity,* and *acceleration. Average speed* is the distance traveled divided by the time required. *Instantaneous speed* is the speed at a particular instant of time, rather than over a time interval. *Velocity* is a vector quantity, meaning that it has a direction associated with it; it is speed with direction. *Constant speed* implies that the rate at which the distance varies with time is unchanging. *Constant velocity* is constant speed in a straight line. *Acceleration* is also a vector quantity and means the rate at which velocity changes with time. It is determined by dividing the change in velocity by the time. Acceleration may result from a change in *speed,* a change in *direction,* or both.

A *freely falling body* is one subject only to the force of gravity. It falls with constant acceleration. The *velocity* that an object attains and the *distance* it falls in a given length of time are the same as for any other object, regardless of weight.

Exercises

1. In Aristotle's view, the fall of objects toward the center of the earth was natural motion, needing no further explanation. Is there an analogy in the present day world view? Is there any kind of motion that we hold to be natural in the sense that it needs no explanation?

2. Give an example of an object that undergoes acceleration while traveling at constant speed.

3. Is it possible for an object whose acceleration is zero to be moving? Explain.

4. Classify the following statements "True" or "False":
   a. An object can have a constant velocity even though its speed is changing.
   b. An object can have a constant speed even though its velocity is changing.
   c. An object can have zero velocity even though its acceleration is not zero.
   d. An object subjected to a constant acceleration can reverse its velocity.

5. Two messengers on bicycles start from rest and attain a velocity of 5 km/h west. Do the bicycles necessarily undergo the same acceleration? Explain.

6. A racehorse running west in a straight line goes from 10 km/h to 30 km/h in 5 seconds. A second horse speeding south in a straight line goes from 20 km/h to 40 km/h also in 5 seconds. Do they have the same acceleration? Why or why not?

7. A motorcyclist traveling on an interstate highway passes kilometer markers every 60 seconds. What is the motorcycle's average speed in kilometers per hour?

8. David killed Goliath by putting a rock in a sling and whirling it around his head before releasing it. If that rock traveled in a circular path in the sling at constant speed, was its acceleration constant? Why or why not?

9. If the speedometer of your car correctly indicates a constant speed of 50 km/h, does this always imply that the car is not accelerating? Explain.

10. Two limousine services, the Red Flash and the Blue Streak, pick up passengers at an airport. The chauffeurs leave at noon and drive in the same direction at constant speeds. The Red Flash driver goes 80 km/h for 4 hours, while the Blue Streak travels for 3 hours at 40 km/h. How fast must the Blue Streak chauffeur drive during the next hour to catch up with the Red Flash limousine at 4 P.M.?

11. Two of the world's fastest land animals are the East African cheetah and the pronghorn antelope of the western United States. Imagine that these two speedsters somehow meet and the cheetah begins chasing the antelope for dinner. The antelope goes from 30 km/h to 50 km/h, and the big cat goes from 50 km/h to 70 km/h. Compare the changes in their speeds.

12. A commercial airplane travels at a constant velocity of 600 km/h east. The pilot decides to decrease the velocity to 240 km/h east, and he does so uniformly over a period of 20 seconds. What is the acceleration? (Use m/s^2, 1 km = 1000 m; 1 h = 3600 s.) Is the acceleration positive or negative? Explain the meaning of your result.

13. Suppose an American-made automobile goes from 0 to 60 km/h in 5 seconds, while a Japanese-made one goes from 0 to 60 km/h in 10 seconds. Both travel east in a straight line. What is the change in velocity in each case? What is t in each case? Compare the cars' accelerations.

14. A cattle truck travels 50 miles in a certain time interval t, and a tractor-trailer travels the same distance in half the time. How does the speed of the cattle truck compare with that of the tractor-trailer for the 50 mile trip?

15. You are driving down the highway at 60 mi/h, which is about 27 m/s. Each broken yellow stripe separating the lanes is 3 meters long. How much time does it take for one stripe to pass your eye?

16. A world-class sprinter runs the 100 yard dash in 9.5 seconds. At this speed, how fast would this person run the mile? (1 mile = 5280 feet)

17. A certain punter kicks a football with a "hang time" (length of time the ball is in the air) of 4.2 seconds. If a teammate runs downfield 45 yards during this time, what is his average speed? Express the result in feet per second.

18. In our discussion of falling bodies (see Tables 2–1, 2–2, and 2–3), we saw that the baseball dropped from rest was traveling 10 m/s downward at the end of the first second. However, at the end of the first second it had traveled a distance of only 5 meters, not 10 meters. Explain.

19. A heavy medicine ball starts from rest and rolls down a ramp with an acceleration of 5 m/s^2 parallel to the ramp. How much speed is it picking up each second? What is its speed after 3 seconds?

20. If we increase the incline of the ramp in Exercise 19 so that it is more nearly vertical, what happens to the acceleration? What is the acceleration of the ball when the ramp is exactly vertical?

21. In Figure 2–6, how far does the object fall during the fifth second? During the third second? What is its average speed during the time interval between $t = 3$ s and $t = 5$ s?

22. In Figure 2–6, what is the average speed of the object for the first 2 seconds of fall? The first 3 seconds?

23. Suppose that rather than dropping our baseball from rest we throw it off the Eiffel Tower with an initial velocity of 5 m/s downward. How would the velocity–time graph appear for this situation?

Deviation from motion in a straight line requires an external force.

3

GALILEO, NEWTON, AND THE FIRST LAW OF MOTION

Motion characterizes the universe and everything in it. Planets orbit the sun, galaxies zoom through space. Cities are alive with moving cars, trucks, motorcycles, people. Motion can be curved, like a golf ball in flight or racing cars going around a track. Or it can be the linear motion of a rocket blasting off its launch pad. When a musician plucks a guitar string or a tailor sets a sewing machine needle going, the result is back-and-forth motion. The salt you sprinkle on food has invisible atoms that jiggle about in crystal lattices. Trees sway in the breeze, windmills go around and around, atoms zip along in giant accelerators. No matter where we look, motion, seen or unseen, is present.

These different kinds of motion, and countless other examples you could name, all have this feature in common: they result from *forces,* which we can think of as pushes and pulls. In Chapter 2 we learned how to describe the movement of things in terms of the distance-and-time concepts of speed, velocity, and acceleration. The next step is to find out how forces affect the way things move.

The stuff of the universe moves according to some rules of the game, formulated by Isaac Newton, called the *laws of motion.* We are going to find out about these laws in this chapter and the next. First, though, we need to go back a little bit, to Galileo, who did the spadework for Newton.

GALILEO, THE FATHER OF PHYSICS

One of the major articles of faith in Aristotle's time and for centuries afterward was that an object at rest was in its natural state of motion. The earth was considered motionless at the center of the universe, and this was natural. It therefore fol-

Perpetual Motion: The Impossible Dream

Motion, Nature decreed at the outset of time, shall not be given away for free." That, however, has not stopped inventors over the last several hundred years from trying to make a machine that could operate continuously without an outside energy source. While numerous such attempts are displayed in museums, a successful perpetual motion machine has never been constructed, neither one designed to accomplish useful work nor one contrived to just continue in motion, running nothing. The differing designs proposed for the type of perpetual motion machine that can do work invariably violate either the first or second law of thermodynamics, both of which we shall study in this course.

Theoretical barriers aside, however, the financial gain that would obviously fall to a breakthrough inventor in this area keeps people trying. After all, practical applications of a perpetual motion machine that can drive other machinery or otherwise accomplish work are almost without limit. At one point, around the turn of the century, so many people tried to mine the gold in this particular hill that the United States Patent Office got fed up. They simply handed applicants a printed circular that read:

The views of the Patent Office are in accord with those scientists who have investigated this subject and are to the effect that such devices are physical impossibilities. The position of the Office can be rebutted only by the exhibition of a working model.

They are still waiting.

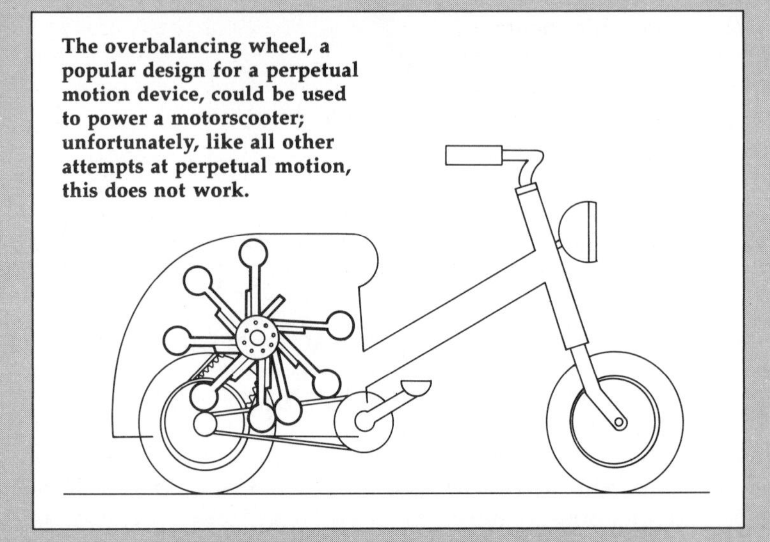

The overbalancing wheel, a popular design for a perpetual motion device, could be used to power a motorscooter; unfortunately, like all other attempts at perpetual motion, this does not work.

Source: Arthur W. J. G. Ord-Hume, *Perpetual Motion, the History of an Obsession.* New York: St. Martin's Press, 1977.

lowed that nature intended for all objects to be at rest; motion of any kind was due to some outside influence. Aristotle's interpretation coincided neatly with later views of the Roman Catholic Church, which taught that humanity, as the focus of Creation, should occupy the most important spot in the universe—its center. Hence, the science of Aristotle and the religion of medieval times joined to support this world view. Almost surely, some people doubted or raised questions. But Aristotle's reputation, combined with the iron hand of the church in Rome, served to squelch any notions contrary to accepted doctrine.

In the latter part of the sixteenth century came Galileo Galilei to present the strongest challenge yet to Aristotle's beliefs about motion. Born in Pisa in 1564, and thus a contemporary of Shakespeare, Galileo was one of the most interesting characters in the annals of science. His father, Vincenzio Galilei of Florence, was a musician and mathematician who had a difficult time making a living with his skills. Because, like all fathers, he wanted a better life for his son, he persuaded

Galileo to become a physician so he could make a lot of money (some things haven't changed). Try as he might, Galileo just didn't have his heart in medicine. He eventually drifted toward physics (it was called *natural philosophy* then), proving to be quite adept at devising various types of experimental apparatus. He taught for several years at the University of Padua, where he performed many of the experiments on motion pertinent to our study.

Galileo was a free spirit with an irrepressible zest for living. He combined his commitment to science with an equally strong appetite for wine and women. His remarkable gift for writing was often used in tossing elegant insults at his opponents. He was very much the Renaissance man. Galileo is perhaps most famous for his use of the telescope, which got him into severe trouble with the Church. His observations of the heavens led him to accept Copernicus's idea that the sun, not the earth, was the center of the solar system and the object about which the earth and the other planets revolved. This model is sometimes called the Copernican heresy; it directly conflicted with official church doctrine. For a time, Galileo evaded serious trouble with the Church in spite of his views, partly because of his friendship with the influential Cardinal Barberini, who later became Pope Urban VIII.

But the Church was serious about the Copernican heresy. Giordano Bruno, a friend of Galileo, was a writer who extended Copernicus's idea by suggesting that the stars might be spread throughout an infinite universe. In 1600, Bruno was burned at the stake as a heretic. Eventually the Inquisition engulfed Galileo himself. His friend the Pope stood aside, and the scientist was forced to recant his views formally in 1633. Convicted of heresy, Galileo was held under house arrest in his home near Florence until his death in 1642.

Galileo didn't much like the way the philosophers of his time contended that words were endowed with supernatural powers. He wrote, "If their [the philosophers'] opinions and their voices have the power to call into existence the things they name, then I beg them to do me the favor of naming a lot of old hardware I have about my house 'gold.' "

Source: Stillman Drake, *Galileo*. New York: Hill and Wang, 1980.

The Concept of Inertia

As we saw in Chapter 2, Galileo provided a better interpretation than Aristotle for the motion of falling bodies. At a more fundamental level, Galileo attacked the Aristotelian concept of natural and violent motion. Recall that the accepted view of motion was that it was natural for an object to be at rest. Sustaining motion required some sort of external push or pull. For example, if a bartender slides a mug of beer to a customer along a flat, smooth bar, the mug will eventually stop. To maintain the mug's motion, according to Aristotle, the bartender must continue to push it, because it is natural for the mug to be at rest. If the bartender stops pushing it, the mug is left alone and, in this view, is thus free to resume its natural, stationary state.

Galileo's genius was in realizing that, as the mug slides across the bartop, it is *not* left to itself, even if no one shoves it.

> **Galileo and the Telescope**
>
> Have you ever looked through a *cannocchiale?* How about just an *occhiale*, then? A *cannone*, perhaps? *Batavica dioptra? Perspicillum?* Surely, a *telescopio?* All of these were names used for the telescope at one time.
>
> Apparently, Galileo did not invent this tool of science. It may have been first conceived by a German or Dutch workman. But Galileo, merely having heard about it, built one and perfected it. He was the first to turn the tube skyward.
>
> Source: Edward Rosen, *The Naming of the Telescope*. New York: Henry Schuman, 1947.

Something else is involved. The bar itself "pushes" in a sort of way that we have come to call *friction*. Galileo believed that friction makes the mug stop. He felt that if the friction were completely removed—say, by making the mug and the bar counter perfectly smooth—then there would be nothing to stop the mug. Whereas Aristotle would assert that the mug is left alone when the bartender ceases to push it, Galileo would say that the mug would not be truly left alone until the frictional effects of the bar were eliminated. Genuinely left to itself, then, once the mug was set in motion it would go on forever (given a long enough bar), with no external push necessary.

This tendency of objects to maintain their state of motion is called *inertia*. If the object is at rest, the natural thing for it to do is to remain at rest until some external push or pull causes it to move. If it is moving, its inertia keeps it moving until something else causes it to stop or somehow change its motion. You might think of inertia as the stubbornness of an object, a resistance to altering its state of motion.

The way we use the word *inertia* in everyday life is similar to its meaning in physics. For example, organizations are reluctant to change; reformers speak of having to "overcome inertia" in their attempts. Economists similarly bemoan the inertia of a sluggish economy.

Galileo tested his ideas about inertia by experiments in which he observed the motion of balls rolled up and down inclined planes. (See Figure 3–1.) He saw that a ball allowed to roll down a slightly inclined surface will gradually pick up speed—accelerate—while a ball on a steep plane will gain speed more quickly. Gravity pulls the ball down the ramp, and the steeper its slope, the greater the effect of gravity on the ball's speed. Similarly, a ball rolled up an incline will slow down, again because of gravity, which, in this case, holds it back. The rate of the decrease in speed depends on the incline: the steeper the ramp, the more quickly the ball decreases its speed. Galileo reasoned that if there were no slope (that is, if the plane were perfectly flat), the ball would neither speed up nor slow down. Gravity would have no effect and the ball would continue to roll on at its same speed. But, in reality, a rolling ball does stop on a flat surface. Galileo attributed this to friction between the ball and the plane. Friction interferes with the motion and slows the ball down.

Figure 3-1. Motion of balls on an inclined plane.

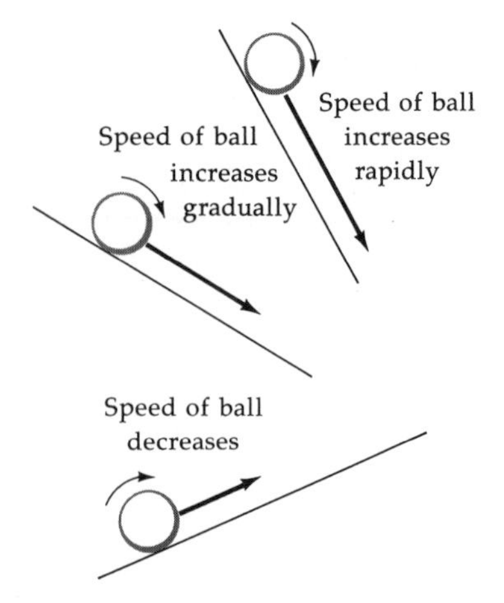

Galileo took great pains to make his inclined planes and rolling balls as smooth as possible to eliminate as much of the troublesome friction as he could. Figure 3–2 illustrates another of his experiments. Here a ball released from a certain height rolls down the plane, across the horizontal part, and up the other side (the corners have been rounded to prevent bouncing). If he could completely remove the effects of friction, how far up the righthand side would the ball roll? Far enough, our

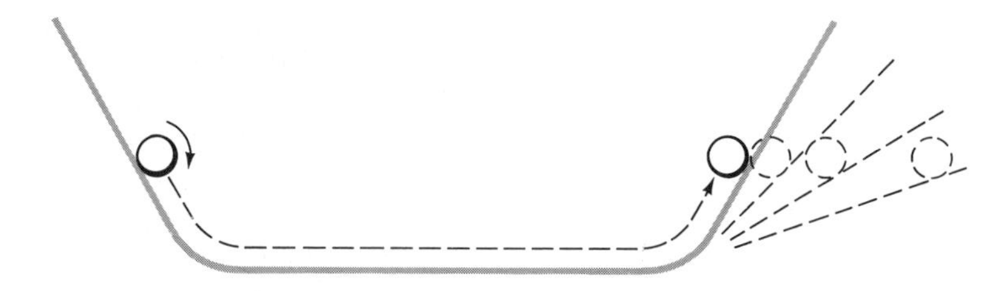

Figure 3-2. Galileo's inclined planes.

intuition tells us, to attain its original height on the lefthand side. The accelerating effect of gravity on the downhill run is exactly matched by its decelerating effect on the uphill run, and the ball regains the height from which it started. This is, in fact, what happens.

Now we will change the situation by lowering the slope of the right arm, bending it more toward the horizontal. Let us repeat the experiment. Again the ball reaches its original height, although this time it ends up farther away from the starting point than in our first try. If we perform the experiment again and again, each time making the right side of the incline slope a little less than before, we find that the ball always ends up at its original height but travels a longer horizontal distance than before. Our surface must extend farther and farther each time. So, the question arises, how far will the ball travel if we make the right side of the apparatus perfectly horizontal? The answer is, of course, that the ball will travel forever if the plane is frictionless and infinitely long, because there is nothing to stop it. If there is no uphill ramp, then gravity has no opportunity to slow the ball down. Once released, it will continue forever and, contrary to the teachings of Aristotle, nothing will be required to maintain its motion.

SIR ISAAC NEWTON

On Christmas Day, 1642, a few months after the death of Galileo, a premature baby boy was born on an English farm in Lincolnshire to a family named Newton. The mother, a widow for three months, named her son Isaac. At the age of three, Isaac Newton went to live with his grandmother, who was largely responsible for his upbringing and early education. He was a sickly boy, rather shy, and was undistinguished in his studies, giving no hint of the great contributions he was soon to make.

Entering Trinity College, Cambridge, at the age of eighteen, he began the study of mathematics. The summer of 1665 saw the Black Death sweep across England and the continent, closing the university and forcing Isaac back to the Lincolnshire farm for some eighteen months. It was during this period that Newton, at the age of twenty-three, conceived most of the

Newton and Interdisciplinary Thought

Newton, a founding father of physics, also had a bit of the biophysicist lurking in him. He anticipated this discipline by considering that "puzzeling problem, By what means the muscles are contracted and dilated to cause animal motion." He thereby paved the way for biology and physics to merge.

Source: A. R. Hall and M. V. Hall, *The Annus Mirabilis of Sir Isaac Newton, 1666–1966.* Cambridge, Mass.: MIT Press, 1970.

The Royal Society of London

The Royal Society of London for the Improvement of Natural Knowledge, established in 1660, is one of the oldest scientific organizations in the world. Its founders prided themselves on being people who valued "nothing but useful knowledge." The society honors selected scientists with medals, works to improve education in science, and assists high school science teachers who wish to do research. Membership has included many famous scientists, such as Newton, Faraday, Boyle, and Rutherford, whose contributions to physics you will learn about in this course. It was researchers from the Royal Society who photographed the solar eclipse of May 29, 1919 and completed the subsequent calculations that verified Einstein's general theory of relativity.

The society's motto is *Nullius in Verba*—"Take nobody's word for it." This motto reflects the basic spirit of experimental science, which has remained unchanged since the seventeenth century.

Source: Sir Harold Harley (ed.), *The Royal Society: Its Origins and Founders*. London: The Royal Society, 1960.

ideas that were to make him famous and whose development were to occupy his mind for the rest of his life. In these incredible months, he deduced the binomial theorem, which you may have studied in high school algebra, and invented the differential and integral calculus. He also made important studies in the field of optics, formulated the law of universal gravitation, and developed the laws of motion.

When the plague lifted and Cambridge University reopened, Isaac Newton was appointed to the Chair of Mathematics. Britain gave him its highest scientific honor, naming him a Fellow of the Royal Society of London at the age of thirty. He spent his life at Cambridge pursuing his two intellectual interests, physics and theology.

It is intriguing that two such great scientific intellects as Galileo and Newton should have such contrasting lifestyles. Whereas Galileo lived with the flair and bravura of the Renaissance man, Newton was every inch the Puritan. He took little interest in anything but his studies. He generally avoided people. The classic absent-minded professor, Newton would become so absorbed in his work that he missed meals, lost track of time, and showed a genuine naiveté about practical day-to-day living. Nevertheless, he and Galileo were scientific kindred spirits. Newton was quoted as saying, "If I have been able to see farther than some, it is because I have stood on the shoulders of giants." And Galileo was chief among the giants.

Building on the foundation that Galileo had laid, Isaac Newton formulated the descriptions of the motion of material bodies into the three laws that form the cornerstone of classical mechanics. We will examine Newton's first law now and then study the other two laws in the next chapter.

THE FIRST LAW OF MOTION

We can state Newton's first law this way:

> Every object continues in its state of rest, or of motion with constant speed in a straight line, unless it is compelled to change that state of motion by forces impressed upon it.

Sometimes called the *law of inertia,* the first law of motion is a precise statement of the results of Galileo's experiments. The key idea is this: An object will continue to do whatever it is doing until some external influence causes it to change. The first part of this law gives no problems. Obviously, an object at rest must have a force applied to it to get it moving, whether it is the pull of gravity, as in the case of our baseball in free fall, or a push of some type, such as when a hockey player slaps the puck across the ice. We would be greatly surprised if the puck, sitting still, suddenly of its own accord began sliding across the ice. In fact, we would begin hunting for the force that had caused this to happen.

It is the second part of the law—the point where Aristotle and Galileo parted company—that might cause us trouble. The notion that the puck would continue to slide forever at a constant speed in a straight line, given a big enough ice rink, seems strange because we have never seen it happen. But a little reflection shows us that the hockey puck eventually stops because there *is* some force impressed upon it, friction. Friction is greatly reduced on ice, as anyone who has slipped on a frozen puddle knows, but it is still present.

Note that the first laws says that the natural state of motion (by "natural" we mean free of external influence) is constant speed in a straight line, or more simply, constant velocity. Whenever the velocity of an object changes, whether its speed increases or decreases or its path deviates from a straight line, there must have been a force impressed on the object. It might be gravity, friction, a push, a pull, you name it; without such a force, the object's state of motion will remain unchanged.

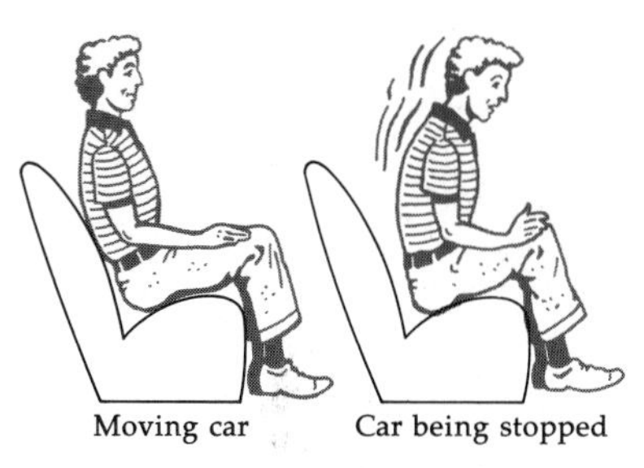

Figure 3-3. Are you thrown forward?

We can see many examples of Newton's first law at work in our everyday world. Everyone who has ever ridden in a car has occasionally been thrown against the dashboard when the driver slammed on the brakes. (See Figure 3-3.) To say you were "thrown against the dashboard" implies that something pushed you out of your seat toward the front of the car, but of course there was no such force. Newton's first law explains what really happened: You were moving along with the car, and when the forces that resulted from applying the brakes caused the car to stop, you kept right on moving, because the brakes were applied to the *car*, not to *you*. Your inertia kept you in your state of motion, moving forward until you hit the dashboard. Your discomfort was caused not by some force acting on you but precisely by the *absence* of any such force. A seatbelt ties you to the car so that when the brakes are applied, the force that stops the car also stops you.

Another example of inertia is the sensation of being pushed back in the seat when the car begins to move. (See Figure 3-4.) In this case, you are at rest and inertia causes you to tend to remain at rest—hence the feeling that your head is being snapped backward as the car lurches forward. Have you ever seen the trick in which a tablecloth is jerked out from under the dishes, leaving them in place on the bare table? The dishes are at rest, no force is applied to them, and hence, according to the law of inertia, they will remain at rest.

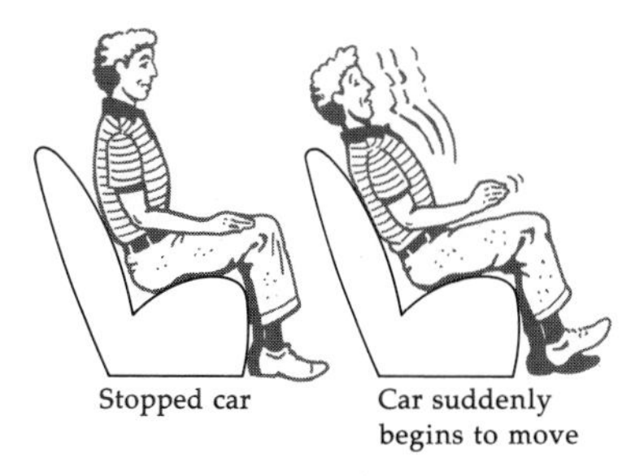

Figure 3-4. Are you pushed backward?

Let us emphasize once again the importance of straight-line motion in Newton's first law. Deviation from motion in a straight line requires an external force. For example, Aristotle considered the (nearly) circular motion of the planets around the sun, or the moon around the earth, as "natural" motion for heavenly bodies. The view of the ancients was that earthly and celestial objects were somehow different, subject to different natural motions. One of the triumphs of Newton's study of mechanics was the demonstration that in fact the stars and planets obey the same laws of motion as billiard balls and rocks. Thus, the curved paths of the planets around the sun imply some kind of force, which we now call gravity, pulling them out of the straight-line paths the first law says they would follow in the absence of such a force. We will study this force of gravity in a later chapter.

Mass

Now that we have the idea for inertia under control, we need some way to measure it. The quantity used to do this is called *mass.* Let's have a look.

Inertia is the resistance of an object to changing its condition of motion. *Mass* is the name given to the measure of the inertia. A heavy object such as a Sherman tank has a greater mass than a buckshot pellet, and we are sure that the tank would be harder to push from rest.

Perhaps it is simpler for you to think of mass as expressing the *quantity* of *matter* in some object. A medicine ball with fifty times the mass of a baseball has fifty times as much matter. It is also fifty times harder to move from a standing start, or to stop once it is rolling. In agreement with what we have already said, the inertia of the medicine ball is greater than that of a baseball by a factor of fifty.

It is important for us to distinguish between *mass* and *weight*. Mass is an *intrinsic* property, which means (among other things) that it is the same regardless of its location. Our Sherman tank has the same mass on the earth as on the moon, even though gravity is much less there. Weight, on the other hand, is the name given to the gravitational pull on an object. Its value depends on where it is measured.The weight you read on your bathroom scale is much greater on the earth than on the moon.

An object has mass even though its weight may be zero. For example, there is a spot between the earth and the moon where the competing pulls of gravity of the two exactly cancel. If we were to put our Sherman tank at this particular place, its *weight* would be *zero,* but it would have the same *mass* it always had. Obviously it has the same amount of matter. Also, its resistance to changing its motion would be the same. We would have to push just as hard to move the tank from rest across the surface of the earth, or the moon, or at that special point of zero weight in between.

The metric unit of mass we will use most often is the *kilogram* (kg). It is quite common to speak of mass and weight interchangeably—for example, to say something "weighs" 10 kilograms. This is incorrect. Remember that mass and weight are related but not identical. Mass measures inertia and quantity of matter. Weight measures pull of gravity.

Learning Checks

1. What is inertia? How is the scientific use of this word similar to its use in everyday language?
2. What is the "natural state of motion" as viewed by Aristotle? by Galileo? by Newton?
3. You are riding down an elevator at constant velocity. Explain the "sinking" feeling in your stomach when the elevator suddenly stops.

SUMMARY

Galileo's studies of motion led him to break with the Aristotelian ideas of natural and violent motion. He formulated the concept of *inertia,* the tendency of an object to maintain whatever condition of motion it has whether moving or at rest. Inertia may be thought of as the resistance of an object to changing its motion. Newton built on Galileo's work, summarizing mechanics into three *laws of motion.* The *first law,* the law of inertia, says that if left to itself, an object will remain at rest or in motion at constant speed in a straight line.

Exercises

1. Compare and contrast Galileo's and Aristotle's explanations for a ball rolling gradually to a stop on a pool table.
2. Galileo studied "falling" bodies that were not actually falling but rolling down smooth inclined planes. Why was this more convenient for him?
3. It is often remarked that Galileo's inclined planes had the effect of "diluting" gravity. What is meant by this?
4. A common classroom demonstration involves covering the top of a beaker with a card and placing a coin on top of the card. When you propel the card forward, by flipping it with a fingertip, the coin drops into the glass. Explain what happens in terms of physics.
5. Suppose a psychologist is testing the intelligence of a chimpanzee by giving it a large stone tied to the end of a string. The stone has food on it and the monkey must lift it by the string to get the reward. If the chimp pulls the string up slowly it can lift the stone. But if it jerks the string suddenly, the string breaks. Explain.
6. An archaeologist at a dig has to push a heavy boulder out of the way. How can this person tell if it is friction or inertia that must be overcome?
7. A carpenter tightens the head of a hammer by slamming the butt end down on a hard surface. Explain.

4

NEWTON'S SECOND AND THIRD LAWS OF MOTION

A familiar modern image: Newton's third law at work. The Space Shuttle *Columbia* is shown here during lift-off in November 1982.

As we saw in Chapter 3, Newton's first law of motion tells us what an object will do if left to itself. That is, if at rest, it will remain at rest, and if moving, it will continue to move in a straight line at constant speed. In this chapter we will examine the other two laws of motion, both of which have to do with the behavior of objects when forces act on them.

THE SECOND LAW OF MOTION

We can state the second law in this way:

> The acceleration of an object is directly proportional to the net force impressed on the object and inversely proportional to the mass of the object. The acceleration is in the direction of the net external force.

Whereas the first law describes what a body does if nothing happens to it, this one gives us some quantitative information about what happens if a force acts on it. Because in the absence of any force the motion is one of constant velocity, it follows that if a force changes that state of motion, the result must be *acceleration*. Recall that acceleration tells how the velocity and time change together, arising from a change in speed or change in the direction of motion, or both. Thus, a jet plane speeding up from 400 km/h to 600 km/h is accelerating because of a change in speed, whereas the moon going around the earth is accelerating because its direction of motion changes. In both cases there is the action of a force.

The second law tells us that the acceleration resulting from a force impressed on a body is directly proportional to the net force and inversely proportional to the mass of the body on

Epitaph Intended for Sir Isaac Newton:

Nature and Nature's laws lay hid in Night
God said Let Newton be! and all was Light

Alexander Pope

The Myth of Sisyphus

Sisyphus, a figure from Greek mythology, was ordered as a punishment to roll a boulder up a hill and topple it down the other side. Unfortunately, whenever he neared the top, the heavy weight of the boulder forced him back, and the boulder rolled to the bottom. Each time this happened, Sisyphus had to begin again, eternally experiencing Newton's second law firsthand.

which the force acts. It also tells us that the direction of the acceleration is the same as that of the net force. We can write this symbolically as

$$\mathbf{a} = \frac{\mathbf{F}}{m}$$

where **a** is acceleration, **F** is the net force, and m stands for mass. The boldface type indicates that **a** and **F** are vectors. We have talked about acceleration as a vector, and we will discuss the vector nature of forces as we go along. (See also Supplement 1.)

To see how the proportionalities work, let us return to our medicine ball and baseball. We can push the baseball with some constant force to give it an acceleration of, say, 5 m/s^2. But pushing the medicine ball with the same force will not generate nearly this much acceleration. In fact, since the medicine ball is fifty times more massive, its acceleration will be one-fiftieth, or 0.01 m/s^2. Put another way, to give the heavy medicine ball the acceleration of 5 m/s^2, we must furnish fifty times as much force as we do with the baseball.

Another way of writing the equation for the second law is

$$\mathbf{F} = m\mathbf{a}$$

This form of the equation says we can calculate the net force by multiplying the mass of an object by its acceleration. If the mass is in kilograms (kg) and the magnitude of acceleration is in meters per second per second (m/s^2), the unit for the magnitude of force is called the *newton* (N). This means that a net force of one newton east will accelerate a one kilogram object 1 m/s^2 to the east. In the British system, the units of mass and force are the *slug* and the *pound* respectively. (One slug = 14.6 kg; one pound = 4.4 N.)

Force and Acceleration

The equation $\mathbf{F} = m\mathbf{a}$ says that the strength of the force on the body dictates how much it will accelerate. Let us illustrate this with a familiar example, a child riding in a toy wagon. The mass m is the total mass of the child and the wagon. The child is going to accelerate from rest to some desired straight-line speed, say, 2 m/s, by applying a force—pushing against the road with one foot.

Suppose the child wants to reach a velocity of 2 m/s to the right as quickly as possible. In the language of Chapter 2, the change in velocity from 0 to 2 m/s to the right is to take place over a very short time interval, t. Since t is *small*, our definition of acceleration on page 15 says that the acceleration is *large*. So we know from Newton's second law that if **a** is large, **F** must be large—the child has to push hard.

On the other hand, if the child wants to be more leisurely in reaching 2 m/s, the time interval t will be larger. The acceleration **a** is correspondingly smaller, requiring a smaller force **F**—the child need not push so hard. These results are common to our experience. Newton's second law expresses in a quantitative way what we observe.

Newton's second law explains how your car can always beat a big truck away from a traffic light. The truck has a much greater mass, so for an approximately equal force, it does not get nearly as much acceleration as the smaller car. The forces produced by the engines of the two vehicles will not be equal, of course, but the truck engine's force is not large enough to make up for the great difference in the masses. So you build up speed more quickly, leaving the truck breathing your exhaust.

Learning Checks

1. For a given mass, what happens to the acceleration as the force increases?
2. In the example of the toy wagon, suppose two children are riding and one pushes with the same force as when alone. How will the acceleration of the two compare to that of the lone child?
3. How is the direction of the acceleration related to the direction of the net force?

Falling Bodies and the Second Law

We pointed out in Chapter 2 that, in the absence of air resistance, falling objects near the surface of the earth have the same acceleration, regardless of their weight. The force that causes this acceleration is gravity, which we shall discuss in Chapter 5. For now, you need only to know that the force of gravity on an object is proportional to its mass. That is, the gravitational force on a 10 kilogram object is ten times that on an object having a mass of 1 kilogram.

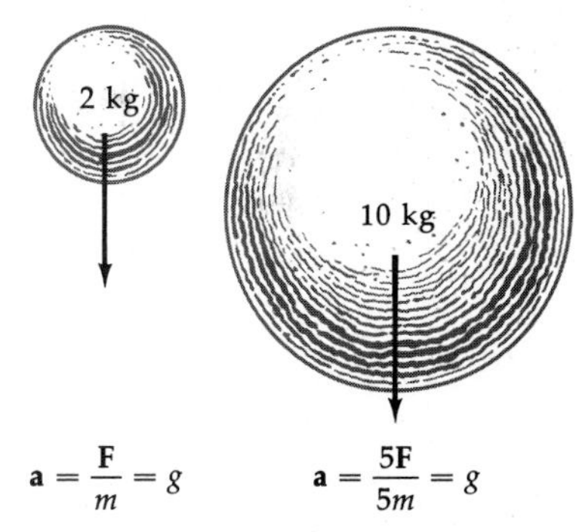

Figure 4-1. Acceleration is the same for all falling bodies regardless of their weight.

Imagine that a warehouse worker drops two steel beams from the top level down to the ground floor. One beam has a mass of 10 kilograms, the other 2 kilograms. (See Figure 4-1.) Since the 10 kg beam has five times the mass of the other, the force due to gravity on it is five times stronger than on the 2 kg beam. However, because it has five times the mass, the 10 kg beam also has five times the inertia of the smaller one, meaning it is five times harder to accelerate. These two effects—the greater *force* on the one hand and the greater *inertia* on the other—exactly cancel, so that the *acceleration* of the larger beam is the same as that of the smaller. Thus, even though the *force* due to gravity is greater on the larger beam because of its greater mass, the *acceleration* is the same for both. The greater mass of the 10 kg beam means that its inertia, and hence its resistance to being accelerated, is correspondingly greater. So, if released together, the two beams will also land together.

Because the acceleration is the same for all freely falling bodies, it is given a special symbol, g, and called the *gravitational acceleration.* Close to the surface of the earth, g is almost constant and has the numerical value 9.8 m/s^2. *Weight* is simply the special name given to the force that causes this gravitational acceleration. The weight of an object is its mass multiplied by the acceleration of gravity:

$$W = mg$$

We can see that this equation has the same form as Newton's second law, $\mathbf{F} = m\mathbf{a}$. The weight W is the force that arises because of the gravitational attraction of the earth for a body of mass m, and g is the acceleration produced by this gravitational force. Mass and weight are seen to be proportional.

Learning Checks

1. Does gravity pull with the same size force on all freely falling bodies? What about the acceleration of these objects—are they the same?
2. What is weight?
3. How is **a** in Newton's second law related to g in the equation for weight?

FORCE AS A VECTOR

Newton's second law refers to the *net* force on mass m, which means the vector sum of all the various forces that may be acting. Force is a vector quantity. The physical result of applying a force depends on the direction of the force vector. The second law tells us what this result is: the body accelerates in the direction of the net force.

For example, two longshoremen are instructed to move a large shipping crate. They can't see each other because the crate is too tall, so they attach grappling hooks to either side and pull in opposite directions. Each exerts a force of magnitude 5 newtons. One tugs toward the east, the other toward the west. We can represent this situation by a simple diagram, as in Figure 4-2. As you can see, since the forces are in opposite directions, the *net* force acting on the crate is zero, even though the *individual* forces have a combined magnitude of 10 newtons. Since the net force is zero, Newton's second law tells us that

Figure 4-2. Force vectors: opposite directions.

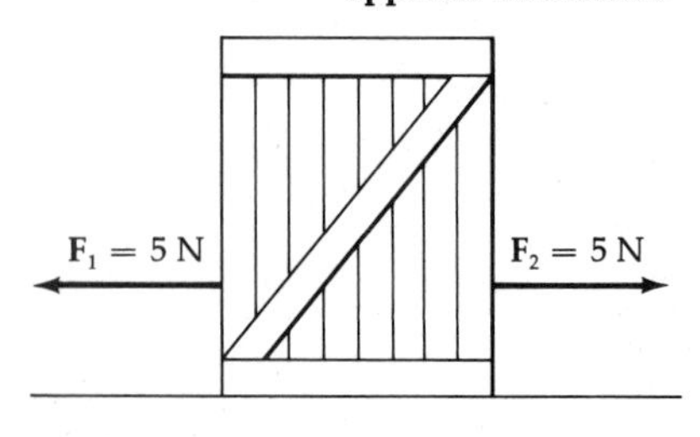

the acceleration is zero, which in turn says that the velocity does not change. Since the crate is at rest, that is, its velocity is zero, it doesn't move when acted on by these equal but opposing forces. This is just what common sense would tell us.

Now let us change things a bit by having one longshoreman move around to the same side of the crate as the other. Again each pulls on the crate with 5 newtons. Now, however, both are pulling to the east, say, so the picture becomes that in Figure 4–3. It should be obvious that this time the net force (neglecting friction) is 10 newtons east.

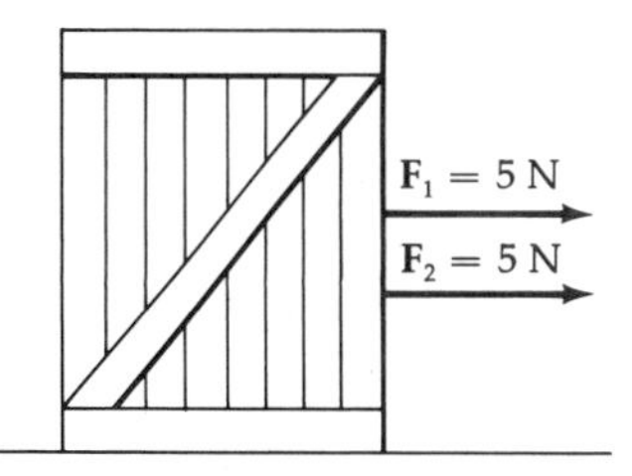

Figure 4-3. Force vectors: same directions.

The crate accelerates to the east by an amount depending on its mass *m*, according to the second law. This illustrates the vector nature of force: even though both people are exerting the same *amount* of force in the two cases, the results are very different simply because the *directions* of the forces are different.

Determining the net force in these two cases is quite simple because the forces are made to act along the same straight line, either in the same direction or in opposite directions. (Supplement 3 deals with more complicated situations involving vectors.)

Weight Off Your Mind

We live with the weight of our bodies from the first instant of life. Though many of us are careful weight-watchers and regularly check ourselves on the scale, our everyday experience of weight in the physics sense—that is, of the gravitational pull that the earth exerts on us—is so deeply ingrained we almost forget about it. Now that humankind is venturing off the earth, however, we *can* forget about it. Away from our planet, the forces of gravitational acceleration we ordinarily feel are lacking. In space, there are variation of body weight and weightlessness, new sensations for future space travelers.

How does it feel to be weightless? You can get a quick sense of this condition here on earth by jumping off a diving board or a chair. During the jump, you feel weightless because no force is supporting you. Have you ever been in an elevator when it dropped quickly, giving you that momentary stomach-lifting sensation? The only difference between

the elevator drop (or the chair-jump) experience and that of astronauts is that your weightlessness state was short-lived—it lasted perhaps a second or two—while the space dweller encounters it for much longer periods.

What effects does an extended state of weightlessness have on the human body? We don't know the whole story yet. Both animal and human physiological studies have been and are being conducted by aerospace medical research laboratories. Indications of disturbances in orientation and muscular coordination could cause some difficulty for orbital workers, but human adaptive abilities are great. "We've adapted very well," U.S. astronaut Pete Conrad reported from *Skylab*. "If you are just reading, you float free and wind up anywhere, on the ceiling, on the floor, even in the corner, sometimes ricocheting, and it doesn't seem to bother you." Researchers in this area have concluded that, though tumbling in a weightless state seems to produce severe disorientation, softly floating about in a padded cabin during a maneuver can be very exhilarating—a weight off your mind.

Sources: E. T. Benedikt (ed.), *Weightless Physical Phenomenon and Biological Effects*. New York: Plenum Press, 1961.

T. Page and L. W. Page (eds.), *Space Science and Astronomy: Escape from Earth*. New York: Macmillan, 1976.

Learning Checks

1. What is net force?
2. In the expression $\mathbf{F} = m\mathbf{a}$, $\mathbf{F}$ and $\mathbf{a}$ are both vectors. Why is m not a vector?

Friction and the Second Law

We must keep in mind the notion of net force even when there is only one apparent applied force, because friction is usually present. In discussing situations involving force, it is useful to neglect friction simply to make things easier. But in reality friction is alive and well, and we cannot ignore it.

To illustrate how friction enters into the net force in Newton's law, let us return to our example of pushing a wagon. The neighborhood child is moving along at a speed of 2 m/s in a straight line on a level road and wants to maintain that constant velocity. The acceleration is zero since the velocity is not changing, so according to the second law, the net force must also be zero. However, experience teaches that the child still needs to apply a small force by pushing the wagon; otherwise, the effects of wind resistance, friction in the wheel axles, and so on slow it down. This "extra" pushing supplies the force needed to overcome these various frictional forces. So the vec-

Plenty of Friction

Friction is a force of serious concern to many industries. In several industrial applications, it can have a devastating consequence: the wear and abrasion of material that can cause financial loss.

Textile manufacturers, for instance, pay a great deal of attention to friction, not only in their machinery as yarns are converted into fabrics, but in their products, too. They study the effects of friction between identical fibers, different fibers, natural and synthetic fibers, and a fiber and nonfibrous material, such as metal or plastic. They want to identify the longer-lasting, more friction-resistant fabrics.

Metallurgists also study the influence of friction on the hard materials with which they must work. An understanding of the frictional behavior of metals is essential to the operation of such equipment as a high-speed friction saw, which must cut through armor plate, carbon steel, plate glass, and specially hardened plastics.

a

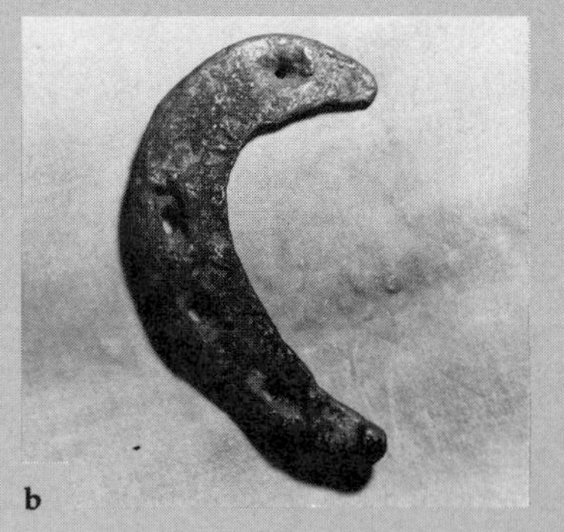

b

Nearly all the tools of humankind are subject to frictional wear—from the hardest and most modern, such as earthmoving equipment (**a**), to the oldest and most fragile, such as prehistoric scrapers fashioned from rock (**b**). The wear on these latter artifacts offers anthropologists and archeologists clues as to what kind of work prehistoric humans did.

Sources: H. G. Howell, K. W. Mieslkis, and D. Tabor, *Friction in Textiles*. New York: Textile Book Publishers, 1959.
American Institute of Physics, *Friction*. New York: AAPT Committee on Resource Letters, 1964.

tor sum of the *applied* force, which acts to move the wagon forward, and the *frictional* force, which opposes the motion and hence is in the opposite direction, is a net force of zero. The friction makes an applied force necessary. The child has to push to maintain constant velocity.

Anyone who has pushed a car by hand will have a feeling for Newton's laws. A car initially at rest requires a large force to get it up to a velocity of, say, 10 km/h, because of its large inertia. Once that velocity is reached, however, it can be maintained with very little force. Now the inertia tends to make the car continue its state of motion (first law), and zero acceleration means applying just enough force to overcome friction to achieve a *net* force of zero (second law).

Learning Checks

1. If the child in the wagon applies a force just equal to the friction, will the wagon accelerate?
2. The Environmental Protection Agency rates the gasoline mileage of automobiles for both city and highway driving. Why is the mileage always higher for highway driving?

THE THIRD LAW OF MOTION

We can express Newton's third law this way:

> For every action there is a reaction force, both being equal in magnitude and opposite in direction.

Whenever one object exerts a force on a second, the second object exerts a force on the first that is equal in magnitude, or strength, and opposite in direction. If we think of a force only as a "push" or a "pull," it is easy to fall into the habit of supposing that forces are caused exclusively by the action of our muscles. It seems strange to imagine a table "pushing" on the floor. And when a baseball crashes into an outfield fence, the baseball must have exerted a force on the fence. Because the ball bounces off the fence and rebounds toward the infield, the fence must also have exerted a force on the ball. Newton generalized this observation, recognizing that all forces connect two objects. When you push on a wall, the wall pushes back.

The essence of the third law is this: forces always occur in pairs. Which one we call "action" and which "reaction" is a matter of taste. When we walk across the floor (Figure 4–4), we exert a force on it by the action of our feet and legs. The friction between our shoes and the floor provides the connection, and the floor pushes back on us. Lift a box off the floor: you

Figure 4-4. A third-law example: F is the force you apply; f is the friction.

pull *up* on the box, and the box pulls *down* on you. These forces are of the same magnitude and point in opposite directions.

People often get confused about the third law. Students usually ask, "If I pull up on the box, and the box pulls down on me, why don't these forces just cancel? How can anything happen?" The key here is understanding *on which objects* the forces act. Earlier, when we illustrated the cancellation of forces, the two oppositely directed 5 newton forces were both exerted on the *same* object: the crate. But the action-reaction pair that Newton's third law talks about *never* act on the same body, so they cannot cancel. They always act on *different* bodies. Each object involved receives a single force that cannot be canceled by its "partner" force, since this other force is on the other body. For instance, if you throw a baseball, the force you give it is matched by a force the ball exerts on you. Since ordinarily you are standing on rough ground, the resulting additional frictional forces keep you from moving and thus mask the effects of the third-law force. However, if you throw the ball while standing on an ice skating rink, the force the ball exerts on you will send you skidding off in the opposite direction. (See Figure 4–5.)

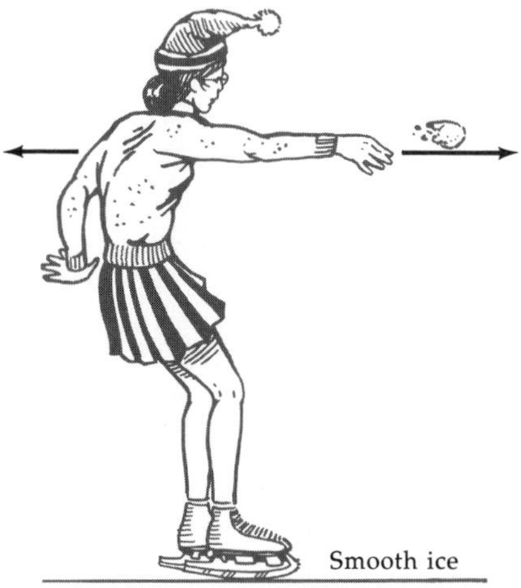

Smooth ice

Figure 4-5. The force the ball exerts on you causes you to recoil.

Further confusion often arises about the third law because the way the two objects involved in the interaction behave is usually quite different. One of them may even seem to be unaffected by the force. For example, in Chapter 2 the baseball that we dropped fell to the ground because the earth pulled on it by the force of gravity. This force is directed straight down, and the baseball, like any falling body, accelerates in the direction of the force. This acceleration results from the force exerted *by* the earth *on* the ball. Now the third law tells us that the ball exerts a force of the same magnitude *on* the earth, and in the opposite direction. And since the earth has a force impressed on it, it accelerates according to the second law. The catch is that the earth's acceleration is extremely small, too small to measure, and we don't notice it. So, for this reason, it may seem strange to say that the ball pulls on the earth with the same force as the earth pulls on the ball.

The key thing to notice is that the *force* is the same, but the acceleration of the ball and the earth are quite different. This has to do not with the force itself but with the masses of the objects involved. (See Figure 4–6.) Recall that the second law tells us that $\mathbf{a} = \mathbf{F}/m$. The acceleration of a body is small if its mass is quite large (for a given force) simply because dividing by a large number gives a small answer. Thus, when the baseball pulls on the earth with the force of gravity, the acceleration of the earth is negligibly small because its mass is tremendous compared with that of the ball. The ball, however, subject to the same force, has a much larger acceleration, which we can measure nicely because its mass is relatively small.

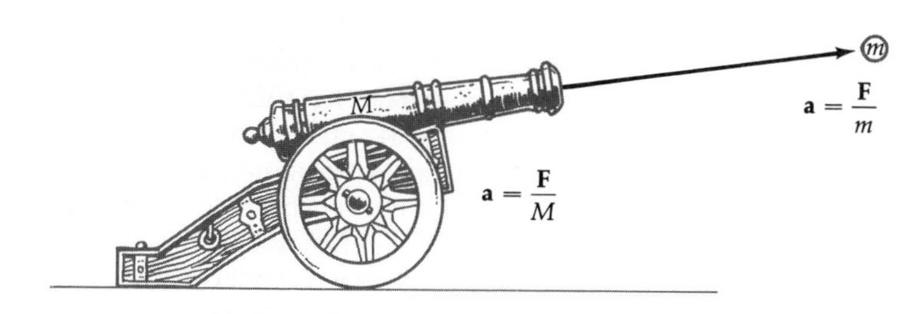

Figure 4-6. Acceleration of cannon and ball are different.

Another example may help clarify things. Anyone who has ever fired a rifle held a few centimeters from the shoulder has observed Newton's third law in the form of the rifle's "kick." Here the force on the bullet-rifle combination is caused by the explosion of the gun powder. We might label the "action force" that part of the interaction that sends the bullet zipping through the air, and the "reaction force" that which recoils the rifle in the opposite direction, "kicking" the shooter. Again, the *accelerations* of the bullet and rifle caused by the same *force* are different because of the masses. The bullet's mass is small and its acceleration is large, whereas the rifle's mass is large and its acceleration is small.

This same principle is involved in the thrust produced by rocket engines, one example of a general type of so-called *reaction engines.* A simple illustration occurs when you blow up a toy balloon, let it go, and it flies around the room. The pressurized air pushes against the balloon and sends it off in one direction, while the balloon squeezes against the air, forcing it out in the other direction. In the case of a rocket, combustion of the fuel expands the resulting gases that exert a force against the walls of the engine, while the walls of the engine exert an equal and opposite force against the gases, ejecting them out the rear.

Aerodynamic engineers build jet engines on a similar idea. A jet engine draws air in at the front. This air is compressed and burned with fuel in a combustion chamber. The engine ejects the resulting hot gases from the rear, propelling the aircraft forward.

Marine biologists are familiar with the squid as a life form that uses action-reaction forces. The squid draws water into a body cavity. Muscle tension forces the water out at high speed, propelling the squid in the opposite direction.

Learning Checks

1. What is the essence of Newton's third law?
2. Suppose a rifle and bullet have equal masses. Describe what happens when the rifle is fired.

Figure 4-7. The straight-line path of a dropped ball and the curved-line path of a thrown ball.

MOTION ALONG A CURVED PATH

We have seen that a baseball dropped from rest will travel in a straight line toward the earth's surface with an accelerated motion, the acceleration being about 10 m/s^2. Now suppose that instead of dropping the ball I *throw* it horizontally. It also falls to the earth, of course, but along a curved path, landing downrange from its starting point. (See Figure 4–7.) Obviously the harder I throw it, the farther away it strikes the ground. Now, this curved-line motion seems more complicated than the one that results from simply dropping the ball. Let us see if we can analyze it with what we know about Newton's laws.

While the ball is in my hand, I apply a force to it through the action of my muscles. This force is directed horizontally. It takes the ball from a horizontal velocity of zero to the velocity it has at the instant it leaves my hand—let us say it is 7 m/s to the east, just to have some numbers to make things more concrete. Let us see what forces act on the ball once it has left my hand with this horizontal velocity of 7 m/s east.

If we neglect air resistance, the only force in the *vertical* direction is gravity. The force of gravity gives the ball an acceleration straight down, just as in the case of the ball that is dropped. What about forces in the horizontal direction? Once the ball has left my hand, I no longer exert a force on it or have any influence on it at all. I could drop dead in my tracks and the ball would still have the horizontal velocity of 7 m/s east, which I gave it by the throwing motion. There are no other possible horizontal forces. So we conclude that the horizontal force acting on the ball in flight is *zero.*

Now, if there are no horizontal *forces* acting on the ball, then Newton's second law tells us that the horizontal part of the *acceleration* is also zero. There is no change in the horizontal velocity. Throughout the ball's trip from my hand to the ground, the horizontal velocity remains at 7 m/s east, the same value it had at the instant the ball left my hand. Gravity does not affect the horizontal velocity. The gravitational force is a vector directed straight down and has no component in the horizontal direction.

What about the vertical component of the velocity? Because gravity accelerates the ball, the vertical part of the velocity will take on the same values we discussed in connection with falling bodies in Chapter 2. Thus, at the end of 1 second the vertical velocity will be about 10 m/s down, after 2 seconds it will be about 20 m/s down, and so on. This means that *the thrown baseball is also a freely falling body,* because it is subject only to the force of gravity. (See Figure 4–8.)

Let us summarize. While in flight, the thrown ball feels only the force of gravity. This causes it to accelerate vertically downward like any freely falling body. The ball also has a hori-

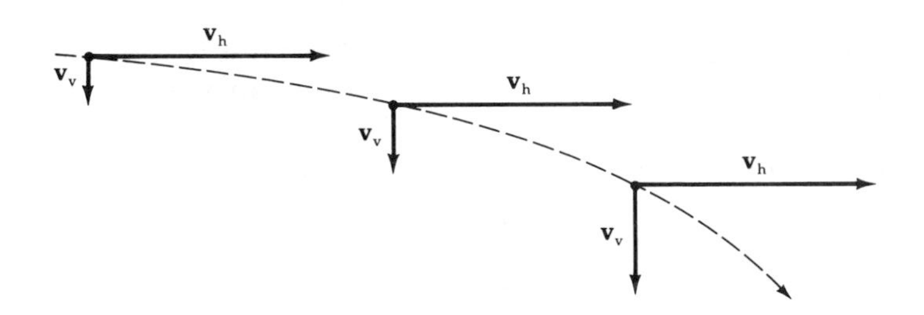

Figure 4-8. Vertical and horizontal components of velocity.

zontal velocity that remains constant throughout its flight. The ball follows a path that results from the combination of the vertical and horizontal parts.

We can carry this discussion one step further. Suppose I throw one baseball horizontally while letting a second one drop from the same height. I release both at the same time. Which strikes the ground first? Probably your inclination is to say that the dropped ball lands first, since it travels only a short

Physics on the Mound

The behavior of tossed baseballs becomes a highly important matter in the business of major-league pitching. Recently, physicists have applied their skills to look into the forces involved in the various kinds of pitches, such as curveballs, fastballs, and knuckleballs.

In one highly publicized experiment (produced jointly by the magazine *Science 82* and the NBC news show "Today"), curveballs thrown by professional pitchers (a lefthander and a righthander) were photographed with the aid of a high-speed strobe light. The balls were thrown through a "corridor" of strung ping-pong balls (as markers) and were photographed against a wall of black paper. In this way, the flight-path of the ball could be determined. The object was to settle once and for all the debate between physicists and batters as to whether curveballs curve in a smooth arc or a sharp break. The photographic data were used to calculate the factors involved in the ball's flight. Result: The downward-curving motion of the ball is gradual, a smooth arc. But if you view it head-on, as batters do, it gives the illusion of "breaking"—a ball falling off a table.

In another interesting experiment, a biophysicist attached electrodes to the fingers of college pitchers. A test ball was coated with electroconductive paint, and an electronic switch was embedded in the catcher's glove to determine the speed of a pitch. When the athletes wound up to throw, they triggered a microswitch that recorded information about how the ball was handled. "The system showed that while the arm may be responsible for launching the ball, the fingers play an important part—the order in which each of them releases the ball is the key to the pitcher's precision. If an index finger leaves the ball even a fraction of an instant before another digit, a smoking fastball may become a slow, hanging curve, and a dizzying strike may become an easy home-run pitch" (Kluger, 1982, p. 68).

Next time you see a pitcher throw a fastball or a curve, think about how researchers set up experiments to uncover the physics behind what he's doing.

The great Baseball Hall of Fame pitcher Sandy Koufax threw a powerful curveball. Hitters said it looked like it rolled off a table. Physicists have shown that the curve doesn't really "break" in this manner but, rather, travels downward in a smooth arc.

Sources: William F. Allman, "Pitching Rainbows, the Untold Physics of the Curve Ball," *Science 82*, October 1982, pp. 32–39.

Jeffrey Kluger, "The Human Machine," *Science Digest*, June 1982, pp. 64–71.

distance straight down, while the other travels a greater distance and lands down the way a bit. However, common sense fails us here. In fact, *both strike the ground at the same time.* Once the two balls have left my hands, they are both subject to the same force, gravity. Neither experiences any horizontal acceleration. So, because they travel the same vertical distance and undergo the same vertical acceleration, it follows that they must strike the ground simultaneously.

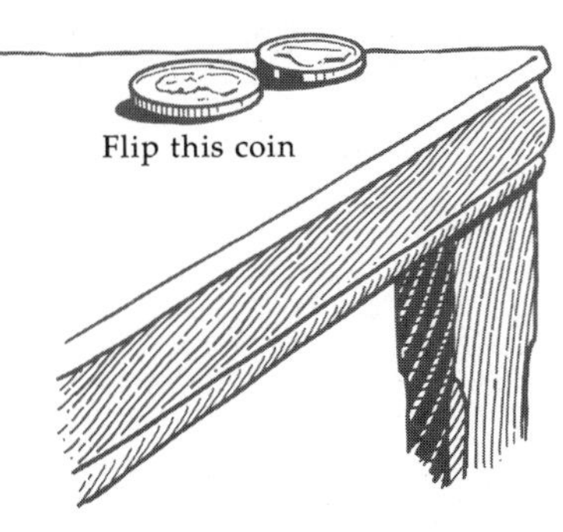

Figure 4-9. The two coins will hit the ground at the same time.

Check this out for yourself. Arrange two coins at the edge of a table, one behind and off to one side of the other, so that when the rear coin is flipped, the one in front falls off the table. (See Figure 4–9.) Notice that you get a single sound when they strike the ground together.

Learning Checks

1. Does an object in flight have to fall straight down to be in free fall? Explain.
2. Suppose you throw a ball straight up. After it leaves your hand, what is its acceleration? (Neglect air resistance.)
3. Answer Question 2 for the ball thrown straight *down.*
4. For any body in flight, what is its horizontal acceleration? Its vertical acceleration? (Neglect air resistance.)

Circular Motion

We have just seen that the thrown baseball represents a situation in which the direction of motion (along a curved path) and the direction of the acceleration (straight down) are not the same. An important special case of this kind is motion in a circle.

Let us consider a simple illustration, a rock tied to a string and spun in a circle. (See Figure 4–10.) Obviously this is an instance of accelerated motion, since the rock doesn't move in a straight line, so there must be a force involved. The question arises, what is the direction of the force?

Figure 4-10. Circular motion of a rock whirled around on a string.

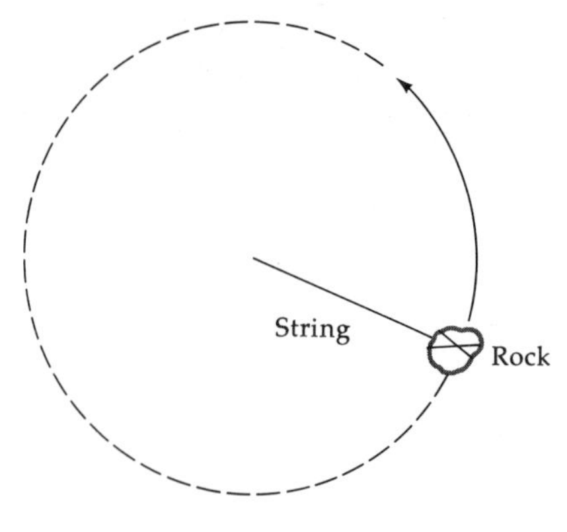

We can better understand the answer by asking another question: What would happen if we were to let go of the string? Somehow the string provides the force that confines the rock to its circular path. If we "turn off" that force, you can see that the rock would fly off in a straight line. This agrees with Newton's first law—with no force involved, the rock moves at constant speed in a straight line.

But what happens when the rock moves along while we are holding the string? During each instant of time, it wants to keep moving in a straight line. But the force arising from the tension in the string continuously tugs it out of that straight-line motion and along the circular path. (See Figure 4–11.) The string is always perpendicular to the path of the rock, so the force (and hence the acceleration) is always *toward* the center of the circle. This force has a special name, *centripetal* force, which means "toward the center."

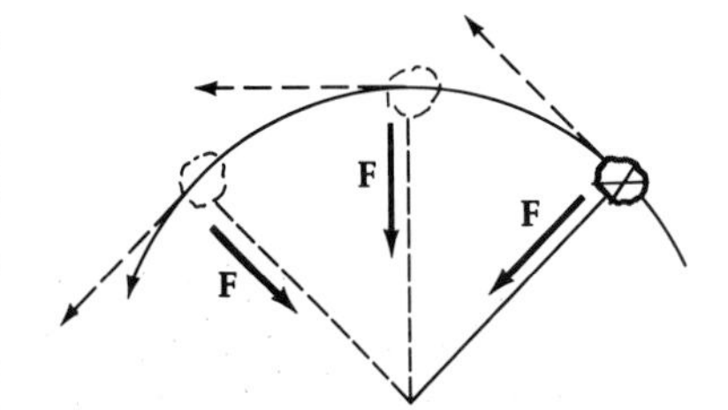

Figure 4-11. In the absence of a force, the rock would fly off on a straight line.

There is a popular misconception about circular motion, an idea that a force pushes the rock outward and causes it to fly away when you release the string. The name given to this is *centrifugal* (away from center) force. But such a force on the rock does not exist. As we have seen, it is the *absence* of any force that allows the rock to pursue its natural straight-line motion once the string is released. And the force that holds the rock along the circle is *toward*, not away from, the center.

It is true that as you whirl the rock around on the string, you can feel the rock tugging on your hand because of the tension in the string. This force is centrifugal, true enough, but it is *not* a force on the rock. In fact, this centrifugal force is the pull of the rock on the string. It is the third-law partner of the centripetal force that keeps the rock in its circular path.

The same is true of an orbiting satellite, whether the moon or an artificial rocket-launched object. The moon stays in its orbit around the earth because of gravity, which exerts a centripetal force. If we could somehow "switch off" gravity, the moon would fly off in a straight line—not because of some centrifugal force pushing it away from the earth, but simply because there is *no* force to make it deviate from straight-line motion.

Amusement parks often have a "centripetal force" ride that goes by various names but operates on the principle we have discussed. In one version, you stand with your back to the wall of a large drum that spins about a vertical axis. When the drum is rotating fast enough, the floor drops out from beneath your feet and you are pinned to the wall by the centripetal force of the drum against your back. There is no force on you to push you against the drum wall. The only net force acting on *you* is the centripetal one. The centrifugal force is not on you—it is the force that you exert on the wall.

CHECKPOINT ON FORCES

Now let's see how far we have come. In Chapter 2 we saw a way of describing motion with speed, velocity, and acceleration. In Chapters 3 and 4 we have learned from the work of Galileo and Newton the rules governing motion. The essence of the rules is this: Whenever a body *accelerates* it does so because there is a *force*. Without a force, an object just goes right on doing its thing, either sitting still or moving with constant velocity. The more massive a body, the smaller the acceleration it gets from a given force.

> *Don't fight forces, use them.*
>
> Buckminster Fuller

We began developing the idea of forces by thinking of them as pushes or pulls. Now you can see that we have expanded that notion quite a lot. The concepts of "push" and "pull" tend to make us think of muscular actions by animate objects, but of course forces are not restricted to these alone. We can think of a force on an object as anything that can cause the object to accelerate—that is, change its velocity. Friction is a force—it slows down sliding objects. Gravity is a force—it causes objects falling straight down to increase their speed, or objects thrown straight up to slow down and eventually reverse their direction of motion. And it keeps satellites in orbit around the earth, preventing them from flying off into space.

We will come across a few other kinds of forces as we progress, but they all have this in common: they are capable of producing changes in the velocity of the object that they act upon.

SUMMARY

Newton's first law deals with the motion of objects in the absence of any net force; the second and third laws tell us what to expect when forces do act. The *second law* relates the *net force* on an object to its *acceleration:* the acceleration imparted to an object by a force is directly proportional to the force and inversely proportional to the mass. The *third law* shows that forces occur in pairs: every action force is accompanied by an oppositely directed reaction force of the same magnitude. The paired forces act on *different objects.*

Force is a *vector* quantity. The net force on an object is the vector sum (resultant) of all the individual forces.

We can study the curved-path motion of an object such as a thrown baseball by splitting it into *horizontal* and *vertical* parts. The motion is just that of a freely falling body, since the only force (neglecting air resistance) is the vertical force of gravity. The horizontal component of the motion is characterized by constant velocity because there is no force, and hence no acceleration, in the horizontal direction. Two objects starting from the same height, one thrown horizontally and the other dropped from rest, will strike the ground at the same time.

Exercises

1. In the expression for weight, $W = mg$, what is g?
2. If you know that a particular object is not accelerating, does this assure you that there are no forces acting on the object? Explain.
3. Do Newton's third-law forces cancel? Explain.
4. Distinguish between *mass* and *weight*.
5. Explain why Newton's first law is a special case of Newton's second law for zero net force.
6. Imagine that a soldier tries to push a Sherman tank from rest to a velocity of, say, 20 km/h north, in the absence of friction. Will the force required when the tank is pushed across the surface of the earth be greater than, less than, or the same as that on the surface of the moon? Explain.

7. In Problem 6, suppose that the soldier lifts the tank to a height of 1 meter off the surface. Will the force required to do this on the earth be greater than, less than, or the same as that required on the moon? (Hint: Consider *all* the forces.) Explain.

8. A freight train and a motorcycle collide at a railroad crossing. Which experiences greater *force*? The greater acceleration? Explain.

9. Two stones simultaneously leave the top of a tall building and plummet straight down, one having been fired from slingshot and the other dropped from rest. Once they are in flight, which has the greater acceleration (if either)? Explain. (Neglect air resistance.)

10. Describe the experience of a hunter firing a rifle if the bullet has twice the mass of the rifle.

11. You drop a cannonball and a feather in a vacuum and allow them to fall to the ground. Is the *force* the same on both? If so, why? If not, which experiences the greater force?

12. In Problem 11, are the *accelerations* the same? If so, why? If not, which is greater? Which object strikes the ground first?

13. A racquetball player serves the ball so that it is traveling horizontally when it leaves the racquet. During its flight toward the front wall, what is the ball's *horizontal* acceleration? Its *vertical* acceleration? Explain.

14. A golfer drives a ball off the tee. It follows a curved path and lands in the fairway. Describe the golf ball's acceleration while it is airborne.

15. A weightlifter exerts an upward force on a barbell. Newton's third law says that there is a "reaction" force to the upward force on the barbell. Which object exerts this reaction force? Which feels it? In what direction does it act? Explain.

16. A frustrated geologist kicks a boulder. Why does this hurt?

17. A donkey is urged to pull a cart. The donkey, thoroughly schooled in Newton's third law, gives this as his reason for refusing to try: "If I were to pull on that cart, it would pull back on me with exactly the same size force. Thus, since I can never exert a greater force on the cart than it exerts on me, I can never get it moving, so it is pointless to try." How would you respond?

18. Gordie Howe pushes a hockey puck across frictionless ice from rest to a velocity of 10 m/s east. Gordie then stops pushing. Describe the motion of the puck.

19. In Figure 4–12, the two blocks are connected by a rope that cannot stretch. The magnitude of **F** is 50 newtons. What is the acceleration of both blocks? What is the net force on the 10 kg block? (Neglect friction.)

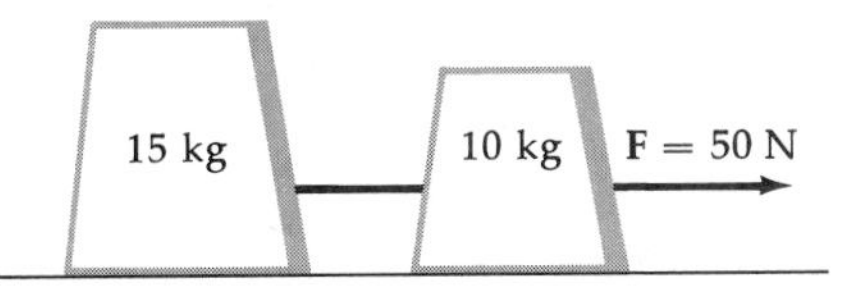

Figure 4-12.

20. A force **F** is applied to a 10 kg block accelerating it to the right at 5 m/s^2. If the magnitude of the friction force is 2 newtons, what is the magnitude of **F**?

21. Two stevedores drag a crate along a loading dock. One of them applies a 3 newton force to the east and the other a 4 newton force to the west. What is the magnitude of the resultant force? In which direction will the crate move?

22. A fisheries expert stands on a cliff 5 meters above the surface of a lake. To stir up some fish, this person throws a stone horizontally out over the water and into the lake. It splashes into the water at a point 25 meters from the base of the cliff. What was its velocity just after being released? What was its horizontal velocity after 1/2 second? Explain. (Neglect air resistance.)

23. A gun having a muzzle velocity of 300 m/s fires a bullet straight up. How far will the bullet have traveled after 1 second? Suppose you could somehow switch off gravity. How far will it travel in this case? (Neglect air resistance.)

24. A boy pulls a toy wagon at a constant velocity of 2 km/h east. He applies a force of 5 newtons east. What is the force of friction on the wagon?

25. Suppose you weigh 130 pounds. Identify the forces (a) as you fall freely toward the earth and (b) as you stand on the earth. Be careful to note which forces are *on* an object and which are exerted *by* an object.

26. Suppose that a wrecker pulls a car, giving it an acceleration of 6 m/s^2 east. What would be the acceleration of another car having twice the mass of the first if the wrecker applies the same force? Answer for a third car of one-third the mass of the first.

27. On the moon, objects weigh about one-sixth of their weight on earth. If a bowling ball on earth has a mass of 7.3 kg and weighs 73 newtons, what would its mass be on the moon? What would its weight be there?

28. Even though a thrown ball has a vertical acceleration, its motion is not vertical but is along a curved path. Give another example in which the motion and the acceleration are in different directions.

5

THE UNIVERSAL LAW OF GRAVITY

Sports and physical laws are associated in endless ways — some subtle, some very obvious, as in the parachutist's reliance on gravity during a free fall.

> *How many times it thundered before Franklin took the hint! How many apples fell on Newton's head before he took the hint! Nature is always hinting at us. It hints over and over again. And suddenly we take the hint.*
>
> *Robert Frost*

Gravity is something we take for granted. It is so much a part of everyday life we don't think much about it. But it is quite an important topic. Gravity makes apples fall and holds the earth in its orbit around the sun. Exciting and controversial topics, such as black holes and ideas about the origin of the universe, are tied to the gravitational interaction. Many of the problems of landing astronauts on the moon and traveling to other planets are associated with gravity. So it is a phenomenon that determines a great deal about our universe. In this chapter we are going to study this interaction and how it came to be discovered.

THE FORCE OF GRAVITY

The gravitational interaction plays an important role in our study of physics for several reasons. First of all, gravity is one of the four so-called *basic interactions* that govern the relationships between things in nature (later we will study the other three—electromagnetism, the nuclear force, and the weak force). Of the four, gravity is the interaction that dominates on the cosmic scale, determining the motions of the stars, the planets, and moons of the planets. Even so, it is a very weak force compared to some of the other basic interactions.

A second important reason for studying gravity is its position in the history of physics. Sir Isaac Newton formulated the law of gravity in 1665. By demonstrating that the force holding the moon in its orbit is the same one that pulls a falling apple toward the earth, he significantly improved his era's understanding of nature. Newton changed the world view. Aristotle and others held that celestial bodies, the moon included, were fundamentally different from earthbound objects. Newton provided convincing evidence that all bodies everywhere obey the same set of laws.

This discovery was crucial in the development of our modern world view. To the ancients, heavenly objects were special, even somewhat spiritual. A circle, thought to be the perfect shape, specified their motion. People thought them to be composed of materials different from those on earth. Heavenly objects, in this view, existed in a realm far removed from that of imperfect, corrupted terrestrial bodies. Today, these ideas sound strange. We are not at all surprised that the moon and earth contain the same stuff, or that the elements found here also make up the stars. Newton's law of gravity forms a key step along the journey the human mind has taken from the old world view to the new.

Newton created a new science of mathematics and astronomy—celestial mechanics—a reaching out into the void to measure and understand. When first revealed, Newton's work in this area sparked sermons and philosophical arguments of the most vehement kind.

A third important reason for studying the law of gravity is that there is a mathematical expression for the force. Having this formula means that we can test the theory and make predictions with it. Gravity is one of the few forces for which this is true. Friction, for instance, which is an example of the electromagnetic force, is too complicated to be expressed in a simple mathematical formula. We understand gravity much better than friction.

We can express the universal law of gravitation in this way:

> Every particle in the universe attracts every other particle with a force directly proportional to the product of the masses of the two particles and inversely proportional to the square of the distance between them.

For example, if two objects having masses m and M are separated by a distance d, the law says that they will attract each other by a force F_g (the subscript g stands for gravity) that may be written symbolically as

$$F_g = G\frac{mM}{d^2}$$

In this expression, G is a proportionality constant called the *universal gravitational constant.* The operative word is *universal.* As far as we know, G has the same value everywhere.

We have written F_g as a *scalar,* not a *vector,* even though we know that forces are vector quantities. The reason is that the gravitational force always points along a line joining the two masses. This means that we only have to keep track of force directions along this line rather than forces in arbitrary directions. Our equation for F_g, then, is the *magnitude* of the gravitational force.

Let us look a little more closely at the way F_g depends on distance. The force gets smaller as the *square* of the distance gets larger. We can see this by looking at the diagram in Figure

5-1. Suppose that the two objects having masses m and M are initially situated a distance of 2 centimeters apart. Each object will feel the force of gravity due to the other one. Now suppose we move one of them—M, say—so that it is 4 centimeters from m, twice as far away as before. Then we find by experiment that the gravitational force is one-fourth of its initial value. The distance has *increased* by a factor of two, so the force has *gone down* by a factor of four, since $2^2 = 4$. If the mass M is moved farther away to a distance of 6 centimeters, three times the initial distance, we find that the force at this distance is only one-ninth of its original value ($3^2 = 9$). As we continue to separate the masses by greater and greater distances, the force of gravity between them diminishes further. Physicists like to say that F_g "falls off" as the square of the distance.

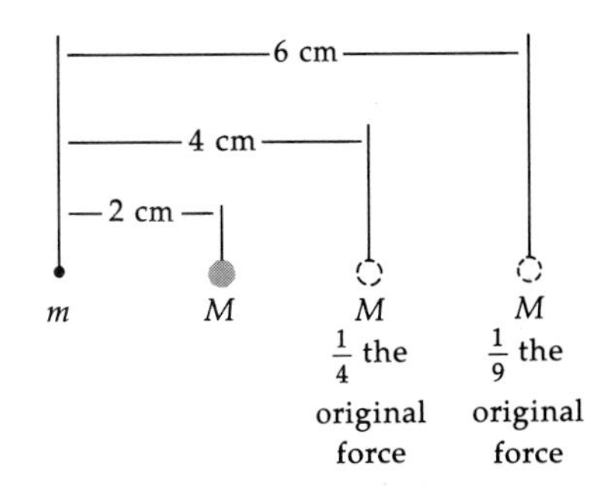

Figure 5-1. The force of gravity "falls off" as the square of the distance.

Gravity is what we call an "inverse-square" law. This can be seen in Figure 5-2, where we trace the pellets fired from a shotgun shell as they spread out after leaving the gun. The first square, situated one unit of length away from the explosion, has an area large enough to enclose all the pellets. As we follow them along straight lines to a distance of *two* units out, we see that it takes a square whose area is *four* times (2^2) that of the first one to contain all the pellets. Put another way, the number of pellets contained in a given area is only one-fourth what it is at a distance one unit out. We can continue this game: at a distance of *three* units, the required square is *nine* times (3^2) larger than the first one, and the number of pellets contained in the original area is one-ninth. If we think of the strength of the shot as proportional to the number of pellets per given area, then we can see that the strength *decreases* with the square of the distance. It follows an inverse-square law.

Note a few things about the gravity law. First of all, what do we mean by "the distance between objects" for bodies that have some size rather than being point masses? Do we mean the distance between the nearest edges, or some average distance, or what? One of the reasons Newton invented the calculus was to deal with this problem. He was able to show that when he considered a spherical body from the outside, he could treat it as if its mass were concentrated at a single point, the center. So the distance d in the equation for the law of gravity means the distance between centers of spherical bodies. (See Figure 5-3.)

Figure 5-2. Number of pellets per area goes as $1/d^2$.

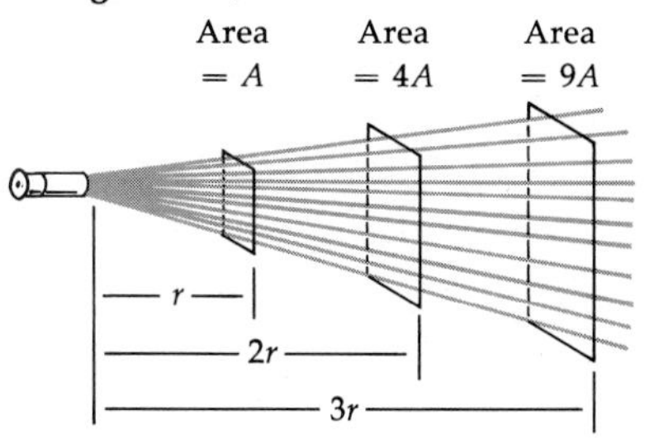

Second, observe that gravity is always *attractive.* That is, the force that the *sun* exerts on the *earth* is a vector pointing *from* the earth *to* the sun. The third law of motion tells us that the *earth* also exerts an equal and oppositely directed force on the *sun.* So we see that the gravitational force is a mutual attraction of two objects. They never repel each other. Hence we speak of the "pull" and not the "push" of gravity. We shall see, in con-

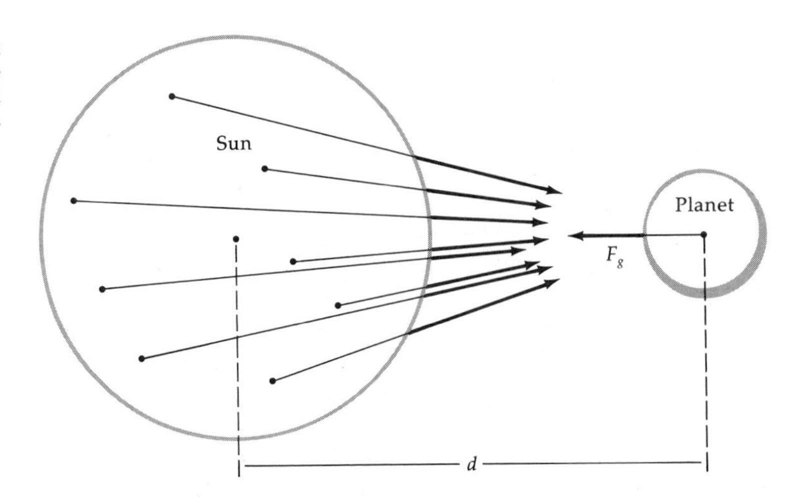

Figure 5-3. All parts of the sun attract the planet gravitationally. The overall effect is the same as if all the sun's mass were at its center.

trast to this, that the forces involving electrical charges and magnets can be either "pushes" or "pulls."

Finally, and most important, the law is a universal one. As far as we can tell, it applies to every pair of objects in the universe—two people, two planets, you and the moon, everything. To say this requires a sort of leap of faith, because obviously we can't go around measuring the force of gravity between every pair of masses in the universe. But every experiment that has been done and all the observations that have been made indicate that gravity works the same way throughout the universe.

You might want to think about this universal phenomenon a little because you probably aren't aware of it. You experience gravity with respect to yourself and the earth, but you don't feel any force between yourself and this book, for example. This is due to the grossness of human senses. It turns out, in fact, that sensitive measuring devices can detect these extremely small forces.

One result of these experiments is that we can determine the value of G, which turns out to be an extremely small number. In fact, G is so small that unless one of the masses involved is quite large, the gravitational force is so small as to be negligible. We don't notice it for ordinary-sized objects. If the masses are in kilograms and distances are in meters, then G has the value 6.67×10^{-11}. Written out, this is 0.0000000000667, a tiny number. We can see why at least one of the masses must be large in order for the force F_g to be very noticeable.

This method of using powers of ten is convenient for expressing very large and very small numbers. The superscript is positive for large numbers and negative for numbers smaller than one. For example, 100 million is written as 10^8, which means "1 followed by 8 zeros," or 100,000,000. The radius of

Geophysics: Gravity and Oil

The techniques of geophysics are used in a number of endeavors. One of the most important is exploring for new oil deposits. With many of the rich, easily found oil fields of America already being depleted, technicians are searching for new finds several miles underground. How can they detect oil at this depth?

Oil is lighter than water; therefore, it moves upward through porous rock formations until it is blocked by denser, nonporous rock. There, it accumulates over the course of tens of millions of years. One method geophysicists use to search for these pockets involves a *gravity meter.* This device is suspended from an aircraft and flown over a region. The high-density rock present will have a stronger gravitational pull than rock of lower density. The gravity meter can detect the slight increase in gravity caused by the dense underground—and perhaps oil-trapping—rock. Thus, the gravity characteristics of an area can be a tipoff to a promising petroleum site. The Arctic Slope of Alaska, whose oil is now being used, was scanned in this manner.

our galaxy is 6×10^{19} or 60,000,000,000,000,000,000 meters. At the other extreme, the mass of a penicillin molecule is about 0.00000000000000005 kg. Notice that between the 5 and the decimal there are 16 zeros; we write this as 5×10^{-17} kg. This shorthand saves a lot of space. We will use this method often throughout the book.

Let us sum up our discussion thus far. Gravity pulls objects together, and never pushes them apart. The gravitational force is extremely weak between bodies the size of billiard balls and freight trains, but on the size scale of the earth and the stars, where the masses are tremendously large, it is the most important interaction in all of nature. Gravity depends on the distance between bodies in a particular way; namely, it obeys an "inverse-square law," so called because the force diminishes as the square of the distance increases.

Learning Checks

1. Why is gravity a key part of physics?
2. State Newton's law of gravity as you would to a friend.
3. What does it mean to say that G is a universal constant?

KEPLER'S LAWS OF PLANETARY MOTION

What led Newton to discover the law of gravity? There is a history to consider here, because the life's work of several people who preceded Newton culminated in this law.

As far back as recorded history goes, and probably even further, people kept track of the movement of planets and stars. They observed the heavens for many reasons, most having to do with religion and agriculture. Much of the mystique surrounding ancient priesthoods, for example, was derived from the priests' ability to predict natural events, such as the flooding of Egypt's Nile Valley or an eclipse of the sun. Ancient people regarded the cosmos with a certain awe. The patterns and regularities in the movements of celestial bodies enhanced this emotion. (Astrology illustrates a holdover of this feeling today.)

Science also encouraged astronomical observations. The motivation here was a desire to understand the universe better through a careful charting of the heavens. Perhaps the most massive study was undertaken by the Danish astronomer Tycho Brahe (1546–1601). Heavily financed by the King of Denmark, Tycho spent the better part of a lifetime, in those pretelescope years of the late sixteenth century, collecting staggering quantities of accurate data about the positions of the planets and the stars. (See Figure 5–4.)

Figure 5-4. Tycho's observatory.

With these data, Johannes Kepler, one of Tycho's assistants, paved the way for Newton's law of gravitation. Kepler, a contemporary and acquaintance of Galileo, had attracted Tycho's attention with his attempts to account for the spacing of the six known planets. Sensing that the planetary positions were somehow related to geometry, Kepler devised a scheme involving the five "perfect solids"—tetrahedron, octahedron, dodecahedron, icosahedron, and cube. These are shown in Figure 5–5. Kepler concocted a model in which these five solids nested inside one another much like a set of mixing bowls. His idea was that four of the planetary spheres could exist in the spaces between these solids, with a fifth sphere inside the whole nest and a sixth around the outside. (See Figure 5–6.) He was able to

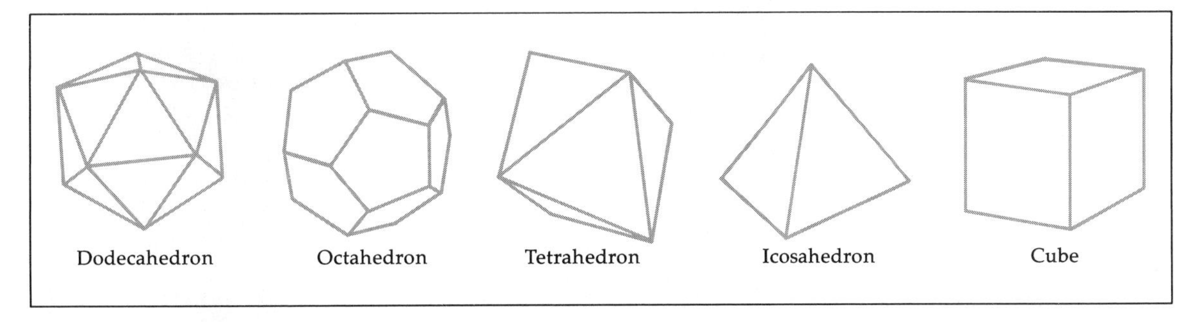

Figure 5-5. The five perfect solids.

arrive at a sequence of the solids that reproduced the planets' relative spacing to within about 5 percent. With his spheres, Kepler remained true to the centuries-old belief in "perfect" circular motion for heavenly bodies.

When Tycho invited Kepler to become his assistant, he had him determine precisely the orbit of the planet Mars. Using Tycho's highly accurate data, Kepler eventually concluded that there was no way to fit Mars' orbit onto a sphere—that is, the orbit was not circular, but some sort of oval shape. His experience with Tycho's remarkable data had a profound effect on Kepler. As a result, he abandoned his earlier mystical notions about celestial geometry.

Guided by his work on the orbits of Mars and the other planets, Kepler formulated three laws of planetary motion. They are:

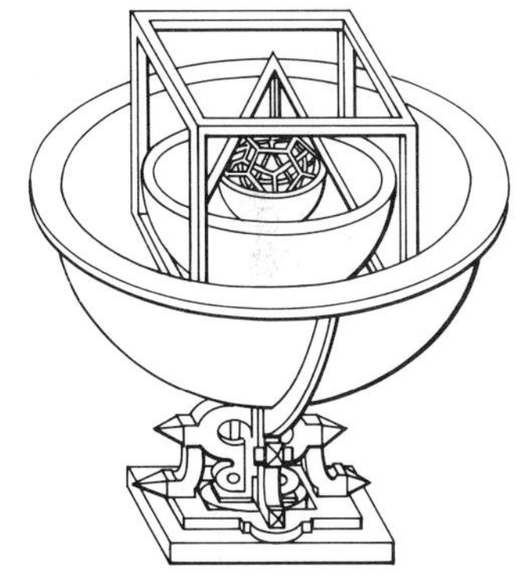

Figure 5-6. Kepler's perfect-solids model of the solar system.

1. Each planet travels about the sun in an ellipse with the sun at one focus.
2. A line from the sun to the planet sweeps out equal areas in equal intervals of time.
3. The ratio of the square of the planet's period of revolution to the cube of the major axis of its orbit is a constant for all planets.

Biophysics: Gravity and the Human Body

Humans have evolved within a gravitational field, a fact that has affected our bodies in a variety of ways. We are, for instance, equipped with highly specialized nerve endings and receptor organs called *gravireceptors.* These are located in muscles, tendons, joints, and our inner ears. They provide the brain with information concerning body position, balance, and the direction of gravitational forces.

Many therapists who do "body work" (that is, employ methods of manipulating the human body for the physical improvement of damaged parts or for reasons of enhancing overall health) realize the impact gravity has had on shaping human form. In this line of work, the body is often visualized as a series of major segments—head, chest, hips, legs—balanced on top of one another like blocks. The ideal body is, then, one in which the "blocks" are in perfect alignment, a condition allowing an efficient use of gravity. As some body work therapists point out, in many people the major segments are not properly aligned, so the body does not work with the gravity field but rather fights against it, causing awkward and inefficient motion. Since gravity is constantly with us from the day we are born, it pays to be in bodily tune with it.

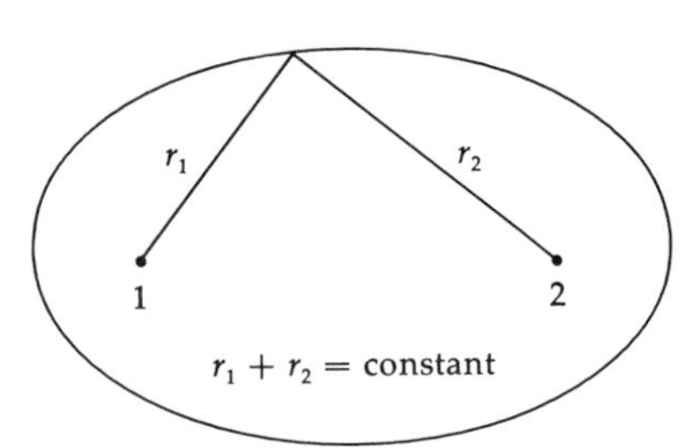

Figure 5-7. The ellipse.

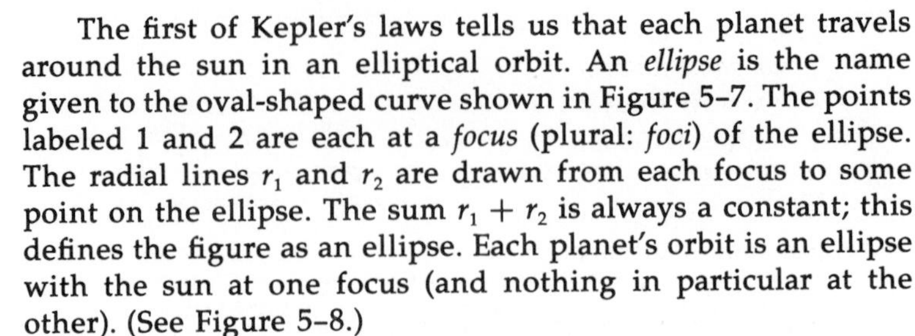

The first of Kepler's laws tells us that each planet travels around the sun in an elliptical orbit. An *ellipse* is the name given to the oval-shaped curve shown in Figure 5-7. The points labeled 1 and 2 are each at a *focus* (plural: *foci*) of the ellipse. The radial lines r_1 and r_2 are drawn from each focus to some point on the ellipse. The sum $r_1 + r_2$ is always a constant; this defines the figure as an ellipse. Each planet's orbit is an ellipse with the sun at one focus (and nothing in particular at the other). (See Figure 5-8.)

The second law gives some quantitative information about motion of a planet at different parts of its orbit. Figure 5-9 shows a planet's position (points A, B, C, and D) at different times of the year. The shaded regions are the areas swept out by the radial line joining the planet and the sun as the planet moves from A to B and from C to D. According to this law, if the planet requires the same amount of time to go from A to B as from C to D, then these areas must be equal. You can see from the picture that the distance along the orbit from A to B must be smaller than the distance along the orbit from C to D. Because these distances are covered in equal intervals of time, the planet must be traveling with a greater speed the closer it is to the sun. So the speed is not a constant but depends on the distance. The farther away the planet is from the sun, the slower it travels.

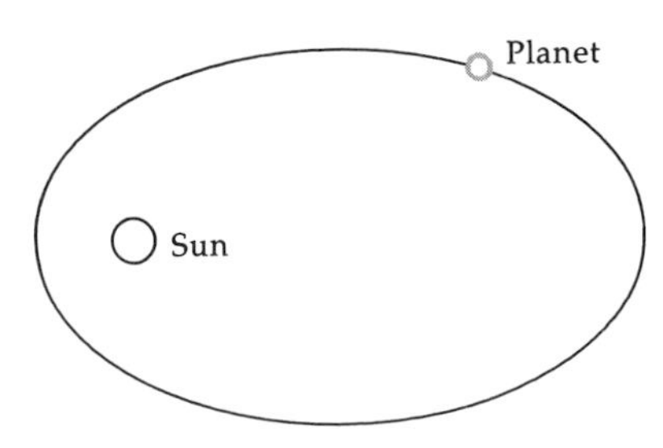

Figure 5-8. The planet orbits the sun in an ellipse with the sun at a focus.

This law is particularly interesting because it represents an early example of the importance of constancy in nature. Kepler noticed that when planets travel in elliptical orbits, the readily measurable quantities such as speed and distance from the sun do not stay fixed. However, the fact that the *area* swept out by the radial line remains constant was an important discovery.

Kepler's third law, formulated some ten years after the other two, is extremely important because it led Newton directly to the equation for the gravitational force. We can write this law in compact form as

$$k = \frac{D^3}{T^2}$$

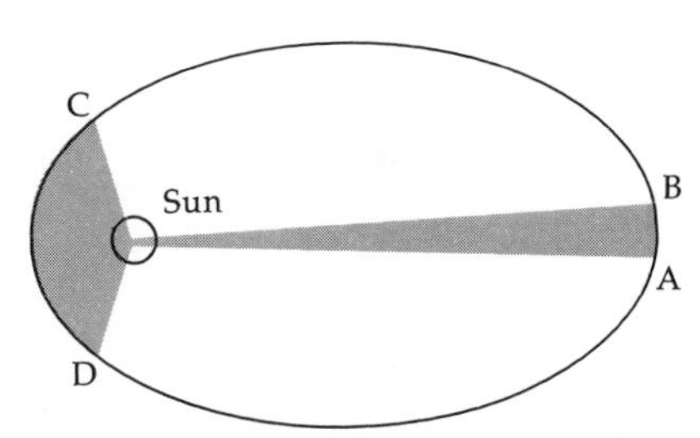

Figure 5-9. The shaded areas are equal for equal time intervals. Therefore the planet moves from distance A to B and distance C to D in the same time, *t*.

where k is a constant; T is the planet's year, which is the time (in earth days) required for a planet to make one complete trip around the sun; and D is the length of the major axis of the planet's orbit. The major axis of an ellipse is the line running across the ellipse through the foci. The minor axis is a similar line bisecting the major axis and perpendicular to it. (See Figure 5-10.)

The remarkable thing about this third law is that the number k is the *same number for all planets in our solar system.* That is, if we know how long it takes Jupiter to orbit the sun, and if we know the dimensions of its orbit, it is easy to calculate k according to this recipe. The surprise is that a similar calculation using the analogous numbers for Mars, Venus, or any other planet in our solar system gives the *same number for k.* This tantalizing fact encouraged Newton to search for some relationship between the properties of the sun and those of the various planets.

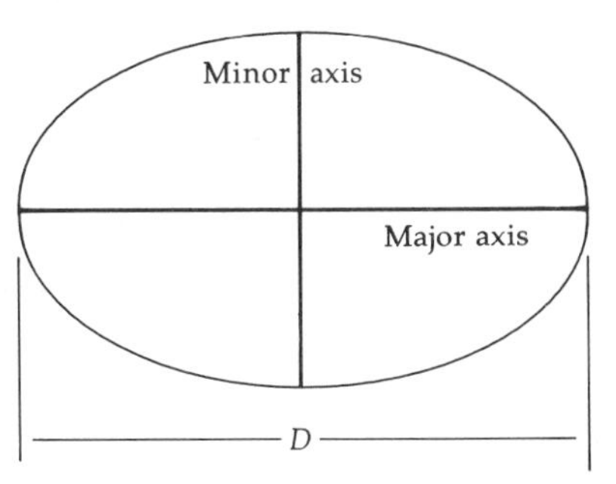

Figure 5-10. Axes of an ellipse.

Here is where Newton enters the picture. What was his reasoning as he developed the law of gravitation? As we know from Chapter 4, Newton deduced that a moving object should coast in a straight line at constant speed until some force caused it to accelerate. The planets' elliptical orbits around the sun implied that some external force was continuously tugging them out of their natural straight-line motion. To Newton, this meant that the force was somehow caused by the presence of the sun, it being the common element linking the planets together in all of Kepler's laws.

Kepler's second and third laws provided the quantitative information Newton needed to give a mathematical form to his law. It was clear from the second planetary law that the distance between planet and sun was related to the acceleration of the planet, since it travels faster when closer to the sun and slower when farther away. The fact that the force depends

Kepler and k

Reflect a moment on the laborious task Kepler performed in discovering the ratio k. One can imagine him almost knee-deep in a huge pile of calculations, trying to distill some simple relationship between T and D. Probably the particular combination of T^2 and D^3 was not the first one he tried. It doesn't leap out at you. Presumably, Kepler tried many combinations before he found the right one. Today we could program a high-speed computer to plow through Kepler's numbers and arrive at the proper equation during the time it takes for a coffee break. But Kepler didn't even have access to a slide rule, much less a computer, since neither had been invented. He did these tedious calculations by hand. It took a long time, to say the least.

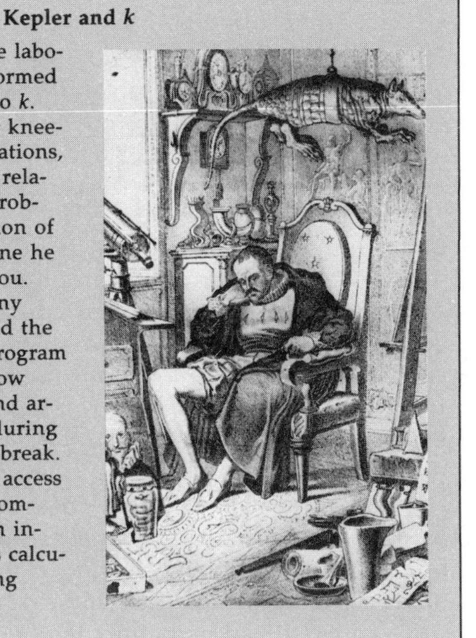

on d^2—and not on d or d^3 or some other form—comes from Kepler's third law. (To show this involves quite a bit of algebra. Supplement 4 shows this development for interested readers.)

Learning Checks

1. Show that a circle is a special case of an ellipse with coincident foci.
2. At what point in its orbit is a planet traveling the fastest? The slowest?
3. What defines an ellipse?
4. How did Newton decide that the planets were subject to a force? What is the direction of this force?

NEWTON AND THE MOON

We learned in Chapter 2 that freely falling bodies are accelerated toward the center of the earth at the rate of 9.8 m/s^2, the value of the constant g. Planets moving in their orbits around the sun also undergo an acceleration that is directed toward the sun. Perhaps Newton's greatest insight was in recognizing that the acceleration of freely falling bodies and that of the orbiting planets are both due to the force of gravity. Luckily for Newton, the moon orbits the earth in a nearly circular path, much like the earth going around the sun. This provided a way of testing his idea. Let us look at the reasoning.

If we suppose the moon's orbit to be exactly circular (it nearly is), then its acceleration is directed toward the orbit's center, which is the center of the earth. The magnitude of this acceleration (see Supplement 3) is $a = v^2/R$, where v is the moon's speed and R is the radius of its orbit. Both of these quantities were known in Newton's day. The speed of the moon in its orbit is $v = 1016$ m/s and the radius is $R = 380{,}000$ km. Putting these numbers in the equation for a gives the moon's acceleration as

"THE BIG BANG? BELIEVE ME, IT WAS VERY, VERY, VERY, VERY, VERY, VERY BIG."

$$a_{\text{moon}} = 0.00271 \text{ m/s}^2$$

Now, look at this from the standpoint of the law of gravity. We have

$$F_g = m\frac{GM}{R^2}$$

where m is the mass of the moon and M is the mass of the earth. Comparing this to Newton's second law for an arbitrary force F

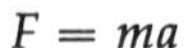

$$F = ma$$

we see that the acceleration caused by the force of the earth's gravity is

$$a = \frac{GM}{R^2}$$

We can use this equation without knowing the mass of the earth, since it tells us that the gravitational acceleration diminishes as the square of the distance increases. At the surface of the earth, where $R = 6300$ km, we know that $a = g = 9.8 \text{ m/s}^2$. The radius of the moon's orbit (380,000 km) is 60 times the radius of the earth. Therefore, the acceleration of the moon due to the earth's gravity should be g divided by 60^2. That is,

$$a_{\text{moon}} = \frac{9.8 \text{ m/s}^2}{60^2} = 0.00271 \text{ m/s}^2$$

which is the value we got earlier. (See Figure 5-11.)

The agreement between these two independent calculations of the moon's acceleration confirmed Newton's idea: The *same kind of force*, gravity, is responsible for the motion of earth-bound objects *and* the paths followed by celestial bodies. This realization was a major step forward in understanding how the universe operates. As we have mentioned, Aristotle believed the planets and stars to have natural motions different from those of rocks and cannonballs. Newton put them all on equal footing. Once he had established that gravity accounts for the motion of falling bodies, the moon, and the planets, Newton had little trouble convincing others that this force was active between *any* two objects: gravity is a universal interaction.

The variation of gravitational acceleration with distance

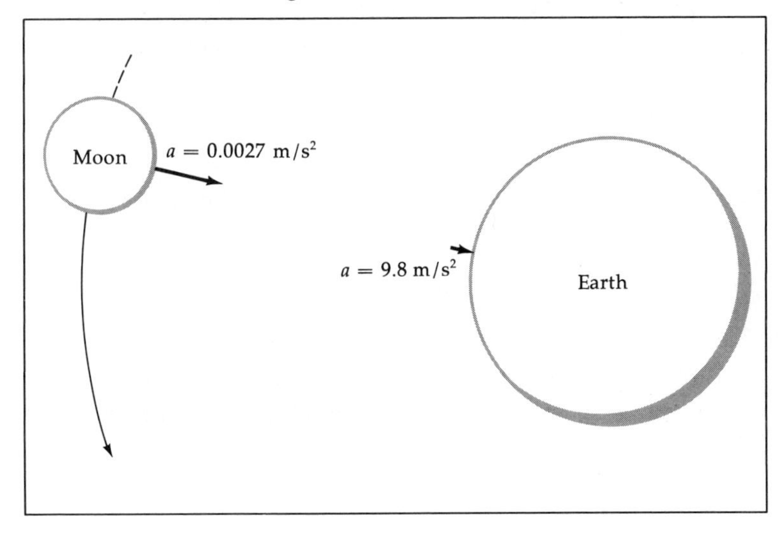

Figure 5-11. Acceleration due to earth's gravity falls off as the square of the distance.

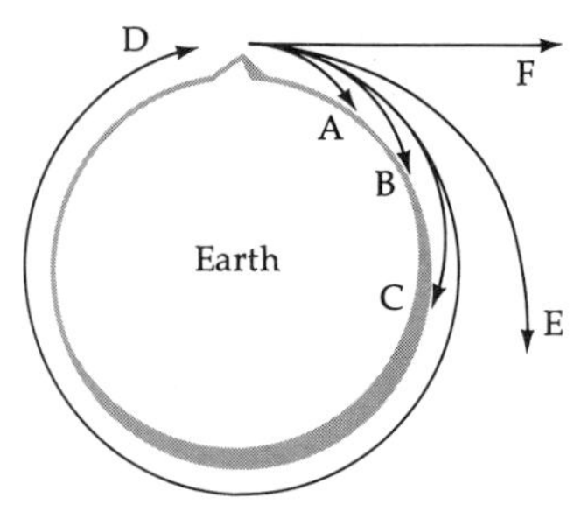

Figure 5-12. Trajectories of stones thrown from a mountaintop with increasing horizontal velocities.

was discussed in Chapter 4. We learned that g has the *nearly* constant value of 9.8 m/s^2 close to the surface of the earth. It may seem rather vague to say that some number is "almost" constant; after all, either it is constant or it isn't. But now that we know about gravity, we can appreciate why this loose talk is justified. We see that the value of g depends on the distance from the center of the earth. At the top of a mountain this distance might be 6305 kilometers rather than the surface distance of 6300 kilometers. The corresponding difference in the values of g at these two locations is quite small, because the 5 kilometer difference in distance is small compared to 6300 kilometers. The difference in your weight as measured in the attic and in the basement is hardly worth talking about. So, there is some justification in saying that g is nearly constant. However, measuring your weight, say, 1000 kilometers above the surface of the earth will make a significant difference; at this altitude g is much smaller than its surface value of 9.8 m/s^2.

Falling Satellites

In our study of projectile motion we saw that an object is in "free fall" whether we drop it from rest, throw it straight up or straight down, or toss it in some arbitrary direction. We say it is in free fall because once it is released the only force exerted on it (again, neglecting air resistance) is gravity, which acts along a line joining the center of mass of the object and the center of the earth. Recall that if we throw the object horizontally, it travels along a curved path and strikes the ground after the

(a) High-capacity communication satellite (COMSTAR D-3) launched for United States domestic telephone communications.
(b) British research satellite (UK-6) designed to conduct studies to promote better understanding of such astrophysical phenomena as quasars, radio galaxies, supernovae, and pulsars.

same interval of time as if it had been dropped. Clearly, if we throw it harder in the same direction, the only thing different is the horizontal distance that it travels. During its flight it is still falling freely: it is still subject only to the force of gravity, and it still accelerates just as if we had dropped it from rest. The greater the speed, the farther downrange it lands. But because the earth is a sphere, the surface of the earth curves and drops out from under the path of the falling object, until at a sufficiently high initial speed the falling object doesn't strike the ground at all. Instead, it circles the earth. (See Figure 5-12.) We thus have a *satellite*, traveling around the earth in a circular orbit with the constant speed that we gave it initially. Thus, our satellite is a body in free fall, and the curved path along which it falls is matched by the curvature of the spherical earth. Although it is still accelerating *toward* the earth, its motion is *around* it. We might say that the satellite "falls around the earth."

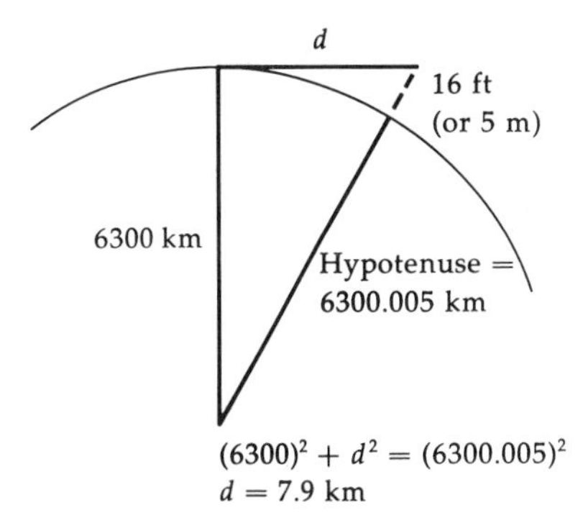

Figure 5-13. Calculation of speed necessary to put a satellite into orbit.

In actual practice, of course, we can't do this by throwing a stone or a baseball because we aren't strong enough to give it a high enough speed. Also, doing it at the surface of the earth isn't such a good idea, because such things as houses, trees, and mountains get in the way. So we use a rocket to replace our pitching arm, and we get our rocket above the trees and the atmosphere (to avoid air resistance) before launching it horizontally. But never mind these engineering details; the basic physics is just as we have described it.

How fast must the satellite be going to get into orbit? It isn't too difficult to calculate this speed because it depends only on the radius of the earth and the acceleration due to gravity. Figure 5-13 shows how to do it. We know from our earlier work that an object in free fall near the surface of the earth travels a vertical distance of about 5 meters (about 16 feet) during the first second after it is released. Knowing the radius of the earth (6300 kilometers), we can determine the horizontal distance corresponding to a drop of 5 meters from a line tangent to the surface of the earth. We find that this distance is 8 kilometers (about 5 miles). Put another way, the earth curves 5 meters for every 8 kilometers of surface, so a ship 5 meters tall would disappear over the horizon 8 kilometers from shore. Thus, the speed of our orbiting satellite must be 8 km/s, or about 29,000 km/h. This speed will vary somewhat according to the radius of the particular orbit. (See Figure 5-14.)

Our own moon is just such a falling satellite, orbiting the earth in a nearly circular path at a distance of some 380,000 kilometers from the earth's surface. A calculation similar to the one we just outlined for artificial satellites shows that the moon falls away from the straight-line path it would travel in the absence of gravity, a distance of about one-eighth of a centimeter in one second.

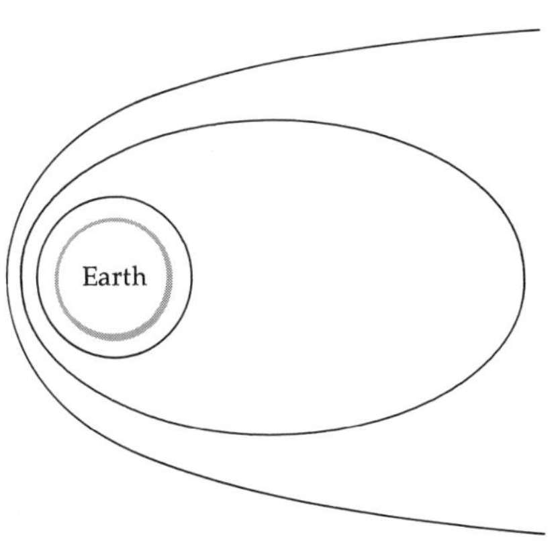

Figure 5-14. Satellite orbits.

If we wish to put our satellite in an *elliptical* orbit rather than a *circular* one, we should give it an initial speed somewhat greater than 8 km/s. It would then orbit the earth in an elliptical path with the center of the earth at one focus of the ellipse. If we give our rocket a still greater initial speed of about 11.2 km/s, or about 40,000 km/h, it would not orbit the earth at all, but would overcome the influence of the earth's gravity and shoot out into space. This velocity of 40,000 km/h horizontal is called the *escape velocity,* because it is that initial horizontal velocity an object must have to escape the gravitational field of the earth. The number we have quoted, 40,000 km/h, is the escape velocity for the earth. Each planet, star, moon, or other celestial body has its own escape velocity that depends both on the mass of the body and its size, or radius.

Learning Checks

1. What does it mean to say that the moon is a freely falling body?
2. What does it mean to say that a satellite "falls around" the earth?
3. What would be the acceleration of a satellite in a circular orbit whose radius is ten times the radius of the earth?
4. We said that the gravitational acceleration g is not strictly a constant. Explain.
5. Could g ever be zero? If so, at what distance from earth?

THE GRAVITATIONAL FIELD

Since Newton's time, ideas about the gravitational force between masses have changed somewhat. It is difficult to imagine two objects exerting forces on each other without touching. This is sometimes called "action-at-a-distance," and some physicists are uncomfortable with it. So they have developed the concept of the *field.* When an apple is released in midair, it falls toward the center of the earth because of the gravitational force. Pictorially, one may imagine that the earth is surrounded by invisible lines of force directed toward its center, as shown in Figure 5–15. The apple falls toward the earth parallel to one of these lines of force that represent the gravitational field of the earth. The *direction* of the field, indicated by the arrowheads, is the direction that an object will move under the influence of the field. The field *strength* is larger where the lines are crowded together. Very near the surface of the earth, where the field strength is large, the field lines are closer together than they are several miles out in space, where the gravitational field is weaker.

Great Lost Discoveries II – Invulnerability

This short science fiction tale by Fredric Brown shows what can happen when someone—no matter how clever—neglects to study basic physics and remains ignorant of fundamental concepts of this science.

The second great lost discovery was the secret of invulnerability. It was discovered in 1952 by a United States Navy radar officer, Lieutenant Paul Hickendorf. The device was electronic and consisted of a small box that could be carried handily in a pocket; when a switch on the box was turned on the person carrying the device was surrounded by a force-field whose strength, as far as it could be measured by Hickendorf's excellent mathematics, was as near as matters to infinite.

The field was also completely impervious to any degree of heat and any quantity of radiation.

Lieutenant Hickendorf decided that a man—or a woman or a child or a dog—enclosed in that force-field could withstand the explosion of a hydrogen bomb at closest range and not be injured in the slightest degree.

No hydrogen bomb had been exploded at that time, but at the moment he completed his device, the lieutenant happened to be on a ship, cruiser class, that was steaming across the Pacific Ocean en route to an atoll called Eniwetok, and the fact had leaked out that they were to be there to assist in the first explosion of a hydrogen bomb.

Lieutenant Hickendorf decided to get lost—to hide out on the target island and be there when the bomb went off, and also to be there unharmed after it went off, thereby demonstrating beyond all doubt that his discovery was workable, a defense against the most powerful weapon of all time.

It proved difficult, but he hid out successfully and was there, only yards away from the H-bomb—after having crept closer and closer during the countdown—when it exploded.

His calculations had been completely correct and he was not injured in the slightest way, not scratched, not bruised, not burned.

But Lieutenant Hickendorf had overlooked the possibility of one thing happening, and that one thing happened. He was blown off the surface of the earth with much more than escape velocity. Straight out, not even into orbit. Forty-nine days later he fell into the sun, still completely uninjured but unfortunately long since dead since the force-field had carried with it enough air to last him only a few hours, and so his discovery was lost to mankind, at least for the duration of the twentieth century.

Source: Fredric Brown, *Nightmares and Geezenstacks.* New York: Bantam Books, 1979, p. 3. Reprinted by permission of the author and the author's agents, Scott Meredith Literary Agency, Inc., 845 Third Avenue, New York, NY 10022.

Formally, the field strength is defined as the force per unit mass at some point in space, or F_g/m. We know that

$$F_g = \frac{mMG}{d^2}$$

Dividing by m, the mass of some arbitrary object, gives

$$F_g/m = \frac{MG}{d^2}$$

which is the gravitational field strength due to a sphere of mass M at a distance d from its center. You can see that this is the expression we had on page 61 for the gravitational acceleration g due to the mass M. This allows a useful interpretation of g: it is a measure of the gravitational field strength at some particular point a distance d from the mass that causes the field.

Obviously, we haven't changed anything fundamental by introducing this notion of a field; rather, the idea is simply a convenient mental construct allowing us to draw pictures (mentally, as well as on paper) of what happens with gravity. One can say that space itself is somehow distorted by the presence

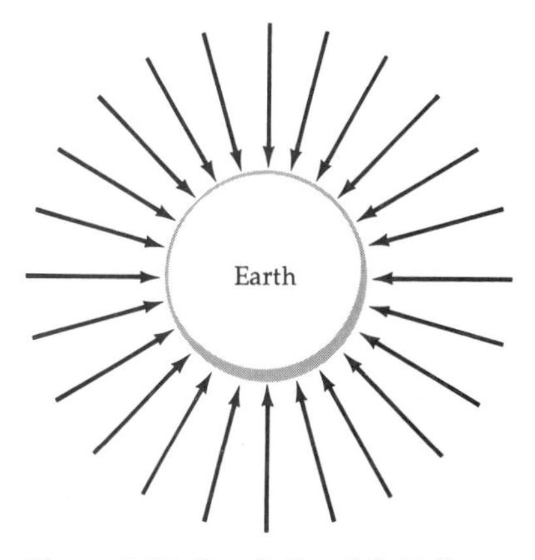

Figure 5-15. Gravitational field lines.

Figure 5-16. Bending of light by gravity.

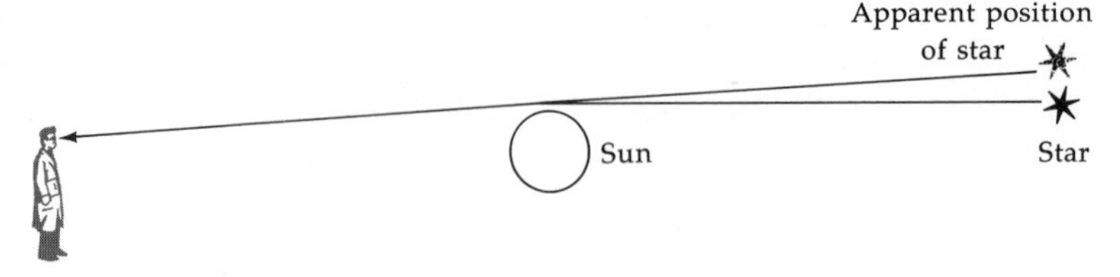

of the mass with which the gravitational field is associated. Any object in the vicinity of this mass will be affected by this space distortion.

BLACK HOLES: GRAVITATIONAL COLLAPSE

Modern theories of gravity, such as the general theory of relativity developed by Albert Einstein in the early part of the twentieth century, are different from Newton's. One of the important features of Newton's gravitational law is that the force of gravity applies only to things that have mass, such as rocks and water and people, objects we are familiar with in everyday life. Einstein generalized this by postulating that *energy itself* is affected by a gravitational field. A beam of light, for example, possesses energy and, if Einstein is right, should be deflected by the gravitational force. Although a beam of light has no mass, we shall see later that it does have an *equivalent* mass because of its energy due to the mass–energy equivalence expressed in Einstein's famous equation, $E = mc^2$, where c is the velocity of light.

For example, Einstein's theory predicts, and experiments have shown, that light coming from a distant star and passing near the sun on its way to the earth is bent by the gravitational field of the sun. (See Figure 5–16.) Because we see the light coming along a straight-line path, we believe the star to be in its apparent position, which is different from its true location.

Now suppose that the gravitational field of the sun were much stronger, perhaps billions of times stronger than it is. Then the light coming from other stars would be severely bent along an ever-tightening spiral until it became totally absorbed. Anything passing by this supersun would also be sucked in—spaceships, comets, anything at all. (See Figure 5–17.) If the gravitational field were strong enough, the light from the sun itself could not escape from the surface. Radio beams could not be bounced off the sun, for they, too, would be trapped by this immensely strong gravitational field. The escape velocity would be greater than the velocity of light.

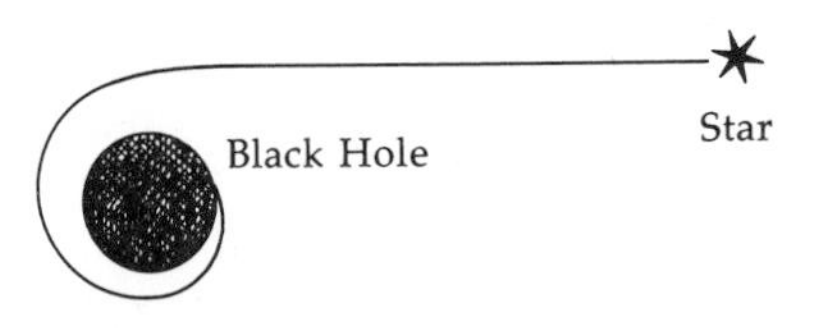

Figure 5-17. Capture of light by a black hole.

Since no light could reach us from this supersun, neither light bouncing off it from another star nor light generated from its own interior, we would not be able to see it. The sun under these conditions would offer no more contrast with the background sky than a black cat at midnight. It would have become a *black hole,* its blackness due to the absence of any light emanating from it.

How might the sun or any other star become a black hole? We know that the force of gravity between two objects becomes greater as they get closer together. An object at the surface of the sun (assuming it could survive the heat) would experience a gravitational force that depends on the radius of the sun, some 645,000 kilometers.

Now suppose that the sun were no longer this large but somehow, without decreasing its *mass,* diminished in *size* until it had a radius of only 3 km. All the matter in the sun would be jammed into this relatively tiny ball of incredible density. Our sun would have collapsed under its own gravity. In this case, the object on the surface of the sun, only 3 kilometers from the center, would experience a gravitational force about 4×10^{10} or 40 billion times stronger than before. Any matter or radiation coming within 2 kilometers of the sun's surface (5 kilometers from the center) would be pulled into it. The sun would be a black hole.

What would happen to the motion of the earth under these conditions? Nothing. Note that we have been talking about the force of gravity at the *surface* of the sun. But since the distance between the center of mass of the sun and that of the earth is not affected by the gravitational collapse of the sun, the mutual force of gravity between earth and sun would be the same as it had always been. The orbital path that the earth travels about the sun would remain unchanged.

THE OSCILLATING UNIVERSE

The most widely accepted mechanism of the origin of the universe is the so-called big bang theory. This theory holds that at the beginning of time all the matter in the universe was gathered into one relatively small, closely-packed sphere of unimaginable density, equivalent to all the matter in our solar system crowded into the volume of a drop of water. This severe crowding generated so much heat that the entire sphere exploded, hurtling everything out through space at tremendous speeds. The universe at that time was inconceivably hot, perhaps a hundred thousand million (10^{11}) degrees Celsius, and filled with an undifferentiated soup of matter and radiation. Within a few seconds the rapidly expanding universe had cooled to a few thousand million (10^{9}) degrees, and hydrogen nuclei began to form in great abundance. (In fact, to this day about 90 percent of the matter in the universe is hydrogen.) At this temperature, any other nuclei would have been blasted apart almost as soon as they were formed.

After about 3 minutes, the temperature had dropped enough so that the helium nuclei could remain stable. From this point for about 700,000 years the universe continued to

A New Energy Source?

Black holes are indeed awesome, not to mention dangerous. That, however, is hardly enough to frighten energy-hungry humans. "In the collapse of a star to form a black hole," John Taylor points out, "vast amounts of energy will be liberated, and while matter is being eaten up by an already formed black hole further energy should be produced. So we expect the black hole to be a source of great supplies of energy."

The energy that some scientists hope to draw from a black hole is gravitational in nature, arising by gravitational contraction. There is, however, a small catch along this primrose path. We have to "tame" the black hole ". . . so that energy generation can be obtained from it in a continuous fashion without constant fear of falling inside . . ."

Source: John Taylor, *Black Holes: The End of the Universe?* New York: Random House, 1973.

expand and cool, until the temperature was low enough for electrons to join with the nuclei to form stable atoms. Later the stars, planets, and galaxies formed as more and more matter began to come together.

The universe is still undergoing expansion, some 12 billion to 15 billion years later, as evidenced by an effect known as the *Doppler shift* in the wavelengths of light coming from the various galaxies. (We will learn about the Doppler effect in our study of wave motion in a later chapter.) This effect shows that the galaxies are moving away from one another at velocities that are proportional to their distances apart. Thus, for example, suppose that galaxy X is moving away from the earth twice as fast as galaxy Y. It turns out that galaxy X is twice as far away from the earth as galaxy Y. This gives credence to the big bang theory, for it suggests that the earth and galaxies X and Y all started from the same point.

For instance, the cluster of galaxies in the constellation Virgo is about 78 million light years away, and, according to the Doppler shift, moving away at about 1200 km/s.* Galaxies in Ursa Major are seen to be about thirteen times as distant (1 billion light years) and moving nearly thirteen times as fast, or 15,000 km/s.

Figure 5–18 illustrates the expansion of the universe. The central dot (G) represents any typical galaxy. The other dots are shown moving away with velocities relative to G indicated by the length and directions of the attached arrows. Notice that the galaxies farthest from G have the longest arrows.

We can get a feeling for this expansion by a very simple experiment. Mark ink spots fairly close together on a collapsed balloon. Now, inflate the balloon and watch how the spots separate. Pick out one spot and imagine we are "riding" on it, watching the other spots run away from us. The result is that, no matter which spot you select, all others are moving away as the balloon expands.

Figure 5-18. Expansion of the universe.

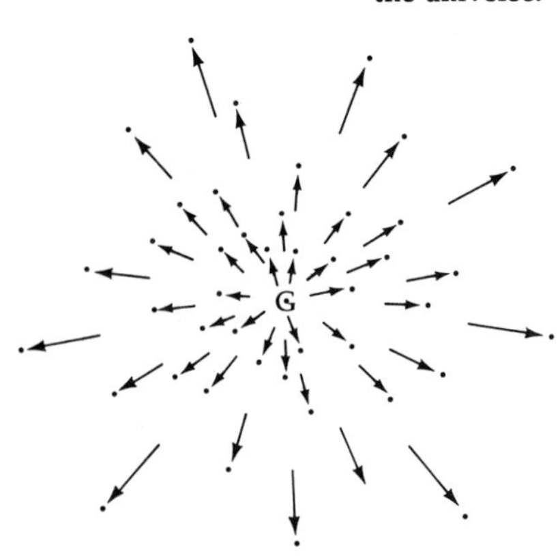

Will this expansion continue indefinitely? Perhaps not, and here is where the gravitational force enters the picture. The force of gravity between, say, two billiard balls is very small, but it is the dominant force affecting the motions of the celestial bodies because of the tremendously large masses involved. Some scientists believe that if there is more mass in the universe than has been detected so far, the gravitational force among the galaxies might be sufficiently great to halt the expansion. Where might this undetected mass be hiding? Some obvious candidates might be the black holes. We can't see them, since light cannot escape from a black hole, and it is conceivable that millions of black holes might be sprinkled throughout the universe. These pockets of high density would contain

*A light year is 10 million million (10^{12}) km, the distance light travels in one year.

large amounts of matter. Some people predict that including this mass in the gravitational equation for the universe just might do the trick.

The scenario goes something like this: as the gravitational force causes the galaxies to slow down, they would eventually come to a stop and then begin moving back toward one another. Since the force of gravity increases with decreasing distance, the attraction among the galaxies would become ever stronger, flinging them together with increasing accelerations. Eventually, they would come together as in the beginning, gravitationally attracted into a crushingly small sphere, reducing all the matter to fundamental particles. The density would generate large amounts of heat, and the big bang would happen once again.

Could it be that this has already happened? Perhaps so, perhaps even more than once. It is possible that the universe has oscillated like this several times. While it is clear that the universe is expanding now, there is no *a priori* reason to assume that the present expansion is the initial one. The remnants of any previous civilization would certainly be destroyed under the atomizing forces at work in creating the primeval superdense sphere, so there is no way to tell whether we are living on the first loop or the fifth.

Learning Checks

1. What is the direction of the earth's gravitational field?
2. How does the gravitational field strength of a body vary with distance?
3. Why is the black hole called that? Why is it invisible?
4. What is the escape velocity of a black hole?
5. Would the motion of the earth in its orbit be affected if the sun were to undergo gravitational collapse? Explain.
6. You observe two galaxies, one of which is twice as far from earth as the other. The nearer galaxy is moving away from earth at only half the speed of the farther. Explain how these facts indicate that the two galaxies and the earth all started from the same point.

CHECKPOINT ON GRAVITY

Earlier we learned how Newton concluded that whenever an object accelerates it does so because of some kind of force. We have now learned about one particular kind of force, gravity. It operates according to one of the most important rules of the game and, as far as we know, is the same throughout the universe. It is the force that ties you firmly to the earth, keeps the moon in its orbit, determines the behavior of waterfalls, and gives the pinwheel appearance to some of the galaxies.

SUMMARY

The most familiar force in all of nature is the *universal force of gravity.* It is one of the four basic interactions, dominating the behavior of objects on the cosmic scale. Newton's discovery of this law was a milestone in the understanding of the motion of terrestrial and celestial bodies. The force obeys an *inverse-square law,* having the form $F_g = GmM/d^2$ for objects of masses M and m separated by a distance d. The distance d is measured between the centers of spherical objects.

The gravitational force always *pulls* the objects toward each other. Unless at least one object has a very large mass, the force is too small to be observed.

Newton formulated his theory of gravity on the basis of *Kepler's laws of planetary motion,* which say that the planets move in ellipses about the sun at a focus; that the line from the planet to the sun sweeps out equal areas in equal time intervals; and that D^3/T^2 is a constant for all planets, D being the length of the major elliptical axis and T the length of the planet's year. By studying the motion of the moon and determining its acceleration, Newton was able to demonstrate that its orbit is governed by the same force as the motion of freely falling bodies on earth.

Satellites are freely falling bodies also, falling toward the earth while moving around it. The path that a satellite takes depends on its initial velocity. With an initial velocity of about 8 km/s, the orbit is circular. It is elliptical for a velocity of just under 11 km/s. Any velocity greater than this will shoot the satellite into outer space. The *escape velocity* for the earth is about 11.2 km/s, or around 40,000 km/h.

The "action-at-a-distance" concept for gravitational force has been replaced in modern times by the notion of a *field.* We can think of a field as the distortion of space caused by the presence of matter.

Einstein's ideas about pure energy being susceptible to the gravitational force have generated the notion of the *black hole,* the final state in the gravitational collapse of a dead star. For a black hole, the gravitational field is so strong that not even light can escape.

The big bang theory holds that the beginning of the universe occurred with an explosion of a superdense matter core. The universe has continued to expand ever since. If more mass than is now known were present in space, gravity would eventually reverse the galaxies' motions and pull them back together.

All the matter in the universe would be crushed together again. The resulting heat would generate another big bang and the expansion would take place all over again.

Exercises

1. Discuss why Kepler's third law was so important in Newton's concept of gravity.

2. When it is winter in the Northern Hemisphere, the earth is closer to the sun than in the summer. How does the wintertime speed of the earth's motion around the sun compare with that in the summertime?

3. Would a mountain climber weigh less at the top of Pike's Peak than at sea level? Explain.

4. Does a typewriter exert a gravitational force on the earth? If so, how large is this force?

5. Draw a diagram showing what would happen to the moon if suddenly the earth's gravitational pull were "switched off."

6. Should Kepler's laws apply to artificial satellites of the earth? Does an earth satellite necessarily travel with constant speed? Explain.

7. In the expression for weight, $W = mg$, is g a universal constant? Explain.

8. Would it ever be correct to draw the field lines of the earth pointing *away* from its surface? What would this mean experimentally? Explain.

9. Speculate on what the world would be like if G were ten times its actual value.

10. Knowing the time T required for the moon to orbit the earth, and knowing D for its elliptical orbit, we can calculate the ratio D^3/T^2 for the moon's motion. Would you expect this number to have the same value as k for the planets? Explain.

11. Surveying is difficult in Northern India because a plumb line does not hang quite vertically, due to the neighboring Himalaya Mountains. Explain.

12. The earth is not quite a perfect sphere, having a slight bulge at the equator and being flattened somewhat at the poles. Will the value of g be the same at the equator as at the poles? Explain.

13. Suppose that you are applying for a job in the telecommunications field and the employer asks you, "What keeps our communication satellite in its orbit around the earth?" How do you answer this question?

14. Since the earth is attracted by the sun's gravity, why doesn't it fall into the sun?

15. Would you predict that the escape velocity for the moon would be smaller than, the same as, or greater than that for the earth? Explain.

16. A space station can simulate gravity by rotating slowly, like a wheel, so that its occupants do not experience the discomfort of weightlessness. How can the rotational motion simulate gravity?

17. Suppose the earth were four times as far from the sun as it is now. By what factor would the gravitational force of the sun on the earth be reduced? By what factor would the acceleration of the earth toward the sun be changed?

18. The radius of Mars is one-half that of the earth and its mass is one-eighth the earth's mass. What is the value of g at the surface of Mars?

19. What would your weight be if you were in a space vehicle 4000 miles above the earth's surface?

20. Imagine that planet X has the same mass as earth but that its radius is twice that of earth. If Jon weighs 40 pounds on earth, how much will he weigh on planet X?

21. We said earlier that galaxies in Virgo, a distance of 78 million light years from earth, are speeding away at 1200 km/s. Galaxies in Corona Borealis are receding at 22,000 km/s relative to earth. What is the approximate distance of these galaxies from us?

22. Suppose the mass of the earth were tripled while its radius remained the same. Would you weigh *more* or *less*? By how much?

23. A billiard ball and a tiny buckshot pellet in outer space are separated by a distance of 2 meters.
   a. Between what points is the 2 meter distance measured?
   b. Compare the force exerted by the ball on the pellet with that exerted by the pellet on the ball as to (1) size and (2) direction.
   c. In which directions will the objects accelerate? Which object will have the larger acceleration?
   d. Do the forces increase or decrease if the objects are placed 1 meter apart?

24. Answer question 23 assuming that F_g falls off as d^3 instead of d^2.

BUTLER

6

MOMENTUM AND A CONSERVATION LAW

Momentum is the product of mass and velocity.

The rules of the game nature plays are often simple yet intriguing. An example we have seen is Kepler's law about the equal areas a planet sweeps out in equal intervals of time, in its orbit around the sun. Simple enough, it is also fascinating, for we are at a loss to explain *why* the areas should be the same. They just are. This is an example of a group of more general rules that, as far as we know, *everything* in nature obeys, from planets to speeding freight trains down to the atoms in a nuclear reactor. Here we find an example of the so-called *conservation laws*, perhaps the most powerful generalizations we have for understanding the universe.

Conservation is another of those ordinary words that physicists use in a peculiar way. When we say that something is conserved, we mean this: if we determine the value of the thing, then wait for nature to undergo its various changes, and then measure the quantity again, we get the same number. In other words, *a conserved quantity is one that stays constant as time goes on.* The nonphysics use of the word "conservation" is often related to this idea. For example, conservation groups are known for their efforts to keep wilderness areas unchanged over long periods of time. Similarly, we "conserve" a natural resource when we keep its amount constant.

In Chapters 6 and 7 we will learn about two conserved quantities called *momentum* and *angular momentum.* Let us start with momentum.

MOMENTUM

Suppose you are standing at a railroad crossing that has two sets of tracks running east and west. On one track a freight train speeds toward you at 50 km/h west and on the other a small handcar approaches with the same velocity. If you had to be hit by one of them, which would you choose? Admittedly, that is not a pleasant choice to make, but make it anyway. Naturally, you would choose the handcar. The reason? Because it is smaller, you would probably say. There is something equally different about a tanker truck and a baseball going 5 km/h north, even though both are traveling with the same velocity.

How might we express that something? One way is to say that the object having the greater mass will hit you with a larger force. Put another way: it requires a larger force to stop the speeding train than it does to stop the handcar. With these ideas in mind, let us return to Newton's law for force.

Newton's second law is

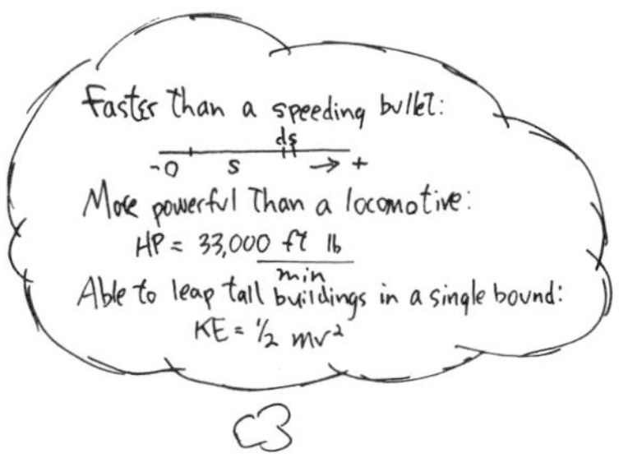

$$\mathbf{F} = m\mathbf{a}$$

where **F** is the net force applied to the body of mass m to give it an acceleration **a**. Since **a** is the rate at which the velocity changes in time, we can write

$$\mathbf{F} = \frac{m(\mathbf{v}_{\text{final}} - \mathbf{v}_{\text{initial}})}{t}$$

where t is the time interval over which the velocity changes from $\mathbf{v}_{\text{initial}}$ to $\mathbf{v}_{\text{final}}$.

The mass of the body remains constant during this change in velocity.* Therefore, we can write this equation:

$$\mathbf{F} = \frac{(m\mathbf{v})_{\text{final}} - (m\mathbf{v})_{\text{initial}}}{t}$$

Here we have expressed the force in terms of a quantity that changes over the time interval. That quantity is $m\mathbf{v}$, mass times velocity, and it has the name *momentum*, labeled **p**. That is,

$$\mathbf{p} = m\mathbf{v}$$

$$\text{momentum} = \text{mass} \times \text{velocity}$$

We can see that momentum provides a quantitative distinction between heavy and light objects moving at the same velocity. We might describe it as mass-weighted velocity. Newton called it "quantity of motion"—perhaps a more graphic description than "momentum" (which comes from the Latin word for "motion"). (See Figure 6–1.)

*This constancy of mass is an approximation for speeds much less than the velocity of light—more about this in our study of relativity, Chapters 8 and 9.

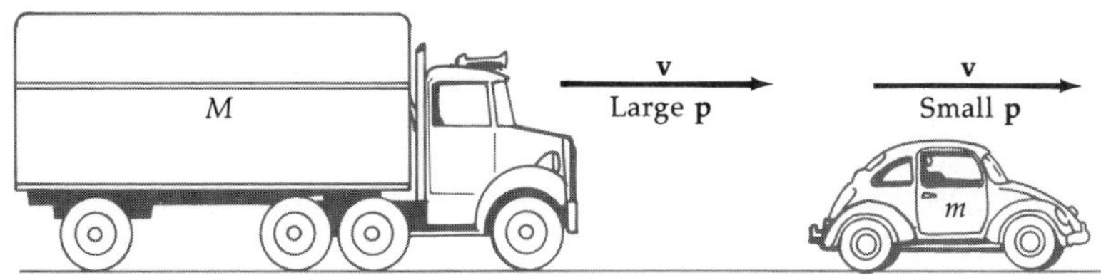

Figure 6-1. Objects traveling at the same velocity have momenta that depend on their masses.

Because momentum is the product of mass and velocity, these two quantities can be combined in different ways to give the same momentum. For example, a 10 kg mass moving with a velocity of 2 m/s west has the same momentum as a 5 kg mass moving at 4 m/s west, or a 1 kg mass zipping along at 20 m/s west. In all of these cases the momentum is 20 kg m/s west. A large object moving at a slow speed may still have a large momentum because of its mass, while a small, fast-moving object may have a large momentum because of its high velocity. (See Figure 6-2.) And, of course, an object that is not moving has no momentum at all, regardless of how large it is. By its definition, momentum must involve some motion.

Notice that we have used the vector notation in our equation by writing **p** in boldface type. Momentum, like acceleration, velocity, and force, is a vector quantity and therefore depends on direction. The direction of the momentum is that of the velocity. If, in our examples above, the 10 kg mass were moving *north* at 2 m/s and the 5 kg mass were moving *east* at 4 m/s, the two would not have the same momentum. The magnitudes of momentum would be the same (20 kg m/s), but the two masses would have different momenta because of the directions of motion.

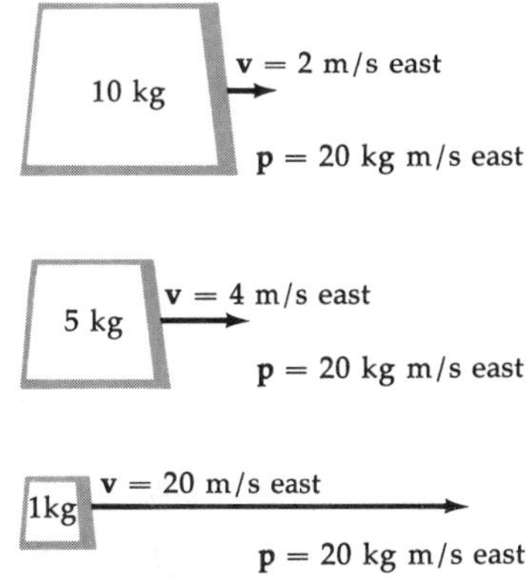

Figure 6-2. The different objects all have equal momenta because of their different velocities.

Learning Checks

1. What does it mean to say that momentum is "mass-weighted velocity"?
2. Compare the momentum of a steamship moving at 30 km/h with that of a tugboat moving at 30 km/h, both in the same direction.
3. Compare the momentum of a bullet traveling at 300 m/s with that of a Mack truck at rest.
4. A hockey puck slides across the ice at 10 m/s to the right while a second identical puck slides to the left at 10 m/s. Do they have the same momenta? Explain.

MOMENTUM AND IMPULSE

Now let us return to that last equation for the force on page 74. Substituting our definition of momentum, $\mathbf{p} = m\mathbf{v}$, we have

$$\mathbf{F} = \frac{\mathbf{p}_{\text{final}} - \mathbf{p}_{\text{initial}}}{t}$$

This form of the force equation tells us that the net force is equal to the rate at which the momentum of the body changes as time goes on. (By the way, this is how Newton originally stated his second law.) It tells us, for example, that more force is required to move our tanker truck from rest to a speed of 5 m/s than to move the baseball from rest to the same speed in the same interval of time. The tanker truck, because of its greater mass, has greater final momentum (that is, its momentum at 5 m/s) than does the baseball. Hence, its *change* of momentum is greater, requiring a greater force. This corresponds to what common sense tells us: it is harder to move (or stop moving) a tanker truck than a baseball.

Physics and Aircraft Safety

The British comedian Spike Milligan brings some humor to the subject of momentum change with the line, "Tank heaven the ground broke me fall." The aircraft industry, on the other hand, has a very somber attitude toward the breaking of falls, especially air-to-ground crashes of planes, which, of course, bear directly on the question of passenger safety.

Because this matter is so important, the Technical Center of the Federal Aviation Administration studies it intently. At a former NASA facility, technicians repeatedly lift small aircraft off the ground with ropes and fling them down at various speeds and angles. More realistic simulations of crashes take place in the Mojave Desert, where larger, remotely controlled aircraft are radio-instructed to plummet from the sky. Upon contact with the ground, a fuselage may survive intact, but a tail may snap or a jet engine may rip off and be driven into a fuel tank. Damage varies with the impulsive forces involved.

All this helps aerodynamic engineers learn about the effects of momentum changes on aircraft in differing crash situations. They can then use this knowledge to design planes to withstand the shock of impact and thus help save the lives of plane-crash victims.

Source: Steve Fishman, "Air Crashes that Save Lives," *Science Digest*, October 1982, pp. 12–14.

The word "momentum" is used in many nonscientific contexts. For example, a marketing executive may speak of the momentum of an advertising campaign. We hear politicians refer to the momentum of a race for public office. While these uses of the word are somewhat vague and less precise than the definition we have given, they nevertheless convey the idea of something difficult to stop—a TV commercial that sparks tremendous sales, a political candidate moving toward a sure win. The use of the word in physics gives the same idea. Our tanker truck and baseball moving with the same velocity are different because of the forces necessary to set them moving or to stop them once they are moving.

We can view this relationship between force and momentum in another way, one that you have experienced many times without realizing it. If we multiply both sides by t, we have

$$\mathbf{F}t = \mathbf{p}_f - \mathbf{p}_i$$

or, in words, the *force* multiplied by the *time interval* during which the force acts, equals the *change in momentum.*

Now let us see what this says physically. As every child knows, it is safer to jump out of a window onto a soft mattress than onto the concrete pavement. The reason is expressed in this equation. When you jump out of the window, you accelerate toward the ground, and at the instant before you crash, your momentum is large. When you strike the ground you naturally stop, so your momentum goes quickly to zero. Thus your body experiences a change in momentum that depends on your mass and your velocity just before you hit. This is the righthand side of the equation. There is nothing you can do about the change in momentum.

However, look at the lefthand side, the product of the force and the time interval. This is called the *impulse.* The impulse involves two factors: the force that causes the change in momentum, and the time interval during which that change takes place. When you strike the concrete, you stop very quickly—that is, t is quite *small,* so $\mathbf{F}$ must be correspondingly *large* in order for their product, the impulse, to equal the change in momentum. The concrete applies this large $\mathbf{F}$ to your body. It may break some bones and cause internal injuries. But now suppose we make t larger; that is, we somehow lengthen the time interval over which your body comes to a stop. We may do this by, say, putting a soft mattress beneath the window. If t is now *large,* then $\mathbf{F}$ may become *smaller,* because it is the *product* of the two that is important. The longer you can stretch out the time it takes to effect the momentum change, the smaller the force your body needs to absorb. The result: you are less likely to break a leg. (See Figure 6–3.)

Figure 6–3. The effects of changing the values of force and time as factors in impulse.

(a) $\mathbf{F}$ is large t is small
The hard surface causes a quick stop, and hence a large force.

(b) $\mathbf{F}$ is small t is large
The net reduces the force by lengthening the stopping time.

Learning Checks

1. Explain in your own words the relationship between *force, momentum,* and *time.*

2. Suppose you have to push a truck from rest to a certain velocity in 10 seconds. By what factor must you increase the force if you are to accomplish this in 5 seconds?

3. Explain what *impulse* is.

4. In terms of momentum and impulse, explain the following:

   a. The use of a net by a trapeze artist.

   b. Diving into a full swimming pool rather than an empty one.

   c. A golfer "following through" with his or her swing.

THE CONSERVATION OF MOMENTUM

We have seen how momentum is related to Newton's second law of motion. This concept has allowed us to look at the second law from a slightly different slant, perhaps enabling us to understand it a little better. This is a good reason for introducing the concept of momentum. But there is a better reason, namely, that momentum is a quantity that is *conserved.*

The law of conservation of momentum may be stated in this way:

> If the net force acting on a system is zero, the total momentum of the system remains constant.

First we look at how the mathematics gives us this result. We wrote Newton's second law in the momentum-impulse form as $\mathbf{F}t = \mathbf{p}_f - \mathbf{p}_i$. Remember that $\mathbf{F}$ is the net force. If $\mathbf{F}$ is zero, we have $\mathbf{p}_f = \mathbf{p}_i$. Since the final and initial values of the momentum are the same, there is no change in the system's momentum.

We must distinguish here between *external* forces, which act *on* a system, and *internal* forces, which result from the mutual interaction among the objects making up the system. The $\mathbf{F}$ in Newton's second law refers to the net *external* force; the *internal* forces all add up to zero because of Newton's third law, the action-reaction one.

For example, if you are riding along in an automobile, the motion of the car is unaffected if you push on the dashboard or throw a ball back and forth with someone else in the car. These interactions are the result of *internal* forces and have nothing to do with *external* forces acting *on* the system—the "system" in this case being the car and everything in it. To say that the momentum is constant means that the initial momentum and the final momentum are the same for *any* time interval during which the net applied force is zero.

Now this is rather remarkable, if you think about it. No matter what takes place inside a system, no matter what sort of havoc internal forces may cause, the overall momentum remains unchanged. Momentum is a vector quantity, so the total momentum is simply the vector sum of the individual momenta of the separate parts of the system. And it is this quantity that forever remains unaltered in the absence of external forces. If the momentum of some part of the system changes, this change must be compensated for by a different momentum for some other portion of the system, in order that the total remain constant.

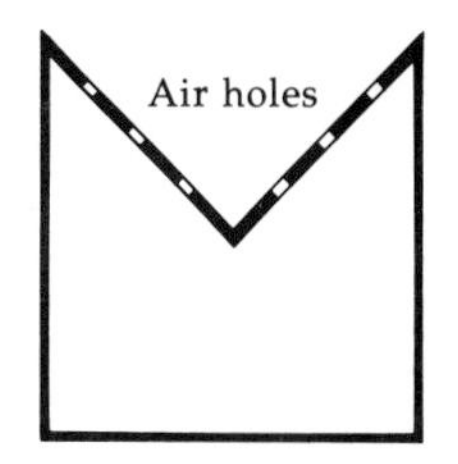

Figure 6-4. Cross section of an air track.

Consider some examples. A convenient device for studying momentum is the *air track,* because it almost completely eliminates friction. An air track is a track designed so that small carts of various masses may slide on a frictionless cushion of air. A cross section of one type is shown in Figure 6–4. The air forced through the small holes in the track serves to lift the cart, removing the force of friction and counteracting the downward pull of gravity, leaving the net force on the cart zero.

Imagine that we place on the track two carts, one having twice the mass of the other. We compress a spring between them and connect them by a rubber band. This system is not moving, initially. Now we cut the rubber band and watch what happens. (See Figure 6–5.)

The uncoiling of the spring pushes the carts apart, the larger one traveling with a small velocity in one direction and the smaller one traveling with a greater velocity in the opposite direction. If we measure these velocities—say, with a meter stick and stopwatch—we would see that the smaller mass moves away with exactly twice the velocity of the larger mass.

Let us put in some numbers. Suppose that our small cart has a mass of 1 kilogram and the larger one 2 kilograms. If the 2 kilogram mass has a velocity of 5 cm/s in one direction after the "firing," then the 1 kilogram mass will be traveling twice as fast, 10 cm/s, in the opposite direction.

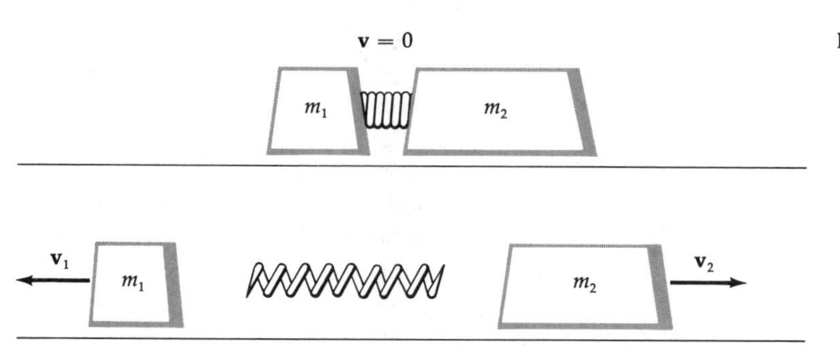

Figure 6-5. Carts on an air track.

Does this check with the conservation law for momentum? As we said, the initial momentum is zero since the two carts begin at rest. What about the final momentum? Well, it is the vector sum of the momenta of the two carts. The momentum of the 1 kilogram cart is

$$\mathbf{p}_1 = 1 \text{ kg} \times 10 \text{ cm/s} = 10 \text{ kg cm/s}$$

and that of the 2 kilogram cart is

$$\mathbf{p}_2 = 2 \text{ kg} \times (-5 \text{ cm/s}) = -10 \text{ kg cm/s}$$

We have taken care of the directions by putting a minus sign on $\mathbf{p}_2$. The vector $\mathbf{p}_1$ points in the opposite direction from $\mathbf{p}_2$; they are *antilinear.* So the final momentum is

$$\mathbf{p}_1 + \mathbf{p}_2 = 10 \text{ kg cm/s} - 10 \text{ kg cm/s} = 0$$

and the momentum is conserved. It hasn't changed, because of the *internal* force of the coiled spring pushing the carts apart. The initial momentum—in this case, zero—has merely been redistributed among different parts of the system.

It is important to notice that the *smaller* mass moves away with the *larger* velocity, and vice versa. This is not only true for carts on air tracks but also for real-life situations. For instance, we may go back to our example of the "kick" that occurs in the firing of a rifle. In your mind's eye, let the bullet take the place of the small air-track cart, substitute the rifle for the larger cart, and replace the rubber band–spring combination by the powder in the bullet and the firing mechanism of the gun. When the gun is fired, the bullet and rifle experience the same force—that which results from the explosion of the powder. The small bullet moves forward with a high velocity and the more massive rifle recoils backward with a smaller velocity. Measurements on this system in the absence of friction would reveal that momentum is conserved. That is, the product of $m\mathbf{v}$ for the bullet is equal in magnitude and opposite in direction to the product $M\mathbf{V}$ for the rifle, keeping the total momentum zero for the rifle-bullet system initially at rest. It is this rearward momentum of the rifle that is responsible for the "kick" the hunter feels.

Let us go a little further with this example. If the hunter holds the rifle a few centimeters away from the shoulder, the velocity at which the rifle recoils will be much smaller than that of the bullet, as we have seen, but large enough to give the hunter quite a jolt. But suppose the hunter holds the rifle snug against the shoulder. Now, the hunter's body is a "part of the rifle," so to speak. The recoil momentum includes the mass of the hunter *and* of the rifle, not just that of the rifle alone. This means that the recoil velocity will be much *smaller* since the recoiling mass is much *larger*—remember, it is the *product* that stays constant. So, it is much safer to hold the rifle tightly against the shoulder, reducing the effect of the "kick."

The main reason conservation laws such as this one for momentum are so important is that we do not need to know the details of the internal force causing the redistribution of the momentum. The explosion of the powder in the bullet due to the impact of the hammer in the rifle may be a complicated process, including such things as the mechanism of the firing device and the chemical reactions within the powder. But these details are irrelevant when we consider momentum. We need only know the initial momentum and observe that no net external force acts on the system. The momentum at any intermediate stage will remain constant, regardless of what happens internally. Put another way: only the beginning and ending are important, not the "path" the system takes during the time between the beginning and ending.

We have discussed situations in which the masses were unequal, but the same principles apply when the masses are equal. Suppose in an ice capades two clowns who weigh the same stand together on the frictionless ice and push each other apart. What will be the result? Since their masses are equal, they will move away from each other with equal speeds. Of course, the greater the force with which they push each other apart, the greater their speeds will be. The point of emphasis in all of these examples is that the velocity of each object depends on the ratio of its mass to that of the other object involved.

Consider this scenario: you are stranded on the ice in the middle of a frozen pond. The ice is perfectly frictionless, so you can't push yourself toward the shore by any means. How can you get off the ice?

Consider this from the standpoint of momentum conservation. Because you are stationary, your momentum is zero. And, since you can't apply any external force (remember, the ice is frictionless), there is no way to change your momentum. It will remain zero. So the way to get off the ice is this: remove your shoe (or jacket or whatever) and give it a heave; you will take off in the opposite direction. Throwing your shoe gives it some momentum. But because the total momentum must remain zero, your body will have a momentum equal to that of the shoe, in the opposite direction. And soon you arrive safely ashore, owing your life to the conservation of momentum.

The reaction engines we looked at in Chapter 4 all operate on the principle of momentum conservation. For example, the momentum of the hot gases ejected by a rocket engine is exactly matched by the momentum of the rocket itself as it races off in the other direction. The overall momentum of the rocket-exhaust system remains the same as at the moment of firing.

Collisions

So far, we have discussed only situations where the original momentum of the system is zero. However, momentum is also conserved when the original momentum is *not* zero. An example occurs when two objects collide. Let us go back to our air track, where we put chewing gum on the carts so that when they hit they will stick together. Suppose for simplicity that both carts have the same mass. One is at rest. We set the other moving at a velocity of 10 cm/s to the right. The carts collide, stick together, and move away with a common velocity. The question is, how fast are they moving?

We can use the law of conservation of momentum to compute this final velocity, even though we don't know anything about how the chewing gum does its job of sticking the carts together. Figure 6-6 shows what happens. The initial momentum of the system is just that of the moving cart, since the stationary one has no momentum. It is

$$\text{initial momentum} = m\mathbf{v} = m \times 10 \text{ cm/s right}$$

The final momentum is the mass of the two carts together, which is $2m$, multiplied by their common velocity:

$$\text{final momentum} = 2m\mathbf{V}$$

Since there is no net force on the carts, the momentum of the whole system is conserved, so

$$\text{final momentum} = \text{initial momentum}$$

If we put in the numbers, we get

$$2m\mathbf{V} = m \times 10 \text{ cm/s right}$$

The m's cancel, and dividing both sides by 2 gives

$$\mathbf{V} = 5 \text{ cm/s right}$$

Figure 6-6. Effect of a collision.

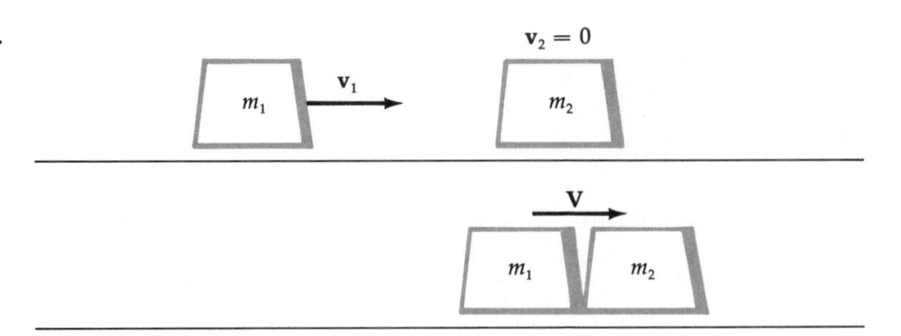

Figure 6-7. Ballistics experiment.

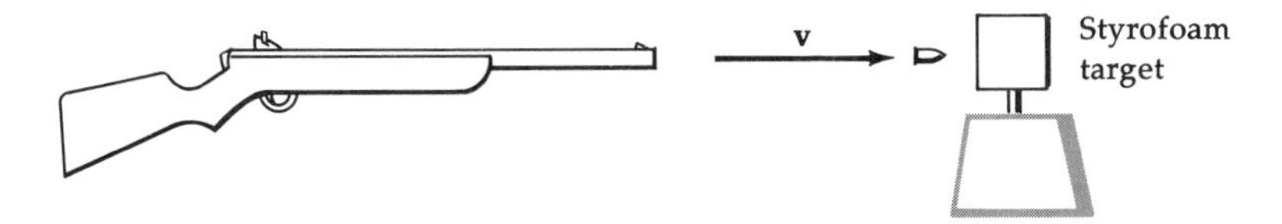

as the velocity of the two carts stuck together. Note also that the two carts move in the same direction that the single cart was going originally, as common sense would tell us.

Suppose we take a case in which the colliding masses are unequal. A practical example is a problem from ballistics, namely, determining the muzzle velocity of a bullet fired from a gun. (The muzzle velocity is the velocity of the bullet when it leaves the gun.) On our air-track cart we mount a piece of styrofoam for use as a target, as shown in Figure 6–7. The mass of the target (cart and styrofoam together) we will call M_T. With the target at rest, the gun is fired and the bullet (whose mass is M_b) is imbedded in the target. The bullet and target together move down the air track with a velocity that we call **V**. We know this velocity will be relatively small because the target is much more massive than the bullet. This means that we can measure **V** easily.

Physics and Police Work

Ballistics is the science that deals with the motion, direction, and behavior of fired missiles, including bullets. Such information can be of great value to those involved in detecting crime. For example, police scientists are often called upon to determine whether a bullet has been fired by a murderer or by a suicide. Such factors as muzzle velocity and direction—concepts from basic physics—play a significant part in these investigations.

Bullets can be "witnesses" that establish an accused person's guilt or innocence in court. During a trial, when a bullet's source becomes important, its markings may be compared to the rifling marks in the barrel of the suspected firearm or to a test bullet shot from the same weapon. This procedure, though popularly associated with ballistics, is, strictly speaking, more a part of "firearms examination." And physics has been influential here as well. It is interesting to note that a physicist was part of the team that developed the special bullet comparison microscope now used in police laboratories the world over.

Source: Eugene B. Block, *Science vs. Crime: The Evolution of the Police Lab.* San Francisco: Cragmont Publications, 1979.

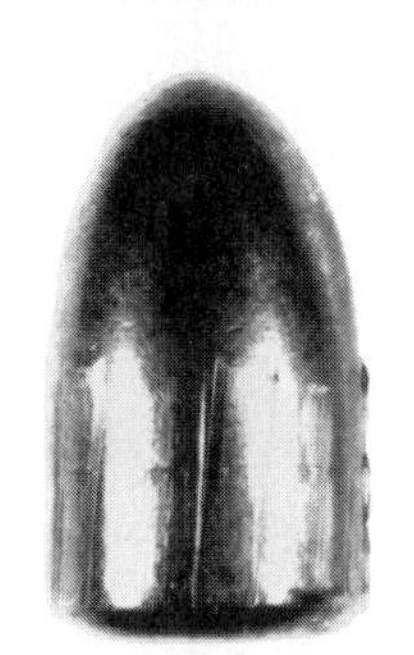

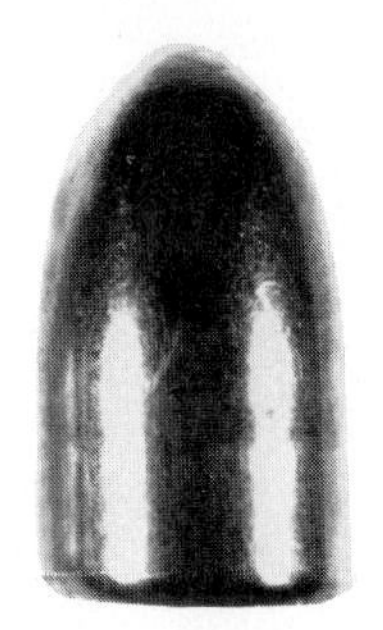

Physics is often helpful in the scientific analysis of fired bullets. The bullet on the left has noticeable striation marks, indicating it has been fired. The bullet on the left, in contrast, has not been fired.

How does the law of conservation of momentum apply here? The initial momentum is that of the bullet alone,

$$\text{initial momentum} = M_b\mathbf{v}$$

whereas the final momentum is that of the bullet and target stuck together. This combination has a mass of $M_b + M_T$, so

$$\text{final momentum} = (M_b + M_T)\mathbf{V}$$

These momenta are equal, giving

$$M_b\mathbf{v} = (M_b + M_T)\mathbf{V}$$

If we divide both sides by M_b, we have the muzzle velocity of the bullet

$$\mathbf{v} = \frac{(M_b + M_T)\mathbf{V}}{M_b}$$

Thus, the conservation of momentum has allowed us to determine the velocity of the fast-moving bullet in an indirect way by measuring the slower, and hence more convenient, velocity of the bullet-target combination.

What about objects that bounce off each other instead of sticking together? Let us return to the carts on the air track. We can arrange them with bumpers so they will recoil rather than become coupled together. Imagine that a cart moving at a velocity of 10 cm/s to the right strikes a stationary cart of the same mass. Momentum is conserved and all of the momentum of the first cart is transferred to the second cart. The first cart stops on impact, and the second cart moves away to the right, also at 10 cm/s.

Learning Checks

1. What is meant in physics by a *conservation law?*
2. Is momentum always conserved? If not, under what conditions is it *not* conserved?
3. In the case of the rifle and the bullet, we saw that momentum is conserved. If we consider the system to be only the bullet, rather than both bullet and rifle, is momentum also conserved in this case? Explain.

SUMMARY

Momentum is defined as the mass of an object multiplied by its velocity and can be thought of as a mass-weighted velocity.

Newton's second law can be stated in terms of momentum: the net external force acting on an object is equal to the rate at which its momentum changes in time.

The change in momentum is equal to the *impulse,* or the force times the time interval during which the force acts.

Momentum obeys a *conservation law:* if there is no net external force acting on a system, then the total momentum of that system will remain constant. This means that even though the momentum can be distributed among the parts of the system in different amounts, its total will not change as long as the system is free of external forces.

Exercises

1. Two patrol cars having the same mass travel on the freeway at 50 km/h in opposite directions. Do they have the same momenta? Explain.

2. What happens if two particles of equal mass traveling at equal speeds collide head-on and stick together? What if they bounce off each other?

3. Do Exercise 2 for the situation in which one particle has twice the mass of the other.

4. Explain this observation: a balloon that has just been inflated flies across the room when you release it.

5. Discuss some instances of momentum conservation in pool shooting.

6. Suppose that a bomb positioned in outer space (free from the force of gravity) explodes into two pieces. Can the two pieces fly away in the same direction? Explain.

7. What do we mean by saying that the explosion of the bomb in Exercise 6 "redistributes the total momentum" of the bomb? Explain.

8. A favorite scene in old slapstick movies is for the star to fall from an airplane and land in a haystack. Explain, in terms of impulse and momentum, why this is better than landing on the ground.

9. In tennis, golf, baseball, and many other sports the importance of "follow-through" is emphasized. Explain this importance, in terms of impulse and momentum.

10. A lumberjack throws a tool to his partner directly overhead in a tree, giving the tool five units of momentum. What is the momentum of the earth in the opposite direction? Do we notice the earth moving in such a situation? Explain.

11. We have said that only *external* forces can change the motion of a body. But when you apply the brakes on a car, the force they exert on the wheels is an internal one. How then do the brakes stop the car? (Hint: think about what happens when you try to stop on an icy road.)

12. An irrigation hose lies in a field. When a farmer turns on the water, the nozzle of the hose jerks backward as the water shoots out. Explain.

13. When a 230 pound linebacker and a 150 pound halfback collide on the football field, which player experiences the greater *force?* the greater *impulse?* the greater *acceleration?* the greater *change in momentum?* Explain each of your answers.

14. Two railroad freight cars are on a siding. One of them is loaded so that its mass is three times that of the other. The heavier car moving at a speed of 4 m/s collides with the lighter car at rest. They couple and move away together. What is their speed? Ignore friction.

15. Answer Exercise 14 if initially the heavier freight car is stationary and the lighter one is moving.

16. Which object has more momentum, one whose mass is m and whose speed is $3v$ or one whose mass is $2m$ and whose speed is $2v$?

17. Two geophysicists at the North Pole shove two identical sleds across the ice. One applies a force of 14 newtons for 3 seconds, while the other applies a force of 21 newtons for 2 seconds. Compare the impulse in the two situations. (Ignore friction.)

18. In Exercise 17, if the sleds each have a mass of 7 kilograms, what is the ultimate speed of each?

19. You stand on a frictionless skating rink and someone throws you a medicine ball that weighs 30 pounds. When you catch it, the ball is moving at 5 mi/h. If you weigh 120 pounds, how fast will you and the medicine ball move across the ice?

20. Imagine three carts of equal mass at rest on an infinitely long air track. Two are connected by the coiled spring–rubber band arrangement described in the text and the third is some distance away from these two. All three are equipped to bounce off each other. Describe what happens when the rubber band is cut: How many collisions are there? What are the speeds of the three carts as time goes by?

21. Compare the momentum in each of these cases:
    a. A 1000 kg object moving at 1 m/s.
    b. A 100 kg object moving at 10 m/s.
    c. A 50 kg object moving at 50 m/s.

Adages such as "What goes up must come down" and "He always lands on his feet" originate from an innate understanding of physics.

7

ROTATION AND ANGULAR MOMENTUM

So far in our study of the motion of objects, we have considered only their overall motion along some path. For example, in studying falling bodies or moving cars, we have treated them as if they were simply point particles, without worrying about any motion other than that of the body moving as a unit. This is sometimes called *translational* motion.

But real objects have shape, they have extension—they are not simply point particles. And they can have another kind of motion in addition to translational motion. A kicked football can tumble end over end as it flies through the air. A pool ball rolls as well as slides across the table. Toss a hammer by its handle and it twists and turns while moving as a whole along its path. These spinning movements carry the general name *rotational motion,* which is the subject of this chapter.

We are going to learn some new terms, important concepts in rotational motion. These have their counterparts in translational motion; you can understand them more easily if you keep the analogies in mind. *Torque* in rotational motion is the analog of *force; moment of inertia* plays the role of *mass;* and *angular momentum* is to rotational motion as *momentum* is to translational motion. We will also encounter a new rule of the game: angular momentum obeys a conservation law that is very much like the one for momentum.

TORQUE

We learned earlier that if two forces of equal magnitude act on a point from opposite directions, the net force on any object at that point is zero. No new motion occurs. (See Figure 7–1a.) But now suppose these forces act on different parts of an extended object—a meter stick, for example, as shown in Figure 7–1b. What about this new situation? Well, the net force is still zero. However, there is motion because the stick will spin, or *rotate,* about a line running through the stick and located between the

Figure 7-1. The net force is equal to zero in both (a) and (b), but a torque is produced in (b), causing rotation.

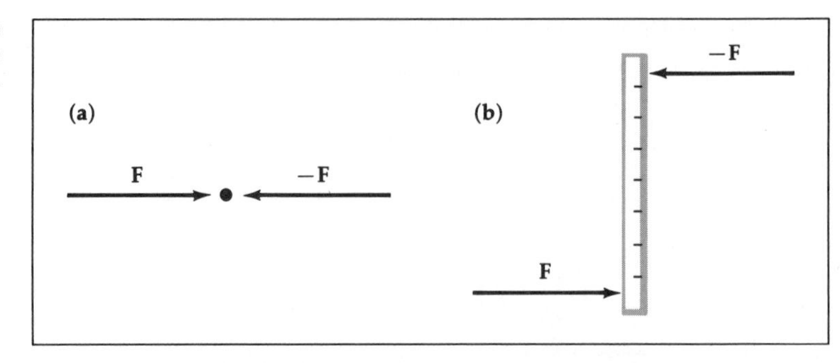

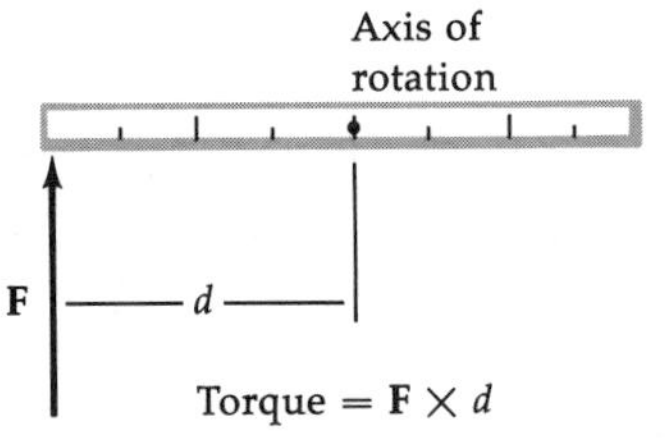

Figure 7-2. Definition of torque.

points of application of the two forces. This line is called the *axis of rotation.* Even though the resultant of the two force vectors is zero, their application to different parts of the stick produces an effect that causes the stick to rotate. This effect is called *torque.* It depends both on the force and on how far from the axis the force is applied. Torque is equal to the product of the force and the perpendicular distance from the axis of rotation to the line along which the force is applied. This is shown in Figure 7-2. The distance d is often called the *lever-arm distance.*

We may distinguish between clockwise torque and counterclockwise torque, depending on which direction the object rotates. Clockwise torque produces a rotation in the direction that the hands of a clock move as you face the clock. Counterclockwise torque causes rotation in the opposite direction. This is illustrated in Figure 7–3.

We can change the magnitude of the torque that produces a rotational motion by varying either the force or the lever-arm distance. A good illustration is the playground seesaw, where children quickly learn that the heavier child must move closer to the middle in order for the seesaw to "balance." (Children are pretty good at physics until somebody tells them how hard it is!) Let us look at the example in Figure 7–4. Jeff weighs 300 newtons and Jon weighs 200. If Jeff sits 4 meters from the middle, then Jon must sit 6 meters away in order for the seesaw to balance. Jeff's weight, a gravitational force of 300 newtons, will produce a counterclockwise rotation of the seesaw:

$$\text{counterclockwise torque} = 300\ \text{N} \times 4\ \text{m}$$
$$= 1200\ \text{N m}$$

Jon, who weighs 200 newtons, will produce a clockwise rotation. At a distance of 6 meters, this torque is

$$\text{clockwise torque} = 200\ \text{N} \times 6\ \text{m}$$
$$= 1200\ \text{N m}$$

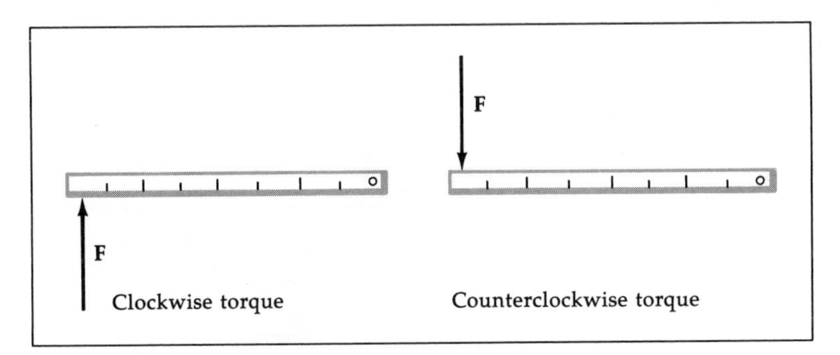

Figure 7-3. Clockwise and counterclockwise torques.

So, as the children experiment to find where they must sit in order to seesaw, they are actually adjusting the clockwise and counterclockwise torques until they are equal. This is a general condition for *rotational equilibrium:*

> If the total clockwise torque and the total counterclockwise torque are equal, the net torque is zero and the system is in rotational equilibrium.

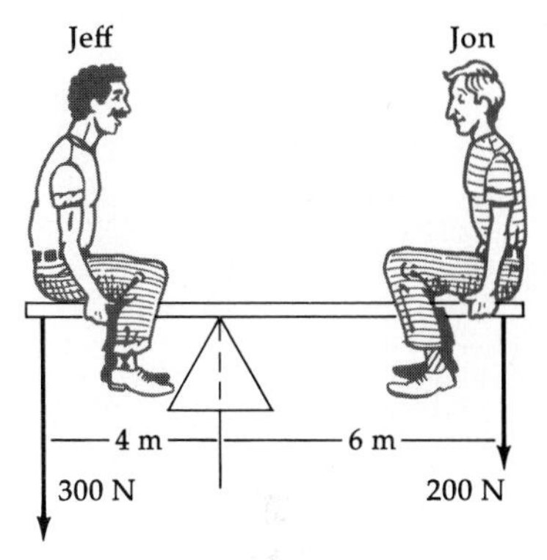

Figure 7-4. Rotational equilibrium on a seesaw.

If either child moves forward or backward, the seesaw will be out of balance because the torques in the two directions will no longer be equal.

Figure 7–5 illustrates the relationship between the force and the lever arm when you use a wrench to loosen a nut. In part (a) of the figure, the force is directed along the wrench. Here there is no torque because the lever arm length is zero. The nut will not budge. Part (b) shows the normal way of using the wrench; here the direction of the force is perpendicular to the wrench, so the lever arm is equal to the length of the wrench. This gives the maximum torque possible for a given force applied to this particular wrench. In part (c), the force is applied at other than a right angle, and hence the lever arm is smaller than in part (b).

Notice that in all these cases the *force* is the same, but the *torque*—and hence the resulting rotation of the nut—is quite different for the three situations, because of the different lever arms.

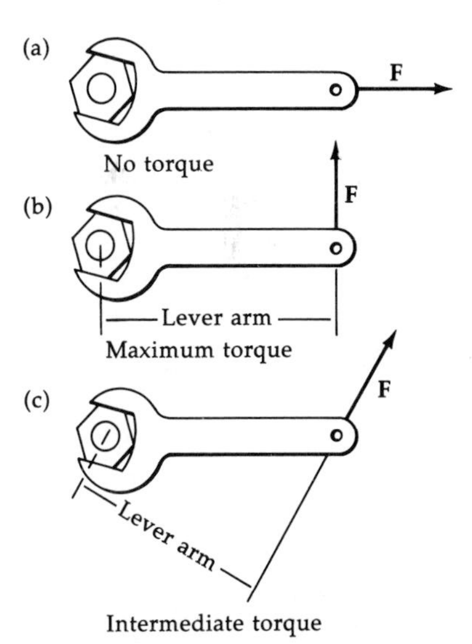

Figure 7-5. Relationship between the force F and the lever arm.

Learning Checks

1. Distinguish between *torque* and *force.*
2. In Figure 7–5(c), why is the lever-arm distance not the same as the length of the wrench?
3. Distinguish between clockwise and counterclockwise torques. Do you suppose from this that torque is a vector quantity? Explain.

Physics and the Art of Motorcycle Maintenance

Motorcycle mechanics, professionals and amateurs, are familiar with the "nuts and bolts" of torque through many articles in their trade magazines. These mechanics know that the components of this twisting operation are distance and force. "More distance, in the vernacular, means 'get a longer wrench' and more force could be translated as 'jump on it.' "

"Torquing a bolt" on a cycle is actually a matter of serious technical concern. In fact, much detailed thought is given to the physics involved: "The torque we apply to a nut or bolt, like income before taxes, is used up in several ways. About 50 percent is used in overcoming friction between the bearing face of the nut or bolt and the surface of the part; approximately 40 percent goes to overcome thread friction, and the rest produces the bolt tension."

Even the selection of the correct torque wrench entails a knowledge of physics. This wrench is a tool gauged in measurements for torque values, such as newton-meters. Such units, as well as concepts like friction, will not be strangers to students who take a basic physics course.

Source: "Applied Torque," *Cycle World*, December 1980, pp. 58–60.

MOMENT OF INERTIA

In our discussion of linear motion, we noted that the mass of a particle is a measure of its inertia, its resistance to changes in motion. A large mass means that the object at rest is difficult to move, or once moving, difficult to stop. A similar idea holds for rotational motion. A wheel not turning requires a torque to start it rotating and, once it is rotating, requires a torque to stop it.

This implies a kind of rotational-motion inertia. It depends not only on the mass but, as with torque, it is also affected by the distance the mass is from the axis of rotation. Physicists define a quantity called the *moment of inertia* (given the symbol I) as the product of the mass and the square of the distance from the rotation axis.* Expressed as an equation,

$$I = mr^2$$

This is for a single particle of mass m. The value of I for a number of particles would simply be the sum of all the values of mr^2 for each particle. Perhaps the term *rotational inertia* would be more descriptive. In any event, the idea is analogous to that of inertia, in that it involves a body's resistance to any changes in its rotational motion.

It is important to understand that the moment of inertia of an object depends not only on the *mass* but even more on the way in which that mass is *distributed*. For example, a bicycle wheel has most of its mass located near the rim, far from the axis of rotation (the spokes don't weigh very much). Hence it

**Moment* as used in physics has come to mean the product of a quantity and its distance from some significant point. For example, torque is sometimes referred to as the *moment of force*.

has a large rotational inertia, because almost every particle making up the wheel has a large value of r. A solid wheel of the same mass would have a much smaller moment of inertia because a larger fraction of the material would be located nearer the axis of the wheel. (See Figure 7–6.)

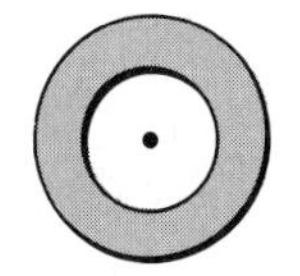

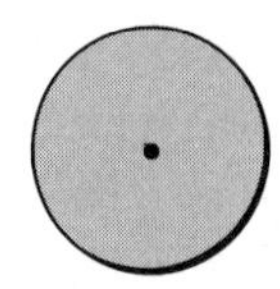

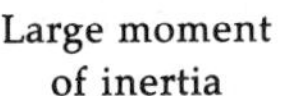

Large moment of inertia

Small moment of inertia

Figure 7-6. Moments of inertia for different distributions of same mass.

Notice, too, that the moment of inertia has nothing to do with whether or not the object is rotating. It depends only on the *mass* and the *distance* and is the same for a given system regardless of whether or not it is rotating. Moment of inertia is a property of the object itself, not of its state of motion.

We find a practical application of rotational inertia in the flywheel connected to the crankshaft of an automobile. The flywheel is quite massive and designed to have a large moment of inertia, so the crankshaft of the automobile continues to turn during those brief periodic intervals when the engine is not firing. Another example: the next time you watch a child trying to balance on a narrow curb, or on one foot playing hopscotch, notice that he or she extends the arms to a position roughly parallel to the ground. This redistributes part of the mass of the body so that the rotational inertia is larger. This furnishes more resistance to rotating (toppling over) and gives the child time to regain balance if he or she begins to lose it.

Physics and Bioengineering: A Fantasy

L. Frank Baum, author of the *Wizard of Oz*, wrote a series of books about the magical land of Oz. One book describes a tribe of people called the Wheelers. They are just like us except that they have wheels for hands and feet and travel on all fours. This, of course, is pure fiction. We do not, in all the real world, know of any animals that have evolved wheels for locomotion. Why is this the case? The biologist Stephen Jay Gould provides an answer:

As its basic structural principle, a true wheel must spin freely without physical fusion to the object it drives. If wheel and object are physically linked, then the wheel cannot turn freely for very long and must rotate back, lest connecting elements be ruptured by the accumulated stress. But animals must maintain physical connections between their parts. If the ends of our legs were axles and our feet were wheels, how could blood, nutrients, and nerve impulses cross the gap to nurture and direct the moving parts of our natural roller skates?

But let us fantasize. Suppose you were hired as a bioengineer to design a pair of circular feet that could be connected to humans. From having read the text, you are now equipped with some of the basic physics involved. What would you do to make the rotational inertia of these feet-wheels large? What would you do to make it small? Which would be better for purposes of locomoting over a flat cement road?

Source: Stephen Jay Gould, "Kingdoms Without Wheels," *Natural History*, March 1981, p. 42.

Learning Checks

1. How does torque depend on mass and distance?

2. A baker turns out a doughnut and a biscuit having the same mass and diameter. Which has the greater moment of inertia? Explain.

3. In Question 2, suppose that you release the doughnut and biscuit at the same time and allow them to roll down an inclined plane. Which wins the race? Explain.

4. Which has the greater moment of inertia, a bicycle wheel that is spinning or an identical one that is still? Explain your answer.

CENTER OF MASS

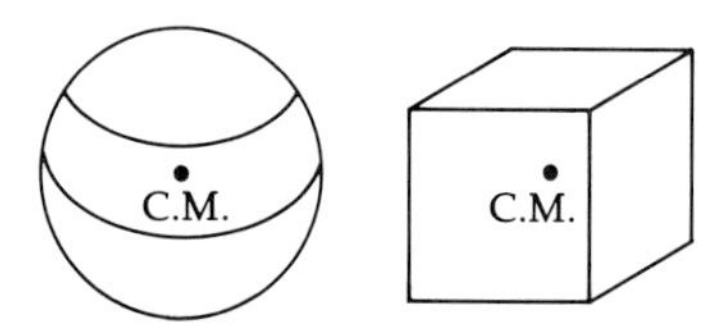

Figure 7-7. Center of mass for symmetrical objects.

We started this discussion by distinguishing between translational motion and rotational motion. If we toss a tennis ball across the room, it follows a parabolic path, and we need only consider this translational motion. If we also toss the tennis racquet, its motion is wobbly as it rotates while also undergoing translational motion. But this apparently random rotational motion is not as chaotic as it appears. The racquet's end-over-end tumbling takes place about an axis that runs through a very special point called the *center of mass.* During the racquet's flight through the air, the center of mass doesn't wobble. It follows the same kind of smooth parabolic path as the tennis ball.

The center of mass is simply the average position of all the particles making up the object. It is the point at which, for translational motion, we can consider all the mass of the body to be located. This means that we can suppose that gravity acts only at this particular point, as far as the translation of the body is concerned.

For a symmetrical body such as a cube or a sphere, the center of mass will be at the geometric center of the object. (See Figure 7–7.) In the case of objects of irregular shape, such as the tennis racquet, the center of mass will be located nearer the heavy end of the object. Occasionally the center of mass will be located at a point not in the body at all. Figure 7–8 illustrates this for a doughnut, where the center of mass is at the center of the hole, and a boomerang, where the center of mass is at a point in the space between the arms.

Figure 7-8. The center of mass can be located where there is no mass.

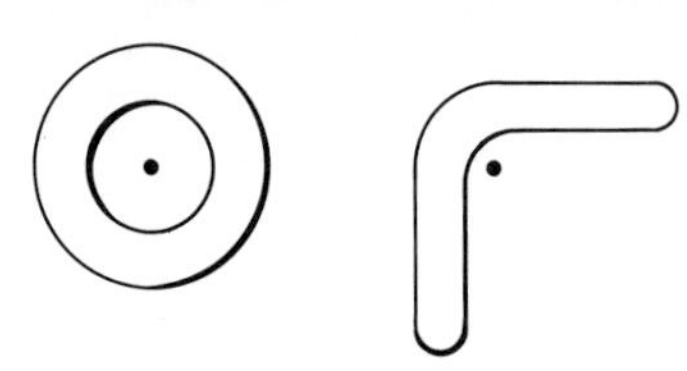

We can locate the position of the center of mass for any object by taking advantage of what we know about torques. Figure 7–9 illustrates the idea. If we suspend the object at any point, it will swing around until the center of mass lies directly below the point of suspension. To see why this is so, remember

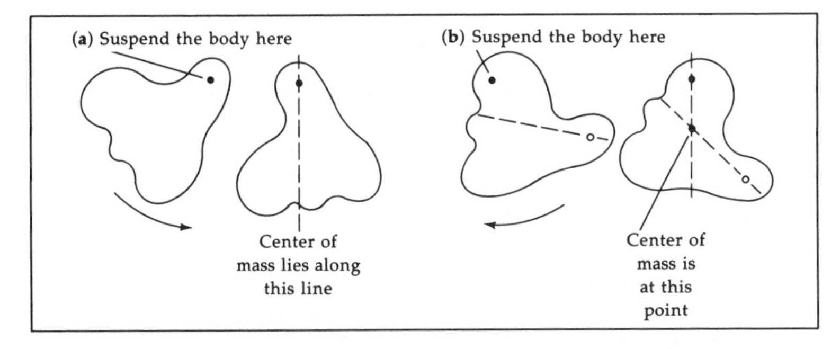

Figure 7-9. Finding the center of mass for an irregularly shaped object.

that the force of gravity acts effectively only at the center of mass. If this is anywhere other than at or below the suspension point, gravity will produce a torque that rotates the body until the lever arm is zero—that is, until the center of mass reaches its lowest position. Here the line of the force of gravity runs through the rotation axis, so the torque is zero. By performing this simple operation for two different suspension points and drawing the lines as indicated in the figure, we can find where the center of mass is located: it will be where the two lines intersect.

The location of the center of mass determines whether an object will stand upright or topple over. The rule is this: the object will remain stationary if a line drawn from the center of mass toward the center of the earth lies inside the object's base. Otherwise it will tip over. Figure 7-10 illustrates these ideas. This rule also comes from the concept of torques. Any edge of the base is potentially an axis of rotation, so if the line through the center of mass lies outside the base, we have a lever arm and hence a torque produced by gravity. This torque rotates the object until the lever arm is zero, and the effect is to lower the center of mass as much as possible. Figure 7-11 shows a book tipped slightly from a position on its edge. The resulting torque causes it to fall over—to the stable position where it has its lowermost center of mass.

(a) Stable (b) Unstable

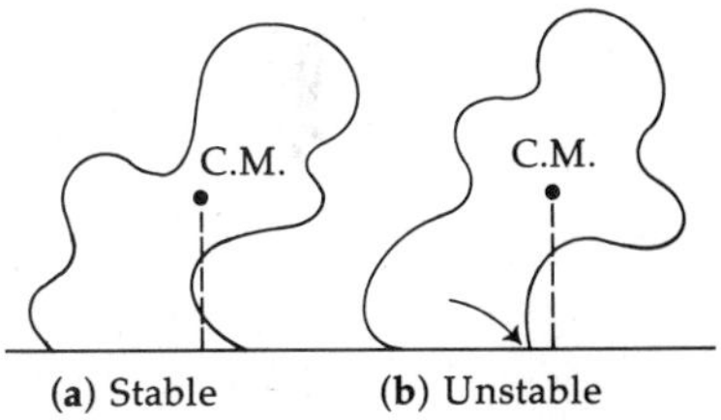

Figure 7-10. Stable and unstable objects. (a) Center-of-mass line lies within the base. (b) Gravity acting on the center of mass produces a torque; the body will rotate about the point marked with the arrow.

You can think of many examples of this stability principle. For instance, it is more comfortable to stand with your feet slightly apart than exactly together. This gives a wider base

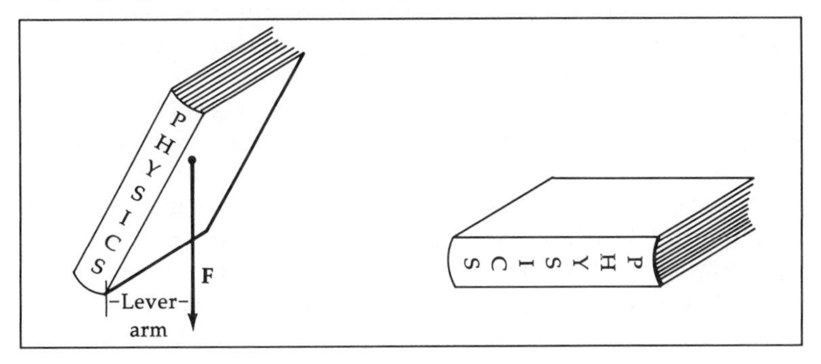

Figure 7-11. The gravity force F and the lever arm produce a torque, rotating the book to the more stable position.

and makes you more stable. A skier can lean forward as much as she wishes without falling, because the long skis provide such an extended base it is virtually impossible for the center-of-mass line to fall outside. You can stand a pencil on its eraser but not on its point, because the point makes such an extremely small base. Thus, many ditchdigging machines have "paws" on extended legs that can be used to vary the base, or stance, of the machine.

Sports cars that sit low to the ground track well on turns because their center of mass is low. Tipping them over would require *raising* the center of mass, which is not likely to happen when the car goes around a curve.

Incidentally, you will often hear the term *center of gravity* used. There is a slight technical distinction between center of gravity and center of mass, but for our purposes the two can be considered the same.

The Center of Gravity of Flying Insects

How do you determine the center of gravity of insects that fly—such as a locust, say? Well, the center of gravity of this insect has been established by simply suspending a preserved specimen from different points by a thread. The center of gravity is given by the intersection of lines AB and CD. Its location depends upon the position of the locust's hind legs.

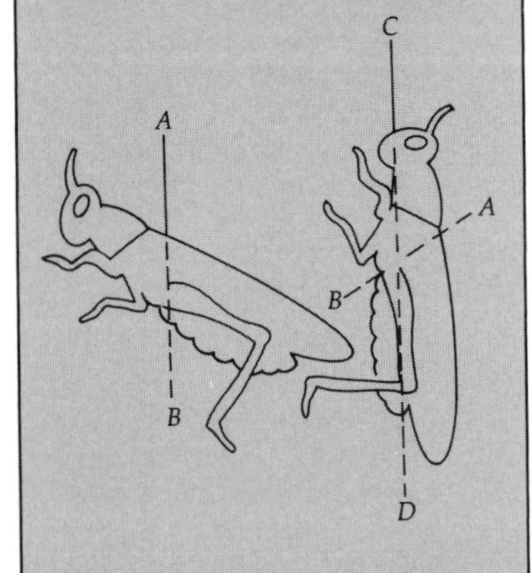

Source: George Duncan, *Physics for Biologists*. New York: John Wiley & Sons, 1975.

Learning Checks

1. Define center of mass.
2. Give an example, other than the ones listed above, in which the center of mass does not lie within the body.
3. What determines whether or not a standing object will fall over?

ANGULAR MOMENTUM

When children begin riding bicycles, one of the first things they learn is to keep moving—if the bicycle stops rolling, it falls over. You may have noticed that an ice skater spins more slowly with arms extended than with them pulled in. By curling his body into a ball, an expert diver can make several turns before hitting the water. A football goes farther if you throw a spiral pass instead of a wobbly one. These and other examples are alike in this respect: they all involve the physical quantity called *angular momentum*.

As the name implies, angular momentum bears a relation to rotational motion similar to the relation of momentum to linear motion. Angular momentum is a vector quantity. To give its most general definition, we would have to use vector properties that are beyond the scope of this book. (We briefly discuss the vector nature of torque and angular momentum in Supplement 5.) However, we can understand the concept of angular momentum if we consider only circular motion.

Suppose that a body of mass m moves with speed v in a circle of radius r. (See Figure 7–12.) Then the magnitude of the angular momentum is

$$L = mvr$$

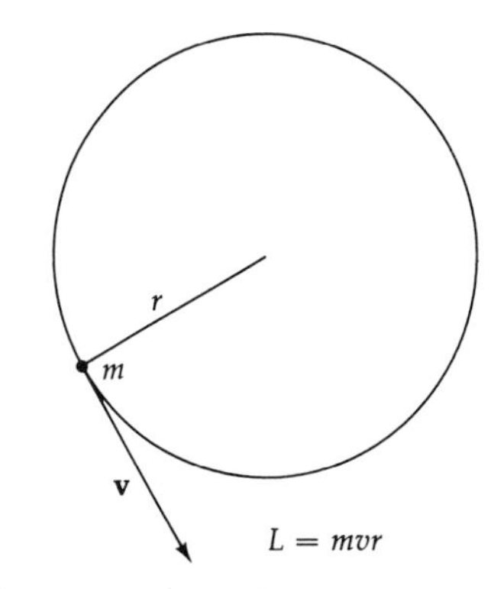

Figure 7-12. Angular momentum for circular motion.

We get the direction of the angular momentum vector in this way: imagine that you turn a righthand screw in the direction that the mass m is moving; the direction of L is the direction that the screw would advance. Another way of getting the same answer is by the so-called righthand rule. If you curl the fingers of your right hand in the direction that the object moves, then your thumb points in the direction of the angular momentum vector. You can see that both methods show that L for Figure 7-12 points out of the page. If the particle were moving in the other direction, L would be directed into the page.

Let us make some comments about this definition. Because the magnitude of the momentum p is mv, then the magnitude of the angular momentum is just the magnitude of the linear momentum multiplied by the radius of the circle. Also, notice that the angular momentum for a given mass depends both on the speed *and* the radius: an object moving with a small speed on a large circle will have a large angular momentum. This means also that a body with a large moment of inertia has the capacity for a large angular momentum, since its mass is concentrated nearer the rim of the object, making r large.

We have seen the close analogy between mass and moment of inertia, between force and torque, and now between momentum and angular momentum. Just as *forces* produce changes in *momentum* for linear motion, so *torques* cause changes in *angular momentum* for rotational motion.

Newton's second law also has an analogous form for rotational motion. As we said in Chapter 4, one way to express the second law is to say that the net force on a body is equal to the rate at which the momentum changes as time goes on. Similarly, we see that the net *torque* on a body is equal to the rate at which the *angular momentum* changes in time.

Let us return to the bicycle. Although the stability of the bicycle is a complicated topic, at least part of the reason the rolling bike stays up is because of the relationship between torque and angular momentum. When the bicycle is rolling, the wheels generate a large angular momentum that tends to keep it upright. The torques produced by the force of gravity are not large enough to make much change in the bicycle's angular momentum, so it keeps on rolling. When it stops, however, there is no angular momentum: even the smallest torque causes it to topple over.

The quarterback who throws a spiral pass is taking advantage of angular momentum. Because of the oblong shape of the football, it will encounter the least resistance as it travels through the air if the leading part of the ball is its nose. As the quarterback releases the ball he gives it a spin about its long axis, generating a sizable angular momentum. Random air currents will not supply enough torque to change the direction of this angular momentum. The ball will tend to maintain its spinning motion in the desired direction and the pass travels farther. The barrels of rifles are spirally grooved for the same reason, giving the bullet a similar spinning motion as it leaves the gun.

Conservation of Angular Momentum

We can write the relationship between torque and angular momentum in an equation similar to the one shown earlier (Chapter 6) involving force and momentum. If we let τ represent the net external torque and **L** the angular momentum, then we have

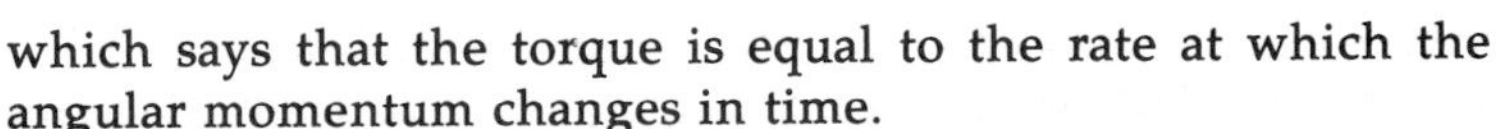

$$\tau = \frac{(\mathbf{L}_{\text{final}} - \mathbf{L}_{\text{initial}})}{t}$$

which says that the torque is equal to the rate at which the angular momentum changes in time.

This gives us a hint that angular momentum can also be a conserved quantity, and indeed this is true. Since changes in angular momentum come about because of torques, it is reasonable to suppose that in the *absence* of any net torque the angular momentum stays the same. The *law of conservation of angular momentum* can be stated like this:

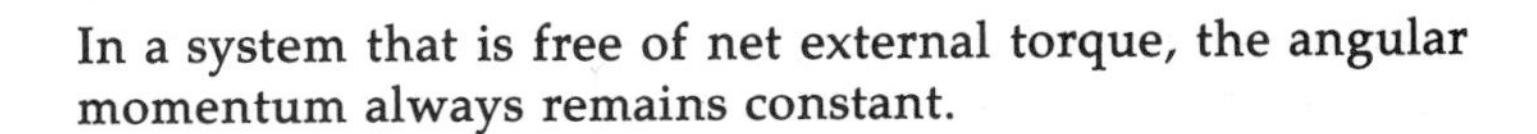

> In a system that is free of net external torque, the angular momentum always remains constant.

Figure 7-13. Conservation of angular momentum.

This law completes the analogies between linear motion and rotational motion. Observe that if we replace the word "torque" by "force," and "angular momentum" by "momentum," we have a statement of the conservation of momentum. Conservation of angular momentum is also a very important concept in physics. Just as in the case of momentum, you need to know only what the angular momentum is at one time in order to know what it is at all times, if the net torque is zero.

An interesting example of the conservation of angular momentum is the ice skater who is spinning with hands outstretched, as in Figure 7-13. Most of the angular momentum is due to the mass in the hands and arms extended a distance r from the axis of rotation, the vertical line running down the center of the body. As the skater draws hands and arms in closer to the body, the radius r decreases. This causes a faster spin. That is, v increases as r decreases in order that the angular momentum be conserved. The conservation of angular momentum dictates that the shorter radius will be exactly compensated by a larger speed of rotation in order that the product mvr remain constant.

You can demonstrate this for yourself with a piano stool or swivel chair. While you sit down with your hands outstretched, have someone give you a good spin. Notice how you speed up as you pull your hands in toward your chest. The effect is even more dramatic if you hold a heavy book in each hand—this gives you a larger moment of inertia when your hands are outstretched.

Recall Kepler's law of constant areas for the planets in their orbits (Chapter 5). In fact, what Kepler had stumbled onto was an example of conservation of angular momentum, although no one recognized its wider applicability until much later. The planets move in their elliptical orbits because of the sun's gravitational attraction. This force produces no torque on the planets, because it is directed along the radial line through a planet and the sun, as shown in Figure 7–14. This is similar to the non-torque-producing force on the wooden bar illustrated in Figure 7–15; you can see that this force, pointing directly toward the rotational axis, cannot change the angular momentum. The planets speed up or slow down as they approach or recede from the sun, keeping their angular momentum constant.

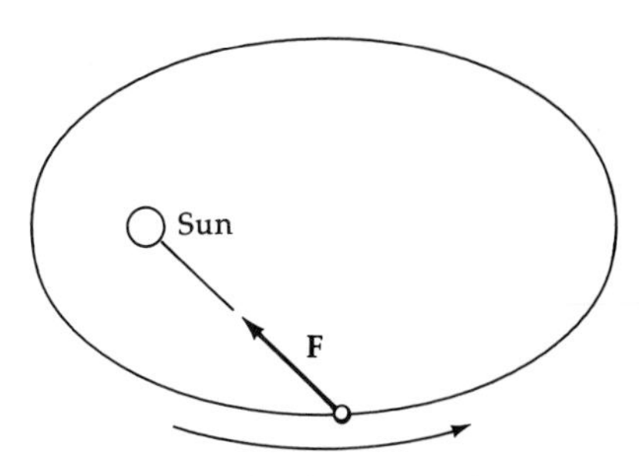

Figure 7-14. The force F produces no torque on the planet.

The earth-moon system is one that conserves angular momentum, because, again, no torques are involved. The earth rotating on its axis has an angular momentum much like that of a spinning bicycle wheel. The moon's gravity produces tidal effects on the earth's oceans, so that as the earth rotates the oceans lag behind somewhat. Friction of the earth's crust against the water in shallower portions of the ocean tends to slow down the rate of the earth's rotation. This effect is quite small—the length of the day increases by about one second every hundred thousand years. Nevertheless, as the earth slows down it loses some of its angular momentum. It isn't lost, really—the conservation rule forbids that—it is transferred to another part of the system: the moon. The moon gains this angular momentum and as a result is gradually creeping farther and farther away from the earth. It is backing away from us at a rate of about 10 centimeters per year.

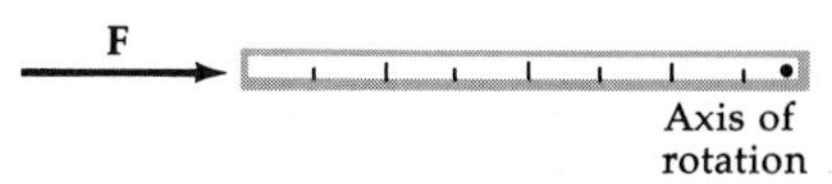

Figure 7-15. The force F produces no torque on the bar.

Learning Checks

1. Explain the relationship between momentum and angular momentum.
2. Convince yourself in qualitative terms that Kepler's second law for planetary motion is the same as the conservation of angular momentum.
3. Explain the gradual increase in the earth-moon separation.
4. Is angular momentum *always* conserved? If not, under what conditions is it not conserved?

GOLF: APPLICATIONS OF THE PHYSICS OF MOTION

Golf is a sport that people of all ages enjoy. For our purposes, it provides excellent illustrations of the fundamental concepts we have developed about motion—momentum, torque, moment of inertia, angular momentum. In essence, playing golf involves two things: (1) selecting the right club for a given situation and (2) developing the skills necessary to use the chosen club properly. We can discuss the first of these now that we have an understanding of mechanics.

Golf clubs are of two types: woods and irons. The woods are longer and have more massive club heads. The driver is the longest and has the largest head. Its function is quite simple: to drive the ball off the tee as far as possible. The fairway woods are shorter and somewhat lighter, but their function is similar—to provide distance rather than pinpoint accuracy.

Because the woods have long shafts and most of their mass is in the club head, they have large moments of inertia. This means that when you swing the driver you generate a large torque, which in turn provides a large force as the club head strikes the ball. This is the whole purpose of the woods—to whack the ball with a large enough force to propel it far down the fairway.

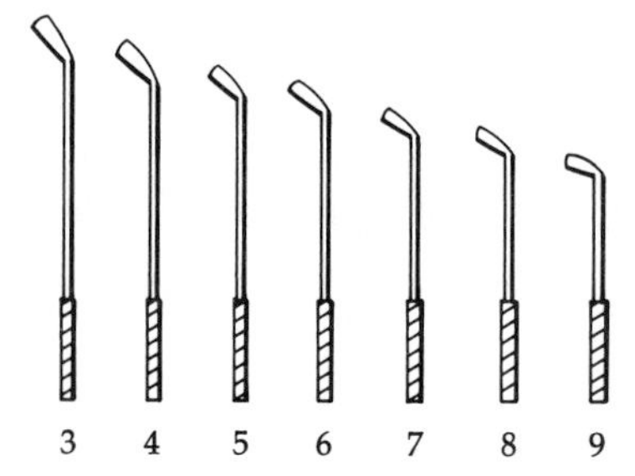

Figure 7-16. The irons of a set of golf clubs.

The irons have shorter shafts and lighter club heads than the woods. In a typical set of seven irons, numbered 3 through 9, the shafts get progressively shorter as the numbers get higher. (See Figure 7-16.) This means that the longer irons—say the 3- and 4-irons—have larger rotational inertia than the shorter 8- and 9-irons. Thus the lower-numbered irons generate a larger torque and send the ball a greater distance than do the higher-numbered clubs. The lengths of the shafts are such that, for a given speed of the club head, the difference in distance obtained from one club to the next is about 10 yards.

Another feature of the irons is the difference in the shapes of the club heads. As the numbers get higher, the angle that the head makes with the shaft gets smaller, beginning at almost 180 degrees with the 3-iron and progressing to slightly more than 90 degrees for the 9-iron. This means that the irons with the higher numbers will loft the ball higher into the air. Let us see what effect this has on the placement of the ball.

As we have seen, the golfer uses the longer-shafted, lower-numbered irons to get a greater distance for the shot. The angle the club face makes with the shaft helps to do this. Because the 3-iron, say, doesn't lift the ball so high in the air, most of the ball's momentum during the flight is nearly parallel to the ground; when it lands it will roll, increasing the distance. (See Figure 7-17.) Close to the green, however, the golfer is more interested in placement than distance. It is undesirable for the

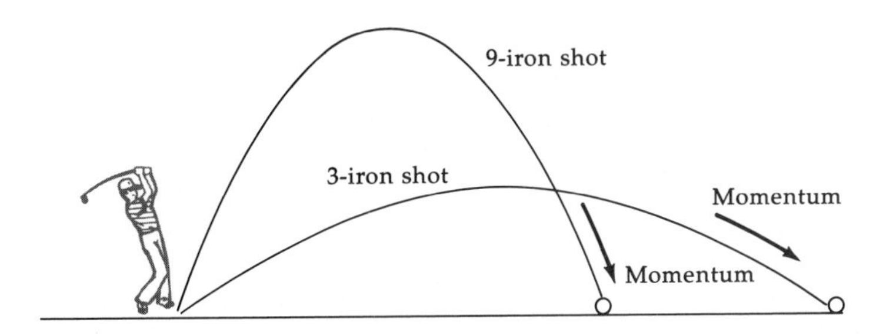

Figure 7-17. A 9-iron gives the ball nearly vertical momentum on landing; the 3-iron produces a nearly horizontal momentum.

ball to roll very far along the ground in this situation. The 9-iron, for example, has a club head whose angle is such that it lifts the ball to a greater height. When the ball returns to the ground, its momentum vector is more nearly vertical, minimizing the amount of roll. This allows the golfer to place the ball more accurately.

> **Learning Checks**
>
> 1. Which has the greater moment of inertia, the driver or the 7-iron?
> 2. For a given club-head speed, which has the larger angular momentum, the 4-iron or the 8-iron?

SUMMARY

As complicated and varied as our physical surroundings may appear, we have seen in this chapter that nature operates according to some rules that greatly simplify the picture. These are the so-called *conservation laws.* We now know two of them: (1) in the absence of any net external force, momentum is conserved; and (2) in the absence of any net external torque, angular momentum is conserved. As far as we can tell, these rules apply to everything everywhere in the universe.

Rotational motion is produced when oppositely directed forces act on different parts of an extended body. We define the *torque* as the product of the force and the lever arm, which is the perpendicular distance between the line of the force and the axis about which the body rotates. Torques may be distinguished by the *clockwise* and *counterclockwise* rotations they produce. A body is in *rotational equilibrium* when the clockwise and counterclockwise torques are equal.

The *moment of inertia,* or rotational inertia, of a body measures its resistance to changes in rotational motion. It depends on the mass of the body and the way in which that mass is distributed. The moment of inertia does not depend on the rotational motion. It is a property of the body itself.

A body's *center of mass* or *center of gravity* is the average position of each bit of matter in the body. For translational motion, one can consider all the mass of the body to be concentrated at the center of mass, so that the force of gravity effectively acts at this point. The location of the center of mass is important in establishing whether or not a standing object will be stable.

Angular momentum is the analog of momentum for rotational motion. For motion in a circle, it is equal to the product of the mass, the speed, and the radius of the circle. In the absence of external torques, angular momentum is *conserved*. This fact accounts for the stability of many rotating systems, such as a rolling bicycle wheel and a spiraling football.

The quantities we have studied in this chapter have closely related counterparts in translational motion. Here is a table that will help you keep these relationships in mind:

Translational motion	*Rotational motion*
Mass	Moment of inertia
Force	Torque
Momentum	Angular momentum

Exercises

1. Waiter A opens the swinging door to his restaurant kitchen by pushing at the door's outer edge; waiter B always opens the door by pushing it in the middle. Opening the door is easier for waiter A. Explain.

2. Explain the fact that steering wheels for buses are made larger than those for Volkswagens.

3. A plumber will often loosen a very stubborn nut by sliding a long pipe over the handle of the wrench. How does this help?

4. One baggage handler carries a 10 pound suitcase in one hand; another is carrying two 5 pound suitcases, one in each hand. Which has an easier time of it? Explain.

5. The effectiveness of the knuckleball is that the batter cannot know which way it will "break." A baseball pitcher throws a knuckleball by purposely holding it so that it does *not* spin. What does the absence of spin have to do with the "fluttering" knuckleball?

6. A space capsule at rest in outer space fires a bullet from a rifle in the nose of the capsule. Describe the motions of the space capsule after the firing.

7. Suppose you are standing on a platform that is free to rotate, holding a bicycle wheel equipped with a handle attached to the axle so that the wheel may spin. You hold the wheel over your head so that its axis is vertical and give it a whirl. What happens? Now, while the wheel is still spinning you turn it over, pointing the handle and axle in the opposite direction. What happens? Explain.

8. Why doesn't the Leaning Tower of Pisa topple over?

9. Young baseball players are often advised to "choke up" on the bat—that is, grip it nearer the middle. Why does this make it easier to swing?

10. We have seen that for translational motion the mass of a body can be considered to be at the center of mass. Is this also true for rotational motion? Explain.

11. Suppose that two identical bicycle wheels are mounted on the same axle. They are then set spinning at the same speed in opposite directions. What is the total angular momentum of the system?

12. Explain how a long drooping pole aids a tightrope walker.

13. Explain the fact that a comet travels fastest in its orbit when it is nearest the sun.

14. Suppose that a rotating ball of hot gas cools down and contracts. What happens to its speed of rotation? What happens to its angular momentum?

15. Some helicopters have two large propellers that both rotate in a horizontal plane. Must they rotate in the same direction or in opposite directions? Explain.

16. Most helicopters have a small propeller in the tail rotating in a vertical plane. What is the function of the small propeller?

17. Two pennies sit on a phonograph turntable, one 5 centimeters from the center, the other 10 centimeters. Which has the larger moment of inertia? By what factor?

18. In Exercise 17, suppose the turntable rotates at a constant speed. Which penny has the larger angular momentum? By what factor?

19. A turtle sits on a red spot on a turntable at rest but free to rotate. The turtle starts walking clockwise as viewed from above. Describe the motion of the turntable. What happens when the turtle stops? While the turtle is walking, is its speed equal to that of the red spot? If not, on what do their speeds depend?

20. A phonograph turntable is turning freely with a turtle standing at the center. Suppose that the turtle walks toward the edge. What, if anything, happens to the speed of the turntable?

21. Suppose that you are going to stop a turning wheel by applying brakes. Is it better to apply the braking force near the axle of the wheel or near its edge? Explain.

Relativity has taught us that time and space are interdependent and has dispelled the long-held notion of time as an absolute.

8

EINSTEIN'S SPECIAL THEORY OF RELATIVITY

Nearly everyone has heard of Albert Einstein. For many people, he serves as the image of a physicist and the prime example of a genius. You will sometimes hear proud parents boast that their child is "another Einstein."

Students often ask what Einstein *really* did. Well, what he really did was revolutionize our world view, our ideas about nature, by considering distance and time in a different way than ever before. The results of this new look are stunning. We are going to learn the bases of these strange ideas in this chapter. Then, in Chapter 9, we will see some of the ramifications of Einstein's theory of relativity.

Let us begin at the beginning, as Einstein did, by thinking about space and time. We have used these concepts to define such quantities as speed and acceleration. Without actually saying so, in our discussion of motion we have assumed that space and time are completely independent of each other. Indeed, this appears so obvious that it hardly seems worth mentioning. All our experience indicates that "space," the three-dimensional world in which we live, is unchanging and detached from our own activities. In like manner, "time" appears to us like an arrow, moving forward, remote and aloof, the "now" for one person and place being the same "now" for any other person in any other place.

Sir Isaac Newton made these same assumptions. However, he did worry about them, and he thought it worthwhile to write down his ideas about them at the outset of his formulation of the laws of mechanics. We quote from his *Mathematica Principia:*

> Absolute, true, and mathematical time, of itself, and from its own nature, flows equally and without regard

. . . And Two and Two Is More Than Four . . .

These excerpts from a parody by Professor W. H. Williams, a specialist in relativity, feature an exchange between Einstein and Sir Arthur Eddington, a noted British astronomer. The humorous verses convey the essence of Einstein's relativity theories.

The time has come, said Eddington,
To talk of many things;
Of cubes and clocks and meter-sticks,
And why a pendulum swings,
And how far space is out of plumb,
And whether time has wings.

. . .

You hold that time is badly warped,
That even light is bent;
I think I get the idea there,
If this is what you meant;
The mail the postman brings today,
Tomorrow will be sent.

. . .

The shortest line, Einstein replied,
Is not the one that's straight;
It curves around upon itself,
Much like a figure eight,
And if you go too rapidly,
You will arrive too late.

. . .

But Easter day is Christmas time
And far away is near,
And two and two is more than four
And over there is here.
You may be right, said Eddington,
It seems a trifle queer.

Source: R. W. Clark, *Einstein: The Life and Times.* New York: World Publishing, 1971.

> to anything external, and by another name is called duration. . . .
>
> Absolute space, in its own nature, without regard to anything external, remains always similar and immovable.

Einstein shattered this comfortable notion of the separateness of space and time just after the turn of the twentieth century. It is a tribute to his genius that he was a major architect of the two scientific revolutions that have taken place in this century, changing the course of history. One of these, the quantum theory of matter, we shall discuss in Chapter 23. Here we look at the other of these twentieth-century revolutions, called the *special theory of relativity.*

Einstein presented this theory in 1905, at the age of 26, while working as a civil servant in a patent office in Berne, Switzerland. (See Figure 8–1.) He published three scientific papers that year. One was on Brownian motion, the haphazard movement of tiny particles in a gas or liquid. Another gave an explanation of the photoelectric effect (Chapter 22), which was an important step in the development of the quantum theory. The third, bearing the rather dull title "On the Electrodynamics of Moving Bodies," introduced special relativity to the world. In the formulation of this theory, Einstein has shown us that the apparent independence of space and time is only an approximation that arises out of the limited range of our experience. We shall see that, in fact, space and time are intimately related. For example, two people moving with respect to one another will make different observations of, say, the size of an object or the timing of two events, the differences depending on their relative speeds.

Such differences seem strange to us. For example, suppose two people have wrist watches that are in perfect working condition. They synchronize their watches and then compare them after one person runs around the block a few times while the

Figure 8-1. Albert Einstein.

other remains still. They would be surprised if the watches showed different times. In fact, if the times were different, they would suppose that one watch had been broken. What Einstein showed was that indeed the runner's watch will run slow, even though both are working. The reasons have to do with a connectedness, an interdependence, between space and time.

Now, if two people were to do this little experiment, they wouldn't notice any difference in how their watches kept time. The mismatch in their readings would be far too small to notice. As we shall see in this chapter, these differences are appreciable *only* when the speed of the moving clock is comparable to the speed of light. Ordinary objects, those with which we are familiar, travel at speeds that are negligible compared with light speed. When we run around the block, the fastest we can go is a few kilometers per hour, whereas light travels at about

300,000 kilometers per *second.* This is almost unimaginably fast—it is about nine times around the earth in 1 second. The discrepancy between the stationary versus the moving timepiece is infinitesimally small. Even if we make the clock comparison using a jet airplane, we can demonstrate the effects of special relativity only a little better, again because a jet airplane can travel at only a small fraction of the speed of light.

So it turns out that our own experience is misleading as it relates to space and time. As a result of our ordinary activities, we have been duped. Simply because we don't travel very fast, we have grown up with notions that are only approximately correct. For objects that *do* travel at speeds close to that of light—certain atomic particles, for example—the predictions Einstein made are exactly correct.

Before looking at Einstein's ideas, we need to understand an important point about relative motion. Suppose you are riding in an airplane and you begin walking up toward the cockpit. If you ask the question, "How fast am I moving?" you can see immediately that you must also ask, "Relative to what?" Relative to the plane, you are walking a few meters per second, but relative to the ground, you are going several hundred kilometers per hour. A bug riding contentedly along on your shoulder would say that you aren't moving at all. The location to which motion is referred is called a *frame of reference,* and you can choose whatever frame you like. In this little illustration, you could choose the airplane or the earth as a frame of reference.

A (Relative) Time and Motion Study

In a well-known science fiction story by H. G. Wells called "The New Accelerator," a scientist discovers how to speed up the processes of his body. Heart, brain, blood all operate at greater speeds, with the result that the outside world seems to come to a standstill. The scientist takes a walk, moving slowly so that the friction of the air won't burn up his clothes. He comes to a park:

The band was playing in the upper stand, though all the sound it made was a low-pitched, wheezy rattle, a sort of prolonged last sigh that passed at times into a sound like the slow, muffled ticking of some monstrous clock. Frozen people stood erect; strange, silent, self-conscious looking dummies hung unstably in mid-stride, promenading upon the grass. I passed close to a poodle dog suspended in the act of leaping, and watched the slow movement of his legs as he sank back to earth.

The relativity of motion is made quite clear in this story. In reality, frames of physiological reference can change. Nowhere is this more apparent than in a rare disease called Cockayne's syndrome, in which a person can age at the rate of 15 to 20 years every year. It is not uncommon, with this disease, for a sufferer to die of old age after only 5 years of life. Metabolic or endocrine defects cause this stupendous speeding up of the aging process. One wonders how, relative to normal individuals, those in this frame of reference view the world's phenomena.

Source: H. G. Wells, "The New Accelerator," *28 Science Fiction Stories.* New York: Dover Publications, 1952.

When Einstein wrote his paper on relative motion, he began with two assumptions:

1. The laws of physics in one frame of reference are the same in any other frame of reference moving with constant velocity relative to it.
2. The speed of light as measured in a vacuum is the same for all observers, regardless of the motion either of any observer or of the light source.

Neither assumption was entirely new. What Einstein did that had never been done before was to consider them together as a starting point for everything about space and time, and then "let the chips fall where they may." The results turn out to be startling. Let us look at this in some detail.

RELATIVE VERSUS ABSOLUTE MOTION

One of the important questions in pretwentieth-century physics was this: Is there a special frame of reference that is always fixed, allowing the measurement of *absolute* motion, not just the relative motions of, say, an airplane and the earth? Recall that in discussing Newton's first law we took great pains to distinguish between straight-line motion and motion in a curved path. But what is straight-line motion?

Suppose you are riding in a car moving at constant velocity and you drop your pencil. As you see it, the pencil travels in a straight line from your hand to the floor. But from the point of view of someone watching from the side of the road, the pencil follows a curved path (called a parabola). As it falls to the floor of the car, the person at the side of the road will claim it moves *downward* as a result of gravity, and *forward* because it shares the constant horizontal velocity of the car. The combination of these motions produces the curved path. Thus, to the person in the car, the pencil falls in a straight line, while to the observer standing on the ground the path is a parabola. Another example is shown in Figure 8-2. What "really" happened? Who is right? Since Newton's laws depend on distinguishing between the two types of motion, the existence of some fixed frame as

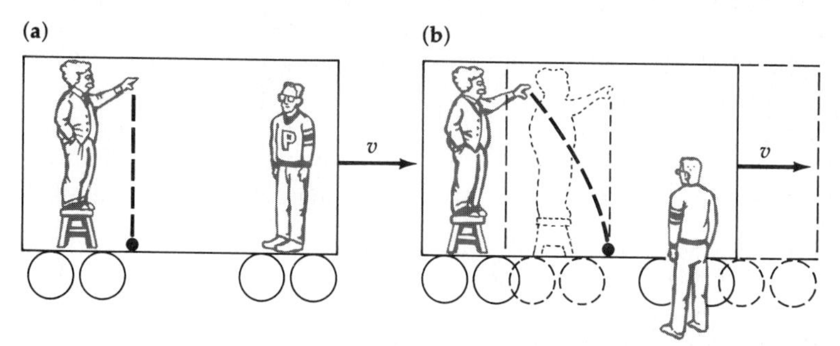

Figure 8-2. Relative motion. In (a), an observer riding in a railroad car sees a falling ball follow a straight-line path. When the car is moving relative to the observer in (b), the ball is seen to follow a parabolic path.

an *absolute* reference for all motion was deemed extremely important for answering such questions.

Scientists "solved" the problem of an absolute frame of reference by believing in a substance called the *luminiferous ether.** The ether (which has nothing to do with ethers you might make in chemistry lab) was thought to fill all of space, even a vacuum. It was a wonderful "material." Scientists could not detect it in any way. It was unaffected by any forces. The ether, so the idea went, was the fixed frame of reference for determining absolute motion. Given an ether fixed in space, the motion of the earth through it must give rise to an "ether wind" similar to the wind you feel when you run on a day when the air is still.

There had been many unsuccessful attempts to detect this ether wind. Most physicists felt that the methods used were not sensitive enough or the measurements not sufficiently accurate to justify abandoning the notion of the ether on the basis of these negative results. So the ether remained.

Then, in 1887 (Einstein was 8 years old) two American scientists performed a series of experiments whose results were interpreted to show that the ether did not exist. Inspired by the work of James Clerk Maxwell on the theory of light waves, A. A. Michelson, a physicist at the Case School of Applied Science in Cleveland, and E. W. Morley, a chemist across town at Western Reserve University, used a technique in which they measured the relative travel times for beams of light moving on perpendicular paths. (Supplement 6 gives a description of the Michelson-Morley experiment.) The sensitivity of these measurements was not seriously questioned. The very accurate results provided dramatic evidence that the ether did not exist.

If there is no ether, then there is nothing with respect to which one can determine *absolute* motion. You can determine only the *relative* motion of two frames of reference. Which one is moving and which is at rest are purely matters of convenience, as long as the velocity of their relative motion remains constant. So, since the motion of one reference frame can be determined only relative to another, it follows that you cannot determine your motion by any experiment solely within the confines of your own frame of reference. This is the crux of Einstein's first assumption: if two observers are moving relative to each other at a constant velocity, neither can determine which is "really" moving. This is often referred to as the *principle of Galilean relativity,* since Galileo expressed a version of it many years ago.

One windy day, two Buddhist monks were arguing about a flapping banner. One said, "The banner is moving, not the wind." The other disagreed, "The wind is moving, not the banner." A third monk settled the issue: "It is neither the wind nor the banner that is moving—it is your minds."

Zen parable

A frame of reference moving at constant velocity is an example of what is called an *inertial* frame, because the law of

*Luminiferous means "light-carrying." Another important reason for needing the ether was the idea that some material was necessary in order for light waves to travel through space. We will discuss this further when we study light in Chapter 20.

inertia (Newton's first law) is valid in such a frame. If you are in an inertial frame of reference, you may legitimately consider yourself to be at rest because there is nothing you can do within your frame to detect your motion. Only by "peeking out" at a frame of reference moving relative to yours can you determine your motion. Even so, you can only determine the *relative* motion between your frame and the neighboring one, and you can choose either of them to be at rest. So the question of which is "really" moving becomes pointless. There is no "really" to it.

A familiar example may help you understand this. Suppose you are riding in an airplane traveling in a straight line at a constant speed of, say, 800 kilometers per hour. As long as this velocity remains constant—that is, no turns, no turbulence, no speed changes—you can't tell you are moving. You may write, play cards, walk up and down the aisle, drink coffee, do all the things you could do if the plane were waiting on the runway. Block out the engine noise, and you might as well be in a long, narrow room. To be technical about it, there is no experiment you can perform within your "capsule"—like dropping your pencil, for example—by which you could detect the fact that you are zipping along at 800 km/h. The pencil drops straight down, just as it does it your classroom.

Only by making an observation *outside* your frame can you detect your relative motion. For example, you might look out the window at the fields passing below. And even then, you could just as easily and legitimately assume that the *airplane* is at rest and it is the *ground* that moves along with high speed. We are conditioned to taking the parochial view that the ground is stationary and the plane is moving, but this is a mere convenience and has no absolute relevance. In flying from Dallas to Atlanta, for example, you could easily suppose that you rise off the ground at Dallas and hover in midair until Atlanta comes beneath you.

If this seems absurd, consider your present situation. You are sitting in a room reading this book. The room is firmly attached to a building that in turn is firmly attached to the earth, which also seems fixed. Without thinking very much about it, you assume that you are at rest. Yet, the earth spins on its axis while orbiting the sun at about 30 km/s. The sun itself is just one more star, moving with a speed of almost 300 km/s through our Milky Way galaxy. And ours is a galaxy among galaxies, moving through the universe at around 600 km/s (well over a million miles per hour). So you are not at rest at all, relative to the nearest galaxy, say. But, Galilean relativity allows you to take the convenient and reasonable view, which is nevertheless parochial, that you are at rest. We do this every day when we speak of the rising and setting of the sun, for instance.

A word of caution: We are talking here about frames of reference whose motion relative to each other is *constant* velocity. It was with this special case of uniform motion that Einstein concerned himself in his assumptions. Hence it is called the *special theory of relativity,* or simply special relativity. When accelerations are involved, things are much more subtle and complicated; this is the realm of *general* relativity, which Einstein published in 1915. We shall be concerned only with special relativity, the motion of reference frames moving with constant velocity.

Learning Checks

1. What was the ether? What purpose connected with motion was it supposed to serve? Why was the notion of the ether abandoned?
2. What is a frame of reference?
3. What is an inertial frame of reference? Is an airplane traveling at 500 km/h in a straight line an inertial frame? How about a rotating phonograph turntable? Is the earth an inertial frame of reference?
4. State Einstein's first assumption in your own words. To what type of motion is this assumption restricted?

THE CONSTANT SPEED OF LIGHT

Einstein's second assumption in formulating special relativity was that the speed of light is always the same, as measured by any observer. The speed of light in a vacuum is a very large number, nearly 300,000,000 (300 million) meters per second, which we may write as 3×10^8 m/s. (Recall from Chapter 5 that we can use this power-of-ten shorthand for writing large or small numbers.) It has become customary to denote this number by the letter c, so hereafter I will write c to mean the speed of light. Einstein says that no matter what the condition of motion of either the source of the light or the observer making the measurement, the speed will *always* turn out to be c.

Now, this seems harmless enough until we compare it with our everyday experience. Imagine a child on a railroad flatcar traveling east at 100 km/h relative to the ground. Suppose that the child rides a tricycle 5 km/h east relative to the flatcar. Clearly the child's speed relative to the ground is just the *sum* of the speeds, 105 km/h. Riding west at the same speed, the child's speed relative to the ground would be 95 km/h, which is the *difference* in the speeds. In general, if the flatcar travels with speed u relative to the ground and the child travels with speed v relative to the flatcar, the child's speed relative to the

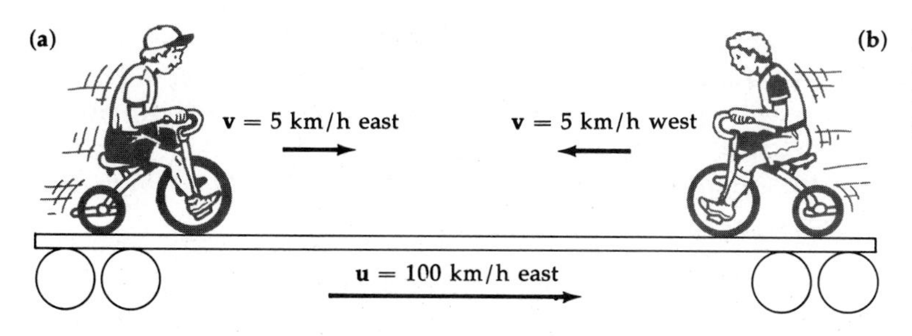

Figure 8-3. Relative speed. Speed of (a) relative to the ground = 105 km/h; speed of (b) relative to the ground = 95 km/h.

ground will be either $(u + v)$ or $(u - v)$, depending on the relative directions. (See Figure 8-3.)

Suppose now that instead of the child on the tricycle traveling at speed v we have a beacon that sends out to the east a flash of light traveling at speed c relative to the flatcar. By analogy with the child on the tricycle, the speed of the light would be $(u + c)$ relative to the ground. However, this violates the theory, because the speed of light must be measured to be c for all observers, including, in this case, one standing on the ground.

As it happens, the quantity $(u + v)$ is only *approximately* correct for the speed measured by the ground observer. It works only when the speeds are very small compared with light speed. Special relativity gives us an *exact* expression, which is always correct, no matter what the speeds are. It is

$$\frac{u + v}{1 + \dfrac{uv}{c^2}}$$

Let us see what this means, by putting in some numbers. For the tricycle rider having $v = 5$ km/h, the term uv/c^2 is extremely small. So when $u = 100$ km/h, the whole denominator is 1.000000000000000429. This is so close to 1.0 we don't need to worry about it, and $u + v = 105$ km/h is very near the exact result.

But now when we replace the tricycle with the flash of light, moving at 1,080,000,000 km/h, we have

$$\frac{(100 \text{ km/h}) + 1{,}080{,}000{,}000 \text{ km/h}}{1 + (100)(1{,}080{,}000{,}000)/(1{,}080{,}000{,}000)^2}$$

This gives

$$\frac{1{,}080{,}000{,}100 \text{ km/h}}{1.0000000926} = 1{,}080{,}000{,}000 \text{ km/h}$$

which is just c, the speed of light back again! This is what the observer on the ground sees for the speed of light, and it is the same speed someone riding on the flatcar measures. Adding the speed of the flatcar has no effect. The light's overall speed is still c, for *both* observers. Notice that even if the speed of the flatcar were also that of light ($u = c$), we would still obtain an overall speed of c.

In summary, then, the speed of light added to any other speed equals the speed of light.

Now that we understand Einstein's two assumptions, let us put them together and see what we can learn. To do this, suppose we imagine an infinitely long straight highway upon which all cars must travel at exactly 60 km/h relative to the earth, no more, no less. Furthermore, the northbound cars are all equipped with signs reading "North" and the southbound cars are all labeled "South." The lone exception is your car, which for some reason is restricted to speeds less than 60 km/h and is not equipped with a speedometer. The only way you can tell how fast you are going and in which direction you are moving is by watching the other cars on the highway. This agrees with the principle of Galilean relativity: if you were to enclose yourself in the car without looking out, you would not be able to detect your motion by any experiment within the capsule, such as dropping pencils or tossing a ball back and forth.

Now suppose you observe that the cars labeled "North" are moving north at 80 km/h relative to your car. Since you know these cars are moving north at 60 km/h relative to the ground, you would conclude that your car is traveling *south* at 20 km/h relative to the ground. You might confirm this by noting that the cars labeled "South" are traveling south at 40 km/h relative to you. Again, if you determined that all cars traveled at 60 km/h relative to you, the "North" cars north and the "South" cars south, you would conclude that your car was not moving relative to the earth. In any event, you could determine your "earth" speed by observing the motion of your frame of reference relative to other frames (the other cars). So far, so good.

Let us go further in this thought experiment by trying to devise an arrangement whereby we might violate the principle of Galilean relativity. Rather than determine your motion with reference to northbound and southbound cars, suppose you try to determine the motion of your car with northbound and southbound beams of light within the car. You might set up a flashlight at the north end of the car and one at the south end, sit in the middle, and observe the speed of the two beams of light relative to you as they zip by. In essence you have replaced the other cars, whose constant speed is 60 km/h, by beams of light, whose constant speed is c. If you observe that the southbound beam of light moves faster than the northbound beam, you must be moving north, and so on. And this would violate Galilean relativity: you would have detected the absolute motion of your frame of reference relative to empty space, without consulting any other reference frame.

Well, it turns out that similar experiments have been tried and the results are always conclusively negative—the two beams of light show no difference in speed whatever, no matter what the motion of the capsule. Had the Michelson-Morley experiment detected the "ether wind," it would have succeeded in violating special relativity, and the result is typical of other experiments. The second assumption of Einstein has been verified many times since then: the speed of light is always measured as c by all observers.

To see what we are in for, let us return to the highway where the cars are traveling at 60 km/h. If the cars behaved like the light, you would always observe that their speed relative to you was 60 km/h no matter what your state of motion. That is, if you were at rest relative to the ground, the other cars would be going 60 km/h relative to you, which seems reasonable enough. But, if you were traveling 40 km/h north, the northbound cars would *still* be traveling 60 km/h relative to you (rather than 20 km/h), and the southbound cars would also still travel 60 km/h (rather than 100 km/h)! If such a state of affairs existed for cars on the highway, you couldn't determine your motion relative to theirs because they would always be going at the same speed relative to you.

This all seems quite ridiculous, for we know that cars on the freeway do not behave like this. And yet, light *does* behave this way. The only difference between the cars traveling at a constant speed of 60 km/h and light traveling at the constant speed of c is that c is such a fantastically greater speed. Therefore, the question arises, would the cars on the highway show such behavior if they traveled at these stupendous speeds? The answer that the special theory of relativity gives is an unequivocal *yes*. This is verified by experiments, not with cars on freeways, but with particles made to travel at speeds very close to that of light. An important point here is that there is nothing remarkable about the light; it is the tremendous speed that makes all the difference. Bertrand Russell, in his excellent book *The ABC of Relativity*, gives a nice example to illustrate the point: If a department store escalator traveled between floors at the speed of light, you would reach the top equally quickly either by standing on the escalator or by running up its steps.

Our common sense is appalled by these ideas, but only because our "common sense" is the product of human beings' having lived for thousands of years in a world where normal speeds are very much less than the speed of light. Part of our way of looking at the world is somewhat a matter of luck, and things that are more or less accidentally the way they are make other things seem like necessities of thought.

Einstein's Second Postulate

H*ow* have contemporary physicists verified Einstein's second postulate? What methods have they used? One based his calculations on an analysis of pulsating x-ray sources in binary star systems. Another studied the relativistic decay of certain subatomic particles. The details of these paths are complex, but results show just how right Einstein's second postulate is.

Source: "New Limit on Constancy of Velocity of Light," *Physics Today*, March 1978, p. 19.

The case of "ordinary" speeds is but one example. We are also used to things being stationary, so that walking to the corner drugstore has a definite meaning. But if things in our world were whizzing about in a random way, then our perceptions and thought patterns would be quite different; indeed, we would have first to inquire where the corner drugstore *is* today. Were we the size of an electron, we would not have the same impression of stability. Were we the size of the sun, the universe would not seem to have nearly the permanence we perceive. So the notion of comparative stability, which forms part of our ordinary outlook, is due to the fact that we are about the size we are and experience the speeds we do. Otherwise, the classical physics that preceded relativity would not be intellectually satisfying, and probably would not have been invented at all.

Learning Checks

1. When does $u + v$ give the correct answer for adding velocities? When is it wrong?
2. If the beam-of-light-in-the-car experiment worked, why would this violate Galilean relativity?
3. How has common sense tricked us about the relation between space and time?

SUMMARY

Einstein's special theory of relativity arises out of the union of two assumptions that can be experimentally observed: (1) the laws of physics are the same in all reference frames traveling at constant velocity with respect to each other, and (2) the speed of light is a constant for all observers. Taken together, these postulates mean that space and time are not separate but in fact depend on one another. We are not aware of this dependence because the speeds of objects in our everyday environment are so slow compared with the speed of light.

Exercises

1. Sometimes you will hear people say that special relativity implies that "all motion is relative." Is this true? Explain.

2. Give two examples from your own experience that demonstrate Galilean relativity.

3. A policeman involved in a high-speed chase fires his revolver at an object in front of the patrol car. Will the bullet travel faster relative to the ground than it would if the gun were fired with the patrol car standing still? Explain.

4. Suppose the policeman in Exercise 3 shines the searchlight mounted on the speeding patrol car at the object in front. Will the light travel faster relative to the ground than it would if the searchlight were used while the car is stationary? Explain any similarities or differences in your answers here and in Exercise 3.

5. One of Einstein's postulates says that the speed of light is the same for all observers. Describe how this contradicts what you observe about the speed of ordinary objects, such as thrown baseballs and rolling streetcars.

6. Suppose you are riding in a golf cart at constant velocity. If you toss a golf ball straight up high into the air, will it land *in* the cart or *behind* it? (Neglect air resistance.)

7. In Exercise 6, describe the path of the golf ball as you see it and as another golfer, standing on the ground, sees it.

8. Physicists like to say that if you are moving at constant velocity relative to a building, say, you can correctly suppose that you are sitting still and the building goes by you with that velocity. Does this follow from Galilean relativity? Explain.

9. We say that the moon goes around the earth. Is this really true? How does it look to someone standing on the moon? Is either point of view "correct" and the other "wrong"? Explain.

10. You are in a canoe, and the fastest you can paddle in still water is 6 km/h. Suppose you are in a stream whose current is 2 km/h relative to the earth. Calculate your maximum speed relative to the earth when going *upstream* and *downstream*.

11. In Exercise 10, how long would it take you to make a round trip of 80 kilometers (40 upstream and 40 back)?

9

SPECIAL RELATIVITY: THE CONSEQUENCES

We don't fully appreciate the consequences of special relativity until we understand the high speeds and vast distances involved in space and time travel.

In Chapter 8 we learned that Einstein came to a startling new view of nature by combining two postulates: (1) we can detect constant-velocity motion only by comparing our reference frame to another, and (2) everybody always gets the same number for the speed of light. Now we are in a position to see the results that flow from these assumptions. We will find that, if Einstein's ideas are correct, then we are compelled to live in a universe where distances measured with rulers and odometers and time segments recorded by watches and clocks are no longer absolute quantities. The answers that come out of such measurements depend on who is doing the observing and, in particular, on how fast the observers are moving relative to one another.

Three results of special relativity form a sort of package deal. Observers in motion relative to each other will find that moving clocks run slow, that moving lengths become shorter, and that events taking place at the same instant in one reference frame will not happen simultaneously in another. We are going to consider each of these ideas in turn. Let us first look at the retardation of moving clocks.

TIME DILATION

To measure time, we have to keep track of some kind of periodic motion, such as the pendulum swing of a grandfather clock or the vibrations of electrons in an atom (the so-called atomic clock). For examining the effects of special relativity on time measurement, we imagine an idealized clock consisting of two mirrors that face each other from opposite ends of a tube and a light pulse that bounces back and forth between the mirrors with speed c. The clock "ticks" whenever the pulse strikes a mirror. Thus, the time between ticks is the time required for the light pulse to leave the first mirror and travel to the second. We give this time the label T. Then the clock's length, which is the distance between the mirrors, is equal to c multiplied by T. (See Figure 9-1.)

Figure 9-1. Light-pulse clock.

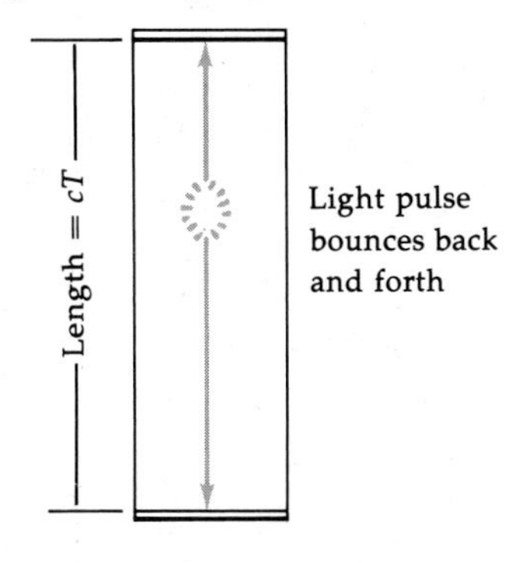

Figure 9-2. Moving and stationary light-pulse clocks.

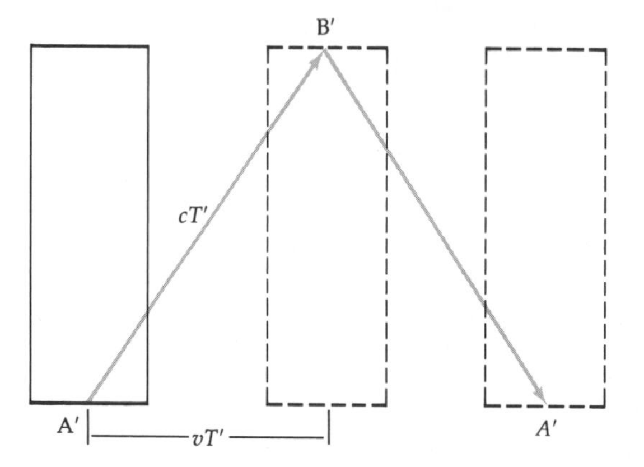

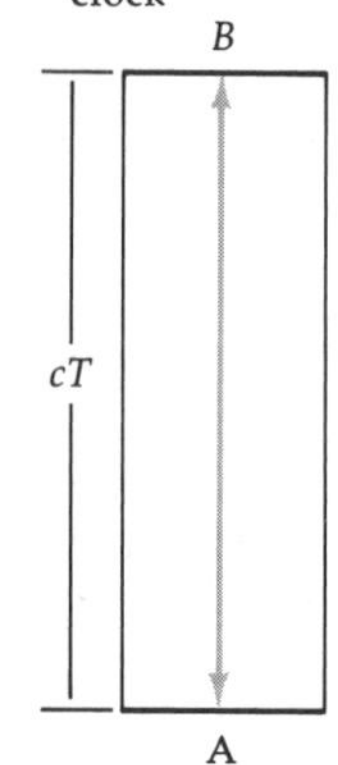

Suppose we make two of these clocks. We place one of them aboard a spaceship and observe what happens when the spaceship moves past us in a straight line with a constant speed v. During the time the light pulse is making its journey from one mirror to the other, the clock is moving relative to us, so we see the light pulse following a zigzag path. (See Figure 9-2.) This path causes the light pulse in the moving clock to travel farther than the one in the stationary clock before striking the other mirror. You can see this by noticing that the line A′B′ is clearly longer than the line AB, which is the length of our stationary clock.

Because we know from Einstein's assumption that the light pulses in both clocks travel at the constant speed c, the moving clock requires a longer time to tick than the stationary clock. We will call this longer time T'. Thus, by the time the light pulse in the moving clock gets to the mirror at B′ to record one tick, the pulse in our stationary clock will have already reached mirror B and started on the return trip back toward the first mirror because it had to travel a shorter distance at the same speed. Therefore, from our point of view, the clock on the spaceship is running slow; that is, more time elapses between ticks. This retardation of the moving clock is called *time dilation.*

We can calculate this time dilation for the moving clock. From Figure 9-2, we can see that during the time T', the clock travels a distance equal to v multiplied by T', because it zooms by us at speed v. The length of the clock is still cT, but the path the light pulse travels has length cT'. These three quantities form a right triangle, shown in Figure 9-3.

Figure 9-3. Right triangle for the light-pulse clocks.

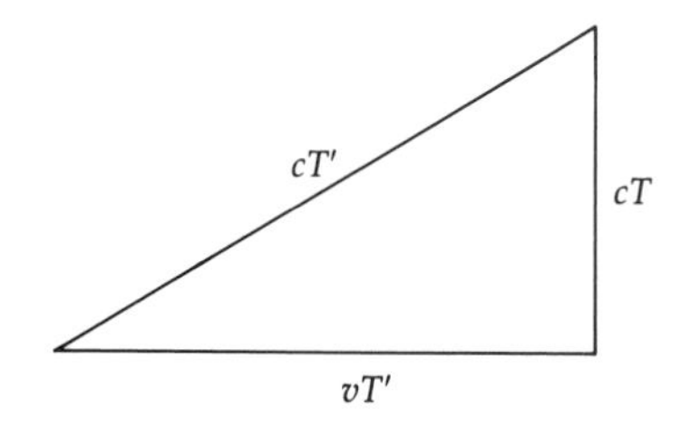

Now, it turns out that there is a very simple rule, discovered a long time ago by Pythagoras, for relating the three sides of a right triangle. It says this: If you multiply the length of each of the two shorter sides by itself and add the results, the number you get is the length of the longest side (the hypotenuse) multiplied by itself. In symbols,

$$(cT')^2 = (vT')^2 + (cT)^2$$

After a few lines of algebra, we can write

$$T' = \frac{T}{\sqrt{1 - (v/c)^2}}$$

This equation tells us how T', the time between ticks for the clock moving at speed v past the stationary clock, is related to T, the time between ticks for the stationary clock. It is especially important to notice that this tick-time T' is measured from the stationary clock. That is, it tells us how a person in *one* frame of reference would count time in *another* frame of reference moving relative to the first. Remember that v is the relative speed of the two frames of reference. If v is different from zero, then T' will be larger than T, so the moving clock runs slow. The greater the relative speed of the two frames, the larger the time discrepancy between them.

Note that the time dilation equation depends on the ratio v/c. For ordinary speeds, this ratio is very close to zero, since c is so large. For even a speed of, say, 300 meters per second, which is over 1000 kilometers per hour, the value of $(v/c)^2$ is only 1×10^{-12}, or 0.000000000001, which is very close to zero on the scale of times we are considering. So ordinarily T' and T are very nearly equal. The time dilation effect is negligible except for very large velocities, so in our everyday activities we don't notice the differences between stationary and moving clocks.

What does this result say? Well, for one thing, it tells us that our measurement of time in our own reference frame will be *different* from our measurement of time in a reference frame moving with respect to us. The motion of our own reference frame will not affect the flow of time that we measure in our own frame—ticks of a wrist watch, oscillations of a pendulum, the aging process—because, according to Galilean relativity, we may assume that we are at rest. We may call this "proper time." On the other hand, "improper time" is the time that *we* observe for a *different* reference frame moving relative to us. This time observation is improper in the sense that two observers moving relative to one another will disagree as to how much time has passed, by an amount represented in the equation for T'. The crux of the matter is that we can no longer picture time as some absolute medium in which we all exist; time depends on the relative motion of the clock and the observer.

Learning Checks

1. What is meant by "time dilation"?
2. Why is it correct to say that the length of the light-pulse clock is equal to c times T?

The Clock Paradox

Suppose we look at this moving clock idea from the standpoint of astronauts on board the spaceship that we observe whizzing by us at speed v. What do the astronauts see? Since their reference frame, the spaceship, is traveling with a constant velocity relative to the earth, they may take the position that *they* are at rest and the *earth* is moving by with speed v. As they see us go by, they notice that the light pulse in our clock bounces between the mirrors along a zigzag path. They therefore conclude, by the same reasoning we used earlier, that *our* clock is running slow. (See Figure 9–4.)

How can this be? We have already seen that it is *their* clock that is wrong. If both clocks keep a record of the number of ticks, then surely by comparing them later on we could tell which one is "really" running slow. It appears that we have a paradox: observers in each frame of reference see that the clock in the other, moving, frame is running slow.

This is often called the "clock paradox." It isn't actually a paradox when we are careful about what Einstein's assumptions really say. The first assumption specifies that the two reference frames move with respect to each other at *constant velocity,* that is, constant speed in a straight line. This means that we and the astronauts will be able to compare clocks only once, namely, when we pass each other. To make a second comparison (to check the number of recorded ticks, for example), at least one of us would have to accelerate in order for us to get back together. That is, the astronauts must either stop their spaceship, turn it around, and head back toward us, or they must move in a curved path in order to get back to make the comparison. In either case, the spaceship has not maintained constant velocity. It has accelerated. *The assumptions of special relativity do not cover instances where acceleration is involved.*

Figure 9-4. The clock paradox.

Astronaut's clock

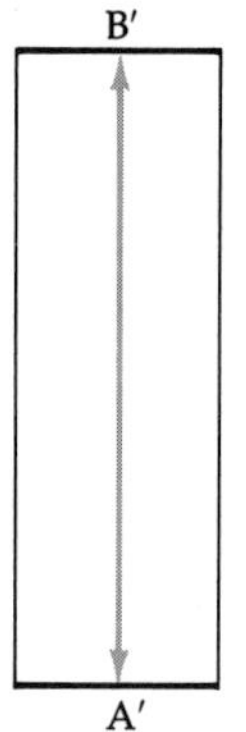

Our clock, as seen by the astronaut

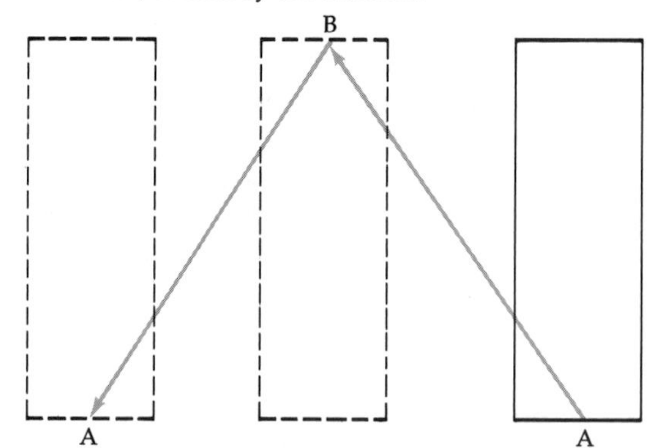

The clock paradox has its origin in the notion that *either* observer can assume that it is the *other* observer who is moving. However, when one of them accelerates, then both agree on *which* one. There is no longer any question as to who is moving. For example, you don't detect the motion of your airplane moving relative to the earth at a constant velocity of several hundred kilometers per hour. However, if the plane accelerates in any way, by changing its speed or going into a turn, then you are aware of that motion. Your coffee spills, your dropped pencil doesn't fall straight down, you can "feel" the acceleration; in short, you *can* detect the accelerated motion without observing anything outside your frame of reference. Special relativity doesn't deal with this case. It turns out that whenever accelerations are involved and both observers can agree which one of them is actually moving, it is always true that the *moving clock* slows down.

Comparison of Times in Two Frames

Let us put some numbers into our equation for T' so that we can see what sort of relative speeds must be involved in order for time dilation to be noticeable. Suppose we put one of our light-pulse clocks aboard a jet airplane and go through the same thought experiment as with the rocketship earlier. We have the airplane moving at $v = 800$ km/h, which is equal to 0.222 km/s. The ratio $(v/c)^2$ is then 0.00000000000055 (recall that $c = 300{,}000$ km/s), and $\sqrt{1-(v/c)^2}$ becomes 0.99999999999972, a number so close to 1 that the time dilation is undetectable.

On the other hand, suppose that we had a rocketship that could travel through space at 80 percent of the speed of light, or 240,000 km/s. (This is over 800 million km/h.) Then $v/c = 0.8$ and $(v/c)^2 = 0.64$. We have

$$\sqrt{1-(v/c)^2} = \sqrt{1-0.64} = \sqrt{0.36} = 0.6$$

Now, the time dilation factor of 0.6 is significantly different from 1. The clock on this spaceship is running slow compared with our clock by a factor of 1/0.6, or 1.667, as viewed by us on the earth. This means that if space travelers moving at 80 percent of the speed of light were gone for 20 years by the earth calendar, they would age at the rate of only 0.6 of this time, or 12 years. Of course, the space travelers would not observe their pulse rates changing as a result of this motion, or their wrist watches slowing down, or any other such effect, because all these "clocks"—including the aging of their own bodies—share the same motion. They occupy the same frame of reference, and hence experience the same proper time.

Learning Checks

1. What is the "clock paradox"? Why is it not really a paradox?
2. As measured from the stationary clock, what happens to the length of a "second" on the light-pulse clock as the speed of the moving ship increases? What would happen to the light pulse if the ship traveled at the speed of light?
3. Why is it true that, if Einstein's assumptions are applied, the time as measured on the two clocks moving relative to each other cannot be compared a second time?
4. In the equation $T' = T/\sqrt{1 - (v/c)^2}$ explain what each symbol means.

LENGTH CONTRACTION

Our sense of length, also, is affected by relativistic considerations.

A companion effect to time dilation is the reduction of the lengths of moving objects as seen from a different reference frame. Again, suppose we have access to a spaceship capable of traveling at 80 percent of the speed of light. We notice that the speed of light c is about 300 m/μs, where μs is an abbreviation for *microsecond*, meaning one-millionth of a second. So we will use this unit to make the numbers come out conveniently. Our rocket can travel 80 percent of this speed, or $v = 240$ m/μs. To perform our experiment, we mark off a course on the ground by placing flags 240 meters apart. As the rocket passes by, it will travel the distance between the flags in a time of 1 microsecond as read on our clock. However, since the clocks aboard the ship run slow, they will only show an elapsed time of 0.6 second according to our time dilation equation.

Now let us analyze these results from the point of view of the astronaut on board the ship. According to Galilean relativity, he can consider himself to be at rest and the earth moving by at a speed of 240 m/μs. Since his clock shows a time of 0.6 microsecond for the passing of the two flags, he concludes that the distance between them is

$$d = 240 \text{ m}/\mu\text{s} \times 0.6 \ \mu\text{s} = 144 \text{ m}$$

Then, from the point of view of the astronaut, the marked course which we measured to be 240 meters long was contracted because of its motion relative to him to a length of only 144 meters!

Suppose that the ship is 24 meters long. An observer stationed on the ground at one of the flags would zoom by the astronaut at 240 m/μs, taking 0.1 microsecond to go from the front of the ship to the back. Since the ground observer's clock runs slow by a factor of 0.6, only $0.6 \times 0.1\ \mu s = 0.06\ \mu s$ would have elapsed on his clock in the time during which he passed the spaceship. From the point of view of the ground observer, the spaceship traveling by him at 240 m/μs and requiring 0.06 microsecond to pass would have a length of $240\ m/\mu s \times 0.06\ \mu s = 14.4$ m, compared with the 24 meter length measured by the astronaut.

This example illustrates a second important general feature of special relativity: *moving lengths contract in the direction of their motion by a factor* $\sqrt{1-(v/c)^2}$. In general, if L_0 is the "rest length," that is, the length an observer in the same reference frame measures, then the length L as measured by an observer in *another* frame moving at speed v will be given by

$$L = L_0\sqrt{1-(v/c)^2}$$

Again, if v is very small compared to c, then L is very close to L_0, which is just what we normally observe.

As these examples show, time dilation and length contraction are intimately related to each other. This is illustrated by the fact that our friend, the factor $\sqrt{1-(v/c)^2}$, appears in both the equation for time dilation and the one for length contraction. Perfect symmetry appears here, for as the time is stretched out, space is contracted in the direction of motion in exactly the same proportion.

It is important to notice that only the dimension *along* the direction of motion is reduced. The other dimensions are unaffected. If you were to speed down the street past a row of buildings, they would appear skinnier and closer together but just as tall as when you were not moving.

Learning Checks

1. Why don't we notice time dilation and length contraction in our ordinary experience?
2. In our example, the ground observer sees that the spaceship is 14.4 meters long rather than 24 meters, which is the length the onboard astronaut measures. Who, if either, is right?
3. In Question 2, would the ground observer and the astronaut notice any difference in the ship's diameter? Explain.

SIMULTANEITY OF EVENTS

There is a third feature of relativity that we must consider, to have the complete picture of a consistent theory. Until the advent of special relativity, no one had thought that there could be any ambiguity in the statement that two events in different places happened at the same time. One might admit that, if the places were very far apart, there might be some difficulty in finding out for certain whether the events were simultaneous, but everyone thought the meaning of the question to be clear.

However, this is not the case. Two events in distant places can appear to be simultaneous to one observer who has taken all due precautions to insure accuracy, while another equally careful observer may judge that the first event preceded the second, and yet a third observer may judge that the second preceded the first. This would happen if all three people were moving rapidly with respect to one another. Furthermore, they would all be equally right. The time order of events is not an absolute; it is a matter that depends on the relative motion of the observers.

Based on our normal experience, we would naturally define two events as simultaneous if they are seen simultaneously by a person who is exactly halfway between them. Figure 9-5 shows stationary observers at points A, B, and M, where M is at equal distances from A and B. If an observer at M sees lightning bolts strike simultaneously at points A and B, then the two strikes took place simultaneously. The observer at A will see the flash there first and then the one at B because of the time required for light to travel between A and B. Similarly, the observer at B will see his flash first, and then the one at A. But all three will agree on the simultaneity of the two strikes because they will allow for the speed of light. No problem.

Such agreement crumbles, however, if two sets of observers are *moving* rapidly relative to one another. Here, our definition of simultaneity is unsatisfactory. Bertrand Russell, in *The ABC of Relativity,* gives an amusing illustration of the difficulty we are in for. In this example, sound is substituted for light, but nothing is changed in principle; it makes things easier to comprehend, however, because of the much slower speed of sound:

> Let us suppose that on a foggy night two men belonging to a gang of brigands shoot the guard and engine-driver of a train. The guard is at the end of the train; the brigands are on the line, and shoot their victims at close quarters. An old gentleman who is exactly in the middle of the train hears

Figure 9-5. Simultaneity of events. M is at equal distances from A and B. Events at A and B are simultaneous if they are observed simultaneously by someone at M.

the two shots simultaneously. You would say, therefore, that the two shots were simultaneous. But a station-master who is exactly halfway between the two brigands hears the shot that kills the guard first. An Australian millionaire uncle of the guard and the engine-driver (who are cousins) has left his whole fortune to the guard, or, should he die first, to the engine-driver. Vast sums are involved in the question which died first. The case goes to the House of Lords, and the lawyers on both sides, having been educated at Oxford, are agreed that either the old gentleman or the station-master must have been mistaken.

In fact, both may perfectly well be right. The train travels away from the shot at the guard, and towards the shot at the engine-driver; therefore the noise of the shot at the guard has farther to go before reaching the old gentleman than the shot at the engine-driver has. Therefore if the old gentleman is right in saying that he heard the two reports simultaneously, then the station-master must be right in saying that he heard the shot at the guard first.

Here is another example, similar to Einstein's original thought experiment. Imagine a railroad train carrying two passengers, one at each end. Two more people are on the ground, also at opposite ends of the train. A flash bulb explodes exactly in the center of the train. If the train is not moving, all four people will agree that the light reached the two ends of the train simultaneously.

Now suppose that the train is moving at a high speed. The ground observers stand so that each is opposite one end of the train when the light from the exploding flash bulb arrives. Let us see what happens from the point of view of the two sets of observers.

First, consider the people on the train. Because the train is moving at constant velocity, they can legitimately assume that the train is stationary and that the ground observers whiz by at a high speed. The flash of light leaves the bulb and begins moving toward the end of the train, as shown in Figure 9–6a. Some time later, the light reaches the two ends of the train simultaneously, because it started in the middle and traveled at a constant speed in all directions, as shown in Figure 9–6b.

But now consider what happens as seen by the ground observers, who see the train moving by them with speed v. The light flashes (Figure 9–7a). Since the light originated at a point closer to the rear observer, and because light travels at constant speed for all observers, the light will reach the rear observer first (Figure 9–7b). It then reaches the forward observer at a later time. The train has moved during this interval (Figure 9–7c).

Hence, the dilemma: the observers on the train claim that the light reached the two ends of the train simultaneously. This

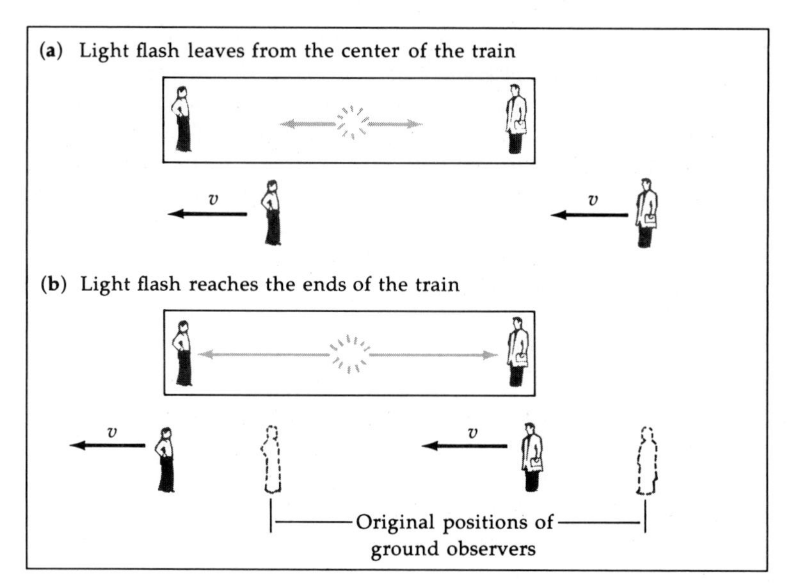

Figure 9-6. The situation as seen by observers on the train.

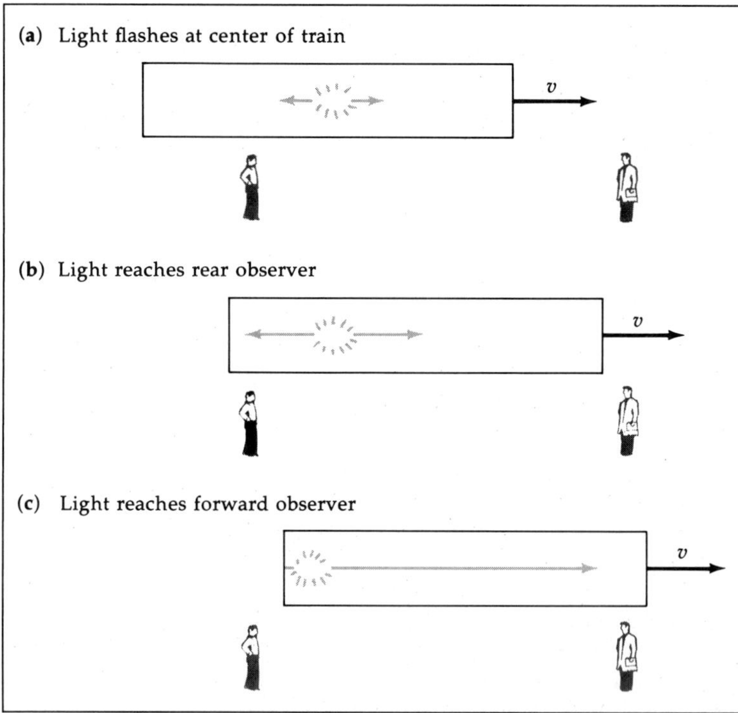

Figure 9-7. The situation as seen by observers on the ground.

is correct because it started at a point exactly halfway between them and traveled in all directions at a constant speed relative to the train. Not so, argue the observers on the ground. They say that because the flash occurred nearer the rear observer and the light traveled at a constant speed relative to the ground, the event of "seeing the light" happened first for the rear observer and later for the forward observer, when the end of the train was opposite him. The observers on the train retort that, while it is true that the rear observer on the ground was closer to the flash, he had moved in the meantime, running away from the light, so to speak, and was really in the same position as the train's rear observer when the light arrived. And so it goes.

Notice here that both sets of observers have adhered to Einstein's postulates for the special theory of relativity: the frames of reference are moving at a constant speed relative to each other, and observers in both frames regard the speed of light as a constant for them. Therefore, both are "right." One cannot say that the two events—seeing the light at the two ends of the train—"really" occurred simultaneously, or that the rear observer "really" saw it first. There is no "really" to it. The time order of events depends on who is making the observations.

Learning Checks

1. What does it mean for two events to happen simultaneously? If all observers are in the same frame of reference, can simultaneity be defined? What if some observers are moving relative to others? Does simultaneity have meaning in this case?
2. In Russell's train-robbery example, how would the station master have observed the sequence of events if the train were not moving? Answer the same question for the old gentleman on the train.
3. Again in the train-robbery example, would it be possible for a third observer to believe that the engine-driver had been shot first?

THE UNIVERSAL SPEED LIMIT

In all that we have said so far, it is apparent that the speed of light plays a central role in the special theory of relativity. The speed of light is also important for another reason: it serves as a universal speed limit that cannot be attained or exceeded by matter. While there are bundles of matter—fundamental particles, they are called—that travel at speeds very close to the velocity of light, as close as 99 percent in some cases, no material object in the universe has ever been observed to travel *at* the speed of light, and there are strong reasons for believing that this will never happen.

Look again at the expression we developed earlier for time dilation

$$T' = \frac{T}{\sqrt{1 - (v/c)^2}}$$

and consider what happens to the denominator as v gets closer and closer to c. Recall that v in this equation is the speed of a spaceship relative to the ground, or more generally, the relative speed of the two frames of reference. Table 9-1 shows several values of the factor $\sqrt{1-(v/c)^2}$ and the time T' for values of v ranging from 10 percent to 99.9 percent of the speed of light.

Table 9-1
Time Dilation

v	$\sqrt{1-(v/c)^2}$	T'
$0.1c$	0.995	$1.005 \times T$
$0.2c$	0.980	$1.021 \times T$
$0.3c$	0.954	$1.048 \times T$
$0.4c$	0.917	$1.091 \times T$
$0.5c$	0.866	$1.155 \times T$
$0.6c$	0.800	$1.250 \times T$
$0.7c$	0.714	$1.400 \times T$
$0.8c$	0.600	$1.667 \times T$
$0.9c$	0.436	$2.294 \times T$
$0.99c$	0.141	$7.089 \times T$
$0.999c$	0.045	$22.366 \times T$

The table shows that as v gets increasingly larger, the denominator $\sqrt{1-(v/c)^2}$ gets smaller and smaller, approaching zero as the speed of the moving clock gets closer and closer to the speed of light. (Refer back to Figure 9-2.) This means that T', the length of time between ticks of the moving clock, gets larger and larger the faster the clock travels. In fact, if v were ever equal to c, then we would have $T' = T/0$. Now, division by zero is something that is undefined in mathematics. However, our tabulation of what happens as we allow v to creep up on c shows us that T' would become infinitely large when v equals c—the light pulse in the moving clock would never reach the top mirror.

Now imagine the impossible: that the spaceship (or any other object) could exceed the speed of light. For the sake of argument, suppose that it travels at twice light speed, $v = 2c$. Then $(v/c)^2 = 2^2 = 4$, so

$$\sqrt{1-(v/c)^2} = \sqrt{1-4} = \sqrt{-3}$$

Now a number like $\sqrt{-3}$ has no meaning in the description of quantities that are physically realizable, like ticks of a clock or speeds or anything else we may encounter in the real world. This is because the symbol $\sqrt{-1}$ means you have to have a number that when multiplied by itself equals -1. Since $(+1) \times (+1) = +1$, and also $(-1) \times (-1) = +1$, there is no such ordinary number that when squared gives -1. Mathematicians call such a number an imaginary number and assign the symbol i to the number $\sqrt{-1}$. So the square root of -4 is $2i$. This is a convenient device in calculations, but such imaginary numbers must not appear in any final results that are to have physical meaning.

Exceeding the velocity of light would lead to an imaginary number for the time T'. Such a result has no more physical meaning than saying you have \$$4i$ in the bank or that your uncle is $52i$ years old.

Changing Mass with Changing Speed

Consider for a moment what this universal speed limit tells us about forces in view of Newton's laws of motion. Recall that Newton's second law is $\mathbf{F} = m\mathbf{a}$, where the force $\mathbf{F}$ applied to the mass m results in the acceleration $\mathbf{a}$. If the acceleration is due to an increase in the velocity v, continuing to apply the force gives the mass greater and greater speed. But we have seen that the speed of a particle cannot increase indefinitely, because it is limited by the speed of light c. So as we continue to apply the force, and v gets closer and closer to c, the speed changes less and less. Hence, the particle is no longer accelerating as it was when the velocity was small.

Nevertheless, we can continue to apply the force $\mathbf{F}$ without limit. Since $\mathbf{a}$ is approaching zero, the inescapable conclusion is that the mass m is changing as the velocity increases. This mass change has been experimentally observed many times for particles such as electrons. So the mass is *not* a constant. It depends on velocity, as do time and length and sequences of events. The way in which the mass depends on the velocity is given by the expression

$$m = \frac{m_0}{\sqrt{1 - (v/c)^2}}$$

It is beyond the scope of this book to derive this equation, but it is not surprising that it has this form—the familiar quantity $\sqrt{1 - (v/c)^2}$ has cropped up once again. The term m_0 in the numerator is called the *rest mass,* appropriately enough, since it is the value of m when $v = 0$, that is, when the mass is at rest.

Notice that as in our discussion with time, the mass m gets infinitely large as v gets closer and closer to the speed of light. For $v = c$, we have $m = m_0/0$, which again is an undefined quantity for all objects whose rest mass is different from zero. Velocities exceeding that of light would give an "imaginary mass," a concept that has no meaning in the ordinary world.

Faster-than-Light Particles

What we have said so far applies to "ordinary" matter, objects having nonzero rest mass. It is impossible to accelerate these particles to the speed of light.

Of course, we know that some things *do* travel at light speed—in particular, light itself. As we shall see in Chapter 20, one way of considering light is to treat it as a stream of particles, called *photons*, that have a rest mass of zero and *always* travel at the speed c. There are other such zero rest-mass particles, also traveling at light speed, that we will learn about later. Newton's laws do not apply to them.

This means there are two kinds of particles that we know about: familiar objects that always travel at *less* than the speed of light, and zero rest-mass particles that always travel at *exactly* light speed—not a hair more or less. We can look at the speed of light as a sort of "wall" through which we cannot shove ordinary matter, a speed limit that particles in our domain cannot exceed regardless of how hard we push.

But now what if there were other particles—extraordinary ones, of a kind that we have never encountered—on the other side of this light-speed wall? Physicists speculate on what they might be like and have even invented a name for them—*tachyons*, from the Greek word for swift, *tachys*. This choice of name recognizes that these particles would always travel at speeds *greater* than the speed of light. To make the distinction, we might call our ordinary, slower-than-light-speed particles "tardyons." (See Figure 9-8.)

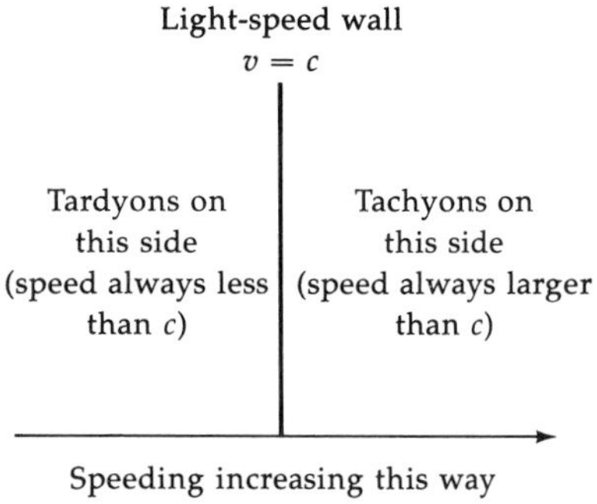

Light-speed wall
$v = c$

Tardyons on this side (speed always less than c)

Tachyons on this side (speed always larger than c)

Speeding increasing this way

Figure 9-8. Tardyons and tachyons.

What would a tachyon be like? Well, for one thing, it would have an "imaginary" mass, imaginary in the sense of being proportional to the mathematical quantity $i = \sqrt{-1}$. Thus, for example, a tachyon traveling at twice the speed of light would have $\sqrt{3}\, mi$ as its rest, or proper, mass.

We said that such a number has no physical meaning, and in our tardyon universe, this is true. But what is to prevent there from being a whole "tachyon universe," a mirror image of ours, on the "other side" of the light-speed wall? We haven't observed tachyons, but that may be because we aren't clever enough or are too hidebound in our tardyon-universe thinking to let the possibility fire our imaginations.

In addition to their imaginary mass, tachyons must have some other properties that seem absurd to tardyons like us. There would be no limit to their velocity, and the faster they went, the less massive they would become. They must travel backward in time. And their reaction to an applied force would be just the opposite of ordinary particles: tachyons would slow down when a force is applied and speed up when they encountered some resistance!

Just as we need to supply tardyons with an infinite force to get them up to light speed, so tachyons would require an infinite force in order to *slow down* to the speed of light. Thus, the speed of light is a minimum speed limit below which tachyons cannot go, so they could not penetrate the light-speed wall, being unable to travel slow enough.

There's a hell of a good
universe next door,
Let's go!

e. e. cummings

Learning Checks

1. What does it mean to say that c is a speed limit in the universe?
2. In the expression for the changing mass, explain what m, m_0, and v are.
3. Distinguish between tardyons and tachyons.
4. Explain how the increase of mass with speed for a tardyon is consistent with Newton's second law.

TIME DILATION AND TRAVEL IN SPACE

When Neil Armstrong stepped onto the surface of the moon in July 1969, the dreams of space travel that had fired countless imaginations for many years became reality. The years since then have seen unmanned flights to the surface of Mars and the vicinities of Venus, Jupiter, and Saturn, and space travel has become one more facet of human existence.

What about the future of manned exploration of space? When we think of traveling to the outer planets and beyond our solar system, we are impressed most of all by the tremendous distances involved. The nearest star beyond our sun is Alpha Centauri, actually a double-star system, about four and one-half light-years away. This means that the light you see coming from Alpha Centauri started on its journey to the earth some four and one-half years ago and covered 43 thousand billion kilometers. This seems far away by our everyday standards, but the distances of other stars and planets stagger the imagination and make Alpha Centauri seem almost like a member of the neighborhood. The edge-to-edge length of our Milky Way galaxy alone is over 900 million billion kilometers.

The vast expanse of space would seem to make space travel to other stars or galaxies impossible, simply because no one could live long enough to make the journey. However, we have seen that because of time dilation, time slows down when objects travel at speeds near light speed, including people and spaceships. At the present, it is not technologically possible to construct a spaceship capable of such speeds. Furthermore, the human body cannot stand the strong accelerations that would be required to build up such high speeds over a fairly short period of time. You may have read of test pilots blacking out when their planes accelerate to several "g's"—that is, several times the normal gravitational acceleration of about 10 m/s^2.

But suppose that it were technologically feasible to build a spacecraft capable of near-light speeds. For traveling to a "nearby" star, say Sirius, 8 light-years away, we could solve the acceleration problem by having the ship steadily accelerate at 1 g. Since there are about 30 million (3×10^7) seconds in a year, the ship's speed after 1 year would climb to about 98 percent of the speed of light. At such speeds, relativistic effects would become important, and the ship's crew would observe a shorter time than would their interested friends who stayed behind. By the earth's clock, the round trip would take about 16 years, but at this same acceleration rate, the ship's occupants would only age by 9 years.

For greater distances, time dilation effects are even more impressive. As an example, suppose a starship traveling with a constant acceleration of g traveled to the center of the Milky Way and back. By the earth's calendar, it would return 40,000 years after departure. But, by the proper time of the starship, only about 30 years would have passed.

Remember, *all* clocks—even the aging process—are subject to time dilation. Suppose twin sisters Arlene and Darlene are 25 years old when Arlene leaves on the trip to Sirius. Darlene, who stays behind, will be 41 years old when the ship completes its round-trip return to earth. But Arlene will have been traveling in a frame of reference where time is kept at a different rate. She will be only 34 when she returns to see her sister. The differences will be real. The "older" twin will really be 41, the younger only 34. Perhaps Darlene will think Arlene found the Fountain of Youth out there in space.

SUMMARY

Einstein's special theory of relativity teaches us that time and space are not separate entities but are related to each other in specific ways. Because we, and the ordinary objects that share our everyday world, travel at speeds much slower than light speed, we are conditioned to believing that time and space *are* independent. But this is only roughly correct. It is an approximation that breaks down as one considers speeds closer and closer to the speed of light.

The consequences of special relativity are that moving clocks run slow and moving lengths contract; both of these changes numerically involve the factor $\sqrt{1 - (v/c)^2}$ for reference frames whose relative speed is v. Also, people who are moving rapidly with respect to each other will not agree on the time order in which events take place.

Light speed is a universal speed limit for all material objects. We can speculate on other objects, "tachyons," which exist on the other side of this wall—they always move *faster* than light, have what we would call an imaginary mass, and travel backward in time.

Exercises

1. How would we "normally" define two events as simultaneous? Why does this seemingly reasonable definition collapse for observers moving at high velocities?

2. What is "proper time"? What is "improper time"?

3. Explain the belief that time dilation holds for *all* clocks, not just light-pulse clocks as used in our thought experiment.

4. It has been said that length contraction and time dilation are two sides of the same coin. What is meant by this?

5. Make up your own thought experiment to illustrate that two events that are simultaneous to one observer are not simultaneous for a second observer moving relative to the first.

6. Suppose you are having lunch with a colleague at work and he or she casually asks about special relativity. How would you go about explaining it—that is, which points would you emphasize, and so forth? Describe your response.

7. Most human beings live less than 100 years. Since the maximum speed one can acquire relative to the earth is c, it is impossible for a person on earth to travel farther than 100 light-years (a light-year is the distance light travels in one year) into space before she becomes 100 years old. Does this necessarily mean that no person from earth will ever be able to travel farther from earth than 100 light years? Explain.

8. When a book is rotated in different directions in a spaceship traveling close to the speed of light, does an astronaut in the spaceship see the book change shape? How about someone watching this from earth? Explain.

9. If an astronaut in a spaceship traveling close to the speed of light shoots a bullet in the direction of his motion, what will a stationary observer conclude? Will the bullet appear to just slowly leave the gun? Explain.

10. Suppose that the speed of light were only 100 kilometers per hour, rather than 300,000 kilometers per second.

   a. Could you drive 600 kilometers in less than 6 hours? Explain.

   b. How long would it take light from the moon to reach the earth (a distance of 384,000 kilometers)?

   c. Suppose that your mass is m_0 grams. If you drive at a speed of 80 kilometers per hour relative to the ground, what will be your mass as measured by someone standing on the ground? (Answer: $m_0/0.6$ or $1.667m_0$.)

11. Relative to the nearest galaxy, you are at this moment hurtling through space at a very high speed. Do you notice any change in your pulse rate or the behavior of your wrist watch as a result of this motion? Explain.

12. In Question 11, would an observer on that neighboring galaxy notice these effects on *your* pulse rate and the time-keeping of *your* wrist watch? Explain.

10

MAKING THINGS HAPPEN: ENERGY AND POWER

Geothermal and petroleum energy are only two of the numerous choices open to us. How we harness energy so that our choices do not harness us is an issue to be continually explored in legislatures and laboratories.

During the last several years, as we have come to realize that oil cannot last forever, energy has emerged as a crucial issue in the United States and the world at large. People constantly talk about the "energy crisis." We read in the newspapers about energy-efficient homes. The auto industry tries to build smaller cars having greater fuel economy. Concerns about energy greatly influence our political relations, especially with South America and the Middle East, which contain vast deposits of petroleum that we need for our industrial society.

Energy, quite literally, is what makes things happen. We will define it more carefully in a few pages, but this idea will get us started. Energy plays a part in every instant of our lives. It cooks our food and freezes water to make ice. It heats our homes in December and cools them in August. Energy keeps our bodies going. It brings us the sounds we hear and the light we see. It cuts our wood, brews our coffee, fuels our cars, powers our communications, runs our mills and factories.

Whenever anything happens, whether a grand and majestic event such as a lightning bolt flashing across the sky, or something mundane such as a student turning pages of a book, energy is being changed from one form to another or transferred from one object to another. As you read this page in a lighted room, many transformations of energy are occurring. The light, called *electromagnetic energy,* that illuminates the page may have begun as the energy of water flowing over a dam, converted in turn to energy that rotated the turbines of a generator, producing the electricity that causes the filament of the bulb to glow. The energy that you exerted opening this book may be traced back through several transformations, from the food you ate to the plants and animals that received energy from the sun.

WORK AND ENERGY

Energy is best defined in terms of another quantity, called *work.* As ordinarily used, the word "work" can mean many things, including what one does for a living, as well as some task that is not particularly pleasant. I am often accused of doing no work while I am at work—can reading books and writing things on paper be work? Mowing the lawn is work, watching TV is not. Most people don't think of a fast game of handball as work, although it can involve more exertion than mowing the lawn.

As used in physics, the word work is always connected with the *motion* of an object through some distance while a *force* is being applied to it. In the simple case where a single force is applied in the same direction that the object moves, the work is just the force multiplied by the distance. If we call the work W, the force F and the distance d, then we have $W = Fd$ (F and d are in the same direction).

For instance, a bulldozer does a certain amount of work in moving a mound of earth across a construction site. As you can see from this equation, the bulldozer does more and more work as it shoves the earth farther and farther along. Figure 10-1 illustrates the case of F and d being in the same direction when you pull a box across the floor. Here the force is "switched on" at the start and "switched off" at the finish.

In the system of units we have been using, the force is in *newtons* and the distance is in *meters.* The product of a newton and a meter is called a *joule,* the unit of work:

$$\text{newton} \times \text{meter} = \text{joule}$$

or, in shorthand,

$$\text{Nm} = \text{J}$$

Take the Word of Science

To nonscientists, many of the words and phrases of science—the jargon common to particular scientific fields—just confuses. A number of very simple words in physics (borrowed from our everyday language), such as *work, force,* and *power,* mean something special to the physicist, usually something different from what they mean ordinarily. Take, for example, the physics definition of *work* given in the text. Here is an expert in the language of science poking fun at the way this word is used technically:

The scientists have decided that a force works, or does work, only when it moves something. I may push and pull in vain at some immovable obstacle, make myself hot and tired by my efforts and yet find that mathematically I have done no work. But if I seize the dangling reins of a runaway horse and pull them, and find that nevertheless the animal continues on its course, I have had work done on me, and I, panting and dishevelled, have done less than no work.

Source: Eric Partridge and Simenon Potter (eds.), *The Language of Science.* London: Andre Deutsch, Ltd., 1966.

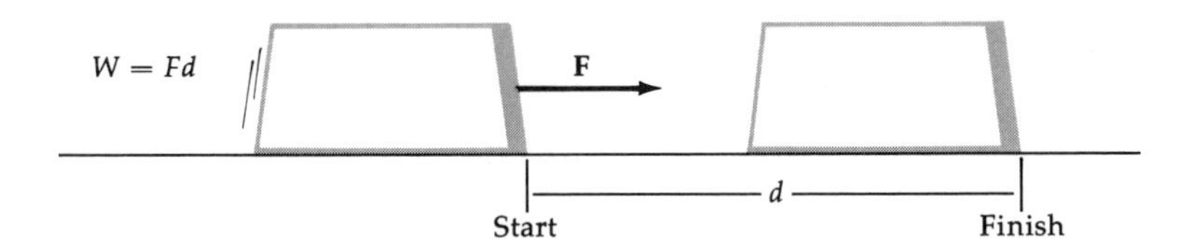

Figure 10-1. Definition of work.

The joule is named for James Prescott Joule, a nineteenth-century British physicist who did some of the important experiments concerning the relationship between energy and heat.

We can also have the force and motion along the same line but in exactly *opposite directions.* Friction is a common example. When a baseball player slides across home plate, he gradually skids to a stop because of friction between his body and the ground. In this case the frictional force does work in slowing the player down. It points toward third base, opposite to the direction the player slides. We can illustrate this simply, as in Figure 10–2, for a box that scoots across the floor and stops. The amount of work the force of friction f does on the box is the product of f and the distance d through which the box moves.

What about the general case where the force and the direction the object moves are not along the same line? Here, it is the *component* of the force in the direction of motion that does the work. Figure 10–3 illustrates this idea for a simple situation, similar to a child pulling a toy wagon along the pavement. The child tugs on the wagon with a force at an angle to the ground. We can picture the force as being divided into two pieces: one, perpendicular to the ground, lifts the wagon slightly, and the other, parallel to the ground, moves the wagon along. It is this parallel bit of the force, in the direction of the wagon's motion, that does the work. This is the only part of the force that counts, as far as work is concerned. If we label the parallel component $F_{\parallel}$, then the work done in moving the wagon is $W = F_{\parallel}d$.

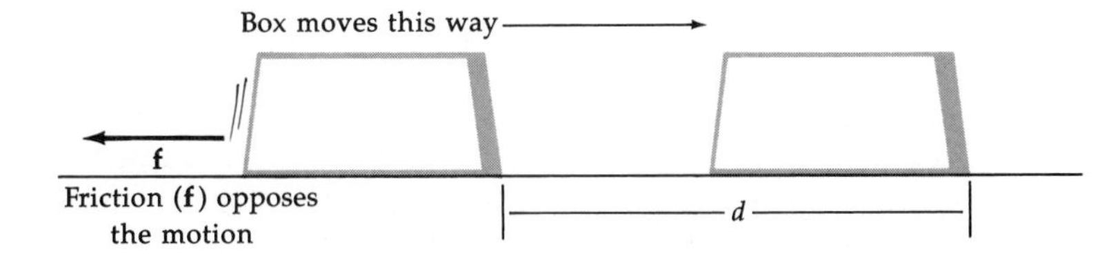

Figure 10-2. Work of friction = fd

Figure 10-3. The component of F parallel to d does the work.

Figure 10-4. **Net work is due to all the forces.**

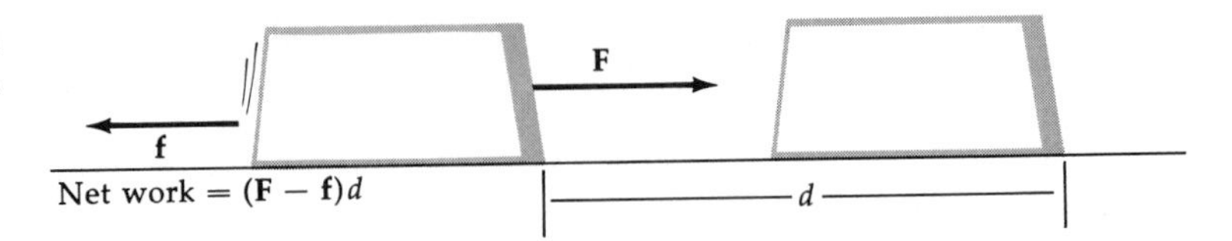

So far we have only discussed the work done by a *single* force. But, of course, we might have *many* forces acting on an object at the same time. Each force could have a component that does work. When several forces are involved, the net amount of work they produce will be that which results from the net effect of all their components along the line of the motion. We simply add the components that are in the direction the object moves, and subtract those in the opposite direction. When the bulldozer moves that pile of dirt, its force is in the direction of motion, while friction against the ground acts in the opposite direction. So, the net work done on the dirt pile is the distance moved multiplied by the difference between the bulldozer's force and the friction force. (See Figure 10–4.)

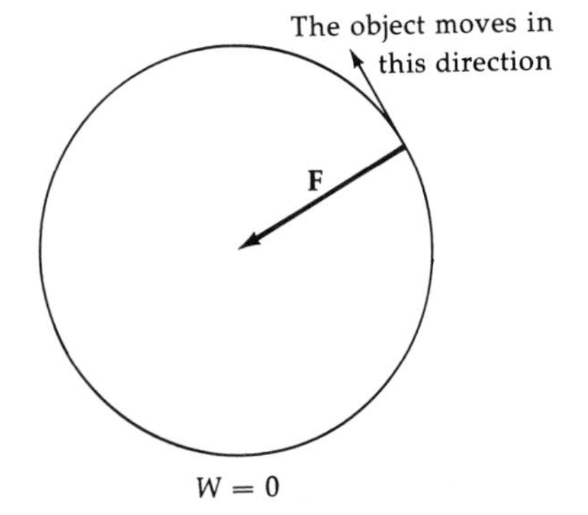

Figure 10-5. F **is perpendicular to the direction in which the object moves, so no work is done.**

If the force is *perpendicular* to the direction of motion, then there is no component of F along d, and no work is done. For example, a rock whirled at constant speed on the end of a string is held in its circular path by a force that points along the string toward the center of the circle (Figure 10–5). The motion of the rock is at each instant directed tangent to the circle, meaning that the force is always perpendicular to the direction of the motion. Hence, this force *does no work* on the rock even though the rock moves through a distance and there is a force applied. But the force has no part, or component, along the direction of the rock's motion. In order for work to be done, the force, or one of its components, *must* be either in the same direction as the motion or in exactly the opposite direction.

Note carefully that the physics definition of work involves *movement*. You may exert a great deal of force by pushing on a building. If you do it for a very long time, your muscles get tired. You sweat and moan and groan. But if the building doesn't move, then you haven't done any work on it, from the standpoint of physics, even though most people would agree that you had "worked" quite hard. If there is no motion, there is no work.

Energy also carries the idea of motion, even as the word is used in everyday situations. When you describe someone as having a lot of energy, usually you mean that the person is active, busy, and makes things happen. "Where does he get all that energy!" is the exasperated cry of the mother whose 5-year-old has left a trail of activity through the house and across the lawn. We generally associate the word energy with motion.

Energy is the capacity to do work. It is not the work itself, nor necessarily the actual doing of the work, but the capacity, the ability to do the work. Gasoline contains the energy necessary to drive your car; the capacity is there, in this case taking the form of chemical energy, waiting to be tapped. Steam provides energy in the form of heat to do work in driving a turbine. The word "energy" itself expresses this idea of work capacity, being derived from the Greek words for "work-within."

In order to better establish the connection between energy and work, let us talk a little more about forces. Recall from Chapter 2 that a force causes acceleration, meaning either a change in *speed* or a change in *direction*, or both. If the force and the motion are in the same direction, as in the case of the box pulled across the floor, then the force causes a change in speed. Suppose that the box is initially at rest and the force F acts until the box has moved a distance d. The box has been *accelerated* because its speed has changed from zero to some value during the time it moved a distance d. Or, if it were initially moving, the final speed when the force is switched off would be greater than the initial speed. In either case, the work results in an increase in the speed.

Suppose that the force and the motion are oppositely directed, as with friction. Here the speed *decreases*. The ball player sliding across homeplate is an example. The force of friction acts to oppose his motion and slows him eventually to a stop.

The key thing to note is this: the force that does work also causes the speed to change.

On the other hand, whirling the rock on a string is a situation where the acceleration is due only to a change in *direction*. The speed remains constant. This is generally true when the

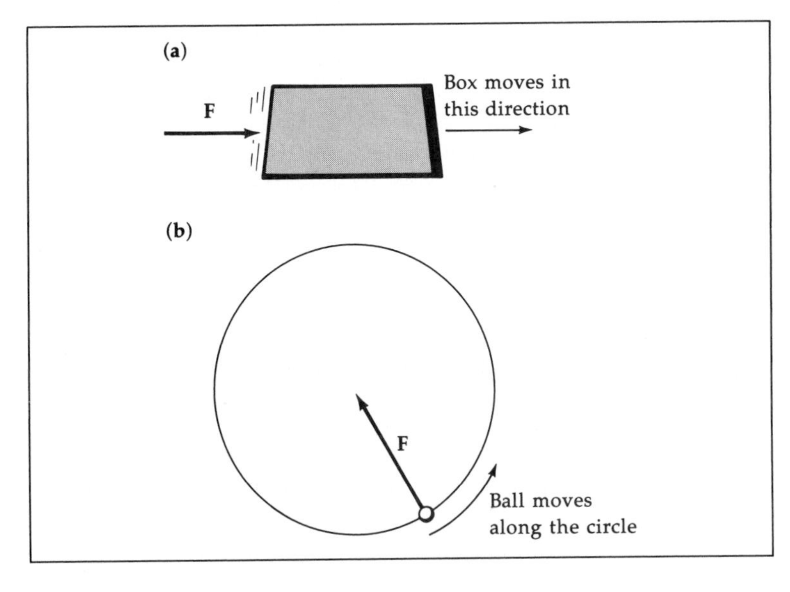

Figure 10-6. Effects of change in speed (a) and change in direction (b). (a) Acceleration is due to a *speed* change; work is done by F. (b) Acceleration is due to a *direction* change; *no* work is done by F.

direction of the force is perpendicular to the direction of the motion, in which case no work is done. These extremes are summarized in Figure 10-6.

For the intermediate case, the force is at an angle to the direction of motion. The object undergoes both a direction change *and* a speed change. The perpendicular portion of the force produces the new direction of motion, while the speed change is brought about by the force's component in the direction of the motion.

Planetary motion is an illustration of this intermediate case. Recall that as a planet skims along its elliptical path, kept in tow by the sun's gravity, it speeds up while approaching the sun and slows down as it moves away. Again, we can view the gravitational force as made up of two perpendicular components. The one along the direction of motion is responsible for the changing speed during the planet's year. The other component, pointing toward the inside of the ellipse, keeps the planet from flying off on a straight line and leaving the solar system. (See Figure 10-7.)

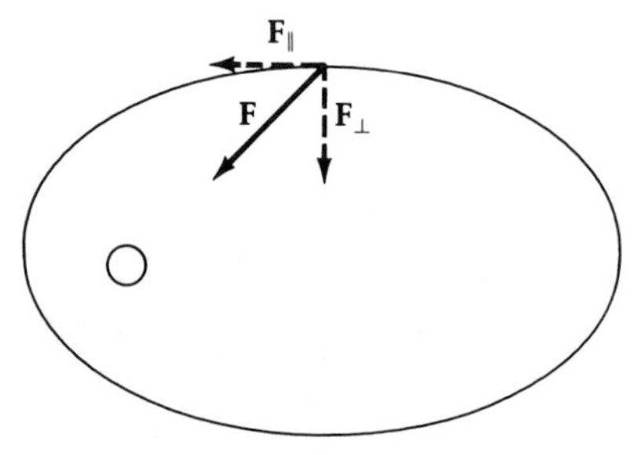

Figure 10-7. Planetary motion. $F_{\parallel}$ changes the planet's speed (does work); $F_{\perp}$ changes the planet's direction (does *no* work).

Learning Checks

1. What is the scientific definition of *work*? How does this definition relate to its everyday use?
2. What is energy? How is it related to work? Are work and energy identical? Explain.
3. A printer's apprentice unsuccessfully tries to lift a large portable printing press. Does the apprentice do any work? Explain.

Kinetic Energy

Whenever any object moves, it has what we call *kinetic energy.* The modifier "kinetic" comes from a Greek word meaning "motion." Kinetic energy is energy of motion. If an object is not moving, then it has no kinetic energy, no matter how large it is or what other properties it may have.

We see many examples of kinetic energy in our everyday world. A circular saw blade rotating at high speed easily cuts through a piece of lumber, but the saw is useless if the blade isn't moving. The kinetic energy you generate when you swing a hammer allows you to pound a nail into a board. A tennis racquet has kinetic energy when you stroke the ball with it.

In all these cases, the kinetic energy of the object enables it to do work. The hammer's kinetic energy drives the nail. The tennis ball has work done on it because of the kinetic energy of the racquet.

If an object has a mass m and it is moving at speed v, then its kinetic energy (KE) is

$$\text{KE} = \frac{mv^2}{2}$$

Notice the units of energy. From this equation we can see that, in the metric system, they are

$$\text{kilograms} \times \frac{\text{meters}}{\text{second}} \times \frac{\text{meters}}{\text{second}}$$

which we can write as

$$\left(\frac{\text{kg m}}{\text{s}^2}\right) \times \text{m}$$

The combination of units in parentheses is a newton, so energy has the units:

$$\text{newtons} \times \text{meters} = \text{joules}$$

That is, *work* and *energy* have the same unit, the *joule* in the metric system.

We have seen that if forces act on an object, any work resulting from their net effect changes the speed of the object. But since the kinetic energy depends on the speed, it follows that if the speed changes so does the kinetic energy. The relationship between work and kinetic energy is this:

> The *net work* done by all the forces acting on a body is equal to the *change* in the body's kinetic energy.

The kinetic energy of a hurricane often leaves disaster in its wake.

That is,

$$\text{work by all forces} = \text{final KE} - \text{initial KE}$$

If v_i is the original speed of a body, and v_f is its final speed after some forces have acted on it, then the net work W done by all the forces is

$$\text{net } W = \frac{mv_f^2}{2} - \frac{mv_i^2}{2}$$

Notice how the speed comes into the equation for kinetic energy. The kinetic energy goes as the *square* of the speed. Thus, the speed appears twice, once each for the *force* and the *distance*, whose product is the work. Recall that the force that does work is also the one that causes the speed to change. The greater the force, the greater the change in speed. The other speed factor appears because the change in speed increases as the distance over which the force acts becomes greater. These two speed factors—one because of the force and the other because of the distance—account for the speed appearing as $(\text{speed})^2$ in the kinetic energy formula.

For example, think about accelerating a motorboat on a lake from 10 m/s to twice this speed, or 20 m/s. The boat's original kinetic energy will be proportional to 100 (because $10^2 = 100$), and its final kinetic energy will be proportional to 400 (because $20^2 = 400$). So, even though the speed has only doubled, there has been a fourfold increase in the kinetic energy. This must come from the work produced by the motor. That is, it has to do four times as much work, and hence supply four times as much energy, to move the boat at 20 m/s as it does at 10 m/s. Baseball pitchers can routinely throw a fastball at speeds of 100 km/h. They must throw four times as hard to make the ball travel at this speed as they do in throwing it at 50 km/h.

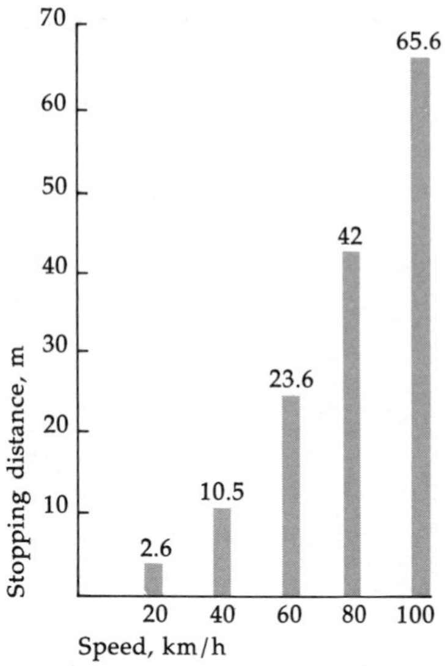

The kinetic energy's dependence on the square of the speed is important in stopping a car. The faster the car is moving, the greater its kinetic energy, so the more work the brakes must do to bring it to a stop. If we suppose that the brakes provide the same force no matter how fast the car is going, then the only way of getting enough work to stop a faster car is for the braking force to act over a longer distance. The "stopping distance," which is the distance required for the car to stop once the brakes are applied, will be greater for higher speeds and will increase as the *square* of the speed. Figure 10-8 shows the stopping distance for various speeds. Notice, for example, that the stopping distance for 80 km/h is four times that for 40 km/h.

Figure 10-8. The stopping distance increases with the square of the speed.

Gravitational Potential Energy

Water pouring over a waterfall can be used to do work, such as turning an electric turbine or a paddle wheel in an old-fashioned grain mill. The heavy ram of a pile driver does work when dropped from an elevated position. The water and the ram speed up as they fall, gaining kinetic energy while plummeting toward the ground. Where does this kinetic energy come from?

We can answer this question by recalling what we know about freely falling objects. Imagine that a house painter standing on a ladder accidentally drops a bucket of paint that falls a distance h to the ground, as shown in Figure 10-9. We will suppose that the bucket's mass is m. When the painter releases the bucket, the force acting on it is just its weight, mg. (Recall that g is the acceleration due to gravity.) This is the net force on the paint bucket (neglecting air resistance) during its flight, so this force, mg, moves the bucket through a distance h. Since the force and the bucket's motion are both straight down, gravity does *work* in this case:

$$\text{work} = \text{force} \times \text{distance} = mgh$$

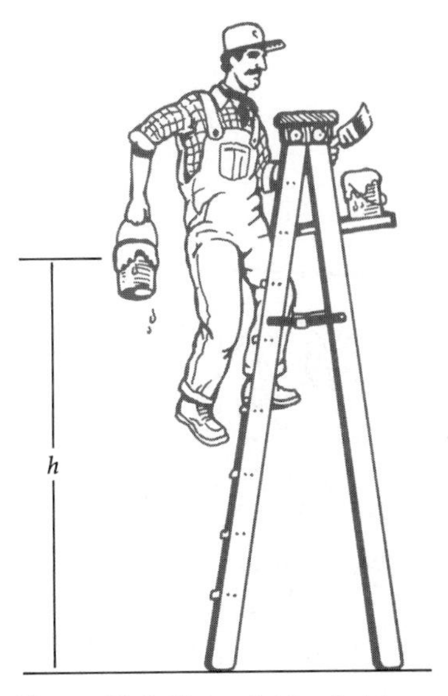

Figure 10-9. The paint bucket has gravitational potential energy.

We have also seen that the work is the change in the kinetic energy. What about the kinetic energy of the bucket? Well, because it starts from rest, its initial speed is zero and so is its kinetic energy:

$$\text{initial KE} = 0$$

As it falls, the bucket accelerates because of the force of gravity. It gradually builds up speed until it reaches its final speed v, at the instant before it hits the ground. Its kinetic energy at this point is therefore

$$\text{final KE} = \frac{mv^2}{2}$$

Since the work is equal to the change in the kinetic energy, we have

$$\begin{aligned} \text{work} &= \text{final KE} - \text{initial KE} \\ &= \frac{mv^2}{2} - 0 \\ &= \frac{mv^2}{2} \end{aligned}$$

So, we have two equations for the work:

$$\text{work} = mgh$$

$$\text{work} = \frac{mv^2}{2}$$

The first comes from multiplying force times distance, the definition of work. The second comes from calculating the change in the kinetic energy. If we equate the two righthand sides, we get

$$mgh = \frac{mv^2}{2}$$

Look at this equation for a moment. The righthand side is kinetic energy, as we have said. So it seems reasonable that the lefthand side, mgh, is also energy of some kind. It is not *kinetic* energy because nothing in the three quantities m, g, and h involves motion. This kind of energy is called *gravitational potential energy.* It is the energy that the paint bucket has as a result of its *position* above the earth's surface. The bucket has capacity for doing work because of the gravitational force tugging it toward the ground. This gravitational potential energy changes to kinetic energy as the bucket falls.

The name *potential* energy implies that the object has energy reserved for future use. We can think of this as stored energy from which work will be gained later on.

Notice that the gravitational potential energy changes as the distance h changes. Moving the object to a greater height increases its potential energy. This means that if we drop it, its kinetic energy upon hitting the ground will also be greater. Of course, this agrees with what we already know. A flower pot will be traveling faster when it hits the ground if it falls off the roof of a building than if it falls from a first-floor window. The taller the waterfall, the greater the energy for turning a mill wheel.

There are other kinds of potential energy besides gravitational. A coiled spring, for example, gains potential energy when work is done to compress it. It can then be released to do work. In its coiled state, the spring has energy stored for potential use at a later time. When you wind an ordinary wrist watch, you are doing work to tighten the mainspring, giving it "spring" potential energy. This energy is then slowly released to turn the watch's balance wheel that controls the time-keeping mechanism.

Slingshots have rubber bands or tubing that store potential energy when stretched. Releasing the tubing unleashes this potential energy, which then does work by shooting a rock through the air. An archery bow has potential energy when drawn.

The chemical energy released when wood or coal or petroleum burns is also potential energy. The energy stored in these fuels is released when bonds within individual molecules are broken.

Learning Checks

1. What is kinetic energy?
2. Does the kinetic energy depend on the direction of the velocity of the moving object?
3. What is potential energy? How does it differ from kinetic energy?

Power

When you drive your car to the top of a hill, the car's gravitational potential energy increases. This energy comes from the work done by the engine. Any car will make it to the top of any hill, which means that each engine will eventually do the same amount of work, but some will do it *faster* than others. That is, some engines deliver their energy in less time. Similarly, you notice a difference when you run around the block rather than walk. You expend the same amount of energy in both cases, but your body does the work in a shorter amount of time running than walking.

We use the term *power* to express the *rate* at which work is done or energy is used. Power is defined as

$$\text{power} = \frac{\text{energy}}{\text{time}}$$

In the metric system we are using, energy is expressed in joules and time in seconds, so the power is in joules per second, which is called the *watt* (W):

$$1\ \text{W} = 1\ \text{J/s}$$

One joule of energy used in 1 second gives 1 watt of power.

We commonly associate watts with electrical power. For example, a 100 watt light bulb burning for 3 seconds uses 300 joules of energy:

$$\begin{aligned} 100\ \text{W} \times 3\ \text{s} &= (100\ \text{J/s}) \times (3\ \text{s}) \\ &= 300\ \text{J} \end{aligned}$$

Automobile engines are routinely rated according to their horsepower, another unit of power. By way of comparison, one horsepower is equal to 745 watts.

Both the horsepower and the watt are units for measuring the same thing, power. You could, for example, rate your automobile in watts and your light bulbs in horsepower. Our 100 watt bulb is also a "0.13 horsepower bulb." Conversely, a 130 horsepower engine is a 100 kilowatt engine (since 1 kilowatt equals 1000 watts).

THE CONSERVATION OF ENERGY

Without doubt, the most important thing about energy is that it obeys a conservation law. Energy is not something that we can easily recognize for what it is, in the sense that we can spot a cup of coffee or a ham sandwich. Neither is it something we can readily observe, such as the speed of a sports car or the spinning of a wheel. Its importance in the physical universe comes not so much from what it is as from the fact that its quantity always remains constant.

Up to now we have discussed only kinetic and potential energy. There are other kinds of energy as well, including chemical, nuclear, heat, radiation, and electrical energy. These others are ultimately either kinetic or potential energy, although it is convenient to retain the various classifications. For example, chemical energy is actually the potential energy due to the positions of atoms bonded together to make up molecules. Heat is a form of energy, the kinetic energy of individual atoms and molecules. Energy can be converted from one form to another, and transferred from one object to another, but when we add up the amounts in the different forms and different places, we always get the same number. This is what we mean by saying that energy is conserved.

Let us focus again on kinetic and gravitational energy to see how the conservation law works. We can use the falling paint bucket as a simple example, where we will ignore air resistance.

Suppose the bucket of paint has a mass of 5 kilograms (weighing just over 11 pounds) and we lift it 4 meters off the ground. By doing so, we have given the bucket a gravitational potential energy (GPE) of about 200 joules (taking g to be $10\ m/s^2$):

$$\text{initial GPE} = mgh = (5\ \text{kg})(10\ \text{m/s}^2)(4\ \text{m}) = 200\ \text{J}$$

While we hold the bucket, its speed is zero, which means it has no kinetic energy. The total energy is therefore 200 joules:

$$\text{initial total energy} = 200\ \text{J}$$

Now we drop the bucket. As it falls, it *loses* potential energy, but it also speeds up, so it *gains* kinetic energy. At the instant it hits the ground, its speed is such that its kinetic energy (KE) is also 200 joules:

$$\text{final KE} = 200\ \text{J}$$

Since its height above the ground is zero, it no longer has any potential energy, so

$$\text{final GPE} = 0$$

Its final total energy is still 200 joules. That is, during the fall, all its original gravitational potential energy has been converted into kinetic energy, keeping the total amount of energy fixed.

What about intermediate points along the way? Except at the start and at the finish, the energy will be partly kinetic and partly potential, but these two added together will always give 200 joules. For instance, after the paint bucket has fallen 1 meter, it will be 3 meters above the ground, so its gravitational potential energy will be 150 joules:

$$\text{GPE (at 3 m)} = (5\ \text{kg})(10\ \text{m/s}^2)(3\ \text{m}) = 150\ \text{J}$$

At this point it has *lost* 50 joules of gravitational potential. However, it is moving with enough speed so that its kinetic energy is 50 joules:

$$\text{KE (at 3 m)} = 50\ \text{J}$$

So the *total* energy is

$$\text{GPE} + \text{KE} = 150\ \text{J} + 50\ \text{J} = 200\ \text{J}$$

The 50 joules, then, has not been "lost" at all. Rather, it has appeared as kinetic energy.

Let us look at another example. Figure 10–10 shows a portion of a roller coaster, similar to one you can find in Disneyland or Busch Gardens or any other amusement park. Imagine that one car of the roller coaster starts from rest in the position shown. We will suppose that the car's mass is 500 kilograms and the vertical distance h in the figure is 50 meters. Then the car has potential energy mgh, which, taking g to be 10 m/s^2, is 250,000 joules. This is also its total energy, because at rest it has zero kinetic energy. Again we are ignoring friction and air resistance.

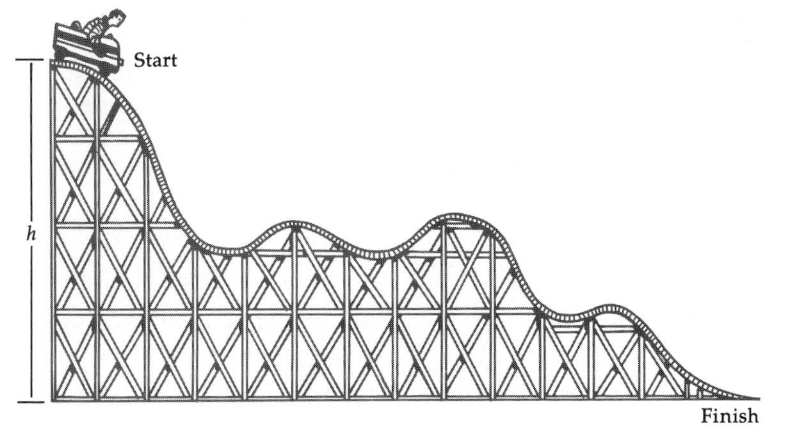

Figure 10-10. The roller coaster.

The Grand Dixence Dam, Switzerland, is the world's highest at 935 feet. It is as tall as a 90 story building, is 7 football fields wide at its crest, and as thick as 2 city blocks at its base. But the impressive Grand Dixence still functions like all hydroelectric dams—it uses the changeover of energy from the potential to the kinetic form.

When the operator releases the brakes, the car begins to move down the track, gaining speed as it loses altitude. At each spot along the way, 250,000 joules remains its total energy, distributed between gravitational potential and kinetic energy according to how high it is above the ground. When the car reaches the finish, it will have lost all its gravitational potential energy. Neglecting friction and air resistance, all of the original 250,000 joules will be kinetic energy, $(mv^2)/2$. So we can calculate how fast it will be going. We have

$$\frac{mv^2}{2} = 250{,}000 \text{ J}$$

where m, remember, is 500 kilograms. Doing the arithmetic, we find that its speed v at the finish is about 32 m/s, or 114 km/h.

Notice that we do *not* need to know the details of the trip. As long as the peak of any given hill is below the starting point, whatever kinetic energy the car loses by climbing that

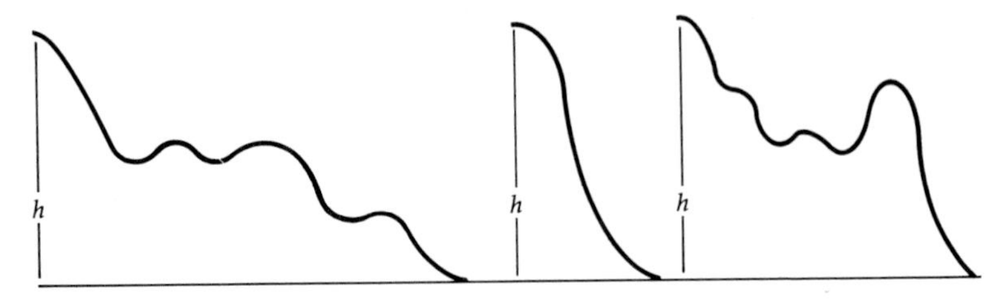

Figure 10-11. Roller coaster designs that give same final speed.

hill is regained when it comes down the other side. So the speed of the car at the finish depends only on the *relative vertical positions* of the start and finish, and not on anything in between. Figure 10–11 shows several roller coaster designs that will give exactly the same final speed.

The shift of energy between kinetic and potential forms has many practical advantages. We have already mentioned the familiar example of the energy obtained by water going over a dam or waterfall. At the top the energy is potential and changes into kinetic energy as the water plummets downward. When the water strikes a water wheel, as shown in Figure 10–12, its kinetic energy turns a mechanism for grinding grain. Hydroelectric dams use the falling water to spin turbines for generating electrical energy.

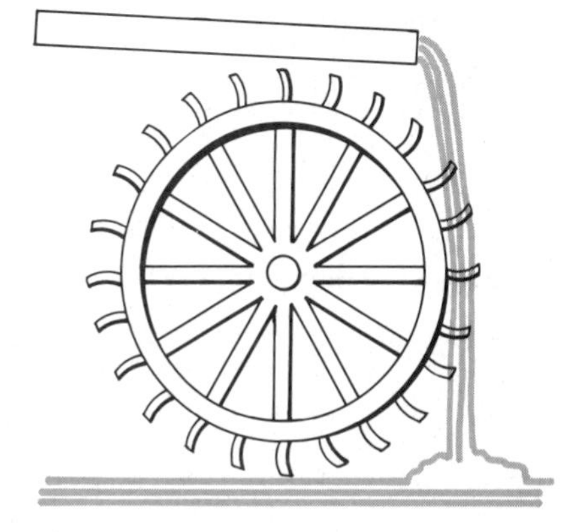

Figure 10-12. Water wheel.

Physicists refer to kinetic and potential energy together as *mechanical energy.* In the idealized situation of a freely falling body, where we neglect the energy-stealing effects of friction and air resistance, mechanical energy is conserved. But in the real world we *can't* ignore friction and air resistance. These generate heat energy, at the expense of mechanical energy. A cannonball fired into the air will have slightly less kinetic energy when it returns to the ground than it had at the moment of firing—a small amount has been "lost" as heat through air resistance. When you drop a medicine ball from the top of a ladder onto a trampoline, it should bounce back to its original height, level with the top of the ladder, if no kinetic or gravitational potential energy were lost. But actually the ball doesn't quite make it to this height. It falls short, because some of its original potential energy has become heat energy due to air resistance and the imperfectness of the trampoline's elasticity. The medicine ball will continue to bounce, each time to a smaller and smaller distance, until it finally comes to rest on the trampoline. All its original potential energy will have been transformed into heat. (See Figure 10–13.)

Figure 10-13. The ball bounces to successively lower heights because of the conversion of potential energy to heat.

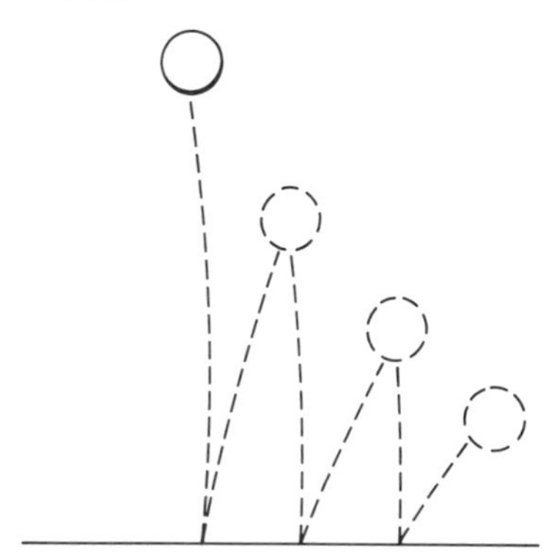

We can carry this line of argument further. Picture a sculptor dropping a wad of modeling clay out the studio window onto the pavement below. The clay just hits the ground with a splat. It doesn't bounce at all. The energy that started out as potential and gradually converted to kinetic energy seems to have disappeared. Mechanical energy is certainly not conserved in this case. In fact, it doesn't survive at all.

As these examples show, in the real world we don't have a conservation law for mechanical energy. And since the real world is the one we must deal with, to have a conservation law for energy we must include the other kinds as well. That there *is* such a universal bookkeeping for all the kinds of energy taken together is what gives energy much of its importance and usefulness.

Heat friction acts on a returning satellite.

For instance, when a space capsule returns to earth toward splashdown, tremendous amounts of heat are generated by friction with the air as the capsule plows through the atmosphere. This was a major engineering problem in the early days of the space program because of difficulties in building a good "heat shield." Heat appears at the expense of the capsule's kinetic energy. The capsule slows down in just such a way that the kinetic energy it loses equals the heat energy produced. Similarly, an orbiting satellite loses speed due to friction as it brushes against the top of the atmosphere. Eventually its speed is reduced so much that it can no longer remain in orbit, and it spirals into a crash landing. This happened to the Skylab satellite in 1979.

Heat from friction also appears in more common experiences. You can feel the heat when you rub your hands together to warm them on a cold day. Here, you are using your muscles to do work in moving your hands back and forth. So the work goes into kinetic energy, which causes the friction that in turn produces heat. If you break a wire by bending it back and forth several times, it may become too hot to touch. The friction of the individual atoms against each other is responsible for this heat. Use a pair of pliers to pry a nail from a block of wood and the nail will feel warm to the touch. Part of the energy you put out was "lost" in the frictional heat of nail against wood.

We can play this game backward, so to speak, by using heat to obtain mechanical energy. The sun's heat lifts many thousands of tons of water vapor into the air every day, increasing the water's gravitational potential energy. This eventually reappears in the energy of falling rain, or the energy of lake water flowing over a dam and generating electrical power, or the energy delivered to a mill wheel by a flowing stream. In short, all the mechanical energy of water stems from the heat of the sun.

Energy taking on a variety of forms is much like money existing in various denominations. If you have a dollar, it may be a single one-dollar bill, or four quarters, or ten dimes, or two quarters and five dimes, or one hundred pennies, and so on. There are a tremendous number of ways that the total amount of one dollar can be distributed among quarters, dimes, nickels, and pennies. You may want it in different forms for different purposes—pennies for the gumball machine, quarters for a locker at the airport. But no matter how the money is divided, it still represents a fixed amount—one dollar.

The conservation of energy is very likely the most important rule of the game. It has much wider-reaching generality, for example, than Newton's laws of motion, which apply only to objects with nonzero rest mass. Newton's laws break down, as we have seen, for bodies traveling at near light speed. We shall learn later that they also fail for bodies very small compared with those in the work-a-day world. Energy conservation, on the other hand, holds, as far as we know, throughout all the universe, for objects of all sizes traveling at all speeds. It applies to living and nonliving systems. It affects atoms and galaxies. And it works in the real nuts-and-bolts world as well as in the idealized models physicists make up in their heads.

Scientists will go to almost any lengths to preserve the notion of conservation of energy. This blind allegiance has paid off handsomely in advancing our understanding of the physical universe. One such instance is the discovery of a new particle, called the *neutrino,* which we shall discuss in a later chapter. The existence of such a particle was predicted in 1931 because certain reactions that take place inside the core of the atom seemed to violate conservation of energy. The neutrino was "invented" in order to make the energy balance come out right. This faith was justified later when the neutrino was observed in some experiments. More important, the discovery of the neutrino led physicists to a deeper and more comprehensive understanding of nature at this fundamental level.

Learning Checks

1. What does it mean when we say that energy is "conserved"?
2. What is mechanical energy? Is it conserved?

SUMMARY

In this chapter we have studied *energy,* one of the most important of all the concepts required for our understanding of the physical universe.

The concept of *work* in physics involves motion. If the force on an object causes it to move, the work is defined as the product of the *distance* moved and the *component of the force in the direction of motion. Energy* is the capacity to do work. It is the source of motion in the universe. *Kinetic energy* is energy due to an object's motion. For an object of mass m and speed v, the kinetic energy is $(mv^2)/2$. *Potential energy* is the energy that an object has because of its position. The most familiar example is *gravitational* potential energy, due to position in the earth's gravitational field. A very important rule, *the conservation of energy,* is illustrated by the gradual conversion of potential energy to kinetic energy for a body in free fall. This conservation law states that the total energy of the universe remains constant.

Exercises

1. When a satellite goes around the earth in a circular orbit, no work is done by the force of gravity exerted on the satellite. Suppose the orbit is elliptical rather than circular. Is work done in this instance? Explain.

2. A compressed spring has potential energy. How do you know this? Why don't we call it "gravitational" potential energy?

3. How much farther will a taxi travel after the brakes are applied when it is going 80 km/h than when it is going 20 km/h?

4. When a car stops as you apply the brakes, what force does the work?

5. A machine that tests for durability stretches a pair of elastic suspenders on a manikin. When the suspenders are stretched out away from the chest, do they have potential energy? How do you decide?

6. Why do the stopping distances of a car when the brakes are applied vary with the *square* of the speed?

7. A stonecutter holds a heavy slab of marble above his head. Does he do any work? Explain.

8. Two armored cars, A and B, have the same masses. Car A travels twice as fast as B. Compare the kinetic energies of the two cars.

9. A mobile library van weighs twice as much as a bookkeeping service van, and they both travel at the same speed. What can you say about their relative kinetic energies?

10. A professional bicyclist rides at 20 km/h when suddenly she slams on the brakes and skids to a halt. What is the final kinetic energy of the bicycle? Discuss what happened to the kinetic energy she had at 20 km/h. Has the gravitational potential energy changed? Explain.

11. In each of the following situations state whether or not any work is being done and explain your answer.
   a. A carpenter holds a tool box by the handle while waiting for a bus.
   b. A satellite travels at constant speed in its circular orbit.
   c. A feather is lifted vertically 1 meter off the ground.
   d. A child pulls a wagon across the ground.

12. A tennis ball when dropped from rest will not bounce back exactly to its original height. This means that the original gravitational potential energy is not fully recovered in that form. Identify possible "losses" (are they really losses?).

13. Suppose the wind is blowing at a speed of 20 km/h and then later at 40 km/h. By what factor has the kinetic energy of the wind increased in the second case?

14. Does a skier at the top of a hill waiting to descend have kinetic or potential energy? What happens to this energy as the skier comes down the hill?

15. Why are there maximum load limits for elevators?

16. A fast-moving baseball will "burn" your hand as you catch it. Explain.

17. An amusement-park roller coaster operates by having a chain-drive pull it to the top of the first hill. Without a motor or other external power source, can it ever reach that height again? Where does it have its maximum potential energy? Its maximum kinetic energy?

18. What energy transformations are involved as you jump on a trampoline?

19. We sometimes say that energy is "lost" when a battery runs down. Is this really true? Explain.

20. A ball of mass m falls vertically 10 meters to the ground. Is it true that (neglecting friction) the kinetic energy of the ball when it hits the ground is equal to the work required to lift the ball back to the 10 meter height? Explain.

21. In which case is more work being done: when a 50 pound sack of potatoes is lifted 3 feet or when a 50 pound crate is pushed 10 feet across the floor with a force of 15 pounds?

22. Suppose you walk up a flight of stairs in 10 seconds and then do it later in 5 seconds. Compare the *work* and the *power* in the two situations.

23. Jon pulls his toy wagon the length of the driveway. Jeff, exerting the same force on an identical wagon, pulls it the length of the driveway and back again. Compare the amounts of work that Jeff and Jon have done.

24. A child plays in a schoolyard swing. At what point in the motion is the kinetic energy a maximum? A minimum? Answer both questions for the gravitational potential energy. At what point is the kinetic energy equal to the gravitational potential energy?

25. Suppose you throw a baseball straight up and catch it as it falls. At what point during its flight is its kinetic energy zero? Where is its kinetic energy a maximum? Answer both questions for gravitational potential energy.

26. When a bullet is fired from a pistol or rifle, the force of the expanding gases acts on the bullet for the full length of the barrel. What effect will this have on the speed of the bullet as it leaves the barrel? Comment on the use of rifles versus pistols for long-range shooting.

27. A particular bale of hay having 1000 joules of gravitational potential energy is dropped from the bed of a truck 2 meters above ground level. During its fall the air resistance amounts to 30 joules of energy. What is the kinetic energy of the bale as it hits the ground?

28. You push a box across the floor by exerting a force of 10 newtons, keeping the box moving at 1 m/s. How much work do you do in 10 seconds? Suppose the speed is 2 m/s. How much work do you do in 10 seconds in this case?

29. How much power is expended in the two situations in Exercise 28?

30. Two bullets are fired into a mattress. If one bullet has twice the speed of the other, how much farther does the faster bullet penetrate?

31. Imagine you are standing on a rooftop holding two billiard balls. You throw one straight *down* at 10 m/s. You then throw the other straight *up* at 10 m/s; this one rises and then also falls to the ground. Discuss energy conservation for these cases. How do the speeds of the two balls compare as they strike the ground?

Photovoltaic cells make it possible to use and store solar energy directly. Scientists continue to search for similar methods of direct energy storage for efficient use by society.

11

ENERGY RESOURCES: CONVERSION, SUPPLY, AND DEMAND

In the last chapter, we focused our attention on kinetic and potential energy. We also touched briefly on the various forms that energy takes in nature—electrical energy, solar energy, and the chemical energy we get by burning fossil fuels. There are others. The main point is this: they are all governed by the law of conservation of energy.

ENERGY CONVERSION

One of the important characteristics of energy is that it can be converted from one form to another. We have seen this in the simple examples of the falling paint bucket and the bouncing medicine ball, both of which make transformations between kinetic and potential energy. For applications to society, the idea is to convert stored energy into a form that produces work or otherwise makes life more comfortable.

Modern technology is based on the simple notion of converting available energy into a form that meets some need. For example, the electricity that runs your household appliances may have originated in a hydroelectric dam. Here is what happens. The sun's heat evaporates water into the air. The water vapor condenses, falls as rain, and fills the lake above the dam. (See Figure 11-1.) As the lake water pours over the spillway, its gravitational potential energy changes into kinetic energy. This fast-flowing water in turn rotates turbines, which move giant coils of wire between the poles of large magnets. We shall see in Chapter 16 how this turbine motion causes electrical current to flow through cables and ultimately to your house. In

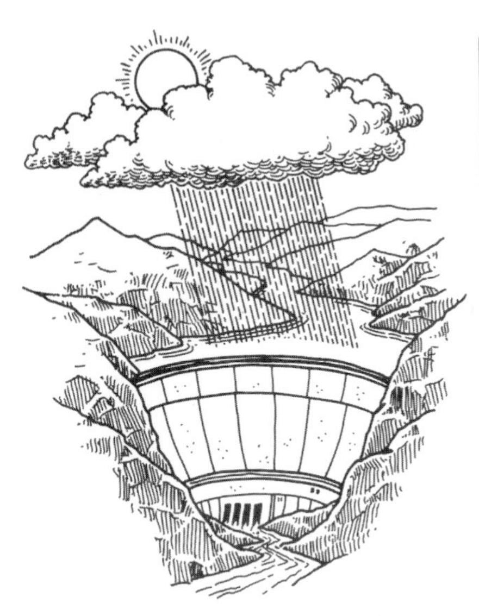

Figure 11-1. The energy of the dam water begins with the sun.

The Push to Microhydro

Many nations depend heavily on hydroelectric systems for their power; Norway, Canada, Sweden, Brazil, and Sri Lanka, for example, obtain over three-fourths of their total power from hydro. So it is not surprising that there are a fair number of gigantic dams in the world, with more under construction. Such massive facilities with their huge power outputs are not trouble-free, however. Sometimes, the dams bring pressing environmental and social problems: Agriculturally valuable land is often flooded; river-running fish, species of wildlife, and rare plants may be threatened; waterborne disease may spread; good residential space can disappear underwater, and local climate may be altered. For these reasons, and for the financial benefits involved, more and more attention is being given to much smaller versions called *microhydro* (or *minihydro*) systems all over the globe. China alone has built over 90,000 small hydro stations since 1968.

Both highly industrialized and developing countries are taking this approach. In the United States, for instance, especially in the Northwest and the Northeast, private entrepreneurs are purchasing and building small, local power plants and dams. These little projects can provide lighting for community schools and parks at lower prices and can even supply power at a price that allows small factories to stay in business and retain their employees.

One small system in Washington state uses a creek dam only 4 feet high. By contrast, the Hoover Dam, one of the seven modern wonders of the world, located between Nevada and Arizona, is 726 feet high. The physics, of course, remains the same for each.

Sources: Steve Ellis, "A Dam Good Idea Makes a Comeback," *Mechanix Illustrated*, September 1982, p. 47.

Abdus Salam, "Unifying the Fundamental Forms of Energy," *Unesco Courier*, July 1981, pp. 22–26.

Comanche Dam, situated between Sacramento and Stockton, California, is a rural dam that provides hydroelectric power to many of the small neighboring towns.

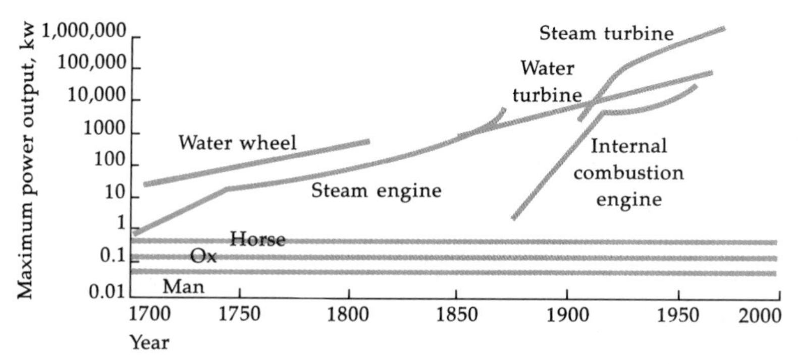

Figure 11-2. Power output of some basic machines. Source: From Chauncey Starr, "Energy and Power," *Scientific American*, September 1971, p. 40.

this way the energy of the sun, after these various conversions, cooks your eggs and boils water for your coffee.

One interesting way of charting the progress of civilization is to track the development of our expertise in using less of our own energy and more of the energy contained elsewhere in nature to produce goods and services. When primitive people learned to use fire to get warm, they took a big step in putting nature's energy to use. Animals supplemented human muscle power for thousands of years. The maximum power output for a human is just under 0.1 kilowatt, but a horse produces nearly ten times this amount. The important factor was that the fuel requirements of the ox and the horse did not deplete the food supply of humans.

Water as a source of energy conversion began with the introduction of the water wheel several decades before Christ. It initially had a power capacity of about 0.3 kilowatt. By 1750 improvements had increased this to 100 kilowatts, or roughly a thousand times that of a human. (See Figure 11-2.) The early Industrial Revolution was founded on the water wheel, and the location of cities and industrial centers was primarily based on the availability of water as a source of power.

Steam power added another dimension to the progress of civilization: mobility. Water power requires a continuous supply of water falling to a lower level, the gravitational potential energy thus being used to do work. Obviously, the water wheel is only useful in conjunction with a waterfall. The steam engine, on the other hand, is self-contained, and hence portable. (See Figure 11-3.) It consists of a heat source and a chamber in which to boil water for making steam. The steam is then used to turn a wheel, say, such as in a railroad locomotive.

Figure 11-3. Simple steam engine.

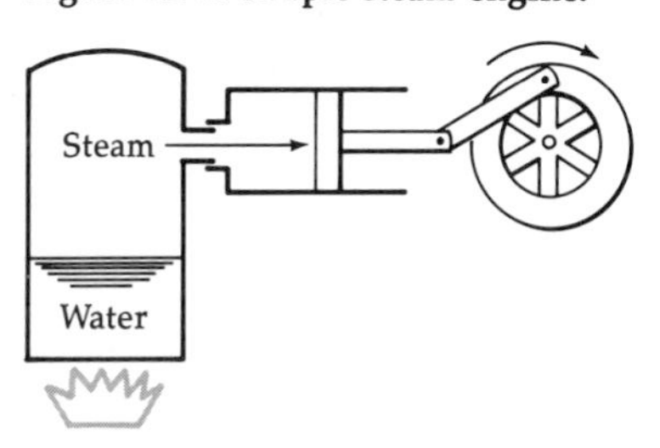

The portability of steam power meant that cities and factories no longer had to be built close to water to have power. Expansion in the use of water power suffered from this geographic limitation. Steam power, on the other hand, was not so limited. It gradually moved from its role as a supplement to the water wheel and windmill to that of a principal source around the middle of the nineteenth century.

The twentieth century has seen electrical power emerge as an important product of energy conversion. Electricity is probably the most convenient form of energy. It can be readily transported by wires to its point of use and then turned into mechanical work, heat, and so on, or stored in batteries to be used later elsewhere.

The ability to harness more and more of our surroundings for the conversion of energy has played an extremely important role in the way the culture has developed. It has allowed a smaller fraction of society to be actively engaged in food production, releasing a larger portion to pursue other goals and interests. We can travel farther and faster than our grandparents ever dreamed. Energy conversion has led to important communication advances, providing us with the opportunity to learn more about the rest of the world and to enjoy a wider variety of experiences than we could ever take full advantage of.

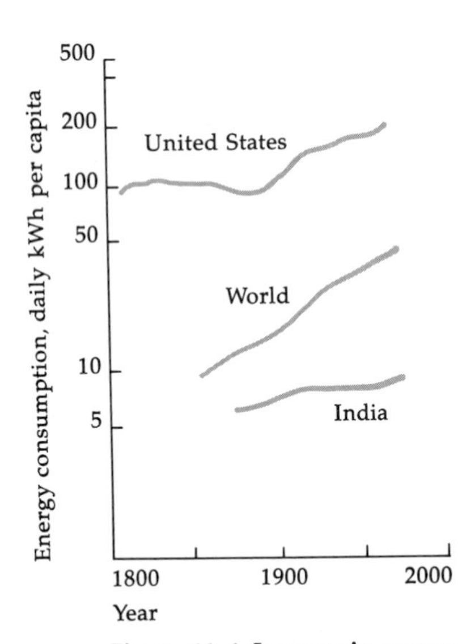

Figure 11-4. Increase in energy consumption. Source: Chauncey Starr, "Energy and Power," *Scientific American*, September 1971, p. 38. Copyright © 1971 by Scientific American, Inc. All rights reserved.

But this advancement to a supercivilization is something of a two-edged sword. It has created a situation that feeds on itself: as the technology for energy conversion has advanced, so also has the consumption of sources of energy. Nature has placed limits on the efficiency with which we can get work from energy; there is inevitably a certain amount of waste heat that is not recoverable. So although the *total* energy remains at a fixed amount, we are continually gobbling up the *usable* energy. The world consumption of energy on a per capita basis has grown from 10 kilowatts each day to its present value of about 50 kilowatts. (See Figure 11-4.) The United States's share is even higher—we have gone from a per capita daily use of 100 kilowatts to around 280 kilowatts at the present.

This, then, is the "energy crisis": consumption of available energy by depleting energy sources that cannot be used again. It may well be the issue that defines our time in history. The problem of energy sources is a serious one; we look at it in the next section.

Learning Checks

1. Trace the energy it took to get you out of bed this morning as far back as you can and discuss any possible losses of energy as waste heat.
2. Consider the energy needed to light your city. Discuss where this energy originated and the transformations and losses involved.
3. In terms of energy conversion, list as many things as you can that are done differently today compared to 50 years ago because of advances in technology. Discuss any problems that this new technology has brought about.

SOURCES OF ENERGY

Modern society has become an energy glutton. We depend heavily on extracting from the earth's resources energy to heat our homes, run our factories, and provide the transportation, recreation, entertainment, and other features of our lives. Energy touches every facet of our physical existence. In its absence, nothing happens.

Nature has clamped a limitation on the possibilities of reusing energy. This barrier is expressed as the second law of thermodynamics, which we shall study in Chapter 14. Accompanying this diminishing supply of available energy is an increasing demand for energy to do the work of modern civilization. In this section we are going to examine this dilemma. We will first look at growth patterns in our consumption of energy, and then we will see what the present status of our resources is and what the future may hold.

Exponential Growth

Growth is a fact of life. Children grow up, prices increase, the population continues to rise. Here we are especially concerned with the growth of our energy demands and, correspondingly, the depletion of energy resources. First we consider some general features of growth rate.

Whenever something increases at a fixed rate, it undergoes what is called *exponential growth.* The term may apply to anything—population, food prices, average age of newlyweds, whatever. For example, when the government announces that the current rate of inflation is 7 percent per year, it means that at this rate prices will be 7 percent higher this time next year than they are now. At first, this doesn't sound too bad—a 75 cent beer would cost 80 cents next year.

However, over longer periods of time, such seemingly modest increases turn out to be not so modest after all. With a 7 percent annual increase, the quantity will double in 10 years. So your 75 cent beer would cost $1.50 in 10 years. Furthermore, its price will double *again* in another 10 years—the cost would be $3. Each decade of 7 percent growth doubles the price. Of course, exponential growth does have its good sides: if you make an investment at 7 percent per year, you will double your money in 10 years.

We can express exponential growth by the concept of *doubling time,* which is the time required for a quantity growing at a fixed rate to double. In our examples, the beer price and your investment both increased at 7 percent per year. Both have a doubling time of 10 years.

There is an easy way to figure out the doubling time for anything that grows at a fixed rate. Simply divide the number 70 by the percent growth rate. The answer is the doubling

How to Request a Wage

A farmer and a prospective field hand were negotiating the terms of his contract for a month. The field hand (an expert, it turns out, at exponential growth) proposed that he be paid 1¢ the first day, 2¢ the second day, 4¢ the third, and so on, so that each day he would receive twice the amount of the previous day. The farmer eagerly agreed, believing such an arrangement to be a steal for him. And so it was, for a few days: the wages he paid at the end of the seventh day amounted to only 64¢ and the cumulative salary for the whole week was a paltry $1.27.

However, as the days went by, the farmer discovered, to his horror, that he had been snookered: the field hand's wages for the fourteenth day—only a week later—were $81.92, and ballooned to $167,772.60 for the twenty-fifth day. On the thirtieth day, the farmer shelled out $5,368,709.12. The total wages for the month amounted to a staggering $10,737,418.23. You can easily check this example for yourself by starting with a penny and doubling the new amount each day. Exponential growth can give enormous numbers rapidly.

Figure 11-5. Exponential growth of bacteria population.

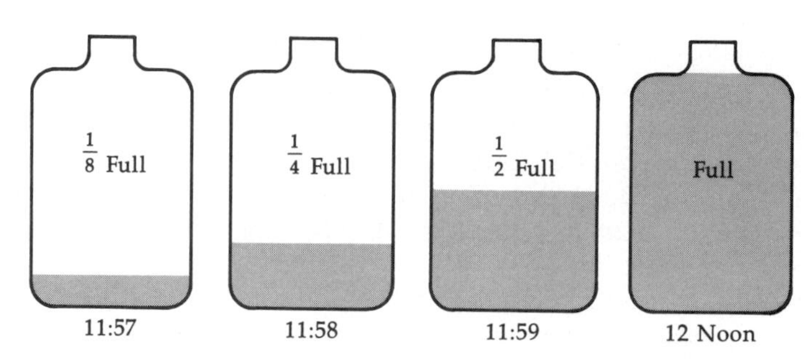

time. If a certain commodity grows at 5 percent per day, then the doubling time is 70 divided by 5, or 14 days. For example, the population of the world currently stands at about 3 billion, with a growth rate of around 2 percent per year. This gives a doubling time of 35 years. At this rate, we can predict there will be 6 billion people on this planet soon after the year 2000.

An important feature about exponential growth is this: In any one doubling period, the increase in the quantity matches the amount that has accumulated in all its previous history. Of course, this is just another way of saying that the amount has doubled, but it brings the effects of exponential growth into sharper focus. Consider the world population, for example: It has taken all of human history to reach our present census of 3 billion people, but only 35 more years will be required, at the present growth rate, to add that many more.

An example will illustrate how quickly a finite commodity can be consumed.* Let us imagine a particular strain of bacteria that doubles its population in 1 minute. Starting with only one, we will have two in 1 minute, four in 2 minutes, eight in 3 minutes, and so on. Suppose that at 11:00 in the morning we put one bacterium in a jar of such a size that it will be full by noon.

Now, we ask ourselves, what time is it when the jar is half full? As you can see from Figure 11-5, it will be 11:59, because, if there are enough bacteria to fill the jar at noon, only half that many will exist one doubling time earlier. Put another way, the bacteria will fill up as much living space in the *last minute* before noon as they did during the preceding 59 minutes, their *entire history.* Even as late as 11:55, the bottle would only be 3 percent full. Undoubtedly the bacteria would believe that the remaining 97 percent of this precious living space would last a long, long time.

But we will suppose that, at around 11:58, when the bottle is one-quarter full, a few forward-looking bacteria become aware of their impending problem and decide to search for more living space. Luckily, they are able to find three more bottles identical to the one they inhabit. Surely their living-

*A. A. Bartlett, *American Journal of Physics.* *46*, 876 (1978).

space problem is solved, because they now have available *four times* the elbow room they have ever known in their history.

No such luck. The original bottle fills up at noon. Then, at 12:01 P.M. there are twice as many bacteria, which means that one of the new bottles is also full. And at 12:02, with another doubling of the population, the remaining two new bottles are also full, and it's all over. By finding three times more space than they had originally started with, the bacteria bought only two more minutes of time. The new space was consumed in two doubling times.

You can see that this idea holds for the increasing consumption of any finite quantity. Even quadrupling the resource extends its life by only two doubling periods.

The Fossil Fuels

These illustrations of exponential growth and consumption of a finite resource provide an excellent framework for looking at the way we are gobbling up the fossil fuels—primarily coal and oil. In essence, we are simply running out of these energy sources. At current levels of consumption even major new finds will not postpone very long their ultimate depletion. Much of modern industry, transportation, and agriculture are geared solely to these fuels—mainly oil—but the time is very near when shifts to other energy sources will have to be made.

"I have a feeling it's too soon for fossil fuels around here."

Let us consider oil first. Before the Alaska oil strike, best estimates of the total available oil in the United States stood at about 190 billion barrels (a barrel is defined as 42 U.S. gallons). For the first 70 years after oil production began around 1860, production of crude oil grew at about 8.25 percent annually, for a doubling time of about 8 years. Production then tailed off to the 1970 level of about 4 billion barrels per year and a zero growth rate.

By 1972, about half of the 190 billion barrel total had been taken out of the ground. If the 1970 production level of 4 billion barrels were maintained—that is, with no increase in production—the remaining oil would last only 23 years. Obviously, if the oil producers heed the call for *increased* production coming from many quarters, even less time will elapse before the complete depletion of this precious resource.

The discovery at Alaska's Prudhoe Bay of a 10 billion barrel field, the largest in the United States, generated a great deal of excitement for the petroleum industry, but it represents only a 3 year supply for this country. The most reliable estimates give about 30 billion barrels as the ultimate crude oil production from Alaska. Even this seemingly rich find, amounting to 17 percent of the 190 billion barrel estimate for the rest of the United States, still buys less than 10 years of time before we run out, assuming no growth rate. Again, if production rates go up, we have even less time.

The world crude oil situation is not much rosier. Estimates range anywhere from 1350 billion to 2100 billion barrels for the total world oil. The production rate since about 1870 has steadily increased at 7 percent per year, giving a doubling time of 10 years. Currently we have produced about one-eighth of the world total. Remember the bacteria-in-the-bottle illustration: we are about two or three doubling times away from the complete depletion of worldwide oil. If no growth in production takes place, there is still only enough oil to last 100 years at the 1970 production level.

The range in estimates between 1350 and 2100 billion barrels—a difference of 750 billion barrels—seems quite large. But because the growth rate is exponential the two ratios give only about 25 years difference in expiration time.

One way of looking at the oil-production picture is to consider how long it takes to produce the middle 80 percent of the total amount. (See Figure 11-6.) For both the United States and the world as a whole, the length of time is around 65 years. This means that in one human lifetime—between 1937 and 2003 for the United States, and 1967 to 2032 for the world—80 percent of the total amount of oil will have been used up. The curve in Figure 11-6 shows that the production rate topped out in 1957 and has been declining ever since, as few new fields have been discovered and the remaining supply has become more and more expensive to tap.

Aside from the growth-rate aspects of obtaining oil, there are also technological problems associated with extracting resources in an economically feasible manner. The oil-well gushers you see in the movies don't quite tell the story, since a relatively small percentage of the total oil flows from a well without technical assistance. The more oil removed from a field, the more difficult it is to remove the remainder, until one fi-

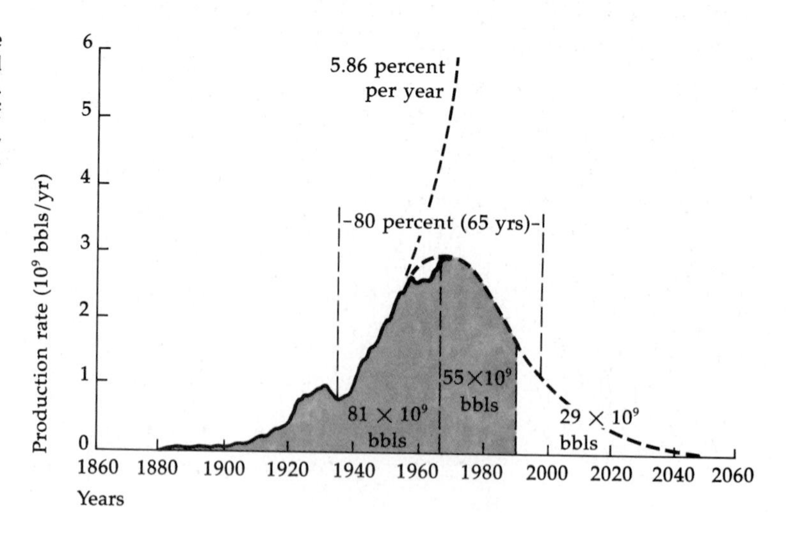

Figure 11-6. Oil consumption in the United States. Source: Reproduced from "Resources and Man," by M. K. Hubbert, with the permission of the National Academy Press, Washington, D.C., 1969.

nally reaches a point of diminishing returns. Some of the oil in the ground is allowed to remain there because it costs more to extract it than it is economically worth. So recoverable resources may be considerably less than total resources.

Much of what we have said about oil also applies to coal and natural gas. The latter usually appears together with oil; its production rate is expected generally to parallel that of oil, implying that natural gas will be in short supply soon after the turn of the century.

The popularity of coal as a fuel in the United States has decreased in comparison with that of oil and natural gas over the last few decades. This trend is expected to reverse in the next few years as production of oil declines. Much more attention has been focused on the production of coal as a result of the oil embargo imposed by the Arab nations near the end of 1973, which demonstrated how closely our ability to purchase energy is tied to the political situation in the Middle East. However, the production of coal is fraught with political problems of its own, particularly those growing out of environmental concerns and the older, more chronic hazards of the traditional mining techniques. Dramatic accidents and the long-range effects of black-lung disease have devastating effects on coal miners' lives. Strip mining, while a relatively easy method of processing coal, carries with it ugly scars and ecological imbalance to the land.

Trains travel day and night out of Wyoming, hauling coal from high rangelands that contain some 50 billion tons of this energy source.

Estimates of the coal supply in the United States range upward from 740 billion metric tons. Since we began mining this coal around the time of the Civil War, about 6 to 13 percent has been taken out of the ground, depending on which estimates you believe. For the first 50 years or so the production rate grew by 6.66 percent each year. It leveled off with no growth for about the next 60 years, when oil and natural gas were meeting our energy needs. Beginning around 1974, when we produced about 600 million metric tons, the rate of increase has shot up to about 11 percent for a doubling time of just over 6 years.

Quite often you hear that we have vast resources of coal in the United States. Huge figures are bandied about describing how long this supply will last. Various politicians call for increasing production of coal. But again remember the phenomenal speed with which exponential growth can dry up a finite resource: even when the bacteria occupied only 3 percent of their bottle, they were only five doubling periods away from disaster. In the case of coal, the current growth in production will deplete the United States's supply in only 40 or 50 years.

ENERGY FOR THE FUTURE

The era of the fossil fuels is almost over. As we have seen, recoverable supplies of oil, coal, and natural gas will be gone very soon after the turn of the century. When we recall that production of nearly 80 percent of the world's total supply will have taken place within the span of one human lifetime, it is easy to see that, viewed against the whole of human history, this era is but the blink of an eye.

At the same time, the fossil fuel age has had a tremendous impact. What has appeared (erroneously, it turns out) to be an unlimited supply of cheap, available energy has spawned gigantic technological advances unparalleled in any comparable length of time. By and large, increasing energy usage has gone hand in hand with the standard of living. Figure 11-7 shows a

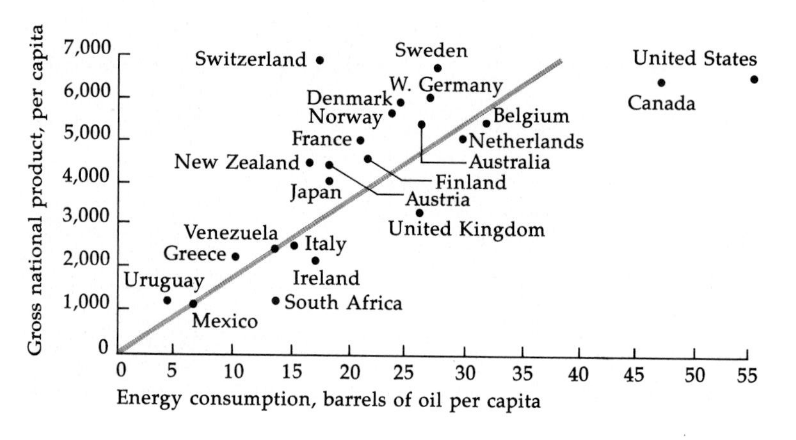

Figure 11-7. Energy consumption in various economies. Source: Earl Cook, "The Flow of Energy in an Industrial Society," *Scientific American*, September 1971, p. 142.

graph of the relationship of gross national product and energy consumption, both on a per capita basis, for the year 1974. The United States, with only about 6 percent of the world's population, uses 35 percent of the world's energy.

With supplies of fossil fuels running out and demands for energy increasing, it is obvious that nontraditional energy sources must be found and utilized if modern society is to continue some semblance of the way we now live. In this section we are going to take a brief overview of some of the possible energy sources for the future: nuclear energy, tidal and wind power, and solar energy.

Nuclear energy refers to energy obtained from deep within the nucleus of the atom. We are going to study the physics of the nucleus in Chapter 25. For now, we can be content with a rough, overall picture. In this view, we can imagine the atom as a miniature solar system, where the sun represents the nucleus and the planets are analogous to the electrons. Making up the nucleus are *protons* and *neutrons*, which contain nearly all the mass of the atom.

Chemical reactions such as the burning of coal involve only the electrons. They generate fairly small amounts of energy. However, if we could get inside the nucleus and observe reactions among the protons and neutrons, that is, *nuclear* reactions, we would find that the energy involved is enormous, many hundreds of times larger than even the most powerful chemical reactions. This energy, the nuclear energy, has been released upon the world, first in a terrifying way, the nuclear bomb, and more beneficially in nuclear reactors.

As we shall see in Chapter 25, the two general ways of obtaining nuclear energy are *fission*, the rupture of a heavy nucleus to form lighter ones, and *fusion*, the joining of very light nuclei. So far fission has been the only practical method; all nuclear power plants involve fission reactions. In both fission and fusion, however, the energy released is in the form of heat that is used to boil water to steam for generating electrical power.

The cooling towers and stacks of this reactor appear deceptively peaceful in the rural Pennsylvania setting. The enormous internal activity, however, is reflected in the heated controversy over the expense and safety of nuclear power.

Wind energy has been used by man for a very long time. Picturesque windmills are the symbol of Holland. Smaller windmills may still be found on farms and ranches throughout the United States, especially in the Southwest, where the wind blows much of the time. Although most such uses of wind power are relics of the past now, this ancient source of energy is staging something of a comeback. While there are few plans for large-scale uses of wind energy, small units can be useful by serving as supplemental sources.

Tidal power is an example of gravitational energy, the most abundant form of energy in the universe. The tides result from the internal gravitational interactions of the earth-moon-sun system. Tidal power may be tapped in a way similar to that of hydroelectric power, by damming a partially enclosed tidal basin and installing generating facilities. Worldwide, the potential for producing electricity from the tides amounts to less than 2 percent of the world's installed electric capacity. However, for those living close by, such an arrangement would provide a nonpolluting and endless supply of energy.

Solar energy may well be the ultimate answer to our energy problems. It furnishes a daily supply of clean, nonpolluting energy that is virtually inexhaustible, expected to last for at least the next few billion years—probably far longer than civilization will survive. Every second about 174 million billion

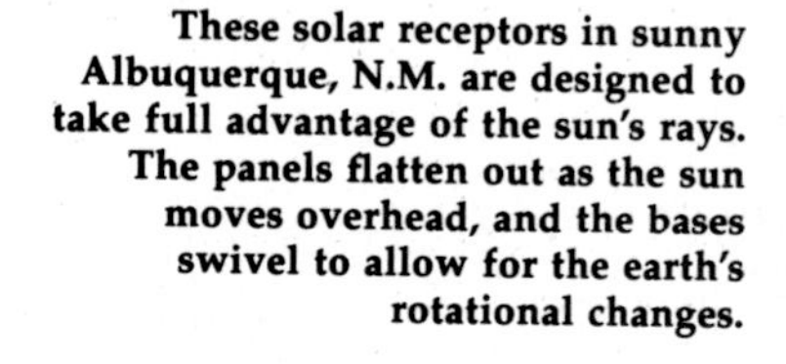

These solar receptors in sunny Albuquerque, N.M. are designed to take full advantage of the sun's rays. The panels flatten out as the sun moves overhead, and the bases swivel to allow for the earth's rotational changes.

Updating Windmills and Sails

Modern versions of the windmill concept would stun the minds of the nineteenth-century Dutch. From experimental wind turbine generators to domelike machines constructed to capture the force of hurricanes, many contemporary windmills have become unrecognizable descendants of their familiar-looking ancestors. Modernization notwithstanding, the use of land-based wind energy systems to obtain power has its problems. One is that the wind blows irregularly, so the captured energy must somehow be stored. That can be costly. However, the Wind Energy Systems Act of 1980 has earmarked $900 million to develop cost-effective wind power systems in the United States.

Seafarers are also rethinking traditional ways of utilizing the wind for commerce. The soaring cost of fuel has stimulated new interest in sails, especially for commercial vessels and huge transport ships. The experimental tanker *Shin Aitoku Maru*, to cite one example, is a cargo ship that sports two computer-controlled sails. These sails are rigid, are of stiff canvas, and are mounted on rotating masts. Computers that detect wind direction and speed trim the sails, which fold when not in use. Fuel savings are approximately 10 percent.

Then there is the idea of combining land and sea methods to create a new wind propulsion system—by substituting special windmills for sails on the decks of ships. In this area, the applied physics of energy and motion enters an innovative stage.

Sources: "Windmills that Harness Hurricanes," *Science Digest*, July 1982, p. 23. "Harnessing the Wind," *National Geographic Special Report*, February 1981, p. 38.

A new wind turbine generator on the Island of Oahu, in Hawaii. With this facility, 18 to 34 mile-per-hour winds can generate the electricity used in 100 average American homes.

The *Shin Aitoku Maru*, a Japanese cargo ship with computer-controlled sails.

joules of energy strike the earth in the form of solar radiation. About 30 percent of this is directly reflected and scattered into outer space. Another 23 percent goes into evaporating water and circulating the earth's atmosphere and water. A small fraction, about 40,000 billion joules per second, is stored by plants as chemical energy during photosynthesis. The remainder is directly absorbed as heat.

Some 35,000 times as much solar energy as humans use falls on the earth annually. In the United States the factor including all losses is about 650 times the amount used. It has been pointed out that power plants receive far more solar energy than the amount of energy they generate.

Roughly 30 percent of the energy used in the world is for space heating. The use of solar energy for this purpose alone could result in tremendous savings in energy. One of the major problems in exploiting solar energy is developing adequate means of storing energy for use during the sunless nighttime hours. For direct conversion of solar energy into electricity, another major obstacle is an economically feasible solar cell, similar to the type used in the manned orbiting laboratory familiar from the space program. Efficiency and mass-production are two problems that must be overcome, and research is being conducted on these fronts.

SUMMARY

One of energy's most important features is its *convertibility* among various forms. Modern technology is based on our ability to change energy found in natural resources into forms suitable for doing *work.* In all its forms, however, energy is still ruled by the *law of conservation.*

Civilization has progressed as we have learned to harness sources of energy other than our own muscles. First came animal power, then water and steam, and finally electrical power, perhaps the most convenient form of energy. However, the advance of energy conversion technology has also increased our appetite for energy, until we are now in the midst of an "energy crisis," which means *consumption* of usable energy by *depletion* of irreplaceable resources.

Our depletion of energy sources is growing in an exponential fashion. An important feature of exponential growth is *doubling time,* the time required for a quantity to double in amount. The doubling time for any quantity growing at a fixed rate is found by dividing the number 70 by the percent growth rate. Because the fossil fuels are a finite resource, the characteristics of exponential growth dictate that we will run out of these energy sources in a fairly short period of time.

Other energy sources will have to be used. Some of these are the atomic nucleus, the wind, the tides, and the sun.

Exercises

1. Suppose that some commodity grows at a rate of 10 percent per month. What is the doubling time? Repeat for a 5 percent growth rate.

2. Our discussion of energy needs and resources implicitly assumed that our standard of living and life style remain pretty much as they are today. Discuss this premise and the possibility of our being forced to adopt new life styles in the face of energy shortages.

3. Many people approach the energy crisis with the notion that they need not worry because science and technology will bail them out. Discuss the reasonableness of this approach, both from the standpoint of purely scientific arguments as well as the interlocking political-social-economic aspects of energy.

4. It has been said that the "energy crisis" is really a "fuel crisis." Comment on this statement.

5. If the rate of inflation holds steady at 7 percent, how many times will prices double during the next 50 years?

6. In Exercise 5, how much will a pair of shoes that costs $40 today cost 20 years from now?

7. The population growth rate for Mexico is about 3.5 percent. If this rate stays constant, how long will it take for Mexico's population to double?

8. Is it true that petroleum is a source of solar energy? Explain.

12

Some Aspects of Fluids

Common experiences often belie complex interactions of physical laws. A simple running tap demonstrates aspects of fluids, pressure, motion, temperature, and gravity, among others.

So far in our study of physics we have focused mainly on the motion of things. We have seen that ideas such as force and momentum allow us to understand and predict how any object will move in a given set of circumstances. Whether it is a golf ball or a toy wagon or a water-filled balloon makes no difference. All we need to know is that the object has mass. Up to this point, we haven't bothered about the details of the object's internal structure.

However, when we do pay attention to what the objects themselves are like, we commonly find matter in three different forms, which scientists usually call *states.* These states of matter are *solid, liquid,* and *gas.* In this chapter we are going to concentrate on the last two, liquid and gas, which together are called *fluids.*

Fluids make up a familiar and important part of our surroundings. We breathe them and drink them. We sail ships and fly planes in them. We use them for fuel in our cars and lawn mowers. Fluids have some simple characteristics and properties that we can easily understand by building on the knowledge of physics we have accumulated so far. A wide variety of phenomena, such as the reason boulders sink and ships float, why a spinning baseball curves, and how a barometer works, can be explained with the principles we are going to study.

Before we focus on these aspects of fluids, though, let us begin by learning how the three states of matter differ. This will serve as an introduction to our study of fluids.

STATES OF MATTER

One of the most sweeping generalizations in all of science is what is known as the *atomic hypothesis.* It says that matter is composed of atoms, tiny particles of about a hundred different

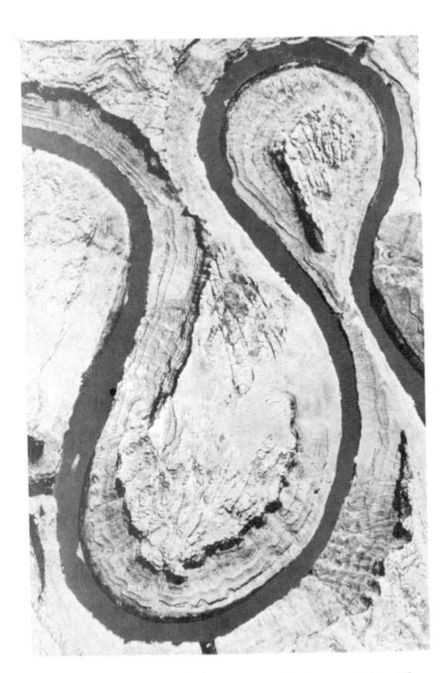

Liquids and gases, integral to the geologic processes of our planet, followed principles of fluid behavior long before humans appeared on earth. The photo above shows a meandering stream developing from the forces involved in flowing water. Volcanic gases erupt quietly or with explosive violence, as shown in the photo on the opposite page.

kinds. The atoms move about randomly when they are far away from each other, they attract each other when they are moderately close together, and they repel when they are forced very close together. The atoms themselves have an internal structure, which we shall study toward the end of the book. For now, however, we can treat them as very tiny billiard balls, each with a diameter of about 10^{-8} (0.00000001) centimeter. If you laid them side by side, it would take a hundred million atoms to form a string 1 centimeter long. Sometimes the atoms get together in specific groups called *molecules,* which we can treat as slightly larger billiard balls.

With this atomic hypothesis, then, we may conveniently distinguish three states of matter simply on the basis of how vigorously their atoms move about and what effect they have on each other.*

In *solids,* the atoms are locked into relatively fixed positions, moving only very slightly from their home base. Thus, a solid has a fixed shape and fixed volume. Attractive forces between the atoms overcome their chaotic random motion and hold them in place, giving the material its familiar rigidity and definite shape. These forces are so strong that the solid resists efforts to deform it in any way. A billiard ball retains its spherical shape in spite of the force of gravity that pulls down on the top part of the ball, trying to flatten it out. *Elastic* solids are those that undergo temporary deformations but quickly spring back to their original shape. A baseball does this when struck by a bat, for example. Of course, if the external force is quite strong, a point will be reached where the internal forces are swamped and the solid breaks or is permanently deformed. If you run over the baseball with a steam roller it won't bounce back to its original spherical shape; it will be forever flattened.

Liquids and *gases,* on the other hand, do not have any definite form or shape. Collectively, they are called *fluids,* from a Latin word for "flowing," because of their ability to be poured and to flow. They shape themselves to the vessel that contains them. Fluids are deformed by external forces. A raindrop, for example, does not stay in its spherical shape when it strikes the ground. It splatters out, because the force of gravity pulls down on each part of the drop. Similarly, a balloon filled with air or some other gas changes its shape as it is squeezed or twisted. Fluids exhibit these properties that distinguish them from solids because the forces between atoms are not strong enough to overcome their random motion.

Gases are at the opposite extreme from solids with regard to their internal make-up. Rather than staying in more or less fixed positions like those of a solid, the atoms of a gas roam

*To avoid cluttering up the discussion, we will use the word "atoms" when what we really mean is "atoms and/or molecules." However, since we are treating both as particles (like tiny billiard balls), no confusion should arise.

about freely and generally are quite far from each other. It is characteristic of the gaseous state that an individual gas atom may travel a long distance (compared with its dimensions) before bumping into another one. In fact, this distance is so large that we can usually consider an individual gas atom to be *isolated,* free from any interactions with its neighbors. So a gas tends to spread itself more or less evenly throughout its container, completely filling it. Its shape and volume are both changeable.

Liquids represent an intermediate state of matter. The individual atoms are more free to move about than in a solid and have no fixed location. However, they are not so isolated as those in a gas, but gather in aggregates that slide and slip over one another. The force of gravity dominates in liquids; they fill the lowermost portions of a container rather than spreading out evenly like a gas. When you pump air into a soccer ball, it fills the whole ball, but a liquid would run to the bottom and form a puddle. Also, whereas the atoms in a gas may be forced together, or compressed, the atoms in a liquid are already close together and strongly resist any forces that would crowd them further. For this reason, we say that liquids are virtually incompressible. A liquid, therefore, has a fixed volume but a changeable shape.

Density

You may have heard it said that iron is "heavier" than aluminum and that both are heavier than water. However, a piece of aluminum the size of a battleship obviously weighs more than an iron cannonball, and the water in the Atlantic Ocean outweighs them both. What we really mean is that iron has a greater *density* than aluminum, and both are more dense than water.

Density is an important property of matter because it allows us to compare objects on the basis of both mass and volume. The density of a substance is its *mass per unit volume:*

$$\text{density} = \frac{\text{mass}}{\text{volume}}$$

In the metric system, the units of density are grams per cubic centimeter (g/cm^3) or kilograms per cubic meter (kg/m^3). For example, the density of water is 1000 kg/m^3, which simply means that a volume of 1 cubic meter can contain a mass of 1000 kilograms of water. In general, the density of anything—solid, liquid, or gas—tells us how much of the substance is contained in a given volume. (See Figure 12-1.)

Figure 12-1. Substances differ in their densities.

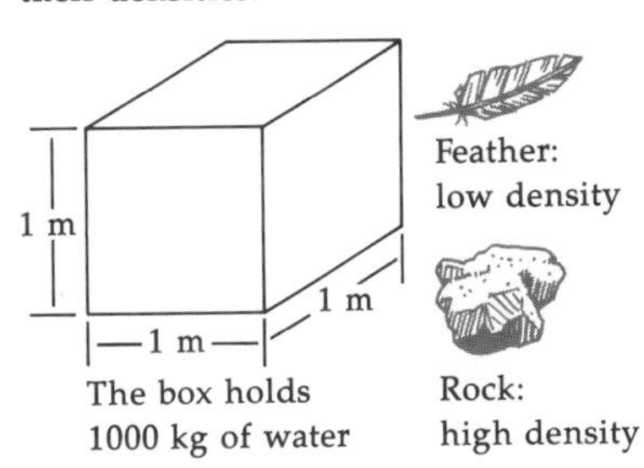

A term closely related to density is *specific gravity,* which is defined as the ratio of the density of a substance to the density of water:

$$\text{specific gravity of substance} = \frac{\text{density of substance}}{\text{density of water}}$$

Thus, the specific gravity gives a direct comparison of the density of something with that of water, a very common and familiar material. Furthermore, since specific gravity has no units, there is no confusion caused by different systems of units. For instance, the specific gravity of iron is 7.8; this tells you that iron is 7.8 times as dense as water.

Pressure

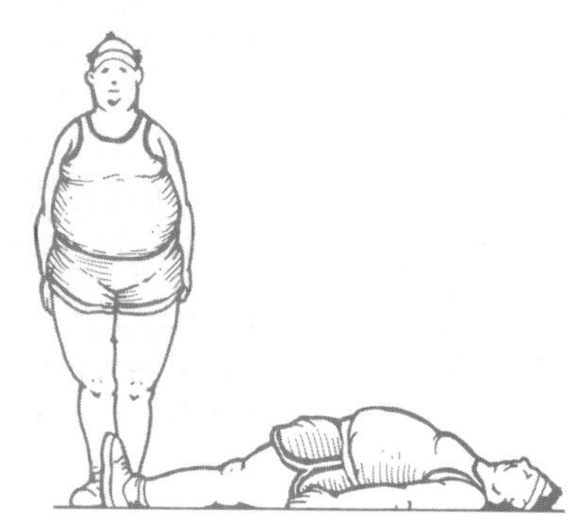

Figure 12-2. You exert greater pressure on the floor when standing than when lying, because the area you cover is smaller.

We have seen that the concept of force is very important in studying the behavior of solid objects. Fluids, however, do not have a fixed shape, so it is more convenient when dealing with them to use *pressure.* Pressure is defined as *force per unit area:*

$$\text{pressure} = \frac{\text{force}}{\text{area}}$$

It tells us how the force is distributed over a surface. (See Figure 12–2.) We will use newtons per square meter (N/m^2) as the unit of pressure.

Pressure and force are not the same thing, as you can see from the definition, even though sometimes you will hear people use them interchangeably. For instance, the head of a thumb tack is rather broad so you can push it into a slab of wood without hurting yourself. The force you apply is spread out over the entire area of the tack head, so the pressure on your thumb is small. But try pushing on the sharp point of the tack with the same force, and you are likely to poke a hole in your thumb. This time the pressure is great, because all the force is concentrated on the tiny area of the point, hundreds of times smaller than the area of the tack head. (See Figure 12–3.)

Consider another example. As fashion cycles change, it is occasionally common for women to wear "spike" heels, high-heel shoes with extremely narrow heels. These shoes can be damaging to floors because of the large pressures exerted. Let us suppose that a 50 kilogram woman wears shoes with square heels 1 centimeter on a side. The area of the surface of the heel is then

Figure 12-3. The same force can yield very different pressures.

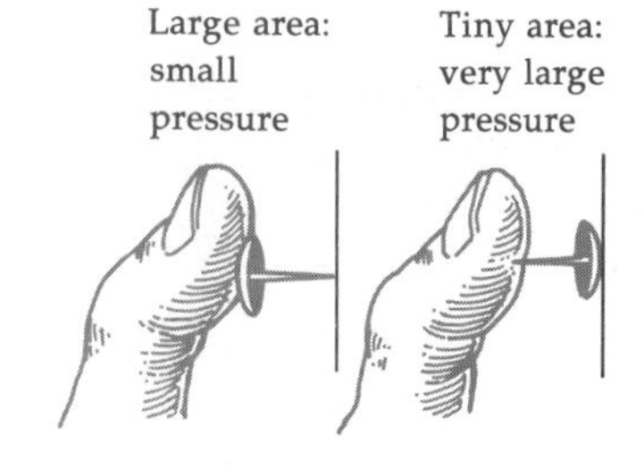

$$A = 1\ \text{cm} \times 1\ \text{cm} = 1\ \text{cm}^2 = 0.0001\ \text{m}^2$$

The woman's weight is her mass multiplied by $g = 9.8\ m/s^2$, or 490 newtons. If she stands on one heel, this weight is the force exerted on that small area, so the pressure is

$$P = \frac{F}{A} = \frac{490\ \text{N}}{0.0001\ \text{m}^2} = 4{,}900{,}000\ \text{N/m}^2$$

Compare this with a broad-surfaced heel, which we may suppose to be, say, 5 centimeters on a side. Then, the area is

$$A = 5\ \text{cm} \times 5\ \text{cm} = 25\ \text{cm}^2 = 0.0025\ \text{m}^2$$

The pressure in this case is

$$P = \frac{490\ \text{N}}{0.0025\ \text{m}^2} = 196{,}000\ \text{N/m}^2$$

which is down by a factor of 25 from that with the spike heels. So, even though the *force* is the same in each case (490 newtons), the *pressure* in the first case is much greater, because in the second case the larger heel distributes the force over a larger area.

A butcher takes advantage of pressure by using a sharp knife to cut steaks. The sharp edge is very fine—that is, it has quite a small area, so when the butcher presses down with a reasonable force, a large pressure is transferred to the steak. The dull edge of the knife is broader with a much larger area. Cutting with the dull edge is more difficult because the butcher can't generate enough pressure.

Learning Checks

1. What is meant by *states of matter?*
2. What fundamental distinctions are there among solids, liquids, and gases?
3. How are liquids and gases similar? How are they different?
4. How is specific gravity related to density?
5. Give metric units for density, pressure, and force.

LIQUIDS

Have you ever wondered how a huge battleship can float, while a chunk of iron sinks rapidly in a pool of water? Do you know why you feel pressure on your ears at the bottom of a swimming pool but not near the top? And perhaps you have noticed that an object apparently weighs less in water than in air. Now that we understand the pressure and density, we can account for all these observations rather simply.

When you swim to greater depths in a pool or lake, the buildup of pressure you feel is due to the water pushing on your body. The pressure in a liquid depends on two things: (1) the *density* of the liquid and (2) the *depth* at which the pressure is to be measured.

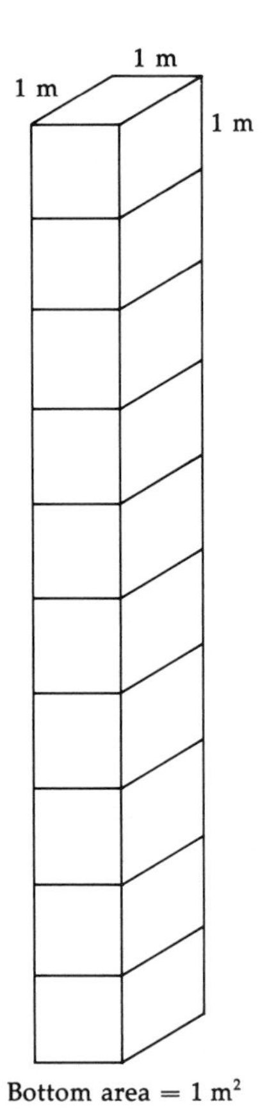

Figure 12-4. Each box contains 1000 kilograms of water and weighs 9800 newtons.

For example, suppose you are swimming at a depth of 10 meters in water whose density is 1000 kg/m^3. You may imagine, then, that there are ten cube-shaped "boxes" of water, each measuring 1 meter on a side, stacked upon your body as in Figure 12-4. Each of these imaginary boxes has a volume of 1 cubic meter (that is, $1 \text{ m} \times 1 \text{ m} \times 1 \text{ m}$), so each contains 1000 kilograms of water. The ten boxes add up to 10,000 kilograms of water, together exerting a force of 98,000 newtons. Since the area of the lower side of the bottom box is 1 m^2, the pressure there is

$$P = \frac{98{,}000 \text{ N}}{1 \text{ m}^2} = 98{,}000 \text{ N/m}^2$$

But notice that this is the same number we get if we multiply the density (1000 kg/m^3) times the depth (10 meters) times the gravitational acceleration g:

$$100 \text{ kg/m}^3 \times 10 \text{ m} \times 9.8 \text{ m/s}^2 = 98{,}000 \text{ N/m}^2$$

So in general, we can say that the *pressure in a liquid is given by the product of the density (D), the depth (d), and the gravitational acceleration (g):*

$$P = Ddg$$

We picked a particularly simple example so the numbers would be convenient. However, the relationship is quite general for all liquids.

This expression for pressure is valid because the density of the liquid is the same at any depth. We made the point earlier that liquids are virtually incompressible over a wide range of pressures. Although they may change their shape, their volume doesn't vary. Imagine what would happen if this were not so. If water could be compressed, that in the lower imaginary boxes would be "squeezed in" by the weight of the upper boxes pushing down. More than 1000 kilograms of water would be forced into the deeper regions, and the density would increase with depth. But this doesn't happen; experiments show that the density in a liquid is *independent* of the depth. We will see later that this is *not* true of gases, which are compressible because there is so much empty space between atoms.

Figure 12-5. Regulator for scuba diving.

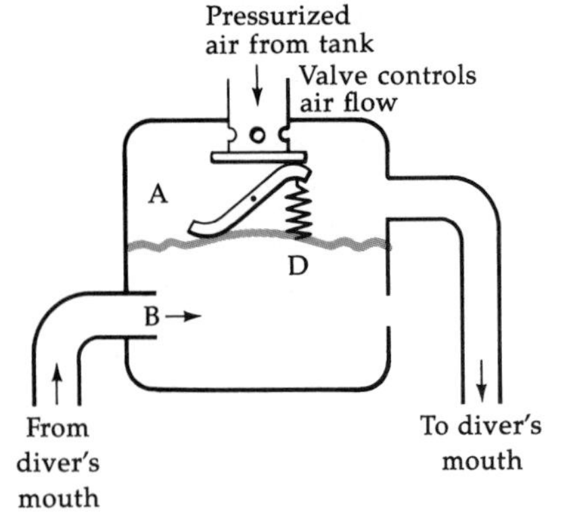

Now you can understand why you feel increasing pressure on your ears as you plunge deeper and deeper into a swimming pool. This underwater pressure causes special problems for deep-sea divers. In order to breathe normally under water, they must receive air at a pressure roughly equal to the pressure of the surrounding water. Shown in Figure 12-5 is a regulator, called a *s*elf-*c*ontained *u*nderwater *b*reathing *a*pparatus (hence the term *scuba* diver), which serves this purpose. Chambers A and B are separated by a flexible watertight diaphragm that controls the flow of air to the diver by manipulation of the valve at the top of the apparatus. Chamber B is open to the water, thus always remaining at the pressure the diver experiences.

The air in the diver's lungs is also at this prevailing pressure. However, when the diver inhales, the lung pressure lowers, in turn lowering the pressure in A. The diaphragm moves up, allowing the valve to move air down and air to flow to the mouthpiece. Exhaling raises the pressure in the lungs and in chamber A, reversing the process and shutting off the air flow. This inhale-exhale cycle can be repeated over and over again. Because chamber B is open to the water, the mechanism insures that air flowing to the diver is always at the pressure of the surrounding water.

Pascal's Principle

We have seen that the pressure exerted by a liquid increases in proportion to the depth. But what about the volume of the liquid? Or the shape of the container? Do they have any bearing on how much pressure a liquid produces?

These questions were answered by Blaise Pascal, a French mathematician of the seventeenth century who studied the dynamics of fluids. He made a very important discovery about pressure in liquids, now known as *Pascal's principle.* We can state it this way:

> Pressure exerted anywhere in a confined liquid is transmitted unchanged to every portion of the liquid and to all the walls of the containing vessel; it is always exerted at right angles to the walls.

Let us see some of the consequences of this principle. First of all, the force giving rise to the pressure in a liquid is exerted equally in all directions, not just downward. For example, if you poke a hole in the side of a container of water, the water spurts out the hole, demonstrating that there is a sidewise force exerted on the vessel walls. If the force were only downward, the water would not escape from the container. In fact, you wouldn't need the container at all, for there would be nothing to change the shape of your column.

Notice the sharp distinction between the way solids and liquids behave in this regard. Architects must know their physics here. The Washington Monument exerts tremendous pressure on its base, but none at all on its sides. It does not collapse under its own weight. If we were to replace this tall spire with one made of water, of course, it would not remain as a spire for more than a fraction of a second. Sidewise pressure would cause it to collapse under its own weight.

Demonstrating Pascal's Principle

Pascal thought up and carried out many experiments on fluid pressure. Here, he describes one:

. . . if a worm were put into a mass of dough, then although it were squeezed between the hands it could never be crushed nor even injured nor distorted, because it would be pressed on all sides. The following experiment will prove this. Into a glass tube, closed at the bottom and half-filled with water, drop three things—namely a small balloon half-filled with air, another quite full of air, and a fly (which can live in luke-warm water as well as in the air); and then push into this tube a piston that reaches the water. If now you press upon the piston with whatever force you please, as, for instance, by piling many weights upon it, the water will press all that it contains; the half-filled balloon will be very noticeably compressed; but the taut balloon will be no more compressed than if it were under no pressure at all, nor will the fly, which will feel no pain under this heavy weight but will move freely and briskly along the glass, and if released from its prison, will fly off immediately.

Source: I. H. B. Spiers and A. G. H. Spiers (translators), *The Physical Treatises of Pascal.* New York: Columbia University Press, 1937.

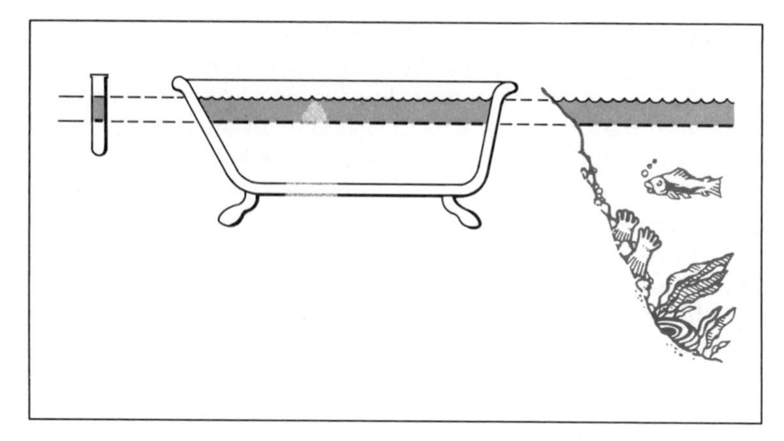

Figure 12-6. Pressure depends on depth, not volume.

Pascal's principle also accounts for the observation that the pressure in a liquid does not depend on the shape of the container or its volume, but only on the depth (and the density, of course). Thus, the water pressure is the same at a depth of 5 centimeters in a slender test tube or a lake or your bathtub. (See Figure 12–6.)

Remember the story of the little Dutch boy who saved his town from flooding by sticking his finger through a hole in the dike? It may seem amazing that he could hold back the entire North Sea in this way, but Pascal's principle provides an explanation. The pressure on him would depend only on the density of the sea water and the distance of the hole below the water surface. The volume of the North Sea has nothing to do with it.

If two or more tubes of various shapes are connected to a common vessel and liquid is allowed to flow freely among the various tubes, the liquid will rise to the same height in each tube. (See Figure 12–7.) Pascal's principle provides a simple explanation.

If the liquid in the tube A were higher than that in the tube B, say, then the pressure at the bottom of A would be greater because of the greater depth. (See Figure 12–8.) This pressure is exerted equally throughout the liquid. Thus, the *upward* force at the bottom of B due to the greater head of liquid in A would

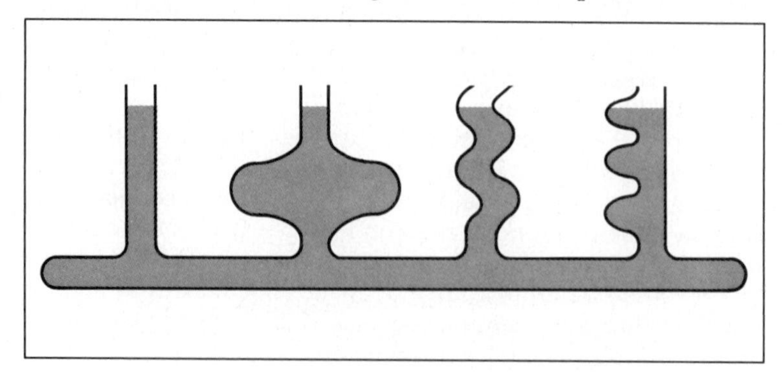

Figure 12-7. Equal pressure in all tubes puts the liquid at the same level.

be greater than the *downward* force due to the smaller head of the liquid in the second tube. The net force on the liquid in the second tube would therefore be directed upward, pushing the liquid up the second tube until the heights of the liquid in the two tubes were the same. By extending this argument to several tubes, you can easily see why the levels in all the connected tubes are the same.

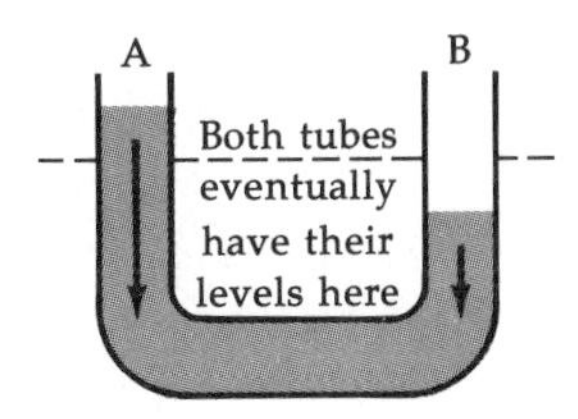

Figure 12-8. Initially, the pressure in A is greater than that in B.

Restaurants use this principle in the large tanks from which the waitress draws coffee. Running up the outside of the metal tank is a "sight glass," a length of transparent glass tubing connected to the tank at the bottom. Because the coffee is free to flow between tank and tube, it will rise to the same level in both, enabling the waitress to tell at a glance how much coffee is left in the tank.

The hydraulic lift (Figure 12-9) is based on Pascal's principle. Any pressure exerted on the small platform is transmitted undiminished to the larger one. This allows one to lift heavy objects by applying a small force. For example, if the smaller platform has a cross-sectional area of 1 m^2, then a force of 10 newtons would deliver a pressure of 10 N/m^2 throughout the liquid. Suppose that the larger platform has an area of 5 m^2. The force the liquid exerts on it is 50 newtons, because force is pressure times area ($10\ N/m^2 \times 5\ m^2 = 50\ N$). So a 50 newton object could be raised by application of a force of only 10 newtons.

Figure 12-9. A hydraulic lift. The 10 N/m^2 pressure is delivered throughout the liquid.

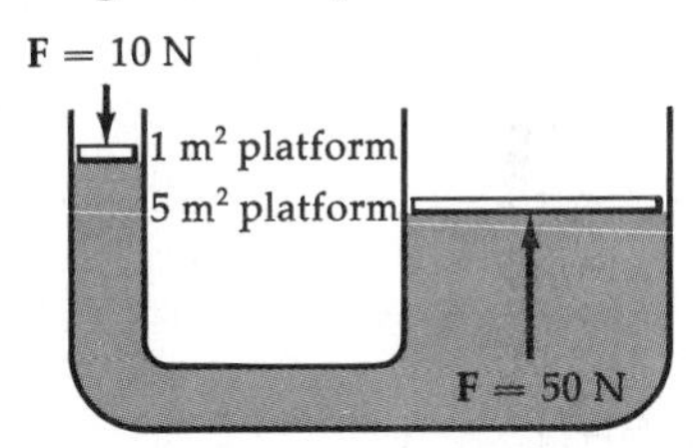

Physics and Engineering: The Hydraulics Industry

Some work requires brute strength, some gentleness. Fluid power systems can accomplish both. Their versatile nature can be seen as they bore into thick cylinder blocks, forge tools, lift elevators, apply brakes, and softly ease damping rods into a nuclear reactor. Their ability to produce high forces at low speeds gives them a distinct advantage over electric power and this, plus the sheer range of their capability of motion, is why the hydraulics industry is among the most rapidly expanding in America.

Now, in a promising practical application, a hydraulic system can be used to *generate* electricity. An ingenious Pennsylvania inventor has devised a way to exploit the weight of moving cars to drive generators. His system, set up at highway toll plazas, works in this fashion:

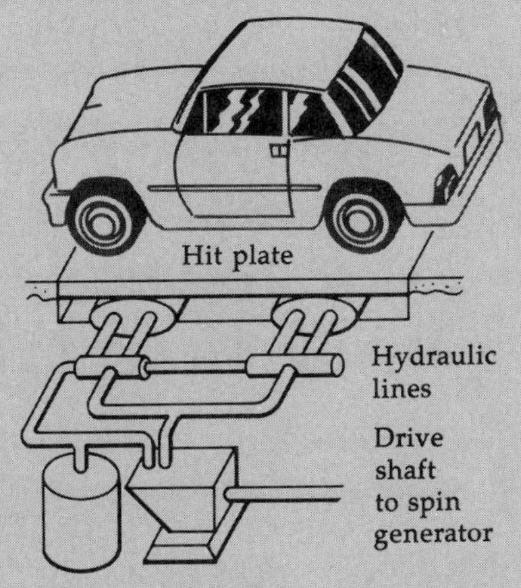

A car passes over a hit plate, a 30-inch-wide metal slab that projects three-quarters of an inch above the surface of the street or roadway. The pressure from the car's weight forces hydraulic fluid out of a rubber chamber attached to the underside of the plate. The fluid flows at high pressure into a nearby building. There it drives a pump-operated motor, which spins a flywheel, which in turn spins a generator, producing electricity. The fluid is then channeled back to the rubber chamber under the plate so that the process can be repeated.

A prototype has been built outside of New York City, and the electricity generated by the moving traffic is expected to cost far less than that obtained by hydro power.

Sources: J. J. Pippenger and R. M. Koff, *Fluid Power Controls.* New York: McGraw-Hill Book Co., 1969.

"How Cars Can Generate Electricity," *Science Digest*, May 1981, p. 29.

Learning Checks

1. Does the pressure at the bottom of a swimming pool depend on the density of water? Explain.
2. What does it mean to say that liquids are incompressible? How is this important in determining the pressure at a given depth?
3. Estimate the density of your body.
4. Estimate the pressure your body exerts on the floor when you stand on both feet; on one foot; on tiptoes.
5. State Pascal's principle in your own words.
6. In what way does the pressure within a liquid depend on the depth?

Buoyant Force: Archimedes' Principle

Archimedes was a mathematician who lived in Syracuse, the capital of a Greek colony in Sicily, during the third century B.C., some one hundred years after the time of Aristotle. He inherited an interest in mathematics from his astronomer father and made many important contributions to pure mathematics as well as to physics. Archimedes' best-known discovery is the law pertaining to the apparent weight loss by bodies submerged in a liquid. This law is called *Archimedes' principle.*

As happens so often in science, the discovery of Archimedes' principle occurred somewhat by accident. The story goes that when Hiero became King of Syracuse, he wished to place a gold crown in the temple of the gods to celebrate his rise to power. When the newly fashioned crown was delivered, Hiero suspected that some of the gold had been removed and replaced by cheaper silver. Hiero instructed Archimedes to determine whether or not this was true, assuring him that if he failed, the king would fix Archimedes' head so he could carry it around in his hands, so to speak. Because silver is less dense than gold, Archimedes knew that the gold-silver combination would have a smaller density than the gold alone. He therefore needed to determine the density of the crown, meaning he needed to know both its mass and its volume. Finding its mass was simple enough, but to get its volume he would have to hammer it into some familiar shape whose dimensions he could measure geometrically. He knew that Hiero wouldn't like that.

Puzzling over this problem of determining the crown's volume, Archimedes went home one night to take a bath. While doing so he noticed that as he sank deeper into the tub, the water level rose, spilling water over the top if enough was displaced by his body. He instantly realized that here was the solution to his problem. Jumping out of the tub, he ran naked through the streets of Syracuse shouting "Eureka! Eureka!" ("I've found it! I've found it!").

Archimedes made a sample of gold and one of silver, each having the same mass as that of the crown. He then filled a vessel with water and successively measured the spilled water after having submerged each of the three items in the filled vessel, one at a time. The silver displaced the most water, the equal weight of gold the least, and the crown caused an intermediate amount of water to spill over. This meant that the crown had a greater *volume* and therefore a smaller *density* than an equal mass of gold, which could only have been due to the presence of silver instead of some of the gold. Hiero had his answer, the crown-maker was duly punished, and Archimedes saved his head.

We get a feeling for the principle Archimedes discovered by considering a thought experiment. Suppose we want to determine the weight of a metal ball when it is submerged in water. Imagine first that instead of the metal ball we submerge a thin balloon of the same diameter filled with water. (See Figure 12–10.) Let us assume that the balloon holds 1 kilogram of water (weighing 9.8 newtons). Since we may suppose that the weight of the balloon is negligible, the situation will be the same as if the water in the balloon were just part of the water in the bucket, so the scale reads zero (Figure 12–10a). We now replace the water in the balloon with the metal ball, which we suppose has a mass of 10 kilograms and weighs 98 newtons. Our kilogram of water was supported by the rest of the water in the bucket with the scales showing zero; the metal ball, which has the same volume as the balloon, will get the same support. Therefore, the scales will read not 98 newtons, but 98 − 9.8, or *88.2* newtons as the weight of the ball submerged in the water (Figure 12–10b). The apparent weight loss is equal to the weight of the water displaced by the metal. This is called *buoyancy*. We often say that the metal ball is "buoyed up" by the water.

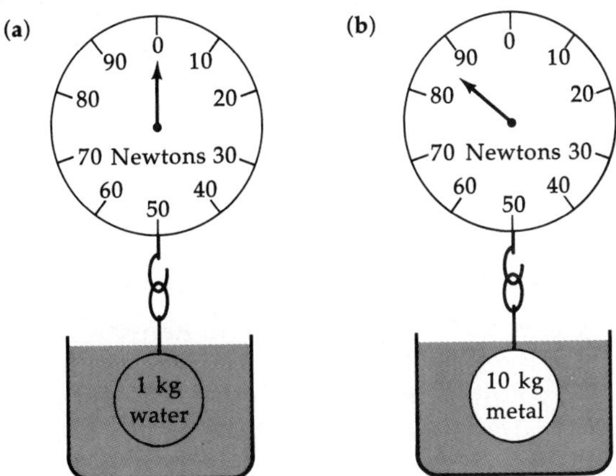

Figure 12-10. Archimedes' principle. The apparent weight of the ball is reduced.

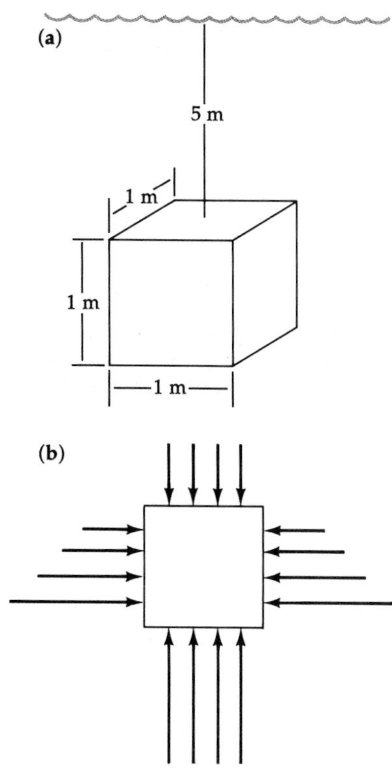

Figure 12-11. The difference in pressure is the same at all levels.

This, then, is Archimedes' principle:

> The buoyant force on an object immersed in a fluid is equal to the weight of the displaced fluid.

The term "displaced fluid," by the way, means just what it says: in order for the metal ball to be immersed, some of the molecules of the liquid have to get out of the way. When they do, the level of the liquid rises by an amount corresponding to the volume of the ball.

Let us see how the buoyant force comes into play. To make the arithmetic simple, let us imagine that we suspend in water a rock in the shape of a perfect cube having dimensions 1 meter on each side, as in Figure 12–11. We suppose it is located so that its top face is 5 meters below the water surface (Figure 12–11a). Now according to Pascal's principle, the water will exert pressure on all the faces of this cube (Figure 12–11b). The arrows representing the pressure have different lengths because, as we recall, the pressure increases with depth. The pressures exerted *horizontally* on the cube from the sides will cancel; that is, each arrow pointing to the left is matched by an arrow of equal length at the same depth pointing to the right. So we need not worry about the sidewise pressures.

What we are really interested in are the *forces* on the two faces. We can calculate the force by multiplying the pressure by the area (recall that the definition of pressure is force/area). The area of each face of the cube is $1\ \text{m} \times 1\ \text{m} = 1\ \text{m}^2$, so we have

$$\text{force on bottom} = 58{,}000\ \text{N/m}^2 \times 1\ \text{m}^2$$
$$\mathbf{F}_{\text{bottom}} = 58{,}000\ \text{N (upward)}$$
$$\text{force on top} = 49{,}000\ \text{N/m}^2 \times 1\ \text{m}^2$$
$$\mathbf{F}_{\text{top}} = 49{,}000\ \text{N (downward)}$$

These two forces are shown in Figure 12–12. So the net force due to the water, called the *buoyant force,* is the difference between these:

$$\mathbf{F}_{\text{buoyant}} = \mathbf{F}_{\text{bottom}} - \mathbf{F}_{\text{top}}$$
$$= 58{,}000\ \text{N} - 49{,}000\ \text{N}$$
$$\mathbf{F}_{\text{buoyant}} = 9{,}800\ \text{N (upward)}$$

The buoyant force is directed upward and hence works against the force of gravity, reducing the apparent weight of the rock.

What about Archimedes' principle? Well, since the rock has a volume of 1 cubic meter ($1\ \text{m} \times 1\ \text{m} \times 1\ \text{m} = 1\ \text{m}^3$), it displaces a cubic meter of water. And since the density of water is 1000 kg/m^3, the volume of displaced water has a mass of 1000 kilograms. But the weight of this mass of water is just exactly equal to the buoyant force we calculated above, namely, 9800 newtons. So we have verified what Archimedes' principle tells us, namely, that the buoyant force is equal to the weight of the displaced fluid. Although we have shown it for a very simple geometry, it is true for any object of any shape.

Notice that the buoyant force does not depend on the depth, but only on the displaced volume. This is because the force arises from the *difference* of forces on top and bottom, not on their absolute value. If we had taken our cube-shaped rock so that its top face was at a depth of 10 meters, we would still get the same final result of 9800 newtons because the bottom face would still be 1 meter deeper than the top face, or at 11 meters in this case.

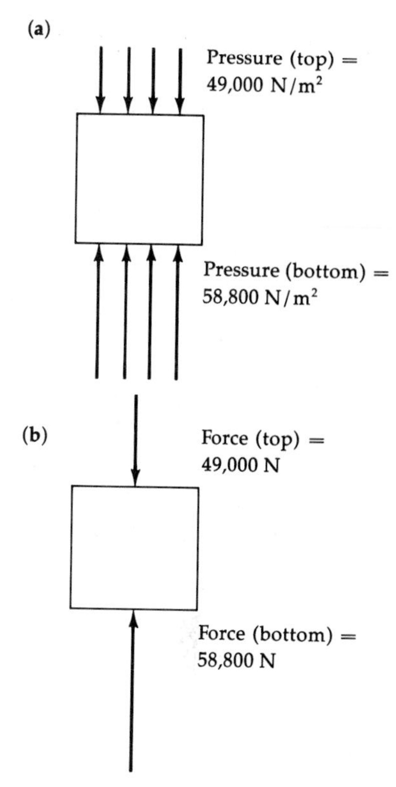

Figure 12-12. The difference between the top and bottom forces equals the buoyant force.

We may use Archimedes' principle to establish some criteria for determining whether or not an object will float in a liquid. It all depends on the relative densities. Figure 12-13 illustrates the possibilities. For example, a rock sinks in water because its density is greater than 1000 kg/m^3. Thus, a given volume of water weighs less than an equal volume of rock, so the buoyant force on the rock—that is, the weight of the displaced water—will be less than the weight (in air) of the rock. The force due to gravity will be greater than the buoyant force, and the rock will sink.

On the other hand, if the density of the body is *less* than that of the liquid, the liquid displaced will weigh more than the body, and the buoyant force will be *greater* than the force of gravity. The net force is thus upward, and the object will rise to the surface.

What about the case in which the two densities are equal? Can you explain why such an object will remain at any depth at which you place it? Naval engineering relies on the answer to this question. A submarine, for example, descends by taking water into ballast tanks. This increases its mass without changing the sub's volume, and hence raises its density above that of the water. When the water is pumped out of the tanks, the reverse takes place: the submarine's density decreases, so that now the buoyant force exceeds its weight, and it rises to the

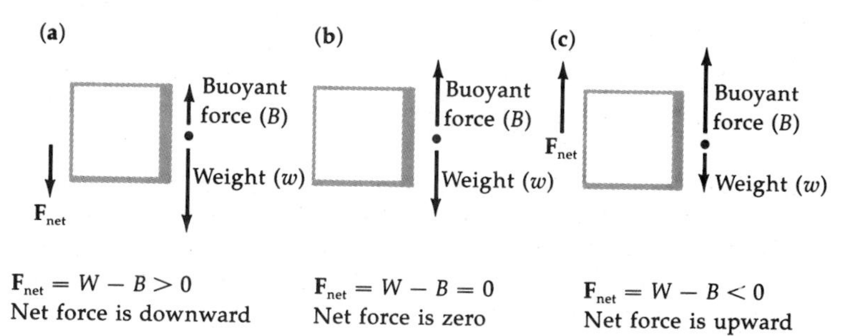

$F_{net} = W - B > 0$
Net force is downward

$F_{net} = W - B = 0$
Net force is zero

$F_{net} = W - B < 0$
Net force is upward

Figure 12-13. Comparative densities. (a) Density of object is *greater* than density of fluid. (b) Density of object is *equal* to density of fluid. (c) Density of object is *less* than density of fluid.

Buoyant Brain

While examples of buoyancy abound in the world, there is one that is quite close to us: our brain. Encircling the brain is a reservoir of cerebrospinal fluid only about two-thirds of a cup in volume. This small amount, however, is sufficient to buoy up the flaccid tissue. If we apply the buoyancy principle to this situation, we can see that the average 1400 gram brain has a mass the equivalent of roughly 20 grams while bathed in the fluid that surrounds it. Thus, your thinking organ floats lightly on the bottom of your skull.

Source: William Oldendorf and William Zabielski, "The Buoyant Brain," *Science Digest*, August 1981, pp. 94–96.

surface. Similarly, a fish controls its density by use of its swim bladder—not, as is often supposed, by contracting and expanding the bladder, but by moving a gas into and out of the bladder to maintain a constant volume at any depth.

The specific gravity of ice is about 0.9, meaning that ice is only 90 percent as dense as water. This means that ice will sink in the water only until 90 percent of its volume is immersed. The other 10 percent rides above the water level. Hence the phrase "the tip of the iceberg": nearly 90 percent of the iceberg is under water.

Although we have used water as our example fluid, these principles hold for other fluids as well. For example, the density of brass is about 8.5 times the density of water, while the density of mercury is 13.6 times that of water. A brass ball, therefore, will sink in a beaker of water but will float in a beaker of mercury. If this seems impossible, try it and see. In general, denser fluids exert a greater buoyant force than less dense ones, and vice versa.

Remember that the buoyant force equals the weight of the displaced fluid. So the greater (or lesser) the density of the fluid, the greater (or lesser) will be the buoyant force on any immersed object. Alcohol is less dense than water—its specific gravity is about 0.8. Bartenders know that the density of a mixed drink depends on the proportions of the alcohol and the mixer. Go easy on that Scotch-and-water if the ice cubes have sunk to the bottom of the glass.

Learning Checks

1. State Archimedes' principle in your own words.
2. What is meant by *buoyant force?*
3. Explain how an object apparently weighs less in water than in air.
4. Explain the fact that you can lift a large rock from the bottom of a lake when you could not lift it from the shore.
5. Does the buoyant force on a submerged rock change as the rock sinks lower into the water? Explain.

Floating Objects

At the beginning of this section on liquids, we asked how it is possible for a battleship to float. We have seen that it is a question of density. Thus, a solid iron ball will sink in water because of its greater density. But if we hammer the iron ball into the shape of a bowl, for example, it will float. We have increased its volume to such an extent that the density of the bowl is now less than that of the water.

Figure 12–14 illustrates the point. In (a), a steel ball having a volume of 1 m^3 and a mass of 1500 kilograms displaces only 1000 kilograms of water, and hence it sinks. In (b), the same 1500 kilograms of steel is shaped into a boat. It now displaces 1500 kilograms of water, and hence it floats. Note that it displaced more while *floating* than while sinking.

Archimedes' principle still holds in the case of floating objects, even though they are not completely immersed. They are partially immersed, however, displacing some water. Hence, they are buoyed up by a force equal to the weight of the displaced water. A floating body, since it is neither sinking nor rising, must be buoyed up by a force equal to its own weight. Therefore, we can say that a floating object displaces its own *mass* while a completely submerged object displaces its own *volume.*

A boat will ride lower in the water with two people aboard than with only one. (See Figure 12–15.) This is because with the greater mass the boat must displace more water, meaning that a greater fraction of it must be immersed. So as more people are added to the boat, it sinks to a deeper level where the mass of the displaced water is equal to the total mass of the loaded boat.

Sea water has a density of about 1030 kg/m^3. Thus, a ship will ride higher in sea water than in fresh water, because it must displace a smaller volume in order to obtain a buoyant force equal to its weight.

In summary, then, we have seen that the pressure within a liquid depends on its *density*—which is the same throughout the liquid—and the *depth.* The pressure at different depths results in an upward force, called the *buoyant force,* acting on any object immersed in the liquid. The magnitude of this buoyancy is numerically equal to the weight of the liquid that has been displaced. Since a floating object must have the force of gravity

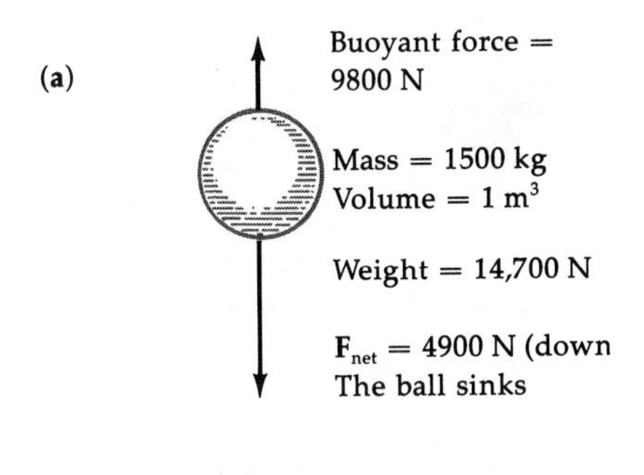

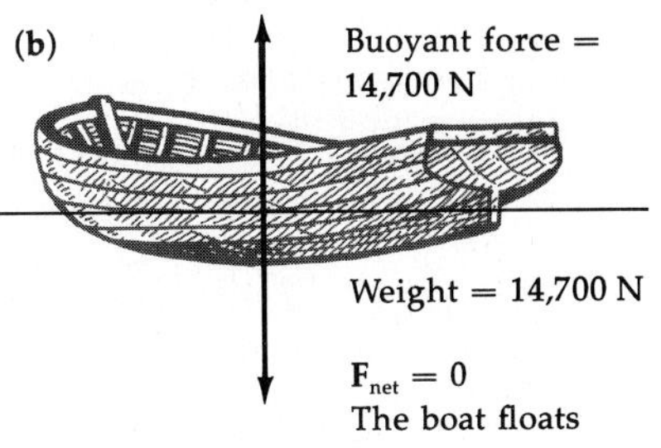

Figure 12-14. The steel sinks or floats, depending on its volume.

Figure 12-15. More water must be displaced in (b).

Physics and Nursing

Physics enters into so many aspects of the clinical nursing profession that an entire course is devoted to the subject. One of the important physics concepts involves the movement of fluids through tubes. It is vital that nurses know about the behavior of blood and other body fluids that must flow through transfusion or drainage tubing. Nurses must be aware of how energy is being expended within a system of flowing fluid, how drops in pressure occur, and how sudden changes in the diameter of a tubing (that is, changes caused by constrictions or kinks) can affect flow. In a hospital operating room, this can sometimes be a matter of life or death.

Source: H. R. Rowe, *An Introduction to Physics in Nursing*. St. Louis, Mo.: C. V. Mosby Co., 1958.

canceled by the buoyant force, it follows that in order for a boat to keep from sinking, it must displace its own mass of water.

Learning Checks

1. A certain object weighs 300 newtons. Is it possible to say whether it will float or sink in water? What other information, if any, is necessary?
2. Explain the statement that a floating object displaces its own *mass* of water, while a submerged object displaces its own *volume*.

Moving Fluids: Bernoulli's Principle

So far we have considered pressure in liquids that are stationary, like a bucket of water or a still lake or a beaker of mercury. What about pressure in moving fluids? Does this introduce any new ideas?

Daniel Bernoulli, a Swiss physicist and mathematician, studied these questions in the eighteenth century, a few decades after Isaac Newton died. Bernoulli investigated the motion of fluids through pipes of varying diameters, such as the one shown in Figure 12–16. The slowly flowing water in the wide-diameter region A will increase in speed as it enters the constricted area B, just as the water current in a broad stream or river speeds up when it passes through a narrow gorge. A familiar example occurs when the water coming gently from a garden hose becomes a high-speed jet stream when you partially cover the outlet end with your thumb. The fluid speeds up because the same amount of water must pass through the same length of the wider region. Otherwise, the fluid would pile up at the entrance to the narrow area. This doesn't happen of course—remember, the fluid is incompressible. Similarly, the speed drops as the water goes from B to C.

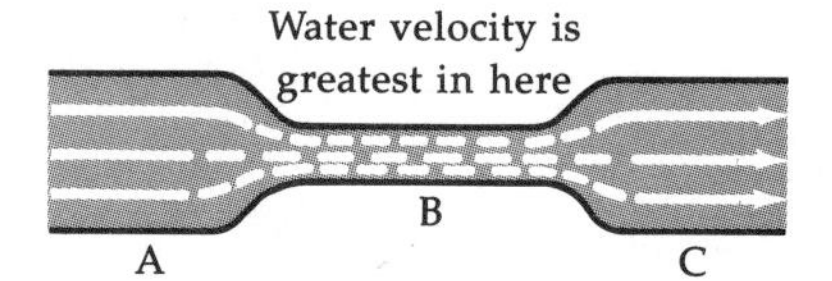

Figure 12-16. Motion of water through pipes of varying diameters.

The effect of the speeds on the *pressure* within the water as it moves from A to B at first seems to contradict common sense. We can measure the pressure in different sections by attaching vertical sight glasses to the pipe at various places, as shown in Figure 12–17. The level to which the liquid rises in the attached sight glasses indicates the pressure, which we find is higher in regions A and C than in B. It might seem that the pressure should be higher in the narrow region B, since water has to "squeeze" through the constriction. The experiment, however, shows that the pressure is greater in regions A and C,

Figure 12-17. Relationship between pressure and velocity.

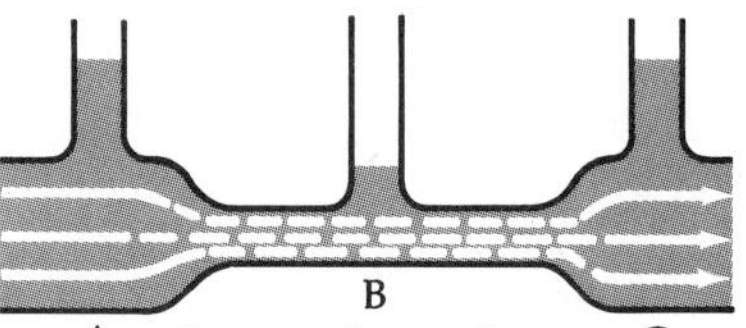

where the water moves relatively slowly, than in the region of high speed, B.

This relationship between speed and pressure is *Bernoulli's principle:*

> The pressure in a moving fluid *decreases* as the speed *increases,* and *increases* as the speed *decreases.*

Let us see if we can account for this effect. To do so, consider Newton's second law. As the water moves from A to B its speed increases; that is, the water *accelerates,* indicating the presence of a *force* acting in the direction of the flow. The only force operating here is that due to the difference in pressure between the broad and the narrow regions of pipe. Because the force is directed from region A toward B, the greater pressure must be in region A, and this is exactly what the experiment indicates. Conversely, as the fluid goes from the narrow to the wide section of pipe, it slows down, or decelerates, implying the presence of a force directed *against* the flow. Again, this leads to a greater pressure in the wider region C, as indicated by the sight glasses. (See Figure 12–18.)

Bernoulli's effect is especially important for gases. We shall see several examples after we have learned some other facts about matter in the gaseous state.

Figure 12-18. Forces due to differences in pressure.

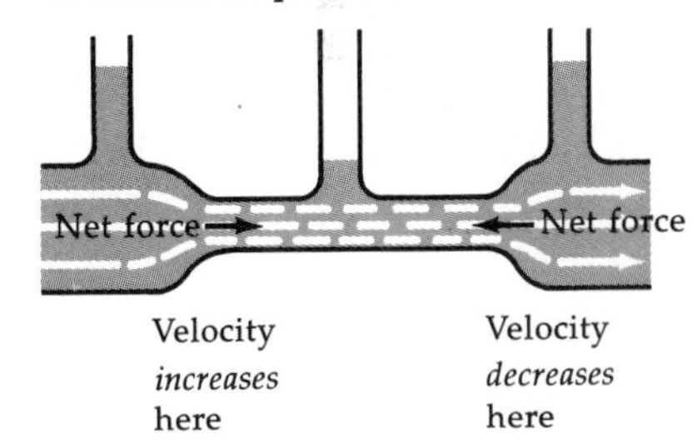

THE BEHAVIOR OF GASES

In contrast to liquids, where the atoms are in contact with each other, gases are characterized by essentially isolated individual atoms, free to move about in random motion, only rarely bumping into one another. Because of the very small atomic masses, the kinetic energy of this random motion overcomes the force of gravity for the individual atoms. As a result, any closed container will be completely filled by a gas distributed

Liquefied Gases

The knowledge of gases that physics provides us with is essential to much of the manufacturing sector of the economy, as gases of one kind or another are used in processes as diverse as steelmaking and the fizzing of soft drinks. Gases converted to liquid state have come to serve as the foundations for entire industries. Liquid nitrogen, for example, is used in the refrigerated cargo compartments of transport trucks that haul frozen foods. Liquid oxygen and liquid hydrogen are used as rocket propellants.

Natural gas (that piped into our homes) is often liquefied for ease of shipping and storing. Liquefied natural gas (LNG), however, has become a controversial fuel. It liquefies only at −162°C (−259°F), which means that specially fabricated, rupture-free tanks must be built to contain it. It burns with intense heat, is highly flammable, and is difficult to control when on fire. Physicists and engineers are employed to perfect the best methods for the safekeeping of LNG and for the handling of possible spills on land and sea.

Sources: Charles S. Simpson, *Chemicals from the Atmosphere.* New York: Doubleday & Co., Inc., 1969.

L. N. Davis, *Frozen Fire.* San Francisco: Friends of the Earth, 1974.

more or less evenly to every nook and cranny of the vessel. Contrast this with liquids, in which the force of gravity dominates, causing the liquid to occupy and shape itself to the lowermost part of a container.

The density of gases is very much smaller than that of liquids. You would expect this, because the atoms are far apart and hence take up a large volume. Air, for instance, has a density of only about 1.3 kg/m^3 at sea level under ordinary temperature conditions. This is about 1/800 the density of water.

Pressure of a Gas

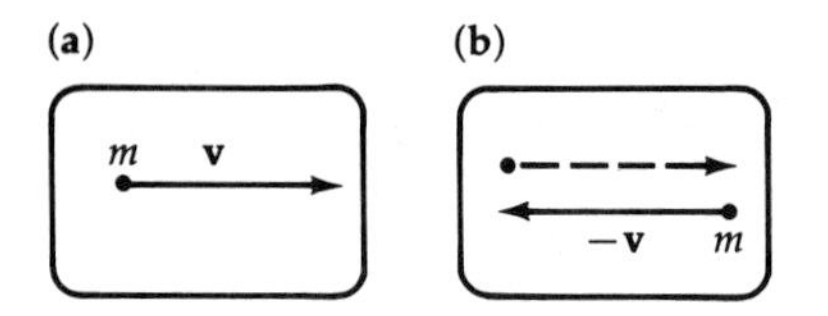

Figure 12-19. The velocity changes direction at the wall.

The random motion of a gas enclosed in a container causes pressure on the vessel walls. Suppose that we have a few atoms of a gas trapped in a closed vessel. We will focus our attention on a single atom of mass m moving with velocity $\mathbf{v}$ in a direction perpendicular to the righthand wall of our box. This is shown in Figure 12–19a. We imagine that upon striking the wall it rebounds with the same speed, so that now its velocity is $-\mathbf{v}$ as in Figure 12–19b.

The momentum of the atom before striking the wall is $m\mathbf{v}$, and afterward it is $-m\mathbf{v}$, so the change in momentum is $-2m\mathbf{v}$. We saw in Chapter 6 that a change in momentum is due to a force—in this case, the force of the wall on the mass m. According to Newton's third law, this force is accompanied by an equal and oppositely directed force that the atom exerts on the *wall.* Every time an atom strikes the wall it will make a similar force contribution, giving a cumulative effect that is the total force on the wall. The pressure on the wall will be this total force divided by the total area. So the pressure of the air on the walls of a balloon, for example, is due to the random motion of the individual gas atoms.

It is clear that the pressure depends on the amount of gas enclosed in the container. Suppose that we double the number of atoms. This doubles the average number of collisions with the wall in a given length of time, which means doubling the pressure. Hence, the *pressure is proportional to the number of gas atoms.*

Figure 12-20. If the velocities are equal, the smaller box has the larger number of collisions.

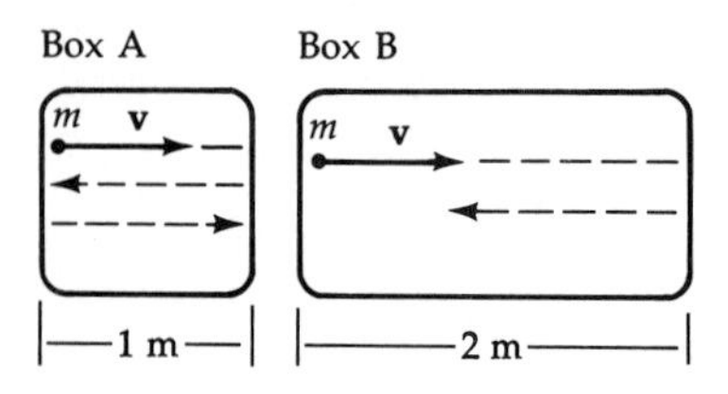

But the pressure depends on other things as well. As our particle bounces back and forth with speed $\mathbf{v}$ between the walls at opposite ends of the container, naturally the length of time between its collisions with the righthand wall depends on how long a path it travels. This is illustrated in Figure 12–20. If box A has length 1 meter and box B has length 2 meters, a particle in box A will make *two* collisions with the righthand wall during the time that an identical particle in box B traveling with the same speed makes only *one.* And, because the pressure depends on the number of collisions of all the particles with the wall, the pressure in box A is twice that in box B (we assume

Gas Pressure and Spacesuits

The public first saw spacesuits in use during the historic 1969 lunar landing and moon-walking activities of the astronauts. In 1973, the suits were again used under actual working conditions when *Skylab* crewmembers exited into space to do external repairs on their orbiting laboratory. Spacesuits are miniature space capsules in which working astronauts must live, and correct gas pressure is a crucial part of this enclosed environment.

Despite the fact that these suits are superbly engineered multimillion-dollar structures, things can go wrong. For example, during one flight, a suit would not reach the required pressure of 4.3 pounds per square inch. (This is adequate even though it is only one-third of normal earth pressure because the astronaut is breathing pure oxygen and not the oxygen-nitrogen mix of earth air at sea level.) It turned out that two plastic pins were missing from the pressure regulator, allowing a locking ring to open, causing a leak. The gas was no longer trapped in a closed container, and the movements of its atoms were consequently affected.

Source: "Some Unsuitable Workmanship," *Time*, December 20, 1982.

A pressurized spacesuit is a container of gas.

that they contain equal amounts of the gas). All other dimensions of the two boxes being the same, the volume of box B is twice that of box A. Therefore, we may conclude that the *pressure is inversely proportional to the volume,* which is one way of saying that the pressure (P) goes up in the same ratio as the volume (V) decreases, and vice versa. In the form of an equation, we may write $P = k/V$, or, the same thing in a different form

$$PV = k$$

where k is constant.

This relationship is called *Boyle's law,* in honor of the seventeenth-century British physicist Robert Boyle, who first clearly presented the pressure-volume dependence to the scientific community in 1660. Boyle's law, which holds for many gases over a wide range of pressures, tells us that for a given amount of gas the product of the pressure and the volume always results in the same number.

There is one more factor that has an influence on gas pressure, and that is the speed of the particle. Suppose we increase the particle's speed in some way. It seems reasonable that at the faster speed it makes more collisions with the walls since it can traverse the distance between them in a shorter time. We can speed up the gas particles by heating them, that is, increasing the *temperature.* We shall look more carefully at the notion of temperature in the next chapter; for now it is sufficient to say that the *pressure goes up or down in direct proportion to increases or decreases in the temperature.* This means that Boyle's law, for example, holds only if the temperature remains *constant.*

Now we are in a position to put together all we know about gas pressure in one neat package. We may write

$$PV = knT$$

where P is the gas pressure, V its volume, n the number of particles of the gas, and T the temperature. This equation is called the *ideal gas law,* because it is rigorously true only for ideal gases, those that obey our assumptions: namely, that the particles of gas do not take up any volume themselves and that they do not interact with each other. Of course, real gases do not strictly obey our assumptions, but most of them come pretty close. (Some of those effects that cause gases to deviate from the ideal gas law are discussed in Supplement 8.)

The important feature of the ideal gas law is that it tells us at a glance the relationship among the pressure, volume, and temperature of a gas. For example, consider what happens when you breathe. To inhale, you expand your lungs, increasing their volume. According to the gas equation, the air pressure in your lungs drops by a corresponding amount, to a lower value than the pressure of the outside air. This pressure difference forces air into your lungs. When you exhale, the reverse process takes place. The lungs contract, decreasing the volume and increasing the pressure. The air is then forced out of your body to the lower-pressure surroundings.

Have you ever noticed that the air pressure in your automobile tires is greater in the summer than in winter? Here we have a situation of constant volume—the volume of the tire. At lower wintertime temperatures the tire pressure is correspondingly lower than in the heat of summer.

As you drive, the friction of the tires against the road heats them up, increasing the pressure of the air inside. That is why it is always a good idea to check the tire pressure after you have driven for a while, when the tires are warm. This gives a better indication of what the tire pressure will be out on the road during a long drive.

You can illustrate the near-ideal behavior of gases by blowing up a balloon, fastening the end, and putting it into a freezer. As the air temperature drops, so does the pressure, collapsing the balloon. Return the balloon to the room, and you can watch it re-expand as its air temperature rises.

Learning Checks

1. What is an *ideal gas?*
2. Explain how momentum is used to account for gas pressure on the walls of a container.
3. In Boyle's law, would you expect the constant k to depend on the mass of the gas atoms? Explain.
4. Why do gases completely fill their container, while liquids fill only the lower portions?

Pressure of the Atmosphere

The earth has an atmosphere because of the competition between the energy supplied by the sun and the earth's gravity. The sun provides energy for the atoms and molecules of gases such as oxygen, nitrogen, and others that make up the air. They move about rapidly and occasionally "leak" out of the earth's gravitational field, escaping into outer space. The earth's gravitational attraction works to counteract this effect of the sun, holding the gas molecules back. On one hand, were there no sun, the air molecules would lie dormant on the surface of the earth. On the other hand, without a strong enough gravitational field, the air would have been boiled away by the sun long ago, depriving us of an atmosphere.

The opposing effects of solar and gravitational energy also account for what we might call the "shape" of the atmosphere. (See Figure 12-21.) At lower altitudes the density of the atmosphere is greater than at higher altitudes, such as the top of a mountain, where the air is relatively thin. So, unlike the case of a liquid, the density of the atmosphere increases with depth, being the greatest at the earth's surface. Correspondingly, the pressure is also greatest there because the rest of the atmosphere is pushing down. As one goes to higher and higher altitudes, the pressure and density of the air gradually decrease. About 99 percent of the atmosphere is below 50 kilometers.

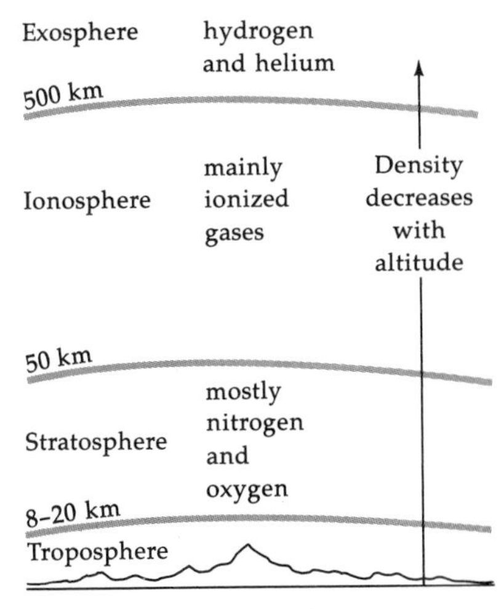

Figure 12-21. The earth's atmosphere.

The decreased atmospheric pressure at high altitudes is the reason airplanes are pressurized. Mountain climbers often carry oxygen with them because the density of the air is so much lower at high elevations. At the top of Mt. Everest, about 9000 meters high, the air is only one-third as dense as it is at sea level—at this elevation a lungful contains only one-third the amount of oxygen.

Physics and Physiology: Movement on the Heights

Adventuresome mountain climbers are not the only people subject to shortness of breath and other unpleasant effects of high altitudes. Helicopters and four wheel drive vehicles have given skiers, explorers, and researchers easier access than ever before to dangerous altitudes.

As the barometric pressure of the air drops with altitude, and the number of gaseous molecules decreases, the human body changes. Breathing speeds up, so that more air is inhaled, the heart beats more rapidly, and more red blood cells are produced. When the body's

responses fail to keep pace with decreasing oxygen, illness results, sometimes in a mild form, sometimes more seriously, as when permanent brain damage occurs.

Acclimation to the heights takes time. Even though athletes trained at altitudes of 6000 to 10,000 feet for the 1968 Olympics in Mexico City (7400 feet), their performances in many events were impaired. A team of researchers is now devoting full time to the study of the physics and physiology of high altitudes. They are erecting electric generators and medical equipment at 20,000 feet, hoping to benefit all those who suffer from altitude sickness.

Source: Eric Perlman, "Walking on Thin Air," *Science 80*, July/August 1980, pp. 89–90.

The normal pressure of the atmosphere at sea level is about 100,000 N/m^2, or 10 N/cm^2. One way of looking at this is to consider a column of air with a cross-sectional area of 1 square centimeter reaching to the top of the atmosphere. It will weigh 10 newtons. This means that every square centimeter of your body constantly experiences a force of about 10 newtons. We aren't aware of this pressure because it is the same both inside and outside our bodies.

Atmospheric pressure is an important consideration in meteorology. This is the reason the weather forecaster gives the pressure in the daily weather report. It is normally quoted as a barometer reading of so many inches or centimeters of mercury, usually 29 or 30 inches, or about 76 centimeters. Let us see what this means.

Mercury, as we have seen, is a very dense liquid, with a specific gravity of about 13.6. One way of making a mercury barometer is to fill a long glass tube with mercury and invert the tube in a reservoir of mercury, such as a shallow pan. The setup looks something like that in Figure 12–22.

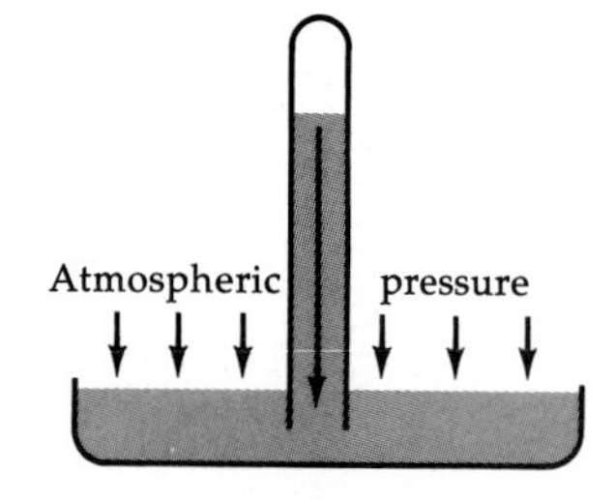

Figure 12-22. The mercury barometer.

Just as water in connecting tubes will adjust its levels until the pressures are equal (according to Pascal's principle), so will the level of mercury in the tube move, as mercury flows between tube and pan, until the pressure of the mercury in the tube is just equal to the pressure of the atmosphere pushing down on the mercury in the pan. One can say, then, that the 76 cm column of mercury supports the column of air many kilometers high. As the pressure of the atmosphere increases or decreases, mercury flows in or out of the tube until the pressures are equalized. So, the next time the weather forecaster on television reports that "the barometer stands at 75.21 centimeters," you will know what is meant.

Using mercury in the barometer is simply a matter of convenience. We could use some other liquid, water, for example. However, since water is "lighter" (less dense) than mercury by a factor of 13.6, we would need a tube 13.6 times as tall in order to accommodate enough water to provide pressure equivalent to that of mercury. In that case the weather forecaster would have to report that "the barometer stands at 10 meters"! Needless to say, such a barometer would be very inconvenient.

The pressure of the atmosphere makes possible the use of various pumps, suction cups, and other devices that depend on pressure differences for their operation. Perhaps the simplest of these is the ordinary drinking straw that you use to sip a soft drink. The popular misconception is that you somehow pull the liquid through the straw into your mouth, but this isn't how it happens. The action of your lips and mouth reduces the pressure at the top end of the straw. Atmospheric pressure on the liquid in the soft drink can is greater than this, so the liquid is pushed up through the straw.

Suction cups, such as those on the ends of the projectiles used in a child's dart gun, work by using the pressure of the atmosphere. When the cup presses against a flat surface, most of the air is forced out from under it. This means that the pressure between the cup and the flat surface is lower than that of the air on the outside; the resulting outside force holds the cup tightly to the surface.

Figure 12–23 illustrates one type of pump, called a *piston pump.* The piston is connected by a shaft to a rotating arm that moves the piston back and forth in the cylinder. As the piston moves to the right from its leftmost position, the pressure in the cylinder is reduced. The greater atmospheric pressure outside the cylinder keeps the delivery valve shut and forces water to flow into the vacancy left by the piston. When the piston starts on its return trip, its motion produces a higher pressure inside the chamber, closing the inlet valve and pushing water through the delivery valve. Repetition of this cycle continues to pump water through the delivery pipe.

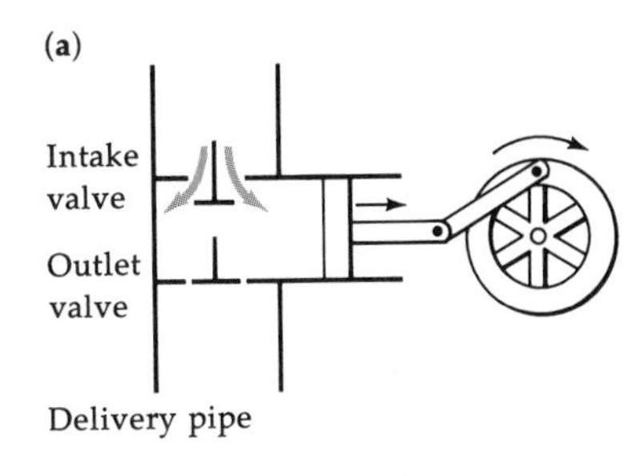

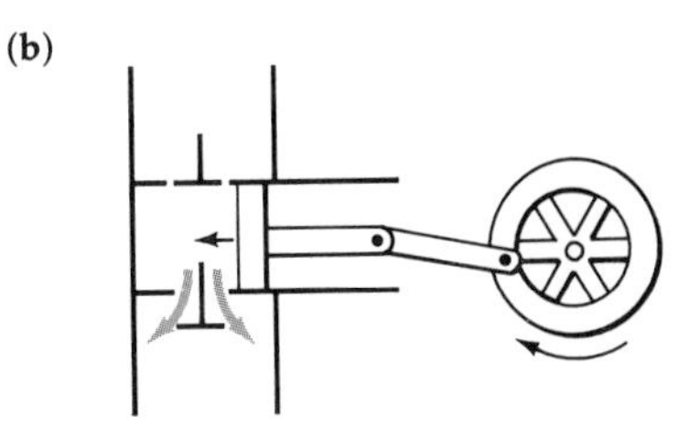

Figure 12-23. The piston pump.

Certainly the most important pump is the human heart, which circulates blood through the body. The heart muscles alternately contract and relax. During the contraction phase, the heart produces its maximum pressure, called the *systolic* pressure, which forces blood out of the heart through the branches of the arterial system. Upon relaxation of the heart, the veins return blood to the right atrium, and the heart is ready to start the cycle again. The reduced pressure during the relaxation phase is called the *diastolic* pressure.

The instrument used to measure blood pressure is called a *sphygmomanometer* (from the Greek *sphygmos,* meaning "pulse"). It consists of an inflatable cuff connected to a manometer, a device that uses the height of a column of mercury to indicate pressures. The cuff is wrapped around the upper arm at the

Example of a living pump. Clams and oysters pump nearly 400 gallons of water through their bodies daily to cleanse themselves of bacteria and pollutants.

same level as the heart and is inflated until the pressure it exerts temporarily shuts off the circulation. As the pressure of the cuff is gradually lowered, the onset of blood circulation can be detected by the characteristic thumping sound heard through a stethoscope placed at the artery in the crease of the elbow. The pressure reading at this time, when the blood is just beginning to flow again, is the systolic pressure. A typical reading is 120 millimeters of mercury. Further deflation of the cuff gives the diastolic reading, of about 80 mm, at the point where the thumping sound disappears. The blood pressure reading is given as the systolic over the diastolic, or "120 over 80" in this example.

Learning Checks

1. When the atmospheric pressure decreases, does the mercury level in the barometer tube go higher or lower? Explain.
2. Your ears "pop" when you go to higher altitudes. Explain.
3. The atmosphere becomes denser at lower altitudes, but the density of liquids is independent of depth. Explain.
4. Estimate the *force* of the atmosphere on your body.
5. Why do you suppose the weather forecaster is interested in regions of high and low pressure?

Bernoulli's Effect for Gases

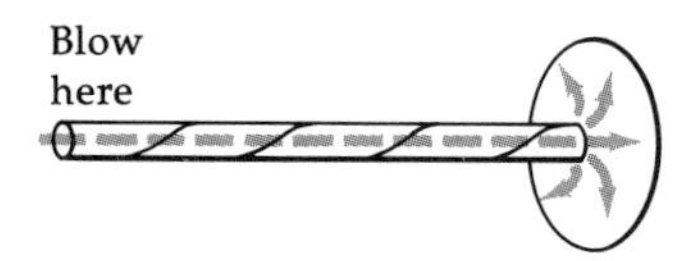

Figure 12-24. Demonstration of pressure-speed relationship.

Moving gases exhibit the same pressure-speed relationship that Bernoulli discovered for liquids. That is, as the speed of the moving gas increases, the pressure within the gas decreases. A nice little demonstration of this effect that you can perform is illustrated in Figure 12-24. Take a cardboard disc, drive a straight pin through the center, and put it into a straw or glass tube as shown. If you blow into the other end of the straw, you might expect the disc to be blown away. This does not happen. In fact, you cannot blow the disc off, and the harder you try, the tighter it clings to the straw.

The air blown into the straw escapes between the disc and the opposite end. The separation between the straw and disc is very small, so the air passing over the top side of the disc has a high speed. Bernoulli's principle tells us that the air pressure in this region will be reduced; hence the greater atmospheric pressure in the room will hold the disc in place.

Figure 12-25. The "lift" for an airplane wing is explained by Bernoulli's effect.

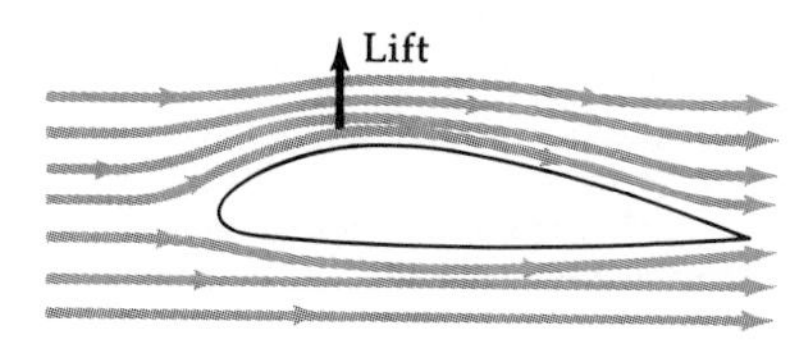

Airplane wings are designed to use Bernoulli's effect in providing the "lift" that keeps the plane off the ground. (See Figure 12-25.) The bottom side is relatively flat, while the top side is curved. Because the air passing over the *top* of the wing travels farther in going from front to back than does the air going *under* the wing, the air speed will be greater on top. This is illustrated by the bunching together of the arrows representing the wind speed. This is analogous to water flowing through the constricted region of a pipe. Hence, the air pressure is greater on the bottom, resulting in a net upward force that lifts the plane. The pilot uses the flaps on the wings to alter the wing profile and thus provide control of the wind speed on top and bottom.

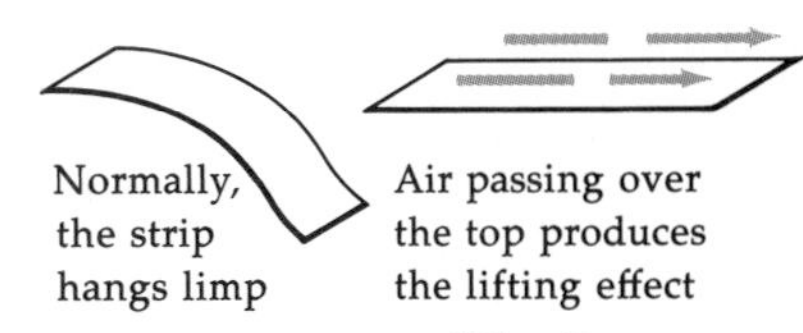

Figure 12-26. Bernoulli's effect.

Another illustration of Bernoulli's effect is shown in Figure 12-26. Hold one end of a strip of paper between your lips and blow across the top. As long as you blow, the strip will remain in an essentially horizontal position. When you stop, the paper relaxes and dangles below your chin. The air moving over the top of the strip reduces its pressure there compared to the still air on the underside. The greater pressure below forces the paper to swing up to its horizontal position, against the effect of gravity.

You can feel the results described by the Bernoulli effect when you pass a large truck on the highway. The shaft of air between your car and the truck is confined to a narrow space and hence moves faster than that on the outside. Thus the pressure is reduced in this narrow region, and you and the truck tend to swerve toward each other.

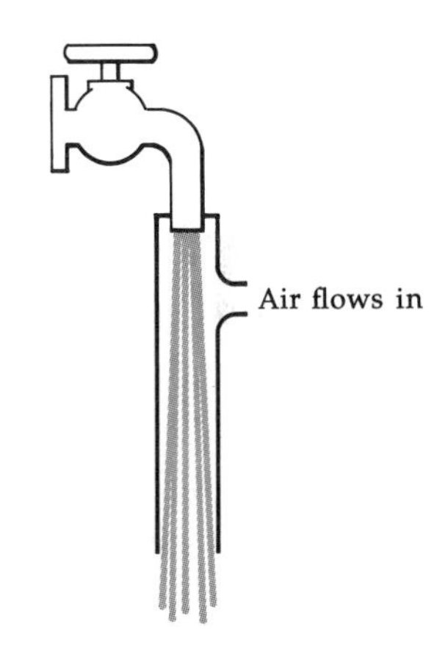

Figure 12-27. An aspirator.

A suction-producing device, called an *aspirator,* makes use of Bernoulli's effect. (See Figure 12-27.) As water passes through the pipe, the still air outside the sleeve is at a higher pressure than inside the pipe, so air rushes into the sleeve. This apparatus is often used to produce small vacuums in the laboratory.

Bernoulli's effect partially explains why a spinning baseball curves. (See Figure 12-28.) As the ball spins while moving through the air, it drags some air with it. On one side of the ball, therefore, the speed produced by the spinning motion adds to the speed of the passing air, while it subtracts on the other side. This difference in overall air speed on the two sides corresponds to a pressure difference, and the ball curves in the direction of the lower pressure. The pitcher can control this effect by changing the amount of spin he puts on the ball, its speed through the air, and the axis about which it rotates.

Figure 12-28. The spinning curveball.

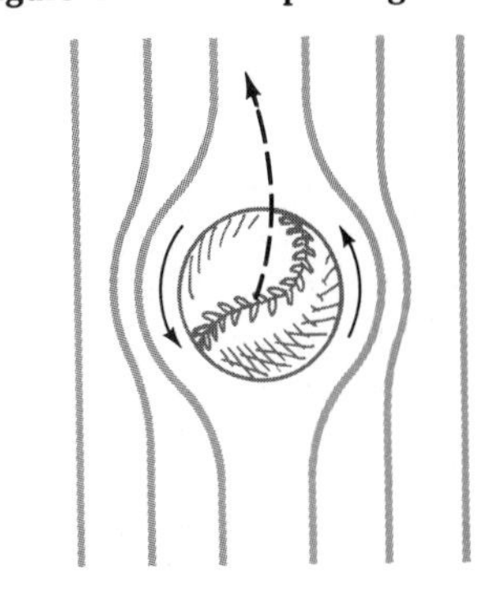

SUMMARY

Matter can exist as solids, liquids, and gases. According to the atomic hypothesis, *solids* are characterized by the individual atoms confined to fixed positions in a lattice. In *gases,* the atoms have minimal interaction with each other and are free to move about. *Liquids* represent an intermediate state—the atoms are relatively free to roam, but are also attracted to one another.

The *density* of a substance is defined as the mass per unit volume. *Pressure* is the force per unit area. For a liquid, the density is the same throughout since the liquid is incompressible. The pressure in a liquid depends on the depth:

$$\text{pressure} = \text{density} \times \text{depth} \times g$$

Pressure in a liquid obeys *Pascal's principle:* the pressure in a confined liquid is transmitted unchanged to every portion of the interior and at right angles to all the walls of the containing vessel.

The force exerted on an object partially or wholly immersed in a liquid is determined by the force of gravity and the pressure, according to Pascal's principle. The upward force of the liquid is called *buoyant force.* It acts against gravity, and so its effect is to reduce the apparent weight of an object in a liquid. According to *Archimedes' principle,* the buoyant force is equal to the weight of the displaced fluid. A completely submerged object displaces its own *volume,* while a floating object displaces its own *weight.*

The pressure in a *moving* fluid, such as water flowing in a pipe, is subject to *Bernoulli's principle:* the pressure in a moving fluid decreases as the speed increases, and vice versa.

For gases, the pressure on the walls of the container is caused by collisions of the individual gas atoms with the walls. For an *ideal gas*—that is, one whose atoms are point particles that do not interact with one another—the relationship among the pressure (P), volume (V), and temperature (T) for a gas of n atoms is

$$PV = nkT$$

which is called the *ideal gas law.* This law tells us that at a constant temperature the product of pressure and volume for a fixed amount of gas remains constant. The equation helps us understand why tire pressure will decrease as the temperature drops.

Atmospheric pressure is due to the weight of the earth's air. At sea level, it is about 10 N/cm^2, which is enough pressure to support a column of mercury roughly 76 centimeters high. Various pumps and suction devices depend on the pressure of the atmosphere for their operation.

Bernoulli's principle deals with moving gases in a way that is analogous with liquids. Again, the pressure is least in the part of the gas that is moving fastest.

Exercises

1. We have seen that the buoyant force on a completely submerged object is the same regardless of its depth in the liquid. Is the buoyant force on a dirigible in the atmosphere the same regardless of its altitude? Explain.

2. For an ideal gas at constant volume, how does the pressure vary with the temperature?

3. A toy boat has a mass of 2 kilograms. What mass of water does it displace when floating in a bathtub? What volume?

4. Archimedes' principle can be applied to gases. Use this idea to explain how a helium balloon rises.

5. Subway commuters are warned not to stand too close to the tracks because they might be sucked under the speeding trains. Is this reasonable? Explain.

6. Explain why the mercury does not run out of a mercury barometer. What would happen if the barometer were on the moon?

7. If you hold your finger over the top end of a drinking straw that has been put in a glass of iced tea, the tea will not run out when you pull the straw out of the glass. Explain.

8. Short-order cooks know that an egg resting on the bottom of a pan of water can be made to float if enough salt is dissolved in the water. Explain the physics involved.

9. Explain Bernoulli's effect in terms of kinetic energy and gravitational potential energy. (See Figure 12–17.)

10. In the story of Archimedes and the king's crown, what would have been the result of Archimedes' experiment if the crown-maker had been honest?

11. How does a life jacket help you float in water? (Hint: think in terms of density and volume.)

12. A fisherman is riding in a boat, holding a fishing tackle box. He accidentally drops the box over the side and it sinks. Does the box displace more water while he is holding it in the boat or while it is in the water? Explain.

13. In Exercise 12, what happens to the water level in the lake when the box sinks—does it rise, fall, or stay the same?

14. As a scuba diver descends into a deep lake or ocean, does the buoyant force increase, decrease, or stay the same? What about the pressure? Explain.

15. Using Bernoulli's principle, explain how a tornado causes the roof to be blown off a house.

16. Why do disaster control officials advise people to open their windows and doors during a tornado?

17. A chunk of iron and a cork of the same volume are held under water. Is the buoyant force on each different or the same? Explain.

18. An overloaded ship barely stays afloat while in the Atlantic Ocean and sinks when it moves into the Hudson River. Explain.

19. Liquid "gunch" has a density of 500 kg/m^3. What is its specific gravity?

20. What is the buoyant force on an object immersed in gunch if the object has a volume of 2 m^2? (See Exercise 19.)

21. A cube of metal 1 meter on a side weighs 48 newtons. Will it sink or float in gunch? (See Exercise 19.) What will be its apparent weight?

22. Suppose that a cube of ice (specific gravity $= 0.90$) is 10 centimeters on each side. A hotel kitchen worker places such a cube in a bucket of water. How far down will this ice cube sink? Explain.

23. If a silver ball floats on a pool of mercury, what fraction of it will be above the surface? (Specific gravities: silver, 10.5; mercury, 13.6.)

24. Consider water in its three phases: liquid, solid (ice), and gas (water vapor). In which phase is the *kinetic energy* of an individual molecule likely to be the greatest?

25. Two paramedics place you on the floor for a practice session in first aid. Estimate the area your body covers as you lie on your back. (Express your estimate in square meters.) From your weight, calculate the pressure you exert on the floor in this prone position.

26. Salt water is denser than fresh water. Compare the pressure 3 meters below the surface of the Great Salt Lake with that at the same depth in a fresh-water lake.

27. Suppose that a rock weighing 98 newtons (that is, a mass of 10 kilograms) has a volume of 0.5 m^3. What will its apparent weight be under water?

28. Explain why the top of a convertible auto bulges when the car moves at high speed.

29. Explain these two observations:
   a. A chimney "draws" better on a windy day.
   b. Cigarette smoke flows out the open window of a moving car.

30. The specific gravity of silver is 10.5. Which has the greater volume, 10 kilograms of silver or 1 kilogram of water?

31. Discuss the "evaporation" of the atmosphere in terms of the escape velocity.

32. Compare the pressure on your head if you stick it 5 centimeters beneath the surface of the water in the bathtub with the pressure at the same depth in a large lake.

33. Would a construction inspector say that, in a hurricane, the pressure of the air is greatest inside or outside a building? Explain.

13

TEMPERATURE AND HEAT

For centuries man has used heat to manipulate metals. As our ability to measure temperature has become more sophisticated, so have our applications, especially in industrial uses.

Our ideas about temperature and heat stem from the most ordinary of everyday experiences. The weather report always includes information and forecasts about the temperature. We pay attention to how warm or cool our homes are. The temperature outdoors helped you decide what to wear today. These and similar experiences show that notions about heat and temperature are very much a part of the way we understand our surroundings.

Everyone has a general feeling about temperature and heat, perhaps even regarding them as the same thing. We know that if a pot of coffee has a high temperature it will feel hot to the touch, and that ice has a low temperature and feels cold. In this chapter you are going to learn some things about heat and the temperature of substances, and how our earlier ideas about energy enter the picture. Then in Chapter 14 you will see that, in addition to the conservation law for energy, nature has some more rules of the game regarding the exchange of energy between objects. These rules are called the *laws of thermodynamics* (literally, "heat movement"). They lead us to some far-reaching conclusions about the universe.

THE CONCEPT AND MEASUREMENT OF TEMPERATURE

Because we know from experience that hot objects have high temperatures and cold objects have low temperatures, it is tempting to say simply that the temperature of a substance tells how hot or how cold it is. This is not adequate, though, because what we mean by hot and cold depends on what we are talking about. For example, suppose you are having a glass of milk and a plate of scrambled eggs for breakfast. If you get interested in the newspaper and let the food sit on the table for a quarter of

Physics and Fashion

We have no fur or feathers to cover our bodies naturally, so we must rely on clothing (and buildings) to protect us from the environment. Researchers have long studied both fully and partially attired people to gain information on the heat loss and retention characteristics of all kinds of clothing. The military originally spurred much of this research with studies of clothing insulation as it related to varying conditions of human activity and exposure. Early investigations used clothed metallic manikins heated electrically to determine heat loss from torso, head, and limbs.

Modern researchers use the thermogram, an image showing differences in infrared heat radiated from the clothing and skin of living models. This technique reveals that specially designed thermal underwear reduces heat loss, but a parka, insulated pants and boots, and a wool cap offer near total heat retention. Styles, then, are often shaped by scientific data. Down-filled clothing, for example, has moved from the realm of outdoor activity to the indoor person's world. The thermogram shows that a down vest protects as well as a business suit jacket.

The combined efforts of zoologists and heating and ventilating engineers may dictate the fashions of the future. Scientists studying the thermal properties of animal pelts are making some amazing finds. Polar bears, for instance, have fur constructed so that radiant energy is transferred to their skin at around 90 percent efficiency. (During cold weather, passive solar collectors, by contrast, operate at under 50 percent efficiency.) Polar bears are therefore kept very warm.

Researchers studying these animals speculate that the hairs of the fur act like little optical fibers that trap the sun's energy and move it down to the skin with no leakage. The skin then absorbs the light. One day this kind of knowledge could have a direct impact on the design and manufacture of winter coats. Thus, the physics of temperature and heat becomes a matter close to our own skins.

Sources: C. E. A. Winslow and L. P. Herrington, *Temperature and Human Life.* Princeton, N. J.: Princeton University Press, 1949.

"An Infrared Look at Personal Insulation," *National Geographic Special Report on Energy,* February 1981, pp. 7–8.

"Polar Bear Energy," in Dick Teresi (ed), *Omni's Continuum.* Boston: Little, Brown and Company, 1982.

an hour or so, you will probably say that the milk has become "hot" and the eggs "cold" when, in fact, they were at very nearly the same temperature—the temperature of the room. Similarly, if the temperature outside were 70° F, you would describe the day as "cool" if it came in the midst of a scorching summer but "warm" if it were in the dead of a frozen winter. So human estimates of temperature often depend on several factors and are not very reliable.

You can get a handle on the concept of temperature by considering what happens when you heat something. If you place a pan of water over a flame, for instance, you know that before long steam will rise from the surface. The water will eventually boil. If you leave the pan on the fire long enough the water will completely boil away.

This process can be viewed within the framework of the energy theory discussed earlier. As the heat is transferred from the fire to the water, there is an increase in the kinetic energy of the individual water molecules. They jiggle around faster and faster, wandering farther and farther from home as more heat is added. Soon some of the molecules are separated so far that they are no longer in the liquid state but now make up a gas, which we call steam in the case of water.

So from this point of view, it is the increase in the activity of the molecules accompanying the addition of energy to the system that is important, for it is the relative amount of motion that determines whether the substance is in the liquid, solid, or gas phase. Physicists define the *temperature* as the *average kinetic energy* of the individual molecules. The kinetic energy in this instance is often called *thermal* energy because it is associated with heating the substance. The *total thermal energy* is just the *sum* of the individual kinetic energies, while the *temperature* is the *average* of all the individual kinetic energies.

Temperature is an *intensive* property, which means that it does not depend on how much material is involved. On the other hand, the total thermal energy *does* depend on the quantity of matter. (By analogy, the average student age in a large university will not change very much from semester to semester even though the enrollment may vary by several hundred students.) Thus, the temperature of a cup of hot coffee will be much higher than that of a lake even though the lake has a greater total thermal energy because of its much larger number of molecules. This idea of temperature as an average quantity will be important when we discuss the flow of energy between two objects.

Perhaps the distinction between temperature and total thermal energy can be made more clear by an example. If you fill a thimble with hot water from a kettle, the water in the thimble will have the same temperature as that remaining in the kettle. However, the water in the thimble will have much less total thermal energy than the water in the kettle simply because of the kettle's greater amount of water. A match and a bonfire have the same temperature, although the bonfire has by far the larger amount of thermal energy.

Temperature Measurement

Because temperature is such an important quantity, scientists for a long time have been concerned with its measurement. As illustrated in the introduction of this chapter, humans are very poor estimators of temperature differences.

Objective ways of measuring temperature take advantage of the fact that the molecules at higher temperatures jiggle around faster. This greater amount of activity means that most materials expand when heated. The most commonly used substance for temperature measurement is mercury, first used in a *thermometer* (from the Greek, meaning "heat-measure") in 1714 by the German physicist G. Daniel Fahrenheit, whose name is associated with the temperature scale used in the United States and a few other countries. Mercury has some properties that make it especially useful for thermometry. Relative to water, it has a low freezing point and a high boiling point, and it expands uniformly over a large temperature range.

Having a substance to record changes in temperature, one must next calibrate the thermometer, that is, associate some numerical values with fixed positions of the mercury. It is only necessary to do this for two points and then to define the num-

Industrial Temperature Measurements

Many industries have unique temperature-measuring problems that have required specialized modifications of conventional instruments. The steel industry is a prime example. In the early part of this century, there was no really accurate way of measuring the temperature of liquid steel (which is necessary to know, if good steel is to be produced). Usually, an experienced steel melter with a practiced eye and cobalt goggles estimated when the proper furnace temperature had been reached by judging the color of the flame. Or the melter could take a small dipper of liquid steel from the furnace, pour it on the ground, and guess its temperature (which had cooled by several degrees Fahrenheit upon being removed from the interior of the furnace). Or a steel rod could be poked into the hot liquid steel, and the time required for the rod to melt could be used to approximate the temperature of the steel "broth." All of these methods were more of an art than a science.

Today, engineer-designed instruments based on traditional temperature scales give accurate readings. The most widely used device—called an immersion thermocouple—consists of a long pole with a customized temperature recorder in its quartz tip. This is inserted through a hole in the furnace door and pushed deep into the liquefied metal. Tailor-made gauges such as this one meet a great variety of industrial needs, and applied physics continues to be vital to their development for new processes.

Source: H. C. Wolfe (ed.), *Temperature: Its Measurement and Control in Science and Industry* (American Institute of Physics). New York: Reinhold Publication Corporation, 1965.

Accurate temperature measurement is a crucial factor of production in some industries. (a) Artificial diamonds are manufactured in giant presses that use a high-pressure/high-temperature system. (b) Steel making requires specially developed instruments for taking the temperature of liquid steel.

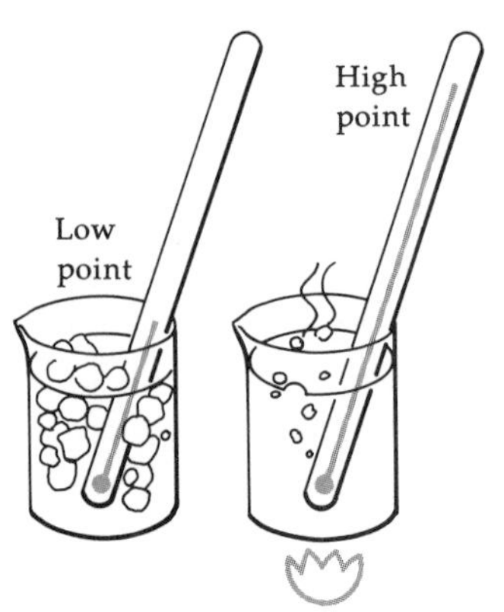

Figure 13-1. Calibration of the thermometer.

ber of units, or degrees, between them. (See Figure 13–1.) Fahrenheit selected the 0° point as the position of the mercury when the thermometer was in a bath of ammonium chloride, ice, and water. For the high-temperature point, he selected the temperature of his wife's body, averaged over several readings. This point he labeled as 96°, which is evenly divisible by 2, 3, and 4, making it useful for calculations. He later redefined the scale so that the freezing point of water was at 32° and the boiling point of water at 212°. Fahrenheit made this choice because the difference between 32 and 212 is 180, which is divisible by 2, 3, 4, and 5.

The Fahrenheit scale does have the convenient feature that in most parts of the world temperatures usually fall in the range of 0° F to 100° F. However, all of the scientific world and most countries (the United States being an increasingly lonely exception) use the *centigrade* or *Celsius* scale, named for its inventor, the eighteenth-century Swedish astronomer Anders Celsius. On this scale, the freezing and boiling points of water are chosen as 0° C and 100° C, respectively. Because these critical values are separated by 100 degree units, the Celsius scale fits nicely into the decimal nature of the metric system of measurement. The gradual move of the United States from the British to the metric system is partially shown by TV weather forecasters, who commonly report temperatures in both Fahrenheit and Celsius units. It is easy to remember that 25° C is a comfortable room temperature (77° F). Figure 13–2 shows the two scales side by side.

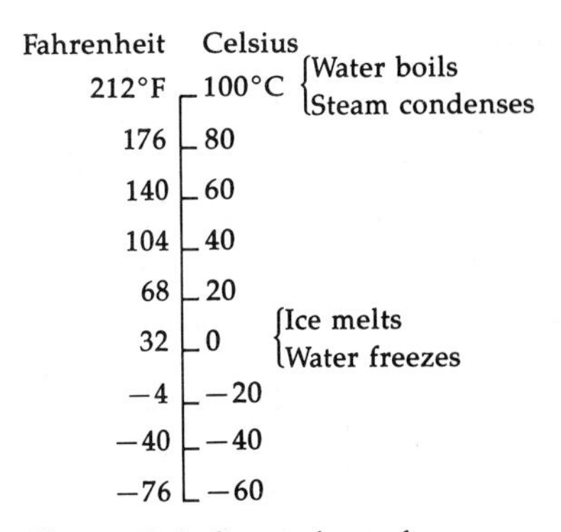

Figure 13-2. Comparison of Fahrenheit and Celsius temperature scales.

Learning Checks

1. What is the definition of temperature?
2. Does it mean anything to talk about the temperature of a single molecule? Explain.
3. Can the total thermal energy of a cold object be greater than that of a hot object? Explain.
4. What is today's temperature in °C?

Temperature and Expansion

We have said that the temperature of a substance is defined as the average kinetic energy of the molecules. Raising the temperature increases the average kinetic energy. This means that the individual molecules travel at greater speeds, because the kinetic energy is proportional to the square of the speed. It follows, therefore, that as the temperature goes up the material may expand, because the individual molecules are moving faster and are more likely to travel farther from "home." By the same argument, most materials contract when they are cooled; the lower temperature means a slower average speed for the individual molecules.

Engineers and builders have to plan for these temperature-related changes. Cracking of roads and sidewalks may be prevented by providing expansion joints—periodic gaps that allow for expansion of the concrete or asphalt. Your dental fillings are made of materials having the same expansion rate as your teeth so they won't break when you eat hot foods.

Different substances expand at different rates. It is generally true that liquids expand more than solids when heated, and gases expand more than liquids. This is reasonable because molecules in solids are more tightly bound than are those in liquids. Gas molecules are the most nearly free, and hence are more likely to occupy a larger volume when we add energy to them.

". . . and the record low for this date is 147° below zero, which occurred 28,000 years ago during the Great Ice Age."

We can conveniently express the rate of expansion by a number that tells how much the size of the object changes for each degree change in temperature. For example, the length of an iron bar increases with temperature by an amount proportional to the original length of the bar and to the change in temperature. We may express this mathematically by writing

$$L = btL_0$$

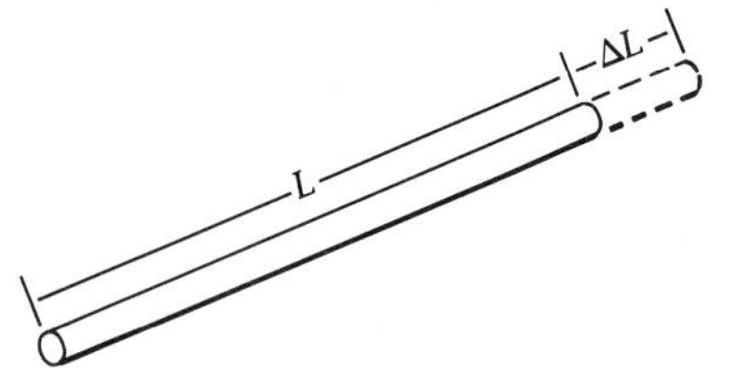

Figure 13-3. Linear expansion. When the temperature of the bar is raised by an amount Δt, the length increases by ΔL.

where L_0 is the original length of the bar, t is the change in temperature, and L is the change in the length of the bar.

The number b is called the *coefficient of linear expansion.* The new length of the bar is $L_0 + L$. (See Figure 13-3.)

Let us take the expansion of a steel rod as an example. For steel, the coefficient of linear expansion, b, is $1 \times 10^{-5}/°\,\text{C}$. This number tells us that a steel bar will increase in length by 0.00001 times its original length for every degree rise in temperature. Suppose that the rod is originally 2 meters long and that we heat it, raising the temperature by 15° C, say, from 25° C to 40° C. Putting numbers into the equation for L, we get

$$\begin{aligned} L &= (0.00001/°\,\text{C})(15°\,\text{C})(2\ \text{m}) \\ &= 0.0003\ \text{m} \end{aligned}$$

So the bar has increased in length by 0.0003 meter, having a new length of 2.0003 meters.

For a gas or liquid, the property of interest is the *volume,* which increases with a rise in temperature. An analogous equation for the change in volume is

$$V = atV_0$$

where now V means the change in volume due to the temperature change t from an original volume V_0. The constant a is called the *coefficient of volume expansion.*

Solids also undergo a change in volume when the temperature changes. This means that the material will expand along all three of its dimensions—the length, the width, and the height. For solids, the value of the corresponding coefficient of volume expansion, a, is approximately three times that for linear expansion. So the volume expansion coefficient for steel would be 0.00003 per degree Celsius.

Figure 13-4. Bimetallic strip. The brass and steel expand at different rates, giving the strip different shapes for different temperatures.

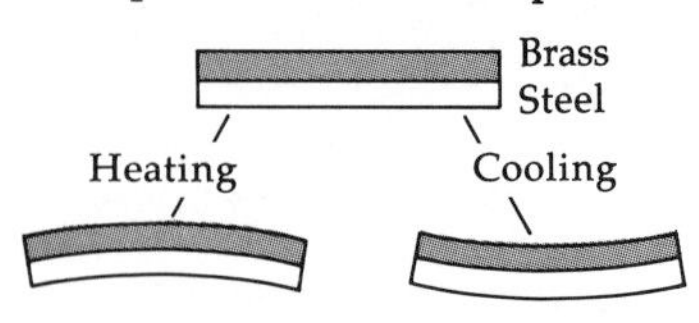

Many practical devices are based on the different rates of expansion of different materials. A common example is a *bimetallic strip.* (See Figure 13-4.) It consists of two dissimilar metals, brass and steel, for example, bonded together as a composite. The coefficient of linear expansion for brass is about twice that for steel, so the brass expands more when heated and contracts more when cooled. This makes the bimetallic strip bend in different directions for rising and falling temperatures.

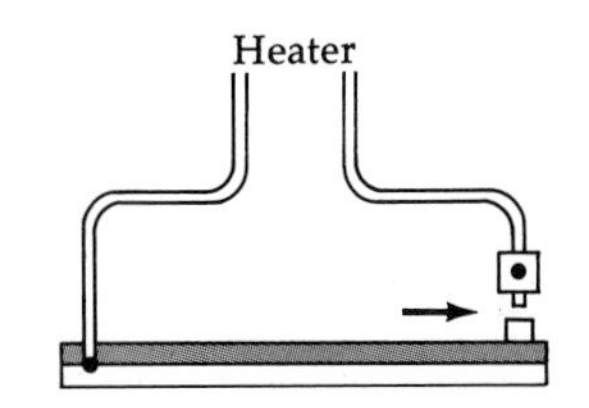

Figure 13-5. Bimetallic thermostat. The strip (arrow) responds to changes in the temperature, switching the heater on and off. When the temperature drops, the strip bends upward, and the heater starts.

Figure 13–5 shows a type of *thermostat* that uses a bimetallic strip. A thermostat is used to maintain a temperature at some desired value. As shown in the figure, when the temperature drops, the brass-steel composite will bend, closing the electric circuit and starting the heater. When the temperature rises above the preset value, the heated strip bends the other way, breaking the contact and shutting off the heater.

Thermometers are often made of a coiled bimetallic strip. (See Figure 13–6.) One end is fixed and the other is attached to a pointer that indicates the temperature on a scale. As the strip bends in response to heating or cooling, the coil tightens or loosens, moving the pointer around the scale.

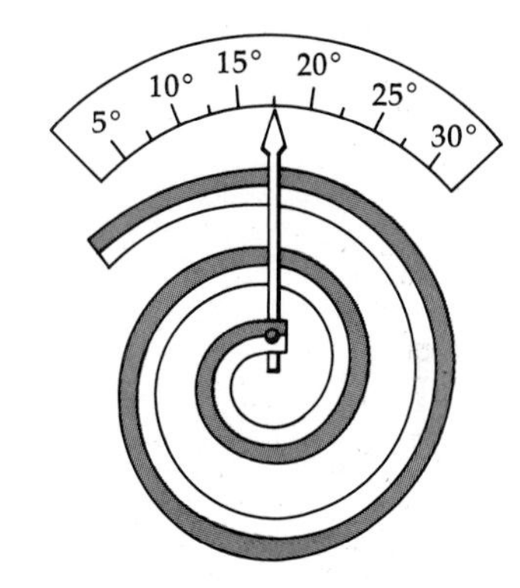

Figure 13-6. Coiled-strip thermometer.

The common mercury thermometer is useful because mercury expands faster than the surrounding glass. What would happen to the column of mercury if the glass expanded faster?

You may have noticed that when you turn on the hot water faucet, the flow of water slows down as it becomes hotter. As the water becomes hotter, the metal valve also heats up and expands, gradually restricting the water moving through the pipe. Once the valve stops expanding, the hot water flows at a steady pace.

The Expansion of Gases: The Absolute Temperature Scale

For gases, the coefficient of volume expansion is many times greater than for solids and liquids. You can understand this by realizing that the gas molecules interact among themselves much less and are therefore relatively free to move around. So when the temperature goes up, the gas molecules occupy a larger volume—much larger than that for a solid.

It turns out that the expansion coefficient has essentially the *same* value for many *different* gases. This is certainly not the case for solids and liquids, whose expansion properties depend very much on the individual substances. Representative values of a illustrate the point: for iron a is 0.000036/° C, for aluminum it is 0.00008/° C, and for gasoline it is about 0.00095/° C. Because of this variety of numbers, it is surprising that the value of a is very nearly the same for the common gases. The French chemist Joseph Louis Gay-Lussac determined a to be 0.00366/° C at 0° C. Notice that this is about 300 times larger than the corresponding coefficient for the average solid.

Now this value for a turns out to have important theoretical significance. We may write 0.00366 as a fraction: 1/273. Putting this into the expression for V gives

$$V = \frac{tV_0}{273}$$

That is, the volume change per degree Celsius temperature change from 0° C is 1/273 of the original volume. If the temperature goes up, then V is positive and the gas expands. If the temperature goes down, t is a negative number. This also makes V negative: the gas contracts to a smaller volume.

For example, suppose that the temperature drops from 0° C to 273° C below zero. Then t is -273, and we have

$$V = \frac{(-273)V_0}{273}$$

which becomes

$$V = -V_0$$

and, adding V_0 to both sides, yields

$$V + V_0 = 0$$

But the new volume is $V + V_0$, so this equation tells us that the new volume is zero! This says that if Gay-Lussac's law holds for all gases at all temperatures, their volume would shrink to zero at $t = -273°$ C. We assume that the gases are ideal; that is, that their molecules are infinitely small in size and do not interact at all with each other. Real gases do not behave this way, of course, and in fact would condense to liquids before collapsing to zero volume.

Nevertheless, this temperature of $-273°$ C is an important number, for, as first pointed out by Lord Kelvin (William Thomson) in 1848, it represents the *lowest possible temperature*, or the *absolute zero* of temperature. We thus have an *absolute* scale of temperature whose degrees are the same size as those of the Celsius scale and whose zero point is at $-273°$ C. That is, 0° C is 273 degrees above absolute zero on this new scale, called the *Kelvin scale.*

Figure 13-7. The absolute temperature scale.

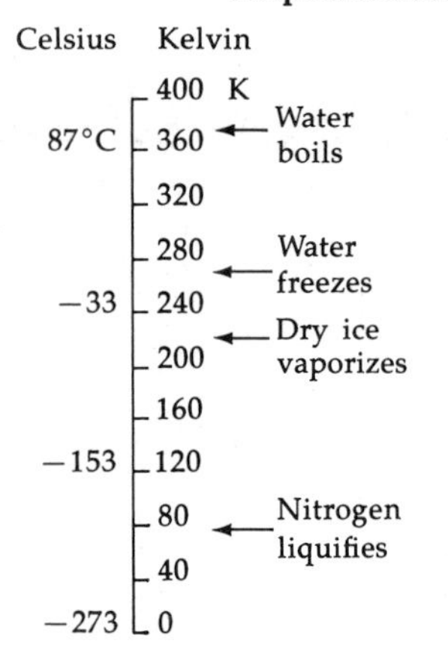

So we may conveniently represent temperature in degrees Kelvin (K) by simply adding 273 to the Celsius temperature. That is,

$$T = 273 + t$$

where T is measured in degrees Kelvin and t in degrees Celsius. On this scale, ice melts at 273 K, water boils at 373 K, nitrogen condenses to a liquid at about 77 K, and dry ice (solid carbon dioxide, sometimes called "hot ice") vaporizes at about 216 K. (See Figure 13-7.)

The Kelvin scale is important because it is an *absolute* scale, not subject to the whims of the person making it up, as are the

Celsius and Fahrenheit and other relative scales. Many important physical relationships are expressed in terms of the absolute temperature. One of these is the ideal gas equation, $PV = nRT$, which we discussed in Chapter 12.

Learning Checks

1. What is today's temperature on the Kelvin scale?
2. How much will the length of a 10 meter steel bar increase if it is heated from 25° C to 35° C?
3. Explain the use of −273° C as the absolute zero of temperature.
4. Since the volume of a gas is related to the thermal motion of the gas molecules, what can you say about thermal motion at absolute zero?
5. How does a bimetallic strip operate?

The Expansion of Water: An Anomaly

Water, one of the most familiar substances in the world, behaves with changing temperature in a way very different from most other materials. This common substance has most uncommon expansion properties.

As we have seen, solids typically expand as they are heated. With many materials, such as metals, continued heating will eventually raise the temperature to the melting point—that is, the point where the solid becomes a liquid. Because of the expansion that has taken place, the liquid occupies a larger volume than it did in the solid form. Put another way, most materials have a greater *density* in the solid phase than they do in the liquid phase. More solid than liquid can be contained in a given volume. Thus, a chunk of solid steel will sink to the bottom of a vat of molten steel.

Water, however, is unusual in this respect. Ice cubes float rather than sink to the bottom of a tumbler of water, which tells us that the density of ice is *less* than the density of water. The solid ice is in a condition of greater expansion than the liquid water, in that a particular amount of it occupies a greater volume in the solid phase.

Suppose we begin with water at room temperature and start to cool it. As we lower the temperature, the water contracts in normal behavior, continuing to decrease in volume, until we reach a temperature of 4° C. This is where the water has its smallest volume and hence its greatest density. On continuing to cool the water below 4° C, we find that it begins to *expand* again, so its density decreases. This expansion continues until the liquid water freezes into ice at 0° C. The ice thus has a smaller density than the water and, for this reason, floats rather than sinks.

This unusual behavior of water explains why lakes and ponds in the wintertime freeze from the top down rather than from the bottom up. Let us suppose the temperature of the air is 0° C on a frosty December morning. The top of a pond, exposed to the air, will cool first. When the surface temperature reaches 4° C, this part of the water will sink to the bottom, because its density is greater than the warmer water beneath it, forcing some of the warmer water to the top. When this water in turn cools down to 4° C it also sinks, again forcing some warmer water up to the surface. In this way, the entire pond will reach a uniform temperature of 4° C before any further cooling takes place. Now when the surface water cools to 3°, 2°, 1°, and finally to 0° C, it remains at the surface, having a smaller density than the 4° water below, meaning that ice begins to form first on the surface. So, even though the pond may be covered with a sheet of ice, there is still likely to be water beneath it, and fish and other aquatic life are not frozen out of their homes.

HEAT

It is clear to everyone that there is some kind of connection between heat and temperature. We speak of "turning up the heat" to maintain a comfortable wintertime temperature. In fact, it is quite common to confuse the two and use them interchangeably. However, heat and temperature are *not* the same thing. Recall that we said that the temperature of a substance is a measure of the average kinetic energy of its individual atoms and molecules. It is an *average* property, one that does not depend on how many molecules there are. So a beaker of water at 25° C and a large pond at the same temperature have the same average molecular kinetic energy.

The thermos bottle that many of us carry to job or school works by blocking the exchange of heat. A coating of silver and a sealed vacuum in the wall of the thermos bottle prevent the quickly moving molecules of a hot liquid from transferring their energy into the air. Thus, the liquid stays hot while the bottle itself feels cold.

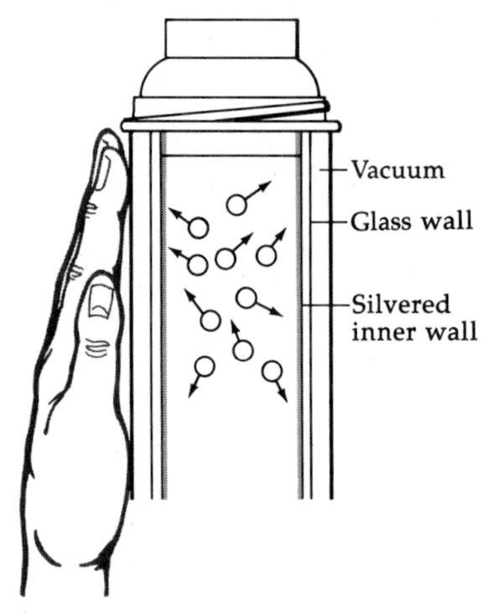

In addition, we may speak of the *internal energy* of a substance. By this we mean the total kinetic energy of all the molecules plus any energy involved in holding the individual atoms together—the energy in the chemical bonds. The internal energy refers not only to the energy of motion the molecules have, the average of which is the basis of the material's temperature, but also to any other energy the substance contains.

The word *heat* is used, not in reference to the internal energy a system contains, but in connection with *changes* in internal energy—gains or losses. The definition is this: *heat is the energy transferred between two substances because of a difference in temperature.* The phenomenon of heat is a very common experience. Ice in a glass of tea melts because of the energy that passes from the warmer tea to the cold ice cubes. This energy is heat. Place your hand in a bucket of ice water. Your hand becomes cold due to the "flow of heat" from your hand to the water—again because of the temperature difference. We might think of the temperature difference as a sort of "driving force" that transfers energy from a warm object to a cold one. And *heat* is the name given to the energy involved in this transfer.

Notice that heat refers to a *process*, not to a condition of the system itself. Thus, something does not *contain* heat. It contains internal energy, and the "flow of heat" means that the system either gains internal energy by contact with something at a higher temperature or loses it to something at a lower temperature.

The flow of heat—that is, the transfer of internal energy from one substance to another—always takes place spontaneously from an object of higher temperature to one of lower temperature. Note carefully that this flow is not *necessarily* from the body containing the greater amount of internal energy to that containing the lesser amount, although it could be. The point is, the energy content doesn't matter. It is the *temperature* difference that counts, not the amounts of internal energy in the two materials. For example, if a cup of hot coffee sits on a large block of ice, the heat flow is from the coffee to the ice, and not the other way around, even though the energy content of the ice is greater due to its larger number of molecules.

The Old Idea of Heat as a Fluid

Because the phenomenon of heat flowing between bodies at different temperatures has been noticed for centuries, it was quite natural for the ancients to develop the notion of heat as a substance, a material stuff, some kind of fluid. The idea of a "heat fluid" goes at least as far back as ancient Greece, in concert with the Aristotelian world view of the time. We mentioned earlier the belief that all the world was made up of various combinations of the "elements" earth, air, fire, and water. So, whenever a substance was heated, it simply took on a little extra "fire."

Scientists have long since abandoned this highly simplistic world view of Aristotle, but the concept of heat as a material substance survived for many centuries. James Black was a Scottish physician of the mid-eighteenth century who dabbled in topics of physics and chemistry that were of interest to him. He pictured heat as a definite physical entity, which he dubbed "calor." He viewed "calor" as an imponderable fluid, capable of interpenetrating all material bodies, increasing their temperatures. For example, when he mixed equal portions of boiling water and ice water, Black discovered that the final temperature of the mixture was very nearly halfway between the initial temperatures of the premixed water samples. His interpretation was that the excess "calor" contained by the boiling water was equally distributed between the two portions during the mixing process.

Benjamin Thompson (1753–1814) was the first to challenge this notion of heat as a material fluid. Thompson was a Massachusetts-born soldier who fought in the American Revolution and afterward lived in England as a Tory exile. He was later titled as Count Rumford in recognition of his reorganization of the German army during a stint as the minister of war in Bavaria.

During all his military activities, Count Rumford maintained a keen interest in scientific problems, especially the nature of heat. It seemed to him that Black's heat fluid could not be a correct interpretation. "Calor," if it existed, would seemingly have had to be created out of nothing in friction processes, for example, which are distinct from heat-producing chemical reactions. While observing the boring of cannon in 1798 at a munitions factory in Munich, Rumford noticed that the boring process generated large amounts of heat, enough to boil water. The more boring took place, the more water could be boiled, almost as if the cannon-borer system contained an infinite amount of heat. Rumford offered the interpretation that the friction of borer against cannon metal was transformed into internal motion of smaller parts of metal and borer, and that it was this motion that was the source of the heat.

Rumford and the Heat Physics of Cookware

Two of Count Rumford's inventions display how clever he was when it came to the practical management of heat. First of all, he invented a portable oven for use in the field by soldiers (a). Instead of an open fire, Rumford suggested a foundation of stone or brick, with a cooking container over the foundation. Food was placed in a pan surrounded by water. This steamed the food, retaining its nutritional value and flavor.

Rumford also designed a double boiler (b). Steam from the receptacle on the bottom passed through a tiny opening; the top pot did not sit in boiling water. This made sauces easier to cook because they did not need to be continually stirred.

The principles of both of these inventions have been applied to cookware commonly found in contemporary restaurants and home kitchens.

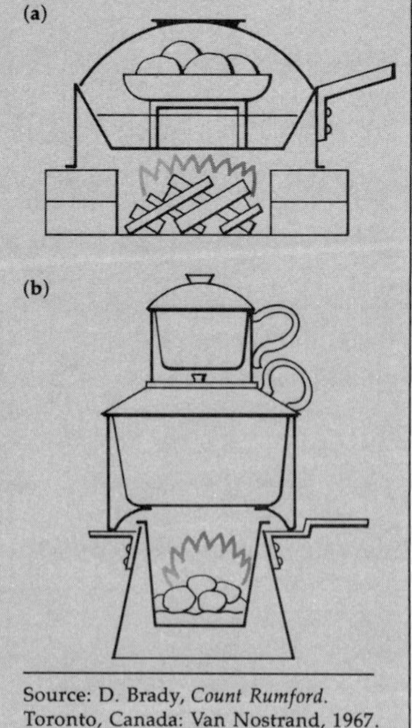

Source: D. Brady, *Count Rumford.* Toronto, Canada: Van Nostrand, 1967.

Learning Checks

1. How are heat and temperature related? How are they different?
2. Is it true that heat always flows from a body of higher energy content to one of lower energy content? Explain.
3. What does it mean to say that internal energy is an *extensive* property while temperature is an *intensive* property?
4. Discuss how we might today account for Black's observation that the final temperature of an ice-water mixture is halfway between the original temperatures of the ice and water.

Heat Capacity and Specific Heat

Although Black's picture of heat as a fluid material does not square with the modern view, many of the concepts he introduced are useful and important today. His observations that cooling a certain amount of water by 1 degree releases more heat than cooling an equal weight of mercury by 1 degree led him to the notion of *heat capacity.* To say that two substances have different heat capacities means that different amounts of heat are required to change the temperature of the two materials.

For example, if you place a pan of water on the electric range and turn on the current, in a very short time the metal coil will be too hot to touch. However, the water in the pan will stay cool enough so that you can stick your hand into it for a much longer time. The water has a greater capacity for absorbing heat without getting hotter—that is, without a temperature increase—than does the metal burner. For the same reason, once the water has been heated it requires a much longer time to cool down to room temperature than does the metal. The water maintains its high temperature due to its greater heat capacity.

A more useful term for comparing heat capacities among various substances is the *specific heat,* defined as the amount of heat required to raise the temperature of a unit mass of the substance by 1 degree of temperature. So specific heat is simply heat capacity per unit mass. The name given to the quantity of heat is a holdover from Black's concept of "calor": we define the *calorie* as the *amount of heat necessary to raise the temperature of one gram of water from 14.5° C to 15.5° C,* that is, by 1° C from a given original temperature. (The temperature range 14.5°–15.5° is specified because it turns out that the specific heat of water is different at different temperatures, although just slightly.) Therefore, the specific heat of water is 1 calorie; the numerical values of the specific heats of some other materials give a direct comparison of their heat capacities relative to that of water as a standard.

Table 13–1 lists specific heats for various substances. From this table we can see, for example, that it takes 100 calories to raise the temperature of 100 grams of water by 1° C, while it takes only 21 calories to raise the temperature of 100 grams of aluminum by the same amount, 1° C. It requires nearly five times as much heat to effect the same temperature change in a given mass of water as it does in an equal mass of aluminum. Another way of looking at this table is to note that, for example, a calorie of heat added to a gram of aluminum will raise its temperature by 1/0.21, or 4.5° C, while a calorie added to a gram of steel will raise its temperature by 0.9° C and that of water by only 1° C. (See Figure 13–8.)

By the way, the "calories" quoted on the labels of food packages for the benefit of weight watchers are actually 1000 calories as we have defined the term, or 1 kilocalorie (kilo = 1000). Sometimes the distinction is made by writing Calorie (with a capital "C") when used in this context. So 1 kilocalorie equals 1 Calorie, which equals 1000 calories.

You can see from our discussion that temperature and heat are very different things. The distinction between them is quite important. Different materials, in the same quantities, absorbing equal amounts of heat will undergo very different changes in temperature. So the temperature is useless as a measure of heat flow.

Table 13-1
Specific Heats for Various Substances

Material	Specific Heat (cal/g ° C)
Water	1.00
Aluminum	0.21
Steel	0.11
Glass	0.16
Sugar	0.27
Wood	0.42

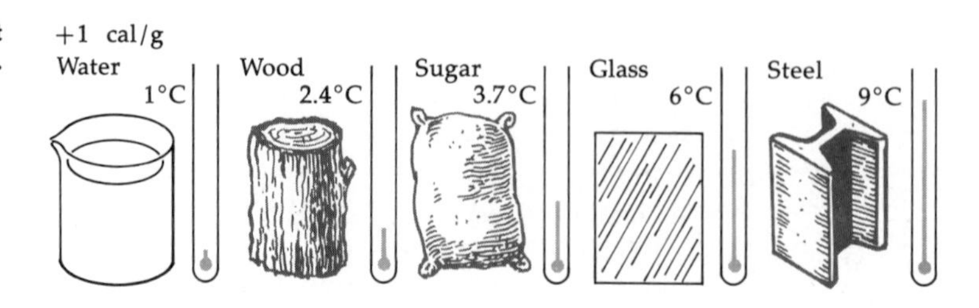

Figure 13-8. Different heat capacities.

Heat Capacity of Water

The fact that water has such a high heat capacity relative to other substances makes it useful as a cooling material, as in the engines of automobiles. Circulating water through the engine block provides a mechanism for removing potentially damaging heat from the motor. Water with its high specific heat furnishes more cooling for a given mass than most other liquids. Power plants and other industrial units also use water as a coolant.

You may have noticed that in the early springtime the temperature outside is high enough for swimming but the water is still too cold. Again, this is an illustration of the high heat capacity of water. The air warms up more quickly than does a body of water the size of a swimming pool or lake, and hence the water is at a lower temperature for a longer time. The opposite effect takes place at the tail end of the water-sport season—swimming in the autumn is usually comfortable as long as you stay in the water, which is likely to be warmer than the surrounding air. The water retains the summer's warmth by virtue of its high specific heat.

The proximity of land areas to large bodies of water has a very pronounced effect on the local climate. For example, Los Angeles and the Texas Panhandle are at about the same latitude. However, the prevailing westerly winds blowing over Los Angeles from the Pacific Ocean keep its wintertime climate much warmer than that in the Panhandle, which is not blessed with the warming presence of a body of water. The high specific heat of water makes the Pacific Ocean a vast thermal reservoir for the entire West Coast.

Learning Checks

1. Explain the distinction between *heat capacity* and *specific heat*.
2. Compare the number of calories required to raise the temperature of a gram of aluminum by 10° C with that required to raise the temperature of 10 grams by 1° C.
3. Why is the climate more temperate near a large body of water than in a landlocked region?

Latent Heats and Change of Phase

We have seen that 100 calories of heat added to 1 gram of water at 0° C will raise the temperature to 100° C, reflecting the fact that at the higher temperature the water molecules have greater average kinetic energy. But now suppose we have a gram of *ice*

Heat Physics and Oceanography

The majority of the solar radiation reaching the earth is captured and stored in the tissues of photosynthetic organisms and in the surface waters of the oceans, which cover 71 percent of our planet. Both these sources are eyed for their energy content.

In the case of the oceans, a concept called *Ocean Thermal Energy Conversion* is being used to develop ways to tap this heat energy to produce electricity. In one version, a system would be constructed off the shores of tropical islands. The 80° F surface waters by the shore would be channeled through a vacuum chamber and turned into steam, which would then be used to turn the blades of a vapor turbine generator to yield electricity.

Source: "Harvesting Ocean Heat," *Science Digest*, January/February 1981, p. 42.

at 0° C and we add heat to it. What happens? Well, we know that eventually the ice will melt, and experiments show that it takes 80 calories to turn that gram of ice into a gram of water, also at 0° C. That is, even though we have added a fairly large amount of heat to the ice—enough to raise the temperature of another gram of *water* from 0° C to 80° C—the temperature of the *ice* has not changed. Rather, all this energy has gone into a *phase change*, carrying the water from its solid form, ice, to the liquid form, still at 0° C. In fact, if we add heat to a mixture of ice and water at 0° C, we find that the temperature will not rise until all the ice has melted. (See Figure 13–9.)

Figure 13-9. The ice-water phase change.

Clearly, then, the heat required for the melting of the ice in these examples has not changed the average kinetic energy of the molecules at all, since there is no change in temperature. But the extra energy cannot have simply disappeared. So the question is, where has it gone?

The answer lies in the arrangement of the water molecules in the two phases. Recall our earlier discussion in which we pointed out that a solid has a regular structure in which the individual molecules are locked into relatively fixed, rigid positions by their mutual interaction. Liquids, on the other hand, are not so well defined as to shape, since the interactions among molecules no longer dominate. During the melting process in which ice is converted into water, the heat absorbed by the ice appears as energy that frees the water molecules from the rigid ice structure. The intermolecular attraction is reduced.

One way to say this is to speak of increasing the potential energy between molecules. Since the temperature is a measure of the kinetic energy—the energy of motion—we know that the heat absorbed in the melting process does not increase the average kinetic energy, because the temperature does not change. What does change is the other part of the internal energy, that is, the *potential energy*—the energy the molecules have because of their positions relative to each other. The molecules absorb energy by getting farther apart, and this increases the potential energy. This is analogous to making the gravitational potential energy of a football larger by lifting it—moving it farther away from the earth.

The 80 calories required for melting 1 gram of ice is called the *latent heat of fusion,* another term due to James Black. It is also the amount of energy that 1 gram of water at 0° C loses to its surroundings on freezing to ice at 0° C, because freezing and melting are inverse processes, two sides of the same coin. Here, some of the energy that the water molecules possess is lost, not by a temperature change but by a decrease in the potential energy. The attractions between molecules halt their rather chaotic motion and force them into the rigid positions of the ice structure. Notice that we have not changed anything about the individual molecules themselves. What has changed as a result of the gain or loss of energy are their positions relative to and attractions for one another.

We illustrate in Figure 13-10 how much energy is involved in the latent heat of fusion for water. It takes as much energy to melt a gram of ice without changing its temperature as it does to raise the temperature of a gram of water from 0° C to 80° C.

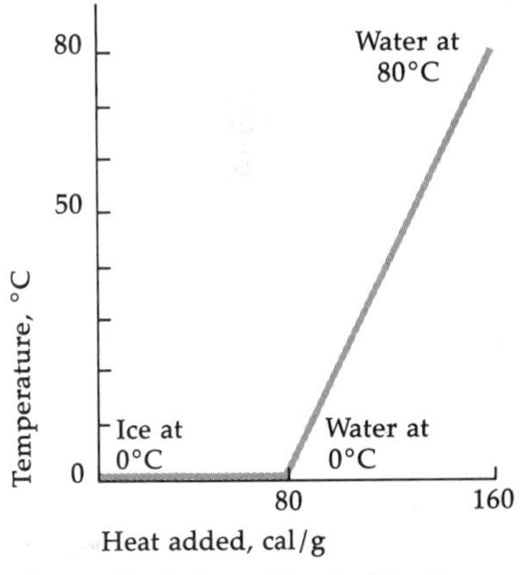

Figure 13-10. Latent heat of fusion.

Suppose, now, that we have our 1 gram of liquid water at 0° C and we continue to add heat. The temperature will rise by 1° C for each calorie of heat added, until we get to 100° C. Here we reach another phase change. Any additional heat will not increase the temperature but will instead go toward converting the liquid water into a gas, steam. Just as in the solid-liquid phase transition, we find that there is a *latent heat of vaporization.* By this we mean a quantity of heat that increases the potential energy of the molecules and reduces the forces of attraction among them until they are far enough apart and free enough of one another's influence to be called a gas. In the case of water, this latent heat is quite large, 540 calories for each gram. That is, it takes nearly seven times the amount of energy to overcome the mutual attraction of the water molecules in the liquid-vapor transition as it does for the solid-liquid transition. A gram of steam at 100° C contains 540 more calories than a gram of water at the same temperature.

Or, put another way, it requires 640 calories to convert a gram of water at 0° C to a gram of steam at 100° C, 540 of those calories—almost 85 percent—being absorbed after the water has already reached 100° C, the boiling temperature. This indicates to us that the forces of attraction among individual water molecules in the liquid phase are quite strong even though the liquid is characterized by relatively random motion of the individual constituent molecules. Only a small fraction of the added heat appears as an increase in kinetic energy (temperature change); by far the greater portion goes into overcoming the intermolecular forces.

Evaporation as a Cooling Process

We have seen that liquids are characterized by fairly strong attractive forces between individual molecules, whereas in the gaseous state one molecule hardly knows the others are present. As illustrated in the case of water, we can change the liquid into a gas by heating, which provides the energy necessary for the molecules to break away from each other. But the liquid can change to the gas very slowly by the process known as *evaporation.* If you have a shallow dish of water sitting out all night, most or all of the water will be gone by morning, having evaporated into the air.

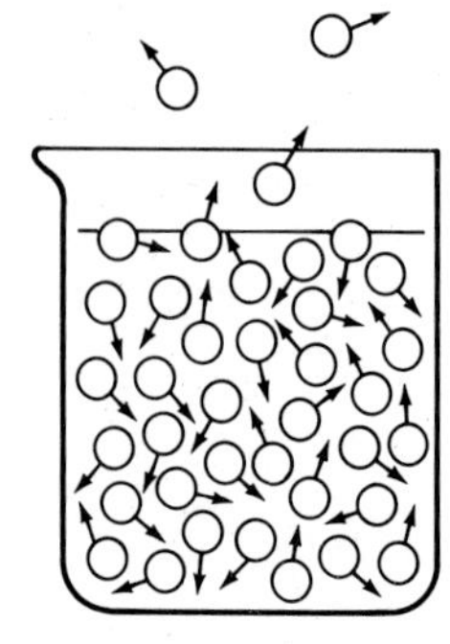

Figure 13-11. Evaporation: A few molecules escape the liquid.

Let us look at what has happened. As the water molecules jiggle about and tumble over one another in the liquid, they will have random directions and various speeds of motion. Most will have the speed that corresponds to the water temperature—its average kinetic energy—but some will be moving faster or slower than this average speed. Occasionally a few at the surface will have enough speed to break away from the liquid and escape into the surrounding air. (See Figure 13–11.)

Notice that it is the fastest molecules that escape; thus the average speed decreases. (This is analogous to the crowd at a party: if the older people leave, the average age of those remaining drops.) Hence, the average kinetic energy decreases and the temperature drops. So evaporation is a cooling process.

When you step out of a shower the bathroom feels much colder than before you got wet. As the water evaporates from your body it lowers the temperature of your skin and you feel cool. The temperature of the room may have remained the same, but the evaporation of the water from your skin wasn't taking place before you got wet.

A breeze on a warm day makes you feel cooler than you would if the air were still, even at the same temperature. The effect of the breeze is to evaporate the moisture continuously from your skin. This is why a fan can make you more comfortable even though it doesn't lower the temperature of a hot room. Television weather reporters often quote a *wind chill factor;* this is the effective temperature that a particular combination of air temperature and wind speed produces as an evaporation cooling effect on the skin.

Figure 13-12. The Japanese dunking bird.

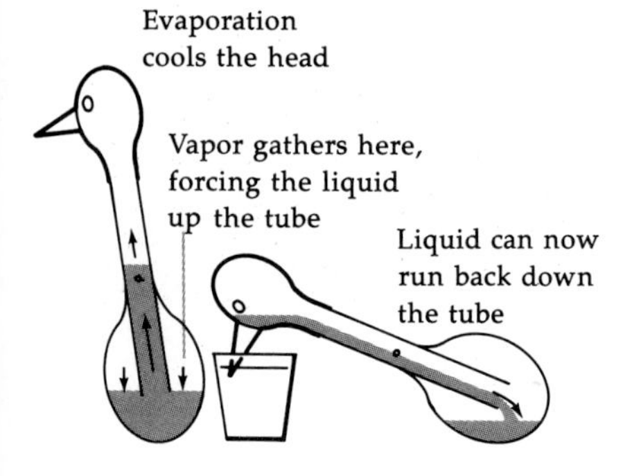

Condensation is the reverse of evaporation. Gas molecules are captured at the surface of the liquid. When fast-moving water vapor molecules strike a cold surface, they lose some of their kinetic energy as they slow down. A large number will coalesce into liquid drops, such as on the cold surface of an iced drink. The window panes in a warm room will "sweat" on a cold day for this same reason.

An amusing application of the cooling effects of evaporation is the Japanese dunking bird. (See Figure 13–12.) The head is covered with felt that is kept moist to make the bird operate. Connecting the head to the lower body is a tube, the bottom of which is submerged in a pool of ether or some fast-vaporizing liquid. The lower part of the bird's body is warmer than the head (because the head is evaporating water), so the ether in the pool evaporates faster than that inside the tube. The increasing pressure of the ether vapor above the pool forces the liquid ether up the tube, until enough has accumulated to make the head heavier than the body. When this happens, the bird swings over on the pivot. The bottom end of the tube is now above the level of liquid in the body and the ether can run back down. This makes the body heavier than the head: the bird straightens up and the process starts all over.

Each time the bird bends down, its nose dunks into a beaker of water, and in this way the head is continuously rewetted. As long as the head remains cooler than the body—that is, as long as water evaporates from the head—the bird will continue to repeat its performance.

Evaporation losses in reservoirs affect the water supply available for crop irrigation, for power production, and for municipal and industrial uses. Thus, evaporation rates are carefully monitored by hydrologists.

Learning Checks

1. What is a change of phase?
2. Explain the terms *latent heat of fusion* and *latent heat of vaporization.*
3. Why is evaporation a cooling process?

Transfer of Heat

You have probably had the experience of holding a metal rod over a flame until the end you are holding is too hot to touch. The heat absorbed by the end of the rod near the flame is transferred along the rod to your hand by a process called *conduction.* (See Figure 13-13.) We may picture this rather crudely in the following way. As the end of the metal rod is heated, the atoms vibrate more and more vigorously about their average location. The atoms and electrons in the hottest portion of the rod vibrate most energetically. They jostle their next-door neighbors, and these atoms and electrons begin to vibrate more rapidly also. They in turn disturb *their* neighbor atoms, and so on along the rod. In this manner, the kinetic energy is passed from one segment of the rod to the next, a sort of domino effect.

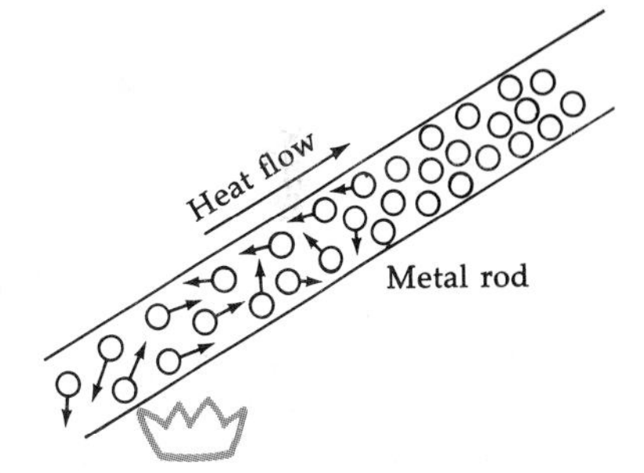

Figure 13-13. Energy is transmitted by conduction along the rod.

Various materials conduct heat to a greater or lesser extent, depending on the structure of the individual atoms and molecules. Solids such as metals conduct heat well because the outer electrons on the individual atoms are relatively far removed from the nucleus and hence are bound rather loosely. As a result, the metal is characterized by a sea of rather free-floating electrons that are tied to no single atom in particular. The kinetic energy generated by a flame is thus more readily distributed throughout the metal by these moderately free electrons. So metals are good conductors of heat. Substances such as glass, plastic, wood, and paper, in which the electrons are bound more tightly, are poor conductors of heat. These are called *insulators.*

Air is a very poor conductor of heat. This makes air an excellent insulator, so long as it remains still. Storm windows provide insulation for homes by trapping a layer of air between themselves and the regular windows, preventing heat from escaping from the house to the outside. Attic insulation made of fiberglass serves the same purpose. It is constructed in such a way as to trap air in little pockets, across which heat is not conducted very well. Styrofoam cups make fine containers for coffee for two reasons—the plastic material itself is a good insulator, and so is the air trapped internally by the construction of the styrofoam.

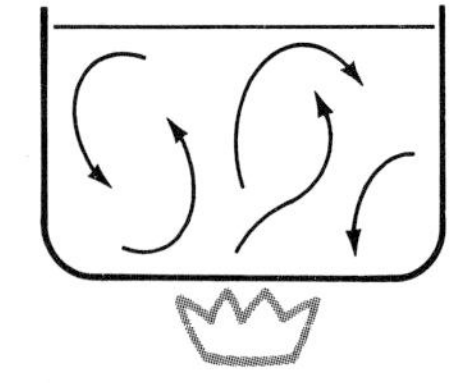

Figure 13-14. Convection in water being heated.

A second way in which heat is transferred is by *convection,* a process in which energy is distributed by currents of liquid or gas. (See Figure 13-14.) When you heat water in a pan on the stove, the water at the bottom gets hot first. It expands, becoming less dense than the cooler water above. This cooler water moves to the bottom and is heated, then it is replaced by cooler water from above it, and the process repeats itself continuously. The turbulence serves to stir the liquid until it is eventually heated uniformly.

In both of these heat transfer mechanisms, conduction and convection, we can account for the energy distribution in a

The Complexity of Simple Convection

Convection is a significant phenomenon that appears in situations easily observed at home (the roiling of a heated soup, the draft drawn up the flue of a fireplace) and easily determined in the world at large (convection flow is behind the great ocean currents and the global circulation of the atmosphere). Less familiar examples also show the vital role of this process: In industry, it plays a part in the drying of paint films; in the lungs, it influences the dispersal of gases; in the earth's mantle, it is evidently behind the slowly migrating continents.

Although extensively studied for over two hundred years, convection has not given up all its secrets. It may seem surprising, but, according to Velarde and Normand,

> *. . . the formulation of a detailed and quantitative account of convection has proved a lasting challenge to the ingenuity of theorists. Indeed, even the simplest system undergoing vigorous convective motion cannot yet be given an exact mathematical description.*

Since much of physics rests on a mathematical foundation, this state of affairs is unique in the field.

Source: M. G. Velarde and C. Normand, "Convection," *Scientific American,* July 1980, pp. 93–108.

mechanical way. Both involve the interaction of more energetic matter with material having less energy, permitting energy transmission by direct contact. There is a third mechanism, however, called *radiation*, that does not involve contact of matter at all. The most important example of this kind of energy transfer is so familiar it is likely to be overlooked: the heat that comes to us from our sun.

Separated from us by some 93 million miles of a vacuum more perfect than any we can manufacture in the laboratory, the sun provides us with most of our thermal energy. Because the transfer of this energy takes place through empty space, it must involve a mechanism that does not require any material. We shall discuss radiation in some detail when we study the properties of light.

"Let's go over to Celsius' place. I hear it's only 36° over there."

SUMMARY

The *temperature* of an object is defined as the *average kinetic energy* of the molecules making up the substance. Temperature is an *intensive*, or average, property.

The measurement of temperature involves observing changes in the properties of substances, such as mercury in a familiar thermometer. Two relative temperature scales are the *Fahrenheit* and *Celsius* scales, the latter being compatible with the metric system.

The *coefficient of linear expansion* is a measure of the fractional change in the length of an object for one degree of temperature change. The *coefficient of volume expansion* is the analogous parameter for changes in volume with temperature. Gay-Lussac's discovery that all gases have virtually the same coefficient of cubical expansion led Lord Kelvin to the notion of the *absolute zero* of temperature. This temperature is $-273°$ C. It represents the lowest possible temperature.

Heat is the energy that is transferred between objects at different temperatures. The flow of heat between objects always spontaneously takes place from high to low temperature. The *heat capacity* of a substance measures the amount of energy necessary to change the temperature by one degree. *Specific heat* is the heat capacity per gram. Water has a high specific heat; metals have low specific heats. Specific heat is measured in *calories*, the amount of heat required to raise the temperature of a gram of water from 14.5° C to 15.5° C.

Latent heats refer to the energy involved in phase changes at a given temperature. The *latent heat of fusion* for water is the heat required to convert a gram of ice at 0° C to a gram of water, also at 0° C. It is equal to 80 calories. The *latent heat of vaporization* is the analogous quantity for converting water to steam at 100° C; it is equal to 540 calories per gram.

Heat transfer can take place by three methods: conduction, convection, and radiation. *Conduction* depends on direct contact, and is effective in materials that have rather free electrons. *Convection* transfers heat by bulk movements of materials, such as the turbulence during the boiling of water. We will take up *radiation* in Chapter 22.

Exercises

1. Is it possible for heat to flow from an object with high internal energy to one with low internal energy? Explain.

2. The mercury in a thermometer expands when it is heated. What happens—does more mercury appear, or does the distance between molecules increase?

3. A recycling center removes the stuck metal lids from jars by warming them under a stream of hot water. Why does this make the lids easier to remove?

4. At the bottom of a waterfall, the water temperature is slightly higher than at the top. Explain.

5. How does the size of the degree on the Fahrenheit scale compare with that on the Celsius scale?

6. When a waitress places a cold metal spoon into a cup of hot coffee, why does the spoon get warmer and the coffee get cooler?

7. Suppose that a pond were somehow cooled at the bottom rather than at the top. Would the pond freeze from the bottom up in this case? Explain your answer.

8. One February, a bridge builder builds two bridges, one in San Francisco and one in Milwaukee. Which, if either, will need the larger expansion joint? Explain.

9. When a metal "doughnut" is heated, does the hole increase in size, decrease in size, or stay the same?

10. Natural gas is sold to homeowners at a certain price per cubic foot. Is the homeowner likely to get a better deal in the winter or in the summer? Explain.

11. If glass expanded faster than mercury when heated, what would happen to the mercury level in a thermometer as the temperature went up?

12. A machinist is asked to assemble a metal ring over a metal cylinder with a snug enough fit so that the ring cannot be removed. How can he use thermal expansion to help?

13. On a warm day, you fill the gasoline tank of your automobile almost to the top. Later you discover that the gasoline has overflowed. Explain what has happened.

14. A chemist pours boiling water into a glass beaker having thick walls and then into one having thin walls. Which beaker is more likely to break? Explain.

15. Suppose that the heat capacity for water were only half its actual value. Would this make ponds more likely or less likely to freeze? Explain.

16. True or false: Any two buckets of water at the same temperature will both contain the same amount of thermal energy. Explain your answer.

17. To keep your feet warm in bed at night, is it better to use a hot water bottle or a chunk of metal of the same size and heated to the same temperature as the water? Explain.

18. A lifeguard stands on a beach, where the sand is uncomfortably hot. To cool her bare toes, she wades into the ocean. Explain the temperature difference between the sand and the water.

19. Why does the temperature in Kansas City vary so much more between winter and summer than the temperature in San Francisco?

20. Why is fiberglass a good insulator?

21. Why do storm windows have two glass panes separated by a thickness of air?

22. Why does a tile floor seem so much colder to your feet than a rug at the same temperature?

23. Oil field workers in Alaska wear down-filled parkas. Why are these coats rendered useless if they get soaked?

24. Can you heat a material without changing its temperature? Explain, and give an example.

25. Water in a canteen will stay cool if the cloth jacket is kept moist. Explain.

26. Why is it useful for your body to perspire when you exercise?

27. A very large percentage of your body is water. Explain how this helps you maintain a body temperature of 37° C when you are in a room either warmer or cooler than this.

28. What temperature on the Celsius scale would correspond to absolute zero if the coefficient of volume expansion for gases were 0.00357/° C, rather than the accepted value of 0.00366/° C?

29. What does it mean, in terms of the temperature, to say that one object is twice as hot as another object?

30. Suppose that metal A is at 0° C and metal B is twice as hot as metal A. What is the temperature of metal B on the absolute scale? On the Celsius scale?

31. Five grams of water cools from 25° C to 15° C. How many calories are released? By how much would the temperature of 10 grams of aluminum be increased by this heat? (See Table 13–1.)

32. Two grams of ice at 0° C are placed in a pan on the stove and heated until all the ice has become steam. How many calories of energy are absorbed by the 2 grams during this entire process?

33. What will be the temperature of a gram of water originally at 10° C after 50 calories of heat are added to it?

34. The heat of vaporization for water is 540 cal/g. Approximately how many grams of ice at 0° C could be melted by this amount of heat?

14

THE LAWS OF THERMODYNAMICS

The science of thermodynamics began taking shape early in the nineteenth century when transformations of energy in heat engines were studied.

In Chapter 13 we studied the concepts of heat and temperature. We saw that the temperature of a substance indicates the average kinetic energy of its individual molecules. And we learned that heat is the energy that moves from one object to another as a result of the temperature difference between them. In this chapter, we are going to focus on this movement of heat from one system to another. Nature has some very important rules—limitations, really—regarding the energy of systems and how the energy flows from one to the other. The formal name given to these rules is the title of this chapter.

THE FIRST LAW

Take a pitcher of cold lemonade from the refrigerator and set it out on the kitchen table. The temperature of the lemonade will begin to rise, since the room air is warmer than the cold lemonade. In the terminology we developed in the last chapter, there is a flow of heat because of the temperature difference between the room and the lemonade. The thermal energy of the room makes the lemonade molecules spread out and move faster. The internal energy of the lemonade increases. The lemonade warms up.

But there are other ways of changing the internal energy of the lemonade that do not depend on a temperature difference, or, put another way, that do not involve any heat. We could stir it. Or we could insert a coil of wire into the lemonade and run an electric current through it. These "nonheat" ways of altering the internal energy are generally classified as *work*. Physicists find it convenient to use only the two categories *work* and *heat* to describe the energy changes of any system. Heat is the energy transferred due to a temperature difference. Work includes all other energy changes.

So we can raise the internal energy of a system either by doing work on the system or by allowing heat to flow into it from some other system at a higher temperature. We can reduce the internal energy by allowing the system to do work on its surroundings or by removing heat from it by the flow of energy to another system having a lower temperature.

The first law of thermodynamics is a rule that nature uses to keep the "energy bookkeeping" straight among internal energy, heat, and work. It is nothing more than the conservation of energy applied specifically to situations involving these three quantities.

Suppose we let U represent the change in the internal energy of some substance. We will label as Q the heat that the system absorbs from its surroundings. Call W the work that the system does on its surroundings. Then conservation of energy tells us that the internal energy will *increase* because of the heat added to the system and *decrease* due to the work that the system does. In an equation, we have

$$U = Q - W$$

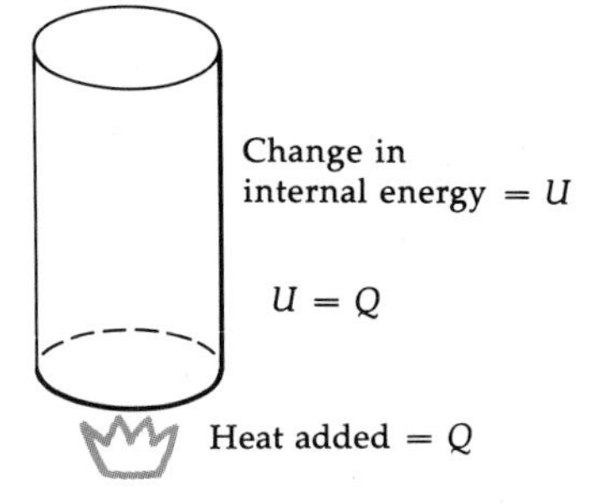

Figure 14-1. Closed cylinder: All the added heat (Q) goes toward changing the internal energy (U).

This equation is a mathematical expression of the first law of thermodynamics.

Let us look at this law as it applies to a very simple system: a closed cylinder full of air. (See Figure 14-1.) We will make sure that the cylinder is tightly shut. This means that since the air cannot expand, it cannot do any work on its surroundings. So W is zero in our equation for the first law, and we are left with

$$U = Q$$

This equation simply says that if we heat the air by, say, putting the cylinder over a flame, all the heat added to the air goes toward increasing its internal energy. And if we treat the air as an ideal gas, then the internal energy is just the kinetic energy of the individual atoms, which increases because of the added energy. So the temperature goes up, just as we would expect.

Figure 14-2. Piston/cylinder: Part of the heat (Q) is expended as work (W) to lift the piston. The remainder changes the internal energy (U).

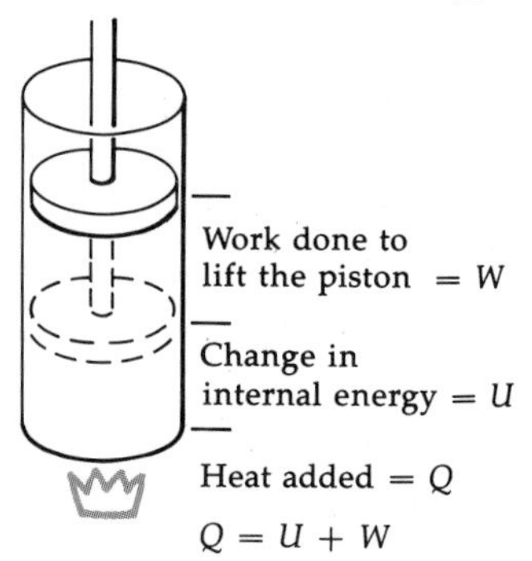

We can change the situation somewhat in order to see the full effects of the first law. Let us take one end off the cylinder and replace it by a piston that is free to move up and down. (See Figure 14-2.) Now when we turn on the flame, part of the heat goes toward raising the piston, in just the same way that a loose-fitting lid bounces off a pan of boiling water. In moving the piston, the air does work that must come at the expense of some of the added heat. So here the heat does two things: It increases the internal energy of the air, raising its temperature; and it does work in lifting the piston.

We can write an equation for this by showing the first law in a different form:

$$Q = U + W$$

That is, the heat added to the system is equal to the change in the internal energy plus the work that the system does. For example, if the internal energy is to increase by 300 calories, and if the air does 50 calories of work in lifting the piston, then the flame must provide 350 calories in the form of heat. As you can see, this is nothing more or less than conservation of energy.

Most people seem to believe [the first law] firmly; mathematicians because they believe it is a fact of observation; observers because they believe it is a theorem of mathematics; philosophers because they believe it is aesthetically satisfying, or because they believe no inference based upon it has ever been proven false, or because they believe new forms of energy can always be invented to make it true. A few neither believe nor disbelieve it; these people maintain that the First Law is a procedure for bookkeeping energy changes, and about bookkeeping procedures it should be asked, not are they true or false, but are they useful.

Henry A. Bent, The Second Law. *New York: Oxford University Press, 1965.*

Learning Checks

1. How is the law of energy conservation related to the first law of thermodynamics?
2. In the expression of the first law, $Q = U + W$, which term is related to the change in the temperature?
3. Consider again the case of the cylinder with movable piston. Suppose that you do work on the system, say pushing the piston down with your hand. What change will this make in the equation $Q = U + W$? What will happen to the temperature in this case?
4. In our example of the lemonade, does the first law have anything to say about which direction the heat will flow?

THE SECOND LAW

The first law of thermodynamics is fine, as far as it goes. Energy is conserved. You can't get something for nothing. But Question 4 of the last Learning Checks raises an important point about which the first law is silent. Granted that within a given system the total energy stays constant, how does this fixed amount of energy shift from place to place?

Here is where the second law of thermodynamics gets into the act. Take the kitchen and lemonade as the "system." We know from the first law that whatever energy the kitchen gives up, the lemonade gains an equal amount. But if the first law were the only consideration, the lemonade could just as well lose energy to the kitchen. The total amount of energy would also remain constant in this case, so the first law would be satisfied. However, our experience tells us that it is the lemonade that warms up by gaining energy from the kitchen. The energy seems not to flow the other way.

> *A good many times I have been present at gatherings of people who by standards of the traditional culture are thought highly educated and who have with considerable gusto been expressing their incredulity at the illiteracy of scientists. Once or twice I have been provoked and have asked the company how many of them could describe the Second Law of Thermodynamics. The response was cold: it was also negative. Yet I was asking something which is about the scientific equivalent of: Have you read a work of Shakespeare's?*
>
> *C. P. Snow*
> The Two Cultures and the Scientific Revolution.
> *Cambridge University Press, 1959*

When we put ice cubes in a glass of hot tea, the cubes melt and the tea becomes cooler. We would be greatly surprised if suddenly the heat flowed the other way, with the ice cubes reforming and the tea heating back up. The first law is no barrier to this. But still it doesn't happen.

In fact, all of our experience points to a pattern in these energy shifts. It is inevitable that the ice cubes melt by absorbing energy from the tea, and the reverse does not happen spontaneously. Hot charcoal will gradually cool as the air around it becomes warm, not the other way around.

The second law of thermodynamics serves as a unifying principle that relates these various forbidden processes. One way of stating the law, appropriate to the examples we have discussed, is

> The spontaneous flow of heat takes place from a region of high temperature to a region of low temperature.

Notice that the second law doesn't provide an explanation for this direction of heat flow, it only summarizes the inevitable results of our experience.

You can picture this transfer of heat from a hot to a cold region as analogous to the flow of water from a region of high pressure to one of low pressure. Figure 14-3 shows two tubes of water at different levels, connected by a passageway with a closed stopcock. When the stopcock is opened, the water flows from the lefthand tube to the righthand tube until the water reaches the same level in both. The pressure difference, which is proportional to the difference in the original levels, is the "driving force" that moves the water through the tubes. In much the same way, the temperature difference between two parts of a system is the "driving force" that moves the heat from the hot region to the cold.

Figure 14-3. Pressure difference as a driving force.

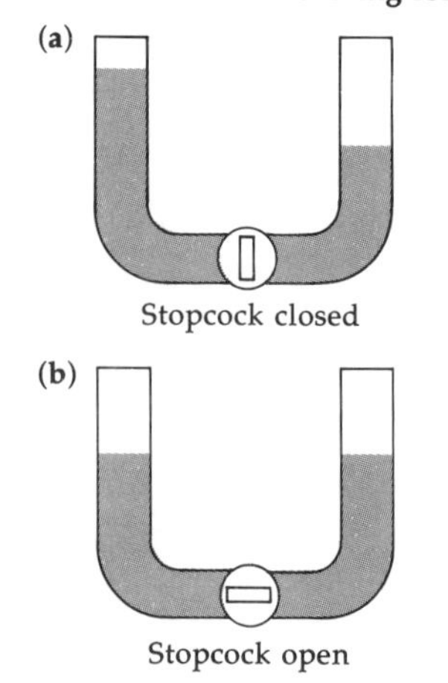

Heat Engines

From a practical standpoint, the second law of thermodynamics has important consequences for putting energy to use—that is, by employing it to do work. In fact, the second law was developed by physicists and engineers as they studied the processes of conversion between mechanical energy and thermal energy. Getting thermal energy from mechanical energy is no big trick—you do it, for example, when you rub your hands to keep them warm on a cold day. In fact, in any process where there is friction—always the case in real-life situations—mechanical energy is converted into thermal energy. A bouncing basketball eventually stops bouncing, its mechanical energy ultimately being dissipated as heat. Conversion of mechanical energy into thermal energy happens as spontaneously in nature as a wagon wheel rolling downhill.

Physics and Technology: An Advanced Heat Engine

Heat engines employed for practical purposes are few and far between. Perhaps the most common example is the steam turbine power unit used in electricity-generating stations. In this mechanism, superheated, pressurized steam is forced into a turbine chamber, where mechanical work is generated as the steam rotates the turbine blades.

Other kinds of heat engines have been devised, but they, too, have little practical value thus far. An exception may be a new design of a Stirling engine created by an industrial group in Sweden. The Stirling engine is a heat engine that works by external combustion, the heating and cooling of a working gas such as helium. The original inventor, Robert Stirling, a Scottish engineer, patented the device early in the nineteenth century. Before it lost out to electric motors and gasoline-powered engines, the Stirling engine was commonly used in such items as water pumps and household fans.

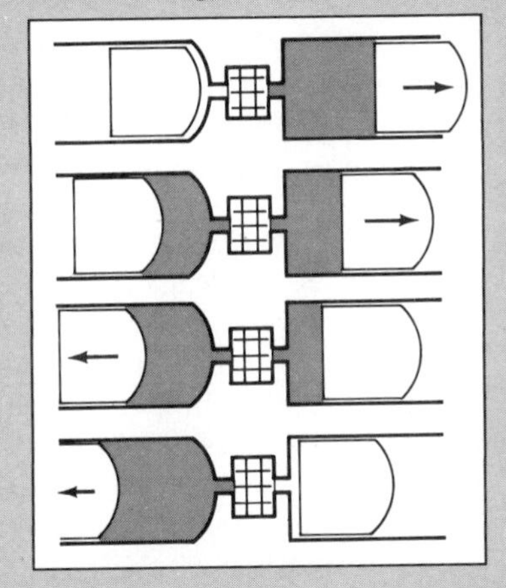

The Stirling engine. The working gas is compressed on the cold side then moves to the hot side where it expands, pushing the piston. It then moves back to the cold side.

Now Swedish researchers, applying modern technology, have developed the Stirling heat engine into a silent, compact power plant that can yield over 7 kilowatts of electricity, as well as hot water for heating. This source of power is free of pollution, can burn most fuels, and needs only slight maintenance. Currently, it is slated for installation in recreational vehicles, luxury motor homes, and yachts. Eventually, the manufacturers anticipate military, agricultural, and regular home uses.

Source: John Free, "Portable Power from a Stirling-Cycle Engine," *Popular Science*, January 1982, p. 40.

But what about going the other way? That is, what about the possibility of converting heat into mechanical energy? This *doesn't* happen spontaneously, any more than a wheel naturally rolls uphill. When you stop your bicycle by applying the brakes, they heat up due to friction. But you cannot turn this process around: the bicycle will not begin to roll again merely by your cooling down the brakes. The dormant basketball doesn't of its own accord begin to bounce when you remove the heat that was generated originally by the bouncing action. Conversion of thermal energy to mechanical energy does not take place naturally.

It is possible, however, to get work from heat, and devices that do so are called *heat engines*. Basically, a heat engine takes advantage of the natural flow of heat from high to low temperature regions by tapping some of this heat for doing work.

Let us take a specific example. Suppose we extract thermal energy from a large hot oven, called a *heat reservoir*. We channel this heat to a gas (generally called the *working substance*), which expands and moves a piston. That is, the expanding gas does work. To continue to get work from such a device, we have to compress the gas again so that it may expand against the piston a second time, and a third time after another recompression, and so on. This recompression takes energy, of course, and we need to make sure that we expend less energy than we obtained during the expansion. Doing the recompression with the gas at a lower pressure will do the trick, because we don't have to work as hard in pushing the piston. The gas pressure

"Frankly, I don't see how we can keep it burning through eternity."

will be lower if we reduce the temperature. So before we push the piston in, we need to remove some heat from the gas.

What if the gas were maintained at the original temperature? If this were the case, the compression phase would require as much work as was obtained in the expansion phase, and the engine would stop—we would not gain any net work. So the engine must be cooled. This means that heat must be removed and ejected into a reservoir at a lower temperature than that of the original hot oven.

In a nutshell, then, a heat engine operates by absorbing heat from a high-temperature reservoir, doing work, and returning the working substance to its original condition by ejecting heat to a low-temperature reservoir. If the reservoirs are at the same temperature, we make no progress, because no net work is generated by the engine. No temperature differences mean no work.

You can see from this illustration that it is impossible to convert the heat *completely* into work, since in that case there is no heat to eject to the cold reservoir. But under these conditions the engine will not operate.

Here again we see the second law in action. The journey from mechanical energy to heat is a downhill run, with 100 percent conversion. But nature will not allow us to convert heat completely into work in a cyclic process. Some waste heat must be expelled.

We can use the water-flow example in Figure 14–3 to understand the importance of a temperature difference for a heat engine. Suppose that in the righthand tube we place a cork that floats on the water surface. When we open the stopcock, the flowing water does work by lifting the cork against gravity. This takes place at the expense of the gravitational potential energy due to the difference in level of the water in the two tubes. As the water flows, less of the original potential energy is available for doing work. Finally, when the levels are the same, there is no more available energy, and no more work is done on the floating cork.

Note that the energy content of the system has not changed, in accordance with conservation of energy. The potential energy lost by the falling water appears as an increase in the potential energy of the cork, now at a higher level relative to the surface of the earth. The *total* energy has not changed, but the *available* energy—that which can do work—has decreased.

The available energy is proportional to the *difference* in the water levels; this difference is the same whether we do the experiment on the top of a mountain, at sea level, or near the center of the earth. There is more potential energy for the *system* at the top of a mountain (relative to the center of the earth, say) than there is at sea level, but this extra energy is unavailable for lifting the cork.

For our heat engine (shown in Figure 14-4), we let T_1 be the temperature of the hot reservoir, and T_2 that of the cold reservoir. The heat Q_1 absorbed from the hot reservoir does an amount of work W, and the engine ejects Q_2 to the cooler reservoir. From the conservation of energy,

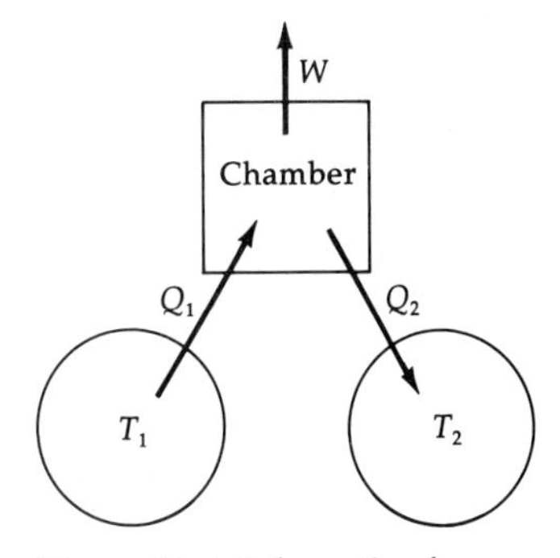

Figure 14-4. Schematic of a heat engine.

$$Q_1 = W + Q_2$$

For instance, the engine might absorb 300 calories from the high-temperature reservoir and eject 200 calories to the cooler one. Then we could obtain 100 calories of work from this engine.

Now, consider what happens as time goes on. Just as in the lemonade example, here we have heat flowing from a hot region to a cold one. As this process continues, the two reservoirs eventually come to the same temperature, and hence there is no driving force to keep the heat flowing. (This is analogous to the water reaching the same level in the tubes in Figure 14-3.) Thus, even though there is still plenty of energy present (since the two reservoirs are at a temperature above absolute zero), there is no longer any *available* energy.

We can express this limitation in terms of the *efficiency* of the heat engine. The efficiency is simply a ratio of the work obtained to the energy supplied in the engine. That is,

$$\text{efficiency} = \frac{\text{work output}}{\text{heat input}}$$

and, in our notation, letting e represent the efficiency,

$$e = \frac{W}{Q_1}$$

In the example, where $W = 100$ calories and $Q_1 = 300$ calories, $e = 1/3$.

In the case of an ideal engine, the work output, or the available energy, is proportional to the temperature difference $T_1 - T_2$, and the heat input is proportional to T_1. Then,

$$e = \frac{T_1 - T_2}{T_1}$$

This efficiency for an ideal engine represents the *maximum* efficiency for *any* heat engine operating between temperatures T_1 and T_2.

As an example, suppose that the hot reservoir is at $T_1 =$ 475 K and the cold reservoir is at $T_2 = 300$ K. Then the theoretical limit to the efficiency of an engine using these two reservoirs is

$$e = \frac{475\ \text{K} - 300\ \text{K}}{475\ \text{K}}$$

$$e = 0.37$$

Multiplying by 100 to express the efficiency as percent gives 37 percent as the maximum efficiency. Any *real* engine operating under these conditions would be *less* than 37 percent efficient because of losses due to friction and imperfections of the equipment.

From the equation

$$e = \frac{T_1 - T_2}{T_1}$$

you can see that even an ideal engine could not be 100 percent efficient unless the cold reservoir were at a temperature of absolute zero ($T_2 = 0$ K), which is an impossible condition. This is just one more way of seeing that it is not possible to convert thermal energy completely into work in a cyclic process.

In summary, then, the first law of thermodynamics tells us that it is impossible to obtain more energy from a closed system than was originally present. The second law maintains that we cannot obtain more work from a system than the amount of *available* energy present, which is always less than the *total* energy.

Put another way, the first law says, "You can't get something for nothing." The second law says, "It's worse than that—you can't break even, either!"

Learning Checks

1. What is the second law of thermodynamics?
2. What is meant by the term *closed system?*
3. What is meant by the available energy in a system?
4. Does an automobile run more efficiently at a higher or lower temperature? Explain.
5. The heat engine in Figure 14–4 is a closed system. What could you do externally to make it operate at a higher efficiency?

ENTROPY: THE SECOND LAW MADE GENERAL

The knowledge we have gained so far about the second law of thermodynamics is due in large measure to the work of a French engineer named Sadi Carnot (pronounced kar-no). Carnot's work in the early third of the nineteenth century was aimed toward understanding heat engines and their efficiencies, and he is credited with fundamental and pivotal contributions to the burgeoning science of thermodynamics. His premature death from an attack of cholera in 1832, when he was 36, undoubtedly cost the fields of physics and engineering one of their brightest stars.

As physicists built on Carnot's work, it became clear that there should be a more general form of the second law, applicable to devices other than heat engines. After all, heat engines are not the only things that produce work. Electric batteries, windmills, chemical reactions, a horse pulling a plow, water falling over a spillway to turn a mill wheel—from all these we obtain work without any temperature difference or heat reservoirs involved.

Nevertheless, there is a common thread that joins all these work-producing devices, and that is the amount of *available* energy. For the heat engine, we have seen that the available energy is connected with the temperature difference of the two heat reservoirs. We shall learn in a later chapter that for electric batteries there is something called the *electric potential difference* that determines the available energy. For the waterfall, it is the difference in gravitational potential energy of the falling water that provides the available energy for the mill wheel to do its work. And so on, for any other work-producing device you can think of. Just as the heat engine runs down as the heat reservoirs gradually come to the same temperature, so the electric battery runs down as the electric potential difference decreases. In each case it is the available energy that "runs out."

We can turn the argument around and say that for an isolated system (that is, one with a fixed amount of total energy), as time goes on the amount of *unavailable* energy *increases.* When all the energy is unavailable, such as in a heat engine when the two temperatures are the same, then we no longer can get work out of the device.

The nineteenth-century German physicist Rudolf Clausius, who studied problems in thermodynamics, invented the word *entropy* as a measure of the unavailable energy. So another way of stating the second law is that *in any isolated system, the entropy always increases.* This form of the law is quite general, for it applies not only to heat engines but to batteries, falling bodies, people, and animals—any situation at all in which energy is transformed.

Note that localized pockets of decreasing entropy are possible. For example, a refrigerator forces heat from the cold and into a warm environment, increasing the temperature difference. This is an entropy decrease. However, this does not hap-

Entropy of the Earth

A point that needs to be emphasized over and over again is that here on earth material entropy is continually increasing and must ultimately reach a maximum. That is because the earth is a closed system in relation to the universe; that is, it exchanges energy but not matter with its surroundings. With the exception of an occasional meteorite that falls to earth and some cosmic dust, our planet remains a closed subsystem of the universe. . . . For life to unfold, the sun must interact with the closed system of matter, minerals, and metals on the planet. . . . This interaction facilitates the dissipation of . . . terrestrial matter that makes up the earth's crust. Mountains are wearing down and topsoil is being blown away with each passing second. That is why, in the final analysis, even renewable resources are really nonrenewable over the long haul. While they continue to reproduce, the life and death of new organisms increase the entropy of the earth, meaning that less available matter exists for the unfolding of life in the future.

Jeremy Rifkin, Entropy: A New World View. *New York: Bantam Books, 1981.*

pen spontaneously but because the refrigerator has a motor doing work to reverse the spontaneous flow of heat. The entropy of the motor increases more than enough to compensate for the decreasing entropy in the refrigerator, giving an *overall increase* in entropy. In this case, then, we must include the motor in the system we are discussing. In all cases of decreasing entropy, it always turns out that there is an increase somewhere else in the universe.

Entropy as Unavailable Energy

Thermodynamics developed out of a continuing study of transformations between mechanical and thermal energy in general, and, as we have seen, heat engines in particular. The ideas involved in thermodynamics pertain to entire systems, not to the individual molecules making up the substance. Indeed, people didn't even know about atoms and molecules when the science of thermodynamics was first taking shape, early in the nineteenth century. When we talk about the temperature or pressure or volume of a bottle of gas or a can of soft drink, we are referring to properties of the system as a whole, the collective characteristics of the entire gas or soft drink. We need make no reference to the billions and billions of individual molecules making up the system.

Now, however, in the latter half of the twentieth century, the notions of atoms and molecules are old hat, common knowledge even among grade school children. So it seems reasonable to try to understand pretwentieth-century thermodynamics in terms of the modern atomic theory, to see how the old ideas fit into the new scheme. In particular, we shall see how entropy as a measure of unavailable energy translates into the modern view.

Let us begin with a familiar example. Suppose that we have a sample of hot air and one of cold air, and we allow them to mix together. Naturally we know that after a while the temperature of both masses of air will settle down to some intermediate value. The hot air will have cooled down and the cold air will have been heated. There is nothing unexpected about this. The flow of heat is from the region of high temperature to the one of low temperature, just as in the case of the heat engine, where the two reservoirs gradually approach the same temperature and the entropy increases.

Figure 14-5. Collision of a "hot" (high speed) molecule with a "cold" (low speed) one.

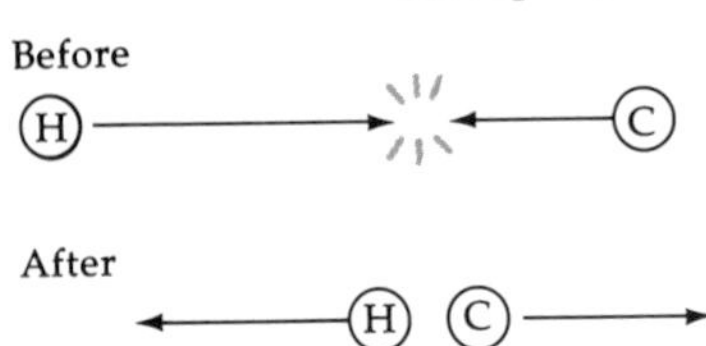

Now let us look at this mixing of the air masses in terms of the individual molecules. The hot air has a high temperature—that is, a high average kinetic energy—which in turn means that the "hot" molecules have higher velocities, on the average, than do the "cold" molecules. When the "hot" molecules collide with the "cold" ones, the vast majority of the "hot" ones will slow down and the "cold" ones will speed up. (See Figure 14-5.) In this way, they achieve more nearly the same speed, and hence the same average kinetic energy. From opposite extremes, they approach the same temperature.

Of course, each group of molecules will have a range of velocities. Among the "hot" molecules, some will be "really hot"—with velocities greater than average—while some will be "cool" in the sense that their velocities will be lower than the average. The same will be true of molecules in the "cold" group. In fact, a very few of the "hot" molecules may actually be traveling slower than a very small number of molecules in the "cold" group.

This means that it is *possible* for a "hot" molecule to become "hotter"—that is, move faster—as a result of colliding with a "cold" molecule, which in turn slows down (that is, gets "colder"). All we would need is a collision between one of the fastest molecules of the "cold" group with one of the slowest of the "hot" group. This is perfectly reasonable and would violate no law of nature.

But the likelihood of this happening is extremely small, simply because of the statistics. Of all the billions and billions of molecules in the hot air mass, only a tiny fraction are moving slower than the fastest of the cold group. Similarly, only a very small number of molecules in the cold group have velocities greater than the slowest of the hot group. So the chances of a collision making the "hot" hotter and the "cold" colder are so tiny as to be negligible.

From this you can see that the *probability* of the hot air mass becoming warmer and the cold air mass cooling down is very small. It isn't impossible, but it is highly unlikely—we can safely say it won't happen. Such a spontaneous occurrence would mean that the entropy of the universe has *decreased.* And the second law of thermodynamics is an expression of the negligible probability of this taking place.

This tendency toward moderation of the velocities of the gas molecules is consistent with the idea of entropy as a measure of the unavailable energy. As long as the energy is more concentrated in the hot air mass, or in the hot reservoir of the heat engine, it is available for doing work. But as it spreads out and becomes more randomly distributed among the formerly "hot" and "cold" molecules, it becomes unavailable. Entropy as a measure of the evenness of the energy distribution applies not only to heat but to other forms of energy as well. The energy in a chemical reaction, for instance, becomes more evenly parceled out among the various chemicals.

On a cosmic scale, many philosophers and scientists believe that the fact of increasing entropy will ultimately lead to the "heat death" of the universe. If we suppose that there is only a finite amount of energy, in all its forms, in the universe, then the persistent increase in entropy means that there is a

Fire and Ice

Some say the world will end
in fire,
Some say in ice.
From what I've tasted of desire
I hold with those who favor
fire.
But if I had to perish twice
I think I know enough of hate
To say that for destruction, ice
Is also great
And would suffice.

From The Poetry of Robert Frost, *edited by Edward Connery Lathem. Copyright 1923 © Holt, Rinehart and Winston. Copyright 1951 by Robert Frost.*

limit to the amount of available energy. When the time comes that there is no more available energy, the entropy will have increased to a maximum value. The unavailable energy will be equal to the total energy. All matter will have reached a uniform mixture, and all parts of the universe will be at one temperature. This means that the universe will have run down, like a watch that has stopped ticking and cannot be rewound.

Entropy and Disorder

Now let us consider entropy from another standpoint. We have seen that hot air mixed with cold air gives lukewarm air and that this corresponds to a moderation of the velocities of all the molecules. When the mixed air reaches its final temperature, the molecules are more disordered than before. We can consider the separation into sets of "hot" and "cold" molecules as a more ordered situation than that of the lukewarm mixture. The increase in entropy has been accompanied by an increase in disorder; hence, we can consider the entropy to be a measure of disorder.

Figure 14-6. Normal sequence of a suit of cards.

What do we mean by order and disorder? It is not a question of pleasant or unpleasant. One person's order can be another's chaos. What we mean is this: *The degree of disorder of a state is a measure of the number of ways that state can be achieved.*

Let us consider a familiar example. Suppose we take an ordinary deck of playing cards: fifty-two cards divided equally among the four suits—clubs, diamonds, hearts, spades. Now we take one of the suits and arrange its thirteen cards in their usual sequence. There is only one way to do this: the deuce must be in the first position, the trey in the second, the four in the third, and so on through the jack, queen, king, and ace in the tenth, eleventh, twelfth, and thirteenth positions, respectively. (See Figure 14–6.) This is a highly ordered state, since there is only one way for it to happen; it has the lowest entropy possible.

Figure 14-7. Switching the Jack-Queen gives two sequences.

Now suppose we want a similar sequence, except that we don't care about the relative position of, say, the jack and the queen. You can see from Figure 14–7 that there are *two* ways to do this: the jack in the tenth position and the queen in the eleventh, or the jack in the eleventh and the queen in the tenth. We have a slightly more disordered state, since there are more ways of getting it than before. Correspondingly, the entropy is higher for this state than for the first sequence.

We could continue this process to more and more disordered states, having higher and higher entropy. For example, if we don't care about the relative positions of the jack, queen, and king, then we can achieve this state in six ways. And so on. Again, we have greater disorder and higher entropy.

The number of ways of getting a particular state, and hence the degree of disorder, increases very rapidly as we relax the restrictions on the position of the cards. Thus, if we allow any *arbitrary* sequence, we get an extremely large number of possibilities. The deuce can be in any of the thirteen positions, the trey in any of the remaining twelve, the four in any of the remaining eleven, and so on. If you do the arithmetic, you will find that there are 6,227,020,800 possible arrangements of the thirteen cards. This is a highly disordered state with a very large entropy.

Suppose that we take the thirteen cards in their original perfect sequence and shuffle them. The disorder is likely to increase, since we are almost certain to break up the initial pattern. This means that the entropy has also increased, since it is a measure of disorder.

Shuffling the deck of cards is analogous to mixing the masses of hot and cold air. The disordered state of the lukewarm air is represented by a randomization of the molecular velocities. The separate hot and cold air is a more ordered arrangement than the disordered state of the lukewarm air. It is just "natural" for the entropy to increase as the air masses mix, just as the shuffling of the deck naturally produces a more random distribution of the cards.

Recall that we said that it is *possible* for the lukewarm mixture to separate into hot air and cold air. Similarly, it is possible by our shuffling of the deck to achieve the highly ordered arrangement of all the cards in sequence. The probability of this happening is extremely small: the odds are one in 6,227,020,800! You might try to compute the odds of dealing a bridge hand containing all thirteen cards of the same suit, from a randomly shuffled deck.

Other processes in nature tend toward higher degrees of disorder. The air molecules in a room don't spontaneously gather all in one corner. Our iced tea example is another illustration: The mixture of melted ice and tea is more disordered, with a higher entropy, than the separated cubes of ice floating in the tea. If you put a layer of blue blocks and a layer of red blocks in a box and shake them up, the two colors mix together. No matter how long or how hard you shake the box, the blocks will not spontaneously return to the ordered, two-layer condition. They *could*, of course—but they won't. The second law insists that order give way to chaos.

The consistency between the old view of the second law, involving the spontaneous flow of heat, and the modern version, involving entropy as a measure of probabilities, is possible because of the tremendous number of molecules in any system. We saw that the probability for a particular arrangement of thirteen cards is one in several billion. Imagine what the likelihood would be for a particular arrangement of the billions and billions of atoms in a balloon full of helium. In shuffling cards, any particular card—the king, say—is equally likely

Entropy and Eating

The ordinary food chain provides an interesting example of entropic disorder. One chemist depicted a simplified food chain consisting of grass, grasshoppers, frogs, trout, and humans to illustrate this point. According to the second law of thermodynamics, at every step of the food chain process, available energy should be converted into unavailable energy, with the result that the environment should undergo a greater disorder. Thus, when the grasshopper chomps the grass, when the frog swallows the insect, when the trout eats the frog, and when a person downs the trout, there is an energy loss. Roughly 10 to 20 percent of the energy devoured stays inside the predator for transfer to the next level in the chain. The remaining 80 to 90 percent is lost as heat to the environment.

Now, consider the question of how many of each species are needed to keep the next higher species from sliding toward maximum entropy: "Three hundred trout are required to support one man for a year. The trout, in turn, must consume 90,000 frogs that must consume 27 million grasshoppers that live off 1000 tons of grass." So for one person to preserve a high degree of order in this system, the energy stored in 27 million grasshoppers (or 1000 tons of grass) must be utilized each year. From this, it can be seen that every living organism maintains its own order only at the price of causing a larger disorder (that is, a dissipation of energy) in the total environment.

Source: G. Tyler Miller, Jr., *Energetics, Kinetics and Life.* Belmont, CA: Wadsworth Publishing Company, 1971.

to land in any one of the thirteen positions. But as you increase the number of cards, you greatly decrease the probability of any one given arrangement.

Similarly, in a room full of air, any one particular air molecule is as likely to be in one part of the room as in any other. For instance, at any given instant it has the same chance of being in the top half of the room as in the bottom half. But it is quite *unlikely* for *all* of the molecules to be in the top half of the room at the same time.

Entropy as disorder is also related to the direction in which time flows. This feature of the second law of thermodynamics sets it apart from the other rules we have considered. For instance, the laws of conservation of energy and momentum do not distinguish between "before" and "after." Indeed, their great power derives from the fact that energy and momentum do *not* change as time goes on, so you could not detect the passage of time just by measuring these quantities.

Entropy, however, does provide some direction to time, because entropy is always increasing. For some isolated system, "after" is distinguished from "before" because "after" has a larger value of the entropy. The direction in which time flows is the direction in which disorder increases.

We all laugh at motion pictures run backward, showing toothpaste returning to the tube or football players running backward from scattered locations to a perfect formation at the scrimmage line. These things seem absurd because they violate our sense of the direction of time. The return of billiard balls to a perfect triangle from all parts of the table does not violate any laws of energy or momentum, so it is not impossible. But such a situation is statistically improbable, violating the second law of thermodynamics.

Note again the statistical nature of the second law. If you watch a motion picture of a single billiard ball, you can't tell if the movie is running backward or forward. We have no sense of time flow here. But as the system becomes more complex—by the inclusion of the other fifteen balls—then the progression from order to disorder becomes quite apparent, and we are convinced about the direction of time.

Learning Checks

1. What is meant by the disorder of a state of a system?
2. How is entropy related to disorder?
3. Cleaning up a messy room represents an isolated region of entropy decrease. Does this violate the second law of thermodynamics? Explain.
4. How is the mixing of hot and cold masses of air analogous to shuffling a deck of cards?

SUMMARY

The *first law of thermodynamics* is a restatement of the conservation law for energy. It says that for an isolated system the heat absorbed is equal to the change in internal energy plus the work done by the system. The *second law* recognizes the pattern that heat always flows spontaneously from high to low temperature. In more general terms, the second law states that the entropy of an isolated system always tends to increase—the entropy being a measure of the energy no longer available for doing work.

Entropy in the modern view is taken as an expression of the disorder, or randomness, of the system. Isolated systems tend toward a more disordered condition. This interpretation of entropy gives us a way of establishing the direction of time flow in the universe.

Exercises

1. Distinguish between *heat* and *work* as used in the first law of thermodynamics.

2. Is "internal energy" the same as "thermal energy"? Explain.

3. Why do we say that the second law of thermodynamics is a *statistical* law rather than an absolute one?

4. Suppose that a particular object does no work, nor is any work done on it. How does the change in internal energy during some process compare with the heat taken in during the process?

5. A health inspector takes two identical jugs of milk from the refrigerator. Both jugs have a temperature of 10° C. One the inspector allows to warm up gradually to room temperature (22° C). The other is heated to 30° C and then allowed to cool to 22° C. How does the final internal energy of the two quantities of milk compare?

6. A quantity of gas is in a container that is thermally insulated, which means that no heat can flow in or out. If we compress the gas by pushing down on a piston, (a) what happens to the internal energy of the gas and (b) what happens to the temperature of the gas?

7. A "perpetual motion machine" is one that, when once set operating, would continue to run with no input of energy. How does this violate the laws of thermodynamics?

8. The second law of thermodynamics places a limit on the conversion of heat into work. Is there any limit to the amount of heat obtained from work? Explain.

9. It has been suggested that the present energy crisis is in fact an "entropy crisis." Comment.

10. Name five processes that always take place in one direction and not spontaneously in the opposite direction.

11. A bakery has three large ovens at temperatures of 200 K, 300 K, and 450 K. Which two would you select to use as hot and cold reservoirs for the most efficient heat engine? What would be its efficiency?

12. In a perfect heat engine, 20 joules of energy are taken up from the hot reservoir and 12 joules of energy are rejected to the cold reservoir during a complete cycle. How much work is done during a cycle?

13. In Exercise 12, what is the engine's efficiency?

14. How many ways are there of arranging four people in four chairs, one person to a chair?

15. In Exercise 14, suppose that a person is restricted to a particular chair. How many ways are there of arranging the other three people? Which is the most disordered state? Which has the greatest entropy?

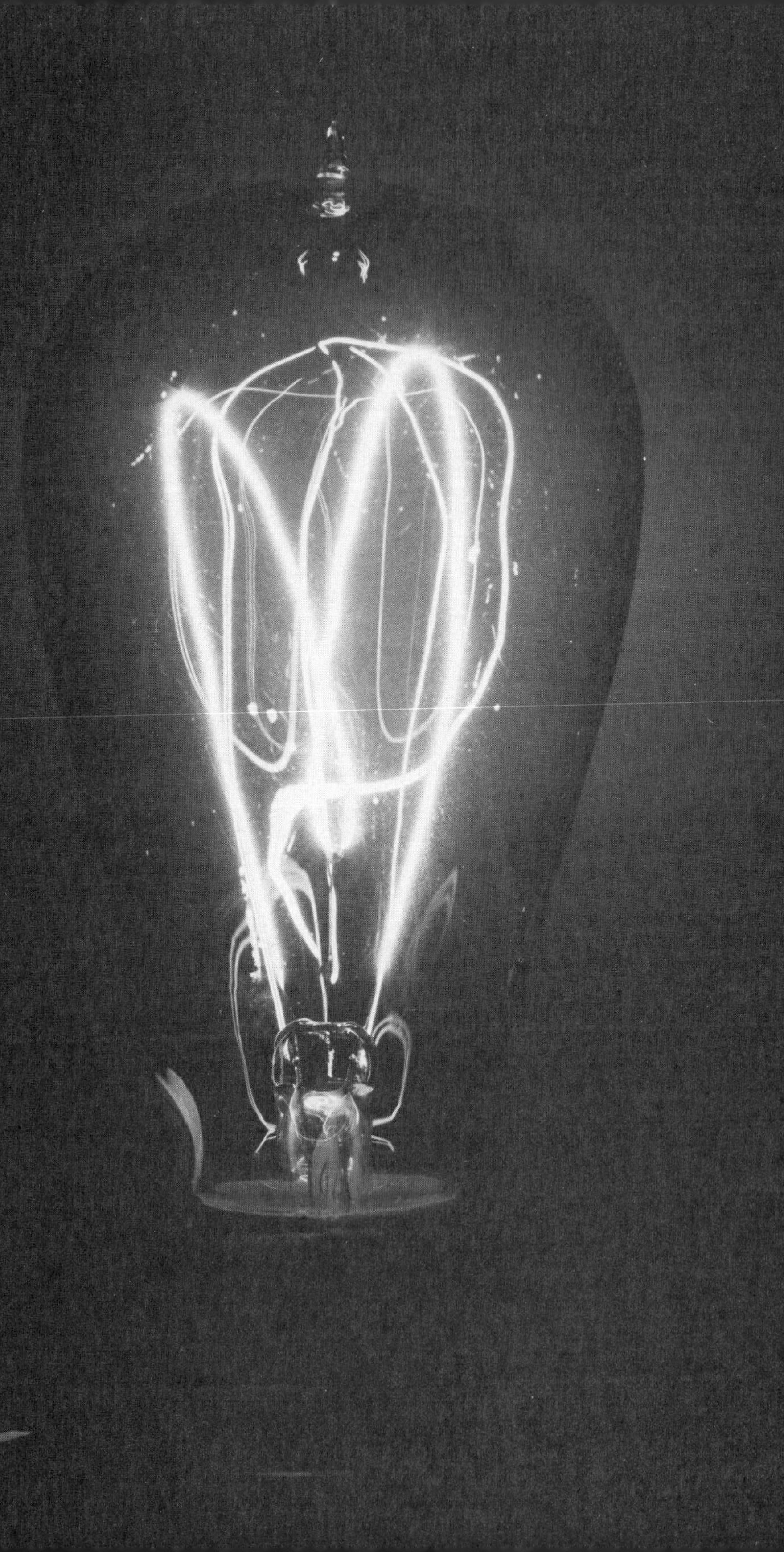

15

ELECTRICITY: CHARGES AND CURRENTS

Electricity has become an integral part of our lives. It is an aspect of electromagnetic interaction – one of the four basic forces of nature.

Electricity is responsible for the most common of everyday experiences. Hardly a minute passes in which modern life is not influenced by various uses of electricity: incandescent lighting, household appliances, telephones, stereos, calculators—the list seems endless. Even the gasoline engines in your automobile and lawn mower depend on an electric spark to ignite the fuel. The furnace that burns natural gas to heat your home won't come on if an electric switch doesn't operate.

These examples of the ways we use electricity are fairly obvious, and doubtless you can think of many more. But there are other, less obvious ways that electricity comes into play in our environment, ways that we don't normally think of as having anything to do with electricity. Along with magnetism, which is the subject of the next chapter, electricity is a component of the *electromagnetic force,* the second of the four basic natural forces we shall study. This force governs nearly every interaction in our everyday world. Electric forces hold together the molecules that make up our bodies and every other object in the universe. Electric signals control our breathing and heartbeat and all other bodily functions. The frictional forces enabling you to walk across the campus to class are electromagnetic. A tennis racquet striking a ball involves the electromagnetic force. Whenever you push or pull or touch anything, the electric force causes the push, the pull, the touch.

In this chapter, we are going to learn about static electricity and electric currents. The next chapter will deal with magnetism and the way it teams up with electricity in many important functions.

STATIC ELECTRICITY

You have probably experienced static electricity by walking across a carpet and getting shocked by touching another person

Figure 15-1. Glass rod, after being rubbed with silk cloth, attracts bits of paper.

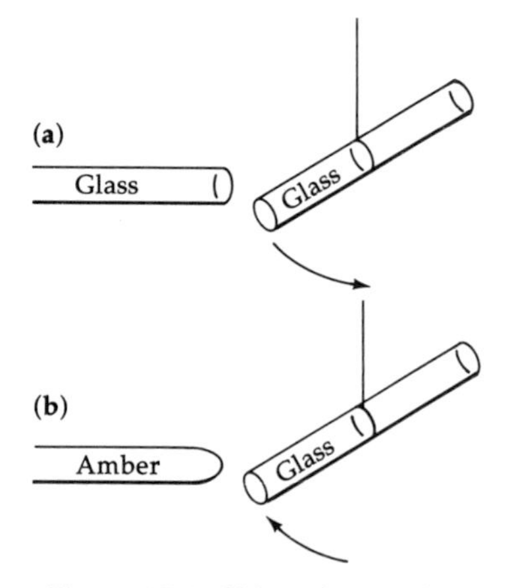

Figure 15-2. Objects in experiment repel or attract one another. (a) Two glass rods *repel* one another after being rubbed with silk. (b) A glass rod, rubbed with silk, is *attracted* to an amber rod that has been rubbed with cat's fur.

Figure 15-3. Schematic of the atom. The nucleus has a *positive* charge, equal in size to the combined *negative* charges of the electrons.

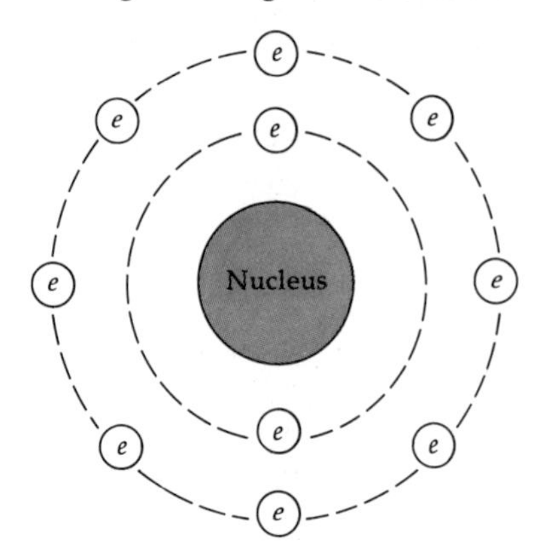

or some metal surface. Comb your hair in the dark sometime, and you might see flashes of light accompanying the crackling sound you hear. Sparks igniting the wick of a cigarette lighter are produced by the flint striking against a rough metal wheel.

Such electrical effects have been known for hundreds of years. As long ago as the late sixteenth century, Sir William Gilbert, a contemporary of Galileo and physician to Queen Elizabeth I, performed simple experiments to study these effects. He found, for example, that a glass rod rubbed with silk attracts bits of dust and paper. (See Figure 15-1.) If a second glass rod is suspended by a string and also rubbed with silk, it is pushed away when the first rod is brought up close. An amber rod rubbed with cat's fur will also attract bits of paper. However, when the amber rod is brought near the suspended glass rod, the opposite effect occurs: the glass rod moves *toward* the amber rod. (See Figure 15-2.)

We can understand these experiments in terms of the atomic hypothesis we mentioned earlier (Chapter 12) in connection with states of matter. This will be covered in more detail in Chapter 24. Each individual atom is composed of *fundamental particles,* two of which are the *proton* and the *electron.* The proton, much the more massive of the two, resides in the central core, or *nucleus,* of the atom. (See Figure 15-3.) The kind of atom, whether it is oxygen or iron or sodium or whatever, depends on how many protons there are in the nucleus. Under normal circumstances, an equal number of electrons swarm about some distance from the nucleus. The electron's mass is only about 1/1840 that of the proton.

The proton and electron have another property, an intrinsic attribute called *electric charge,* that is responsible for the electrical effects we have mentioned. From the behavior of the amber and glass rods, we can conclude that there are two kinds of electric charge. Benjamin Franklin, more famous for his role as a statesman during the Revolution and the early days of the Republic, was also well known as a physicist, particularly for his experiments with electricity. He coined the terms *positive* and *negative* as names for the two kinds of electrical charge. The charge on the proton is positive, that on the electron is negative. Even though the masses of the proton and electron are vastly different, their charges are equal in magnitude and opposite in kind.

Because the electron has so much the lesser mass of the two, it is more mobile. Rubbing the glass rod with the silk cloth strips some of the electrons away from the rod, leaving it with an excess of protons and hence a net positive charge. On the other hand, the amber rod becomes negatively charged when rubbed with the cat's fur. The amber strips some electrons from the cloth, thus gaining an excess of electrons and a net negative charge. Therefore, a negative charge is produced by *adding* electrons, while a positive charge is produced by *removing* electrons. When the number of protons and electrons are equal, the body is uncharged, or *neutral.* The negative and positive charges cancel: they add together, giving zero for the total charge.

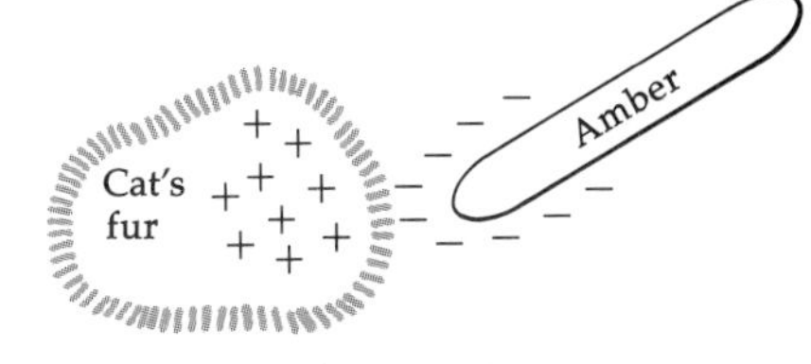

Figure 15-4. The opposite charges on the amber and cat's fur add to zero.

The word *neutral* has much the same meaning here that it does outside physics. Nations that do not take sides during a war, for example, are described as neutrals, as are noninterested parties in a debate or a contest. In physics, a neutral body has no net positive or negative charge.

Notice that when two objects rub together, the resulting friction does not create any charge. Instead, charge is merely transferred from one object to the other. Before you rub the amber rod with the cat's fur, each has an equal number of protons and electrons, and hence each object is neutral. The electrons that appear on the rod are just those scraped from the fur, which is left with a net *positive* charge that exactly matches the *negative* charge on the rod. (See Figure 15-4.) The total charge on the fur and rod taken together is still zero.

We see from this illustration that the total charge is *conserved,* meaning that it doesn't change. This *conservation of electric charge* is a general law of nature, and it is a very important and fundamental rule. Although charges may be shifted from

Electricity Within and Without

Though we tend to think of electricity as a power source for things around us, it is also a force within us. In fact, human beings fairly crackle with low-level voltage. Even the development of the embryo is influenced by organic electric currents.

Electricity powers much that is external to our bodies.

Physiological physicists have been working with the body's electric signals to aid healing processes. Wire electrodes have been inserted in the brain to control severe pain. Electric coils have been used to induce tiny currents into bones and to mend stubborn fractures. The electromagnetic halo around our bodies has been electronically mapped for diagnostic purposes (disturbances in these fields portend illness).

Thus, by mingling the electricity that flows through wires with the electric flows within ourselves, researchers are enhancing the field of medicine. They even hope, one day, to be able to renew ailing organs by simply manipulating our natural internal charges and currents.

Because electricity defines our present to a large extent, and will shape our future to an even larger extent, it is prudent to know something about the basic physics behind this phenomenon.

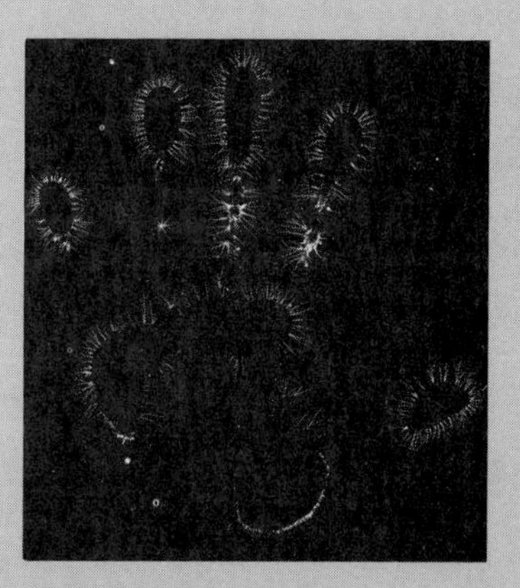

Electricity flows internally throughout our bodies.

Sources: Kathleen McAuliffe, "I Sing the Body Electric," *Omni,* November 1980, pp. 70–73.

Dennis Meredith, "Healing with Electricity," *Science Digest,* May 1981, pp. 52–57.

About Amber

Amber is a hard, translucent yellow or brownish-yellow resin. It was once exuded as a liquid from the trees of prehistoric forests. Amber is found in certain kinds of soil and on some seashores and is used mainly in jewelry. It is easily polished and quickly electrified by friction. The ancient Greeks called amber *elektron* (Latin, *electrum*), which is where our word *electricity* comes from. Gilbert himself never used the noun "electricity." Instead, he invented the word *electrica*, which he defined as, "the things which attract in the same manner as amber does."

Source: Alfred Stills, *Soul of Amber.* New York: Murray Hill Books, Inc., 1944.

place to place through the universe, the total amount of charge never changes. Nature won't allow us to create or destroy electric charge.

In the examples we have been discussing, the excess charges remain stationary on a charged object. We call this *static* electricity, from a Latin word meaning "stationary." *Electrostatics* is the name given to the study of electricity under these circumstances.

The Force Between Charges: Coulomb's Law

Nobody knows exactly what electric charge is. It is an intrinsic quality that cannot be explained or accounted for or defined in terms of anything more fundamental.

Nevertheless, we do know a great deal about the effects of the electric charge. We have seen that, whatever it is, there are two kinds of it, which, following Franklin, we have arbitrarily labeled positive (+) and negative (−). From the experiments illustrated in Figures 15–1 and 15–2, we can also conclude that the same kinds of charge repel each other. The charged glass rod hanging from the string turns away from a second similarly charged glass rod nearby. The experiment with the amber rod demonstrates the existence of the other kind of charge, for the amber rod pulls the hanging glass rod toward it. That is, opposite charges attract each other.

These studies with the charged rods, and other similar experiments, show that some kind of force exists between charged objects. We call it the *electrostatic force.* If the charges are the same, both positive or both negative, then the force is repulsive; the objects tend to accelerate away from each other. If one charge is positive and the other negative, then the objects move together, indicating that the force is attractive. From careful observations like these, one can also see that the force depends on the distance. The two rods do not affect one another unless they are fairly close together. Finally, if the amount of charge on one object is doubled, then the force on the two objects is also doubled. So the force depends on how much charge each object carries.

A French physicist named Charles Augustin de Coulomb was the first to carefully study the relationship among force, charge, and distance, in the late eighteenth century. The result of his work is called, in his honor, *Coulomb's law,* which says:

> The electric force between two charged bodies is directly proportional to the product of their charges and inversely proportional to the square of the distance between the bodies.

We can indicate the charges on the two bodies as q and q'. If they are separated by a distance d, then Coulomb's law gives as the magnitude of the force

$$F = k\frac{qq'}{d^2}$$

where k is a proportionality constant.

The unit of charge is the *coulomb*, abbreviated C. In the mks system, k has the value

$$k = \frac{9 \times 10^9 \text{ N m}^2}{\text{C}^2}$$

This gives the force **F** in newtons. Each electron and proton has a charge whose magnitude is about 1.6×10^{-19} coulomb. Put another way, it takes about 6 billion billion (10^{18}) electrons to equal 1 coulomb of negative charge.

Nature shows us an interesting peculiarity here. As far as we know, charges always appear in whole-number multiples of the size of the electron's charge. Robert A. Millikan, an eminent American physicist of the early part of this century, demonstrated this wholeness of the electric charge in a series of experiments performed between 1906 and 1911. In what is known as the "Millikan oil-drop experiment," he sprayed a fine mist of oil from an atomizer (like that used for perfume) between two horizontal metal plates. The upper plate was positively charged and the lower was negatively charged.

Millikan knew from earlier work that the droplets bore a negative charge as a result of the spraying action, so it followed that they would feel an electric force in the upward direction, opposing the downward gravitational pull of the earth. As time passed, some of the charge gradually leaked to the surrounding air, reducing the electric force and allowing the drops to fall. What Millikan observed was that the electric force, and hence the charge on the drops, decreased in jumps, not in a smooth continuous fashion. Furthermore, the size of the jump always corresponded to some whole multiple of a basic unit of charge, which turned out to be 1.6×10^{-19} coulomb. Millikan's conclusion was that electric charge does not behave like a smooth fluid, but it is "lumpy," and the lumps all have this basic size. If we let e represent the charge on the electron, then a body may carry a charge of $2e$ or $3e$ or $-7e$, but never $(1/2)e$ or $0.32e$, and so on. In recognition of this important discovery, Millikan in 1923 became the first American to win the Nobel Prize in physics.

We can make some interesting comparisons between Coulomb's law and the law of gravitation (Chapter 5). They both depend on multiplying properties of the objects involved in the interaction—the masses in the case of gravity, the charges in the case of electric force. Both gravity and electric

Franklin's Foresight

Benjamin Franklin's imagination soared after his experimental experiences with electricity. In a letter to a friend, for instance, he pretty much predicts our own electrically based world, though he was being humorous at the time: "At a party of pleasure . . ." he suggested, "a turkey [may] be killed for dinner by the *electrical shock*, and roasted by the *electrical jack*, before a fire kindled by the *electrified bottle* . . ." He even foresaw our electrically run toys. In another letter, he talked of demonstrating "an artificial spider, animated by electrical fire to act like a living one and endeavor to catch a fly."

Source: I. Bernard Cohen, *Benjamin Franklin's Experiments*. Cambridge, Mass.: Harvard University Press, 1941.

force decrease as the objects get farther apart; that is, they both obey inverse-square laws, because the size of the forces depends inversely on the square of the distance.*

Although the two laws have this similar appearance, there are important distinctions between them. The forces of gravity and electrostatics represent completely independent phenomena. One big difference is the strength of each force. The electrostatic force is huge, billions and billions of times stronger than gravity. This means that whenever both forces act between two objects, the electrostatic force wins, hands down. Gravity is no match for it.

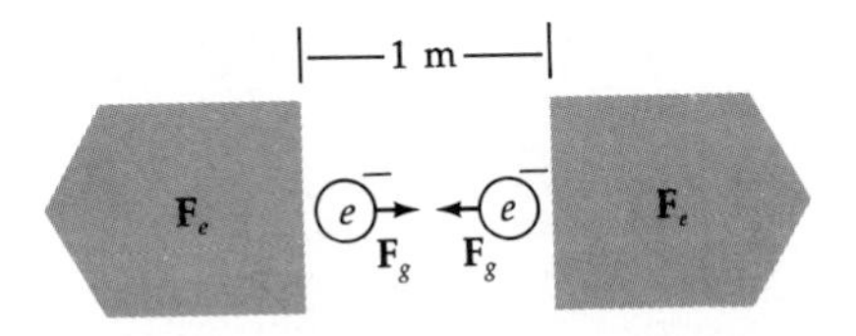

Figure 15-5. The feeble gravitational force (F_g) is no match for the monstrous electrostatic force (F_e).

We can illustrate this by calculating the two forces for a pair of electrons. Let us suppose they are 1 meter apart. (See Figure 15–5.) Because each has a mass (of 9.1×10^{-31} kilogram), the two electrons will be *attracted* to each other by gravity. And since each has a negative charge (of 1.6×10^{-19} C) they will be *repelled* by the electrostatic force. From Chapter 5, we calculate the size of the gravitational force (F_g) to be

$$F_g = \frac{(6.67 \times 10^{-11}\ \text{N m}^2/\text{kg}^2)(9.1 \times 10^{-31}\ \text{kg})^2}{1\ \text{m}^2}$$

$$= 5.52 \times 10^{-71}\ \text{N}$$

On the other hand, Coulomb's law gives as the magnitude of the electrostatic force (F_e)

$$F_e = \frac{(9 \times 10^{9}\ \text{N m}^2/\text{C}^2)(1.6 \times 10^{-19}\ \text{C})^2}{1\ \text{m}^2}$$

$$= 2.31 \times 10^{-28}\ \text{N}$$

Let us compare these numbers. Dividing F_e by F_g gives

$$\frac{F_e}{F_g} = 4 \times 10^{42}$$

This is a tremendous number, 4 followed by 42 zeros. It tells us that the electrostatic repulsion between the two electrons is 4 million-billion-billion-billion-billion times stronger than the force of gravity tending to pull them together. Obviously, we can ignore gravity altogether in the face of its competition. Gravity hardly makes a peep whenever the gigantic electro-

*The fact that both forces behave in the same way with respect to the distance and products of mass or charge leads some people to believe, or at least speculate, that perhaps gravity and electricity are really examples of a more fundamental law. No one has been successful in pursuing this line of thought, but it is tantalizing.

static force is around. Only on the astronomical scale, for the very large masses of stars, planets, and galaxies, does gravity play a major role. For charged objects, it is completely swamped by electricity.

This great size of the electrostatic force accounts for the fact that objects are usually neutral, having equal numbers of positive and negative charges. If we give an amber rod a negative charge by rubbing it with cat's fur, the repulsion of the excess electrons from each other is extremely strong. They will use almost any avenue of escape. This is the reason you don't readily notice electrostatic effects on a humid day. The moisture in the air provides a path for excess charges to "leak" off, restoring the neutral balance. On a day when the air is very dry, charged objects tend to retain their charge, and you are more likely to shock yourself on a doorknob or the handle of the automobile door.

Bolt from the Blue

Originally, lightning terrorized. Mystified by it, the peoples of ancient civilizations mostly believed lightning flashes to be indications of the power or wrath of gods. Those struck by it were often shunned. In many cultures, charms were created to protect against it. During the 1700s, though, as the spirit of science took hold, lightning became the target of rational investigation. Nearly everyone knows about Ben Franklin's famous kite experiment in which he drew what he called "electrical fire" from the sky. Franklin was not the first to capture electricity from clouds in this manner, but he clearly noted in his diary suggestions for carrying out this experiment.

Zeus, the Greeks' mightiest god, is depicted as a heavenly judge wielding lightning bolts.

The physics of lightning given impetus by Franklin has blossomed into many a modernized thunderbolt study project. Out of this research comes valuable aid to meteorologists as well as a range of practical devices. One is a new 50-foot-high lightning trap that can subdue 1,200,000 volts of sudden electricity, thereby possibly saving transformers from being knocked out and preventing storm-caused blackouts.

Natural electricity in the form of lightning can take a heavy toll on the human body. A person struck by lightning is not automatically killed—a bolt can knock you off your feet, scorch your hair or body, tear up your clothing, and leave you temporarily speechless, with one fantastic headache.

Our knowledge of the way this phenomenon works has caused a direct reversal of human attitudes toward it. Now, lightning rods, not charms, protect our homes and aircraft. And, while still awesome to behold, lightning has come to be taken not as an expression of anger but as a symbol of inspiration.

Sources: P. E. Vienmeister, *The Lightning Book*. Garden City, New York: Doubleday & Company, Inc., 1961.

"A New Trap for Lightning," *Science Digest*, January/February 1981, p. 49.

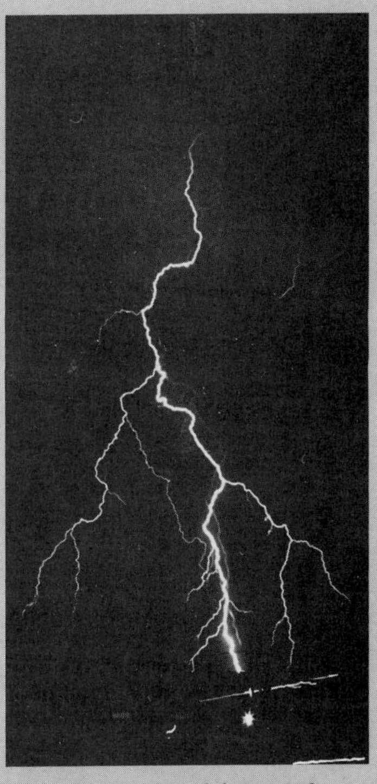

Lightning's colossal force and unpredictability made it the obvious representation of Zeus' power and temperament. Modern science has diminished its fearful mystery but not its mystique.

The other distinction between gravity and electricity is also very critical. Gravity always *attracts* objects to each other, while charged objects are *either* attracted *or* repelled electrostatically, depending on whether their charges are different or the same. It is this feature of the electrostatic force, along with its incredible strength, that accounts for almost perfect balance of positive and negative charges in ordinary objects. If you and a friend were standing at arm's length, and each of you had only *1 percent* more electrons than protons, the repelling force on each of you would be staggering. Such a force could lift a "weight" equal to that of the entire earth. So you can see that there is a tremendous tendency for objects to remain electrically neutral. When you scrape your shoes across the carpet, you become positively charged because you are deficient in electrons. Any electrons on the metal doorknob are attracted by this tremendous electrical force, and the shock you feel is the release of energy as nature restores the balance.

Sometimes this rejoining of positive and negative charges—*discharging,* it is called—is accompanied by a flash of light, a spark. You can see this effect when you take clothes made of certain synthetic materials from the clothes dryer. Those wonderful old Frankenstein-type horror movies wouldn't be the same without a laboratory full of great huge machines churning out sparks and lights and flashes of all descriptions.

Lightning is nature's most spectacular display of electric discharge. A cloud may become charged, perhaps because its water vapor condenses rapidly and hence will be positive or negative relative to another cloud or the earth. This charge imbalance is relieved by the discharge, which we see as a beautiful spark.

Figure 15-6. Electrostatic precipitator.

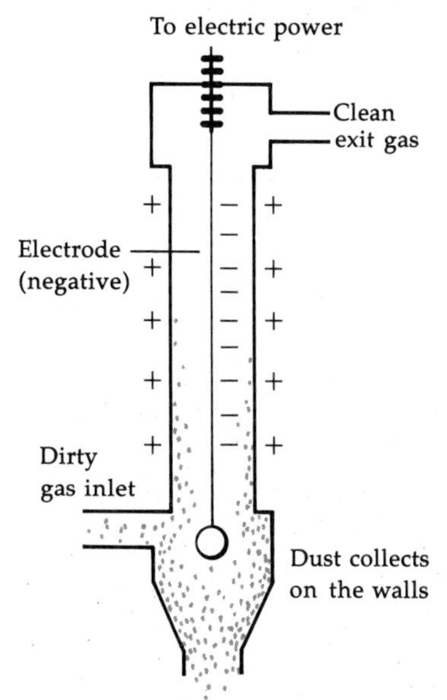

Several applications of electrostatics have importance in modern society. One of these is the *electrostatic precipitator* (see Figure 15-6), important in reducing the air pollution from smokestacks. A thin wire hanging down the middle of the stack is given a very large negative electric charge that causes electrons to come whizzing off the wire. They attach themselves to oxygen molecules in the air. The oxygen molecules, now negatively charged, move across to the wall of the stack, which has a positive charge, and along the way carry with them the dust, fly ash, and other particles in the smoke. This dust can then be collected and disposed of in a nonpolluting way. Electrostatic precipitators can be more than 99 percent efficient in cleaning the smoke from fossil-fuel-burning power plants and other such sources of air pollution.

The familiar Xerox machine also uses electrostatics in the dry-copy imaging process known as *xerography.* A rotating drum coated with selenium is charged in the dark and then illuminated with an image of the page to be copied. The light removes all the charge except where the images of the black areas appear. These images attract a black dust, called the *toner,* which is turn is transferred by electrical attraction to a pre-charged paper that makes contact with the drum. A rapid-heating process affixes the toner to the paper to make a permanent copy.

(a) An electric field created by an ordinary battery with positive and negative terminals. (b) Electric fields are also generated by some forms of life. Each electric discharge from organs located in the rear of this fish makes the tail negative in relation to the head. This particular specimen (*Gymnarchus niloticus*—"naked tail"—from Africa) emanates a weak field for purposes of sensing its environment. Other species generate strong fields to stun prey.

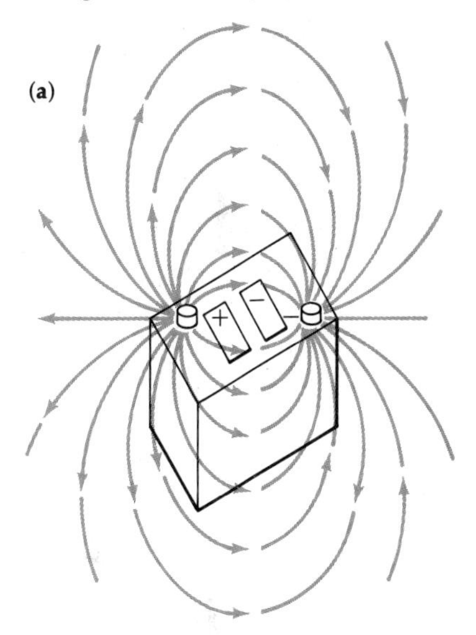

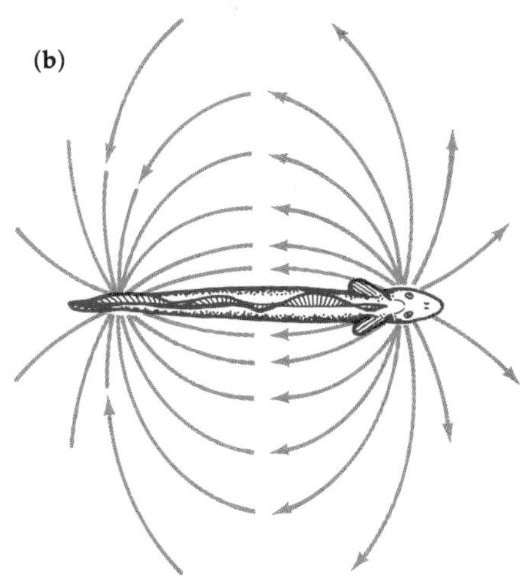

Learning Checks

1. Cite evidence that there are two kinds of electric charge but only one kind of mass.
2. What are the similarities in the *electrostatic* force and the *gravitational* force? What are some differences?
3. If the calculation of the electrostatic force from Coulomb's law gives a negative number, is this force attractive or repulsive? Why? What if F is positive?
4. Imagine two small neutral metal balls 1 meter apart. Is there an electrostatic force between them? If you transfer one coulomb of charge from ball #1 to ball #2, now is there an electrostatic force? If so, is it attractive or repulsive? Explain your answer.
5. In Question 4, if you transfer an additional coulomb of charge from #1 to #2, by what factor does the force change?
6. If a metal ball contains an excess of electrons, predict how they will be distributed. Explain.

The Electric Field

Another anology between electrostatics and gravity is the idea of the *field.* The conceptual problems created by the notion of "action-at-a-distance" are as severe in the present case as in Chapter 5. We have the mathematical expression of Coulomb's law. What does it really mean? Does a positively charged object look about, see how far away a negative charge is, make a rapid calculation of the inverse square of the distance, and then move accordingly? Well, that is hardly a reasonable interpretation. But, nevertheless, how does one charge "know" that the other one is there?

We can't really answer this question, but the idea of the field is a further step along the road toward understanding it. Consider a single positive charge located at some arbitrary position. What we imagine is that this charge, merely by virtue of its existence, *distorts* the space around it. Any other charge (we call it a *test charge*) in the vicinity of the first one is influenced by the distortion; it is accelerated because of the field. The acceleration is toward or away from the first charge, depending on whether the test charge is negative or positive. The magnitude of the acceleration depends on where the test charge is positioned in the field.

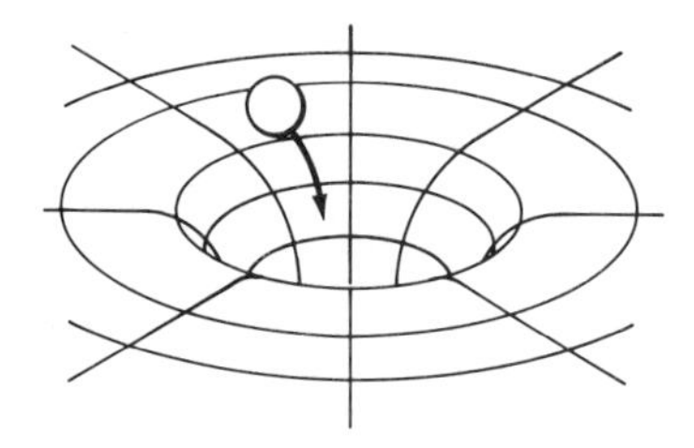

Figure 15-7. The hole in the floor represents a field attracting the ball.

As a crude analogy, imagine a billiard ball rolling across a floor or some other surface. If some region of the surface is not flat, but distorted somehow, then the ball will not merely roll straight across the floor. Suppose, for instance, that the surface curves gradually toward a hole, like the drain in a shower. Then, obviously, the ball rolls into the hole (see Figure 15-7), much like a golf ball rolling into the cup on a putting green. We can imagine that the distortion of the surface represents the field. The behavior of the ball depends on where it is located on the surface.

For the electrostatic field, the original charge produces the spatial distortion. How the test charge responds depends on its location. The *strength* of the electrostatic field tells how much the space has been distorted by the charge. Mathematically, it is

$$E = \frac{kq}{d^2}$$

where q is the charge causing the field and d is the distance from the charge. The field strength is analogous to the steepness of the descent leading into the hole in the floor. The factor k is the same Coulomb-law constant we encountered earlier. You can see that the electrostatic field E has the same mathematical form as the gravitational field g, discussed in Chapter 5.

Carrying the analogy with gravity one step further, we can represent the strength and direction of the field by arrows. Figure 15-8 illustrates these for two fields, one due to a positive charge and one to a negative charge. The convention is that the arrows point in the direction in which a *positive charge* would move if placed in the field. Since it would be repelled by another positive charge, the field lines in Figure 15-8a point outward, and vice versa for the negative charge in Figure 15-8b.

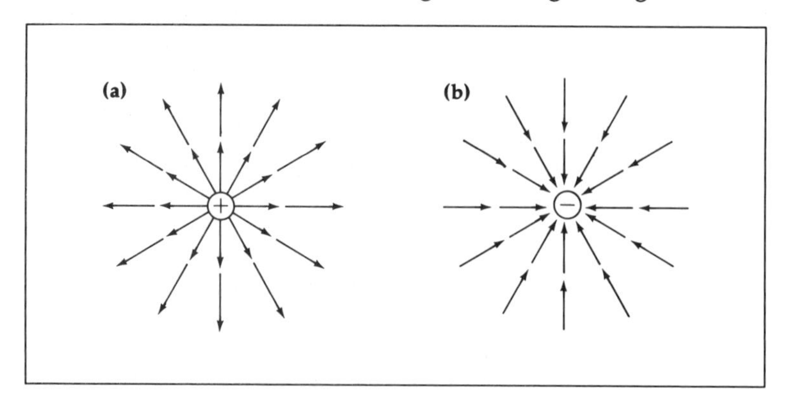

Figure 15-8. Electrostatic field lines for ⊕ and ⊖ charges.

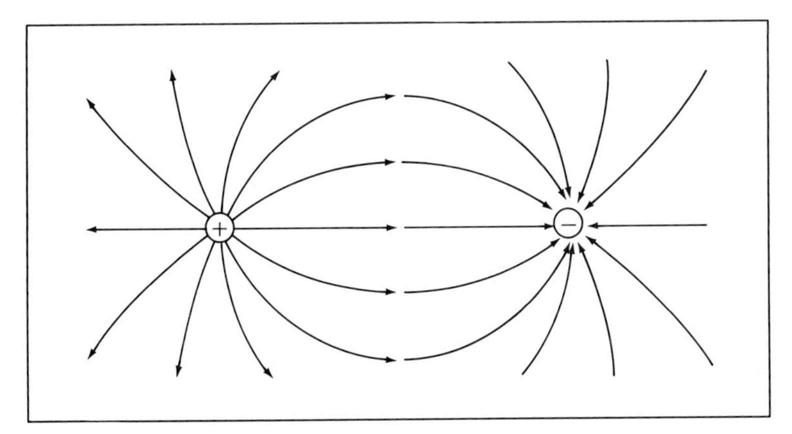

Figure 15-9. Combined field for ⊕ and ⊖ charges.

Up to now we have considered only the field due to a single charge. If two or more charges are present, the resulting field will depend on both of them. That is, it will be a combination of the individual distortions due to each charge. Figure 15-9 illustrates the field due to a positive and a negative charge that we imagine to be clamped in position. Notice that the lines begin on the positive charge and end on the negative charge; this is how a free positive test charge would move in the field. Figure 15-10 illustrates the field between two oppositely charged metal plates.

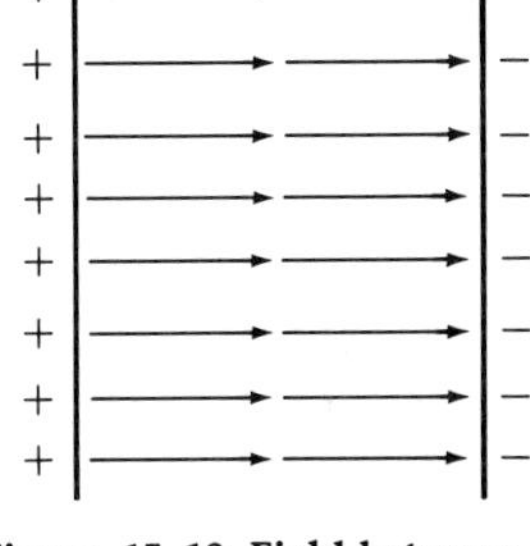

Figure 15-10. Field between oppositely charged metal plates.

Learning Checks

1. How are the *electric* field and the *gravitational* field similar? How are they different?

2. Draw the field due to two negative charges and one positive charge, where all three are on a straight line and the positive charge is halfway between the negative charges.

Electric Potential

Because of the electrostatic force, it takes work to separate a positive and a negative charge, or to press two like charges together. Therefore, we have another similarity with gravity, in which work is required to lift an object off the ground. Recall that in Chapter 10 we described the gravitational potential energy of an object as the energy it has due to its position in a gravitational field. Likewise, we may visualize the *electric potential energy* as the energy of a charged object due to its position in an *electric* field.

When you lift a boulder off the ground, you increase its gravitational potential energy. We can convert this energy into work by letting the rock fall back to the ground through the gravitational field of the earth. In like manner, as shown in

Figure 15-11. Electric potential energy. (a) Pushing the ⊖ charge away from the ⊕ charge increases its electric potential energy. (b) The "pushing energy" in (a) is recovered by letting the ⊖ charge fall back through the field.

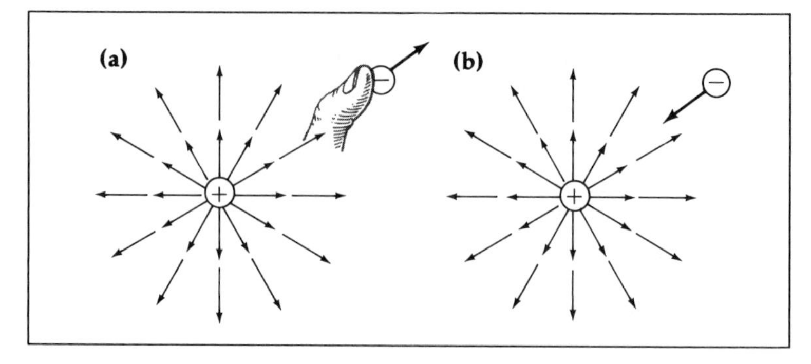

Figure 15–11a, we can increase the electric potential energy of a negative charge in the field of a positive charge by doing work to separate the two. We can reconvert this energy into work by allowing the negative charge to "fall" back through the field toward the positive charge, as shown in Figure 15–11b. Just as the gravitational potential energy of a mass changes with different distances from the earth, so the electric potential energy changes with the distance between the positive and negative charges.

Usually in discussing a charged body in an electric field we speak of its *electric potential* or *voltage.* The electric potential is defined as the ratio of the body's electric potential energy to its charge. In the mks system, energy is expressed in joules and charge in coulombs. The ratio of these, joules divided by coulombs, is called a *volt* (V), named after the Italian physicist Alessandro Volta:

$$\text{volt} = \frac{\text{joule}}{\text{coulomb}}$$

Note the distinction between *electric potential* and *electric potential energy.* They are not identical, and the double use of the word "potential" is rather unfortunate. The *electric potential* is an average property in that it describes how the electric potential energy is distributed among the various charges.

Let us return to the analogy to gravity one more time. You will recall that we were concerned only with the *difference* in gravitational potential energy between two places, not with the absolute potential energies. For example, when the moving company carries a piano from the third floor to the fifth floor, the work required corresponds to the *difference* in potential energy between the two floors. It doesn't matter whether you take the zero of potential energy to be the ground, the center of the earth, or the nearest star. What counts is the difference.

All points on the fifth floor of the building have the same gravitational potential energy. This is shown by the fact that a ball sitting on the floor has no tendency to roll in any direction. Thus, the floor may be called an *equipotential surface,* since every point on it has the same potential energy as every other point. If somehow the floor were tilted, however, the ball would spontaneously roll so as to decrease its gravitational potential energy, and the floor would no longer be an equipotential surface.

Similarly, we may speak of *difference in potential* within an electric field, sometimes referred to as a *voltage drop.* (See Figure 15-12.) An electric charge has no spontaneous tendency to move along an equipotential surface, that is, from one point to another of the same potential. If there is an electric potential difference within the field, the charge will move spontaneously from the higher potential to the lower. This movement of the electric charge leads to the idea of an electric current, which we will study next.

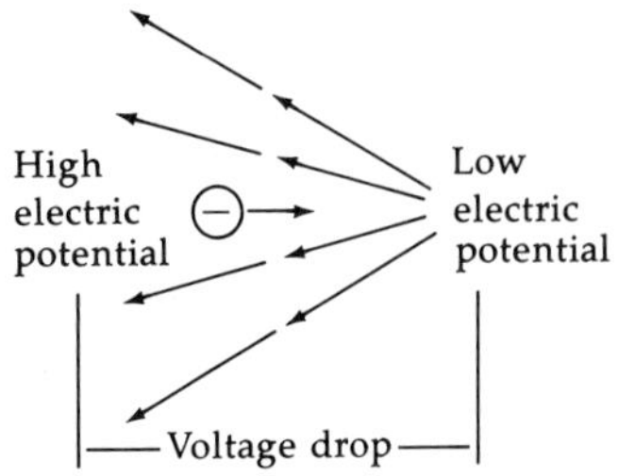

Figure 15-12. The negative charge moves from *high* to *low* electric potential.

Learning Checks

1. What is electric potential energy?
2. Distinguish between electric *potential* and electric *potential energy.*
3. For the electric field in Figure 15-9, which is the direction of decreasing electric potential for a positive charge? For a negative charge?
4. Is any work required to move a charge along an equipotential surface in an electric field? Explain. What is the analogous situation for a gravitational field?

ELECTRIC CURRENT

Some important uses of electricity occur because charges can be made to move along a conductor, such as a wire. A flow of charge is called an *electric current.* By means of electric currents, energy is transferred from place to place, energy that does all the things we commonly associate with electricity—lighting bulbs, playing stereos, and so on. Let us use what we know about fields and electric potential to understand current electricity.

When you scoot across a wool carpet, scraping electrons from your shoes, you become positively charged because of a deficiency in electrons. This sets up a positive electric field around your body. The metal doorknob on which you are about to shock yourself has an electric potential different from that of your hand. As you reach out to grab the metal, the electrostatic force, which increases the closer you get to the door, affects the electrons in the metal; they are attracted to your hand. As soon as the force is strong enough, the electrons jump from the doorknob to your hand and continue to do so until there is no longer a difference in potential. That is, the charges move until the two parts—your hand and the doorknob—are at the *same* electric potential. When there is no longer a potential difference, the flow of charge stops.

This is somewhat like the flow of water in the U tube shown in Figure 15–13. When the stopcock is closed as shown in Figure 15–13a, the difference in the levels of water in the two arms corresponds to a difference in the gravitational potential energy. When the stopcock is opened, the water will flow in the direction of decreasing gravitational potential energy until the water levels are the same, as shown in Figure 15–13b. When there is no longer a difference in the potential energy, the water stops flowing.

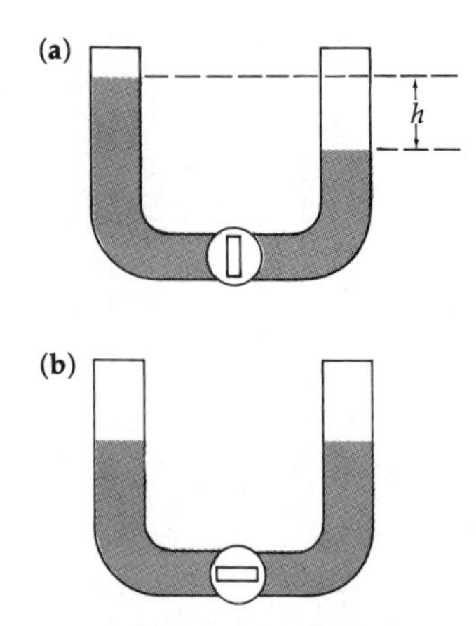

Figure 15-13. Water flows to decrease its gravitational potential. (a) Difference in gravitational potential between the two levels is proportional to *h*. (b) When the levels are at the same height, there is no difference in gravitational potential.

The spark you often see when you shock yourself results from the surge of energy released as the transfer of charge reduces the potential difference to zero. This one-shot process is interesting to watch (and painful to experience!), but it doesn't have much practical value. More useful is the electric current, which results from *maintaining* the transfer of charge as a continuous flow between two points.

An electric current requires sustaining a difference in electric potential between two points, just as a waterfall needs a difference in height between the top and the bottom. A *battery* is a device that maintains this electric potential difference by piling up more negative charges at one terminal than at the other. This makes the terminal having the smaller number of charges positive with respect to the other. If an electrical device such as a lamp is connected between the terminals, a current will flow because of the difference in electric potential. The charges depleted in this way are replaced, usually by a chemical reaction, and the potential difference is maintained throughout the life of the battery.

Figure 15–14 shows a simple version of a *lead storage battery.* The negative terminal is a strip of lead and the positive terminal is lead coated with lead oxide. Both are immersed in a solution of sulfuric acid. The lead at the negative terminal reacts with the acid to produce a fresh supply of electrons. This reaction also dissolves the lead. At the positive terminal, the electrons are captured by the lead oxide, causing it, too, to dissolve.

Figure 15-14. Lead storage battery. Electrons flow along the conducting wire from the negative to the positive electrode.

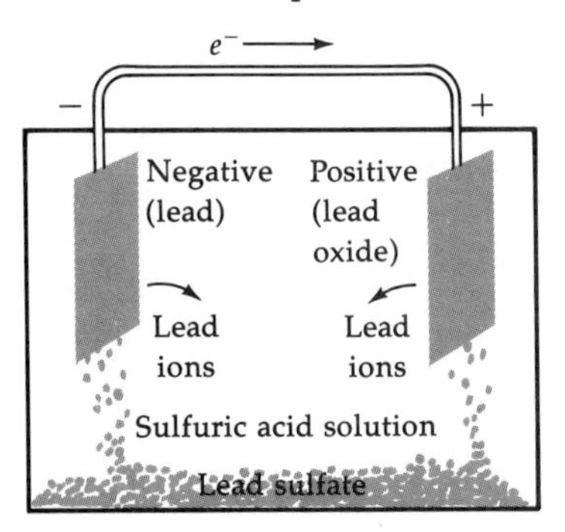

Thus, we have the requirements for maintaining a current: a steady supply of electrons produced at one electrode, and a means of consuming the electrons at the other. The electric current flowing through the connecting wire provides energy for doing work, such as turning a motor or powering electric lights. The battery will keep on producing this current as long as the chemical reaction continues.

Batteries are rated on the basis of the difference in electric potential between the terminals. For instance, the 12 volt battery commonly used in automobiles maintains a difference of

Batteries and Electric Vehicles

"In 15 years, more electricity will be sold for electric vehicles than for light," the great American inventor Thomas Edison predicted in 1915. Why was he so wrong?

The first electric vehicles, built as early as 1837, used crude iron-zinc batteries. Nearly all of the prototype electric cars of today run on deep-discharge storage batteries (up to sixteen may be carried in the car) plus an additional battery that operates the electrical systems (windshield wipers, horn, headlights, and so on). Theoretically, a battery-powered car should be ideal—there are no toxic emissions to pollute the air, there is no danger of explosion in an accident, and there is no tank to keep refilling. Yet, to date, no really successful product has been sold on a mass basis, and many an electric car manufacturer has come and gone in the last 150 years.

The trouble is that the clean, quiet battery of the electric vehicle is also its greatest disadvantage. Electricity itself is not easily portable. Storing sufficient energy to move a 2 ton auto requires a lot of heavy batteries. It takes 500 pounds of batteries, at a minimum, to power a typical compact about 25 miles, which is about what a

The U.S. Department of Energy commissioned General Electric to build this electric car, which carries 1225 pounds of batteries and can cruise at 55 mi/h.

6 pound gallon of gasoline accomplishes. The batteries have to be recharged approximately every 60 miles, a process that involves 4 to 8 hours of the driver's time. Finally, the most unappealing aspect of electric vehicles to consumers is their lack of speed. Some vehicles may be able to go from a standstill to 30 mi/h in 2 seconds, but, in accelerating from 30 to 55 mi/h, they move with extreme sluggishness, weighed down by their batteries.

There are, to be sure, electric vehicles doing serviceable jobs, as post office pickup vans and golf carts. Great Britain's "milk floats" are the best example. These electrically powered, door-to-door milk delivery vehicles have been operating in England's cities for over 40 years. They are the sole long-term cost-effective electric fleet in the world and, as is absolutely necessary, they have a mature support (recharging and maintenance) system.

Researchers have not given up on the electric vehicle, however. New designs offer promise. General Motors has developed an improved zinc–nickel oxide battery capable of storing two times the energy of the lead-acid battery. Gulf & Western has a unique zinc-chloride battery that, on a test drive, has taken a vehicle 175 miles on a single charge. Ford is investigating a battery of sodium and sulfur electrodes that may yield three times the energy of lead-acid batteries. But each of these versions still contains certain engineering difficulties.

In the long run, Edison's vision may come true, but it is obviously going to take more years than anyone foresaw.

Sources: "Energy: Everyone Now into the Act," *Technology Review*, November 1979, pp. 74–80.

David Glass, "Electric Cars at the Crossroads," *Science Digest*, February 1982, pp. 12–14.

12 volts—that is, 12 joules per coulomb. This means that when a coulomb of charge returns to the battery after having expended its energy in doing work, the chemical reactions taking place in the battery resupply it with 12 joules of energy. The charge can then return to the car and again do work.

The path through which the continuous flow of charge takes place is called a *circuit*. Current in a circuit is measured in terms of the amount of charge that flows for a given length of time. One coulomb of charge moving through the circuit each second constitutes a current of 1 *ampere*, abbreviated A and usually called an *amp*. (This unit is named after the French physicist André Marie Ampere.) For example, if you could sit beside a circuit and count 10 coulombs of charge passing you in 5 seconds, the current would be

$$\frac{10\text{ C}}{5\text{ s}} = 2\text{ A}$$

Alternating and Direct Current

In the battery shown in Figure 15–14, the charge always moves in the same direction, negative to positive. This is called a *direct current,* labeled DC. Direct current is the kind generated by all batteries.

The current in houses and buildings is called *alternating current* (AC). Here, as the name implies, the current alternates back and forth, changing directions after a given interval of time. Electrons in the circuit move to and fro about relatively fixed positions. This is accomplished by making one side of the power supply first positive, then negative, then positive, and so on. The most common alternating current is "60 cycle," which means that the current completes one cycle of direction change sixty times each second. This is schematically illustrated in Figure 15–15.

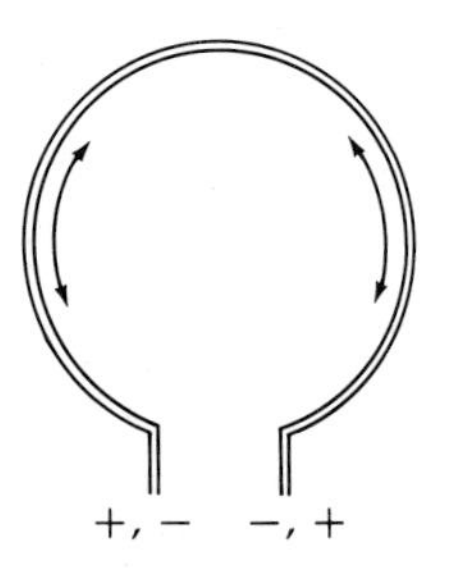

Figure 15-15. Schematic of 60 cycle AC. The current changes direction sixty times each second.

Alternating current is useful because electrical energy in this form can be transmitted over long distances. We shall discuss this topic in the next chapter when we look at magnetism and its relation to electricity.

Learning Checks

1. What is an electric current?
2. What conditions must be maintained in order for there to be an electric current?
3. What is the definition of an *ampere?*

Conductors and Insulators: Resistors

In the battery shown in Figure 15–14, the wire connecting the two electrodes allows electric charge to flow. Materials that readily transmit electric charge are called *conductors.* Silver is an excellent conductor. Other metals, such as copper and aluminum, are also good conductors. Copper is commonly used in household wiring because it is much cheaper than silver.

Many materials do not conduct electricity at all. These are called *insulators.* Some examples are wood, rubber, and various plastics. In an ordinary double-wire lamp cord, for example, it is easier for electric charges to flow through many feet of copper wire than through the few millimeters of rubber separating the two wires.

We can interpret the properties of conductors and insulators on the basis of our picture of the atom. Conductors are made up of atoms whose outer electrons, situated far away from the nucleus, are rather loosely attached to the individual atoms. They form a kind of "sea" of electrons that are relatively free to roam. The electrons in an ordinary wire have large velocities as they zip about, but there is no current, because the electrons move in random directions. A current will result when the electrons are made to move "in step," so to speak—that is, all in the same direction rather than roaming about randomly. We can compel the sea of electrons to move in a particular direction by establishing a potential difference between the ends of the wire—by attaching them to the terminals of a battery, for example. The resulting electric field travels through the circuit, gives direction to the electrons' motions, and pushes them toward the positive terminal. The electrons themselves do not travel very far, but collide with the fixed metal atoms. Their average speed is thus fairly small. But the energy carried by the electric field travels rapidly through the circuit.

Insulators, on the other hand, are composed of atoms whose electrons are more tightly bound. We can picture each electron as being more or less fixed to one particular atom, rather than free to travel about, as those in metals are. Since there are no electrons available to carry charge through the insulator, no current will flow, even if we use a very strong voltage source. For example, a strip of paper attached across battery terminals will not show any measurable current. The glass and amber rods used in the electrostatic experiments are insulators: the charge produced on the rod does not flow through the holder's body to the ground.

Among conductors, the extent to which they carry a current differs quite a lot depending on the properties of the individual material. Some substances readily conduct a current, even under the smallest voltages, while others must have large voltages applied before they will carry an appreciable current.

How well or poorly a conductor carries a current for a particular voltage is expressed by its *resistance*. We said earlier that the individual conduction electrons do not go very far without bumping into atoms. These collisions slow the current down, so materials in which the number of such collisions is high retard the flow of current more than others. That is, they have a large resistance.

Many materials have a very simple relationship among current, voltage, and resistance. George Ohm, the German physicist who first studied this relationship, found that for a given conductor, the amount of current is directly proportional to the applied voltage. If we designate the current as I and the potential difference (voltage) as V, then we can write this relationship as

$$I = \frac{V}{R}$$

where the proportionality constant, R, is the resistance in the circuit.

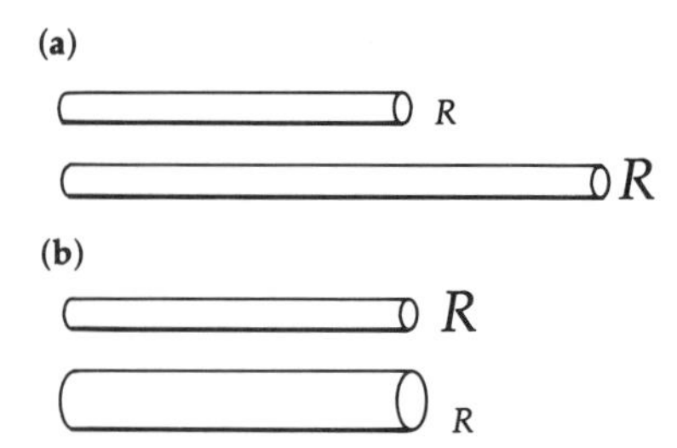

Figure 15-16. Resistance depends on the geometry of the conductor. (a) The longer wire has the greater resistance. (b) The thicker wire has the smaller resistance.

This equation is called *Ohm's law.* It tells us that for a circuit having a fixed value of resistance we can double the current by doubling the applied voltage. Notice also that for a certain voltage the amount of current flowing in a circuit decreases as the resistance increases, and vice versa. This is consistent with the idea of resistance as an impediment to current flow.

The unit of resistance is given in volts per ampere, which is called the *ohm,* abbreviated Ω. That is, if the potential difference in a circuit is 1 volt and the amount of current is 1 ampere, then the resistance in that circuit is 1 ohm:

$$\text{ohm} = \frac{\text{volt}}{\text{ampere}}$$

The size and shape of a material, as well as its atomic makeup, affect its resistance. For example, if you have two pieces of wire that are identical except for their length, the longer one will have a larger resistance. This is reasonable, for in the longer wire there will be more collisions between electrons and atoms—meaning more instances in which the current is slowed down. If 20 volts will drive a 4 ampere current through a 1 meter length of wire, then the current in a 2 meter length of the same wire with the same applied voltage would be only 2 amperes. The values of the resistance in the two wires would be 5 ohms for the shorter and 10 ohms for the longer.

On the other hand, a thicker wire will have a smaller resistance than a thinner wire of the same material. The larger cross-section of the thick wire provides a greater area for charges to move through. Figure 15–16 summarizes these size-and-shape factors of resistance.

Superconductors

Superconductors may render most existing technology obsolete. They can, for example, bring about such marvels as a cubic inch supercomputer, trains that float on energy waves, wire that can transport all the energy for a big city in a single cable, guns with enough power to shoot a satellite into orbit, and instruments that can sense magnetic changes in your brain as you read. Presently, the biggest drawback is that superconductors have to be continuously chilled in liquid helium, which is scarce and expensive. That is why physicists are searching feverishly for ways to make superconductors at room temperatures. This goal has been given top priority in several nations.

Source: Steve Aaronson, "Cold Currents," *Omni,* October 1981, pp. 121–125.

Temperature and Resistance: Superconductors

Varying the temperature of a conductor changes its resistance. This is not surprising, for resistance arises because of the electrons and atoms banging around. Raising the temperature jiggles the atoms around more, so an electron trying to make its way through the metal is more likely to bump into the atoms than when they are relatively quiet. Hence, a given metal has greater resistance at higher temperatures. Conversely, decreasing the temperature lowers the resistance. The electrons can move about more easily, being less likely to bump into an atom making up the metal.

A gradual decrease in resistance with decreasing temperature is characteristic of most metals. But there are some materials whose resistance abruptly plunges to *zero* at very low temperatures, in the neighborhood of absolute zero. Such materials are *superconductors,* because they offer no resistance to the flow of current. Figure 15–17 compares the gradual decrease of resistance for copper with the quick drop-off for tin, which becomes a superconductor at about 3.7 K. Other metals have other characteristic temperatures. Mercury, for instance, has zero resistance at about 4 K, and a tin-niobium alloy at 18 K. Above their characteristic temperatures, these materials show the normal relationship between temperature and resistance.

The advantages of zero resistance in a material are obvious. Once begun, a current would continue to flow forever in principle, and certainly for years in practice. Such a supercurrent would be free of any of the heat losses that occur because of resistance in wires under ordinary conditions. The trade-off, of course, is the expense of creating these extremely low temperatures, certainly not a trivial matter.

A proper understanding of superconductors would take us beyond the scope of this book, but we can get a rough idea. At the characteristic temperature, the electrons interact with the metal atoms in a way that causes the electrons to join up abruptly in pairs. This pairing up prevents the electrons from bouncing off the atoms as they move along. Once the "bumping" is removed, there is no resistance.

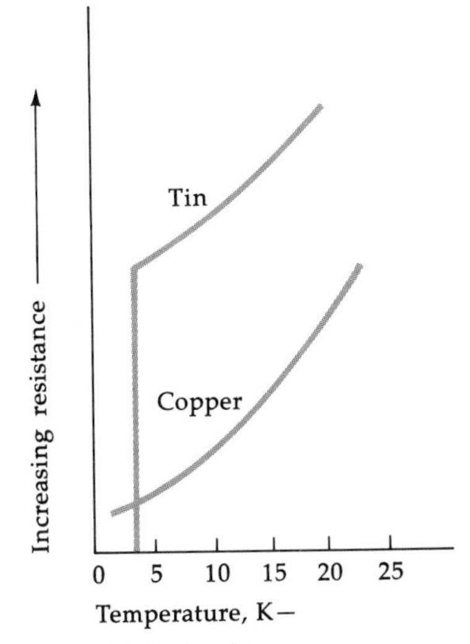

Figure 15-17. Resistance versus temperature for conductors.

Learning Checks

1. Distinguish between conductors and insulators.
2. What is *resistance?* Give its units and tell how it is related to the type of material.
3. If an insulator carries no current, no matter what the applied voltage, what can you say about the value of its resistance?

Parallel and Series Circuits

In any circuit in which more than one resistor is present, there are two ways in which they may be connected, either in *series* or in *parallel.* Figure 15–18 shows a battery circuit in which two resistors are connected in series. The current must pass through both resistors in order that the charge flow from one pole of the battery to the other. Since the current is retarded by both resistors, the *total resistance* in a series circuit is the *sum* of the individual resistances. Thus, for the circuit in Figure 15–18, the total resistance is

$$R = R_1 + R_2$$
$$= 10\ \Omega + 5\ \Omega$$
$$= 15\ \Omega$$

Figure 15-18. Resistors R_1 and R_2 are connected in series.

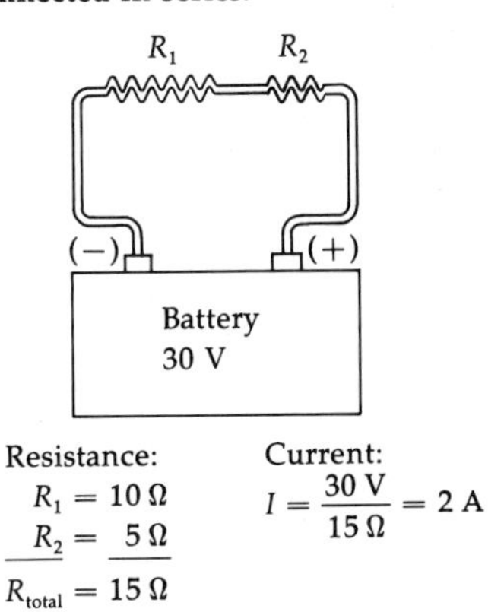

Recall that according to Ohm's law the current in a circuit varies inversely with the resistance. This means that for a fixed voltage the current in a series circuit *decreases* as the number of resistors increases. Suppose, for example, that the battery in Figure 15–10 is rated at 30 volts. Then, if only R_1 were present in the circuit, the current would be

$$I = \frac{V}{R_1} = \frac{30\ \text{V}}{10\ \Omega} = 3\ \text{A}$$

The basic pacemaker — a lifesaving circuit — is a timer for the heart.

With R_2 alone, the current would be

$$I = \frac{30\ \text{V}}{5\ \Omega} = 6\ \text{A}$$

But when both R_1 and R_2 are in the circuit connected in series, the current is

$$I = \frac{30\ \text{V}}{15\ \Omega} = 2\ \text{A}$$

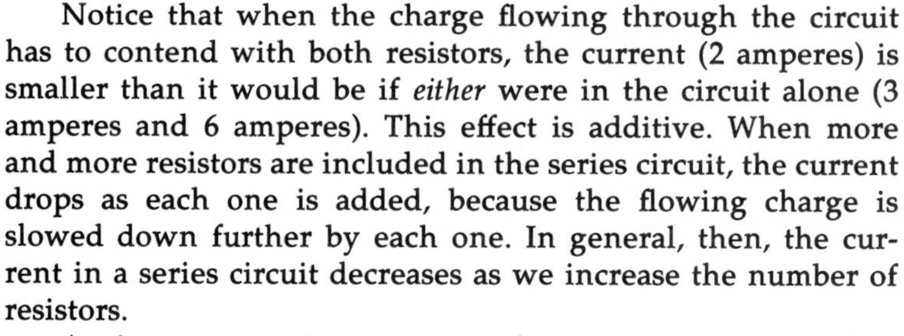

Notice that when the charge flowing through the circuit has to contend with both resistors, the current (2 amperes) is smaller than it would be if *either* were in the circuit alone (3 amperes and 6 amperes). This effect is additive. When more and more resistors are included in the series circuit, the current drops as each one is added, because the flowing charge is slowed down further by each one. In general, then, the current in a series circuit decreases as we increase the number of resistors.

An important point to notice about series circuits is this: All resistors must be connected in order for the current to flow. For example, suppose that in Figure 15–18, R_2 is a light bulb that burns out. This breaks the circuit: there is no longer a complete loop, and the current cannot flow through R_1.

Quite often, strings of Christmas tree lights have bulbs that are connected in series. If one bulb is removed or burns out, they all go out. You must then check them one by one in order to find the culprit.

An example of a *parallel* circuit is shown in Figure 15–19. We may imagine that the charge flows from the negative pole of the battery to the positive pole. When it reaches point A it may take either the path through R_1 or that through R_2. This "choice" of paths means, in fact, that *more* current can flow than if either R_1 or R_2 were alone in the circuit. Put another way, the resistance of two devices connected in parallel is *smaller* than

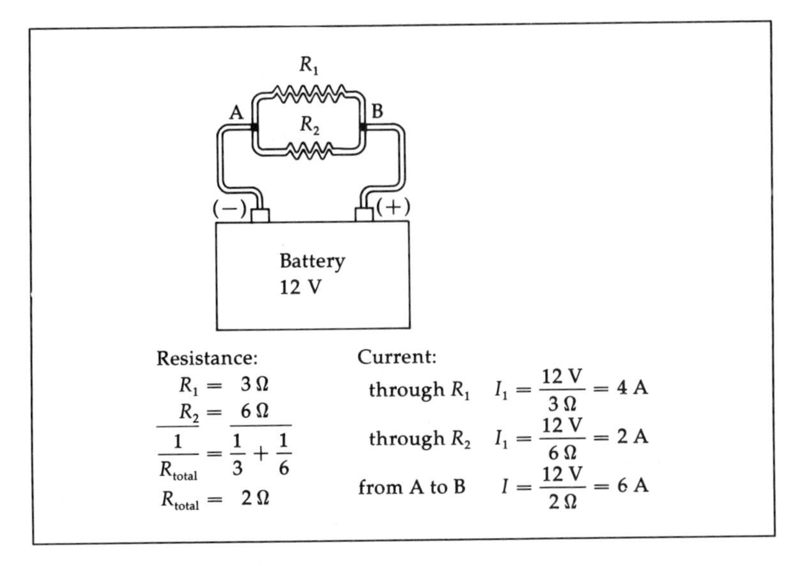

Figure 15-19. Resistors R_1 and R_2 are connected in parallel.

the resistance of either of them taken individually—just the opposite of the effect for a series arrangement.

This may seem like a paradox: the more resistors we have, the smaller the resistance. Suppose we take a familiar situation as an analogy. We can compare charges flowing along a wire to automobiles moving along a turnpike. For the resistance devices, substitute collection booths at a toll plaza—they impede the flow of traffic (that is, the "current") along the highway. The booths are arranged in parallel, like the resistors in Figure 15-19. If only one booth is open, traffic creeps along. But if a second booth opens up, the cars have a choice of lanes and the traffic flows more readily. The more booths available, the faster the movement of the "current" and the smaller the overall resistance.

For the two resistors in Figure 15-19, the total resistance in the circuit, R, is obtained from

$$\frac{1}{R} = \frac{1}{R_1} + \frac{1}{R_2}$$

This gives a value of R smaller than either R_1 or R_2. Taking the values in the figure, we have

$$\frac{1}{R} = \frac{1}{3} + \frac{1}{6} = \frac{3}{6} = \frac{1}{2}$$

so $R = 2$ ohms—compared with R_1 (3 ohms) and R_2 (6 ohms).

Since the total resistance *decreases* as we include more resistors in parallel, it follows that (for a fixed voltage) the amount of current goes up. In our example, the current is 4 amperes with only R_1 in the circuit, but jumps to 6 amperes when we add R_2 in parallel.

One big advantage of a parallel circuit is that even if some of the resistors are removed, the current can still flow through the remaining resistors and the circuit remains complete. This lets you avoid that Christmas-tree-light problem of series circuits, in which they all go out if one does.

Household circuits carry alternating current and are wired in parallel. Typically, a voltage of about 110 volts is supplied to the circuits in which the various appliances and other electrical devices are connected. An example is shown in Figure 15–20, in which a 10 ampere heater, a 3 ampere lamp, and a 2 ampere motor are all in the same parallel circuit. The parallel connection allows each device to operate independently of the others. If only the heater is turned on, the total current is 10 amperes. Switching on the lamp adds an additional 3 amperes, and including the motor adds another 2 amperes. This gives a total of 15 amperes in the circuit.

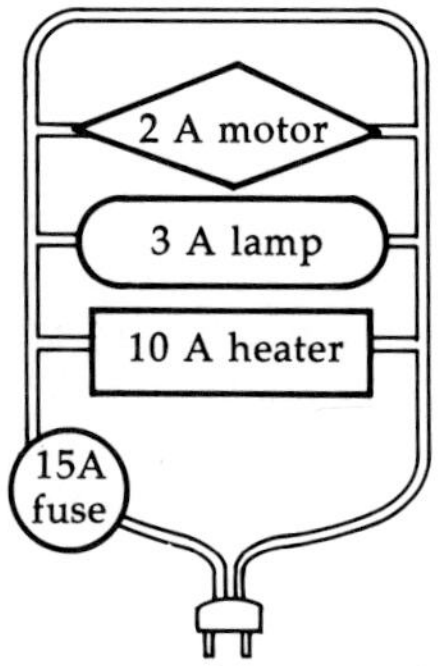

Figure 15-20. Household appliances in parallel. The fuse will blow if the current exceeds 15 amperes.

Adding any additional appliances to this circuit would increase the current to more than 15 amperes. If the current in a typical house circuit is much more than 15 or 20 amperes, the circuit may dangerously overheat and cause a fire. For this reason, circuits include a fuse or other device, connected in *series,* that melts or otherwise breaks the circuit whenever the current exceeds a specified amount. The 15 ampere fuse in Figure 15–20 will blow out if the current in the circuit becomes greater than 15 amperes.

Electric Shocks

Current passing through your body causes electric shocks. This means that if your body is part of a complete electric circuit, then you are liable to get a shock.

It is the *current* that produces the electrical shock, not the *voltage* as such. That is, the amount of voltage has no particular importance except that some voltage is necessary to drive the current through your body. But regardless of how much voltage is present, no current will flow unless there is a complete circuit. This means that different parts of your body must be at different potentials in order to get a shock, for current will flow only as a result of a *difference* in potential—that is, a voltage drop.

For example, suppose you are standing on a dry rug and accidentally touch a frayed electrical cord plugged into the wall outlet. Your entire body will be at a potential of 110 volts, so no current will flow. You receive no shock, since there is no potential difference across your body. But now if you touch,

Table 15-1.
Effects of Electric Current on the Body

Current (ampere)	Effect
0.001	You barely feel this
0.005	The maximum harmless current
0.007–0.015	Muscles out of control—you can't let go
0.05	Pain and exhaustion
0.1–0.3	Uncoordinated contraction of the heart (ventricular fibrillation): death

say, a water pipe with your other hand, you may receive a severe, perhaps fatal, shock. The water pipe is "grounded"—that is, at zero voltage—so now there *is* a difference in potential, the circuit is complete, and current will flow.

You may have noticed that birds can safely perch on high-voltage electric cables. Again, as long as the bird touches only the one cable, its entire body will be at the voltage of the cable. Since there is no potential difference, no current will flow and the bird will not receive a shock.

Recall from Ohm's law that the current in a circuit depends on the applied voltage and on the resistance. Most of the resistance of the human body is in the skin. Dry skin can have a resistance as high as 100,000 ohms or more. But if the skin is wet, either with perspiration (which contains salts, good electrical conductors) or water, then the resistance drops drastically, to perhaps a few hundred ohms.

Table 15-1 indicates the effects of various amounts of current passing through the body. For example, if your skin were dry you would hardly feel 10 volts because only 0.0001 ampere of current would flow across the resistance of 100,000 ohms. Even 120 volts would produce a current of only 0.0012 ampere, barely noticeable. But with moist skin, whose resistance is, say, only 1000 ohms, a 10 volt battery would generate a 0.01 ampere current, giving a fairly strong shock. If your skin were wet, 120 volts would likely be fatal.

Electric Energy and Power

We learned in our study of energy in Chapter 7 that *power* is the rate at which energy is produced or consumed:

$$\text{power} = \frac{\text{energy}}{\text{time}}$$

The metric unit of power is the watt, defined as a joule per second:

$$\text{watt} = \frac{\text{joule}}{\text{second}}$$

We can write the first expression as

$$\text{energy} = \text{power} \times \text{time}$$

which tells us that the energy used by a certain electrical appliance is just equal to the power it consumes times the length of time it is turned on.

Let us look at this in terms of the energy the homeowner or business buys from the utilities company. We have seen that as we add more appliances and other devices that run on electricity in a parallel circuit, we increase the amount of current. Recall that current is the amount of charge flowing for a given length of time:

$$\text{ampere} = \frac{\text{coulomb}}{\text{second}}$$

If we multiply volts by amperes, we get

$$\text{volts} \times \text{amperes} = \frac{\text{joule}}{\text{coulomb}} \times \frac{\text{coulomb}}{\text{second}}$$
$$= \frac{\text{joule}}{\text{second}}$$

which is a watt, the unit of power. So we have

$$P = VI$$

or

$$\text{power} = \text{voltage} \times \text{current}$$

So for a fixed voltage, the more current you use, the greater the amount of power—which means, of course, more energy.

Energy purchased from the utilities company is usually priced in terms of *kilowatt hours* (kWh). (Since a kilowatt is 1000 watts and 1 hour is 3600 seconds, then 1 kilowatt hour equals 1000 watts × 3600 seconds = 3,600,000 joules.) Your electric bill will indicate the price of the energy at a few pennies per kilowatt hour. For instance, if the price is 8 cents per kWh, the cost of burning a 75 watt bulb for 15 hours would be

$$\frac{\$0.08}{\text{kWh}} \times 0.075\ \text{kW} \times 15\ \text{h} = \$0.09$$

or 9 cents.

Learning Checks

1. What is meant by a *series* circuit? A *parallel* circuit?
2. What happens to the other appliances in a parallel circuit if one appliance is disconnected?
3. How does the current in a series circuit change as more resistors are added? How does the overall resistance change?
4. Answer Question 3 for a parallel circuit.
5. How is voltage related to electric shocks?
6. Explain the connection among current, voltage, energy, and power.

SUMMARY

Electricity and magnetism are partners in the *electromagnetic interaction,* one of the four basic forces in nature. We distinguish between *static* electricity, in which the charges are at rest, and *current* electricity, consisting of charges moving through a conductor such as wire.

Experiments with static electricity, such as rubbing a glass rod with a silk cloth, demonstrate that there are two kinds of electric charge. The basic element of negative charge is the *electron,* and that for the positive charge is the *proton.* Since the electron is almost 2000 times lighter than the proton, the overall charge of an object depends on whether there is an excess or deficiency of electrons. Modern applications of *electrostatics* include the precipitator and the dry-copy imaging process called *xerography.*

Coulomb's law expresses the electrostatic force between two charges. It is an inverse-square law in which like charges repel and unlike charges attract each other. The electrostatic force is larger than the gravitational force by a factor of 10^{42}. We can picture this electric force as arising from an *electric field,* the distortion of space due to the presence of an electric charge, analogous to the gravitational field caused by a mass. A charge has some *electric potential energy* due to its position in the field. The electric potential, or *voltage,* tells how the potential energy is distributed among the charges in a field.

Charges in motion make up an *electric current,* which flows from high to low *electric potential.* A battery provides one means for maintaining a current by furnishing a constant supply of electrons produced in a chemical reaction. A current is rated in *amperes,* equal to the number of coulombs of charge flowing past a point each second. Batteries produce *direct current* (DC), in which the charges always move in the same direction.

Household current is *alternating* (AC), in which the direction changes about sixty times each second.

Circuits for the flow of current can be arranged either in *series* or in *parallel*. The current in a series circuit must flow through each item connected to it, while a parallel circuit provides alternate routes for the current. Additional resistors decrease the current in a series circuit and increase it in a parallel one.

Electric shocks are caused by electric current passing through the body when different parts of the body are at different electric potentials. For a given amount of voltage, the size of the current depends on the resistance provided by the skin.

Exercises

1. Figure 15–21 shows an *electroscope*, consisting of a conducting rod leading to two light metal foil leaves. When a charged object touches the electroscope, a current flows briefly along the rod, giving the leaves the same charge. This forces the leaves apart. In the uncharged electroscope the leaves are relaxed. Suppose that you have charged the electroscope by touching it with an amber rod rubbed with cat's fur. Is the charge on the electroscope positive or negative?

2. You bring a charged object near the charged electroscope in Exercise 1. As the object approaches, the leaves come closer together until they eventually relax. When the object is brought still closer (but still without touching the electroscope), the leaves stand apart again. Explain what has happened during this experiment.

3. In Exercise 2, is the charge on the object positive or negative?

4. For a charge q placed in an electric field at a location where the field strength is E, how do you calculate the force on the charge due to the field?

5. Name the forces exerted on two protons located at, say, a distance of 1 meter apart. Which force dominates? How can you tell?

6. The gravitational field lines always point *toward* the object producing the field. Is this true of electric charges? Explain.

7. Sketch the electric field due to three positive charges placed at the corners of an equilateral triangle.

8. Suppose two objects have the same electric potential energy. Do they necessarily have the same electric potential? Explain.

9. What do we mean by "voltage drop"?

10. Suppose that in a circuit you double both the resistance and the voltage. What does this do to the current?

11. Would a music systems analyst advise that it is preferable to connect the speakers of your stereo with long or short wire? Explain.

12. Which draws more current, a 100 watt bulb or a 75 watt bulb? Explain.

13. If a negatively charged particle is placed halfway between two charged parallel plates, one positive and the other negative, in what direction will the particle move?

14. Name three good conductors of electricity and three good insulators. How well do these materials conduct heat?

15. We are not usually aware of the gravitational force or the electric force between two ordinary objects. What is the reason for this in each case? Give an example of a situation in which we *are* aware of each one and explain why.

16. Distinguish between an ampere and a volt.

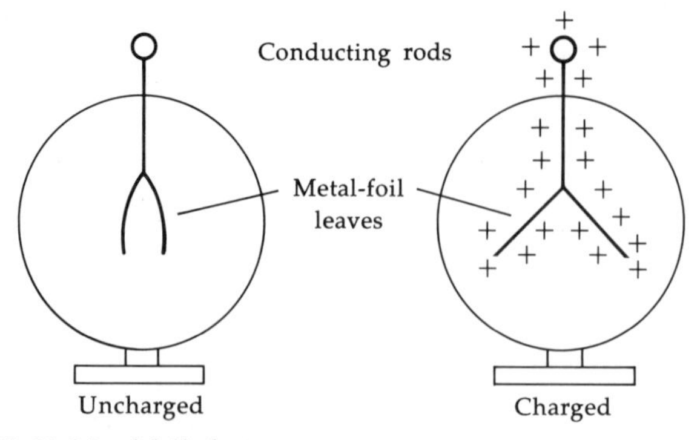

Figure 15-21. Metal foil electroscope.

17. Suppose that you are a security consultant and have available to you the following: a battery, a bell, a pushbutton switch, and plenty of wire. Draw a diagram of an alarm system that sounds whenever the door of a house is opened.

18. Why is it better to use thick wires rather than thin wires for carrying current?

19. A particular circuit has one light bulb. What happens to the current if two more such bulbs are added in series? In parallel?

20. What current does a 30 ohm toaster draw in a 120 volt circuit?

21. How much time does it take for a current of 3 amperes to deliver 75 coulombs of charge?

22. What is the resistance of a device that draws 3 amperes of current from a 12 volt battery?

23. Figure 15–22 shows a circuit that is part series and part parallel. What is the total resistance in the circuit?

24. In a 12 volt battery, suppose that 5 coulombs of charge pass through. How much total energy (in joules) will the battery supply to this charge?

25. A current of 1 ampere flows in a wire. How many electrons are flowing past a given point each second?

26. A flashlight uses two 1.5 volt batteries. What is the voltage applied to the bulb?

27. A radio draws 0.2 ampere and uses a 9 volt battery. What is the overall resistance of the radio?

28. How much current is there in a circuit in which 30 coulombs moves by you in half a minute?

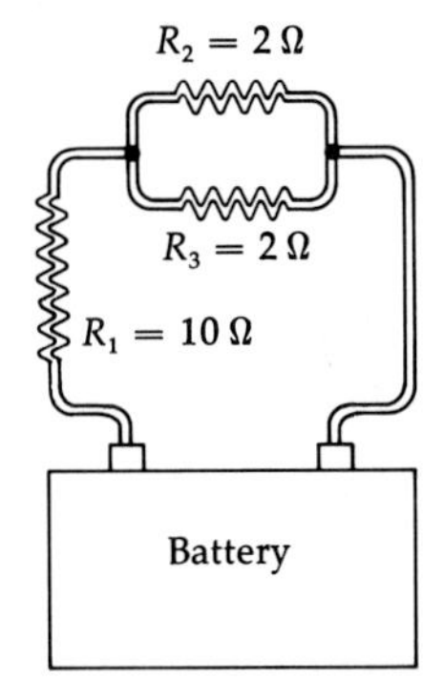

Figure 15-22. Combination parallel/series circuit.

16

MAGNETISM

Magnetic suspension trains may move the populations of the future. With specially designed tracks, they can reach speeds exceeding 300 miles per hour.

In Chapter 15 we learned about one part of the electromagnetic force, electricity. We saw that there are electric charges and electric fields and that the charges could be stationary or moving through a conductor. Now we come to the other half of this force, magnetism. We shall find that there is a close connection between moving electric charges and magnetism.

Almost everyone is familiar with magnets of various kinds, from huge industrial electromagnets to the horseshoe magnets you can buy at the toy store. People have noticed magnetic attraction between objects for thousands of years. Tradition has it that Thales, who lived in Greece in the sixth century B.C., was the first to systematically study these magnetic effects. He found that a particular kind of iron ore called *lodestone* attracted pieces of iron. The sample of ore Thales used came from the neighboring town of Magnesia, so he called it "the Magnesian rock." Iron-attracting materials thus came to be called magnets.

Experiments performed down through the years have demonstrated features of magnetism that are now fairly common. Magnetism can be transferred from one object to another, such as by stroking a piece of steel with a naturally occurring magnetic iron ore. In this way the sample of "normal" steel becomes magnetized.

If a magnetized needle is suspended or pivoted so that it is free to rotate, it orients itself in a direction that runs more or less north and south. (See Figure 16–1.) Furthermore, the same end of the needle always points toward the north pole of the earth. We call this end the *north* pole of the magnet and the other the *south* pole of the magnet. Such a device is the compass, whose invention paved the way for European explorers to venture out more safely into the Atlantic Ocean.

The north and south poles of a magnet are somewhat analogous to the positive and negative charges in electrostatics. We find that the north poles of two magnets repel each other, as do the two south poles, and that the north pole of one magnet attracts the south pole of the other. So, as in the case of electric charges, opposite magnetic poles attract. (See Figure 16–2.)

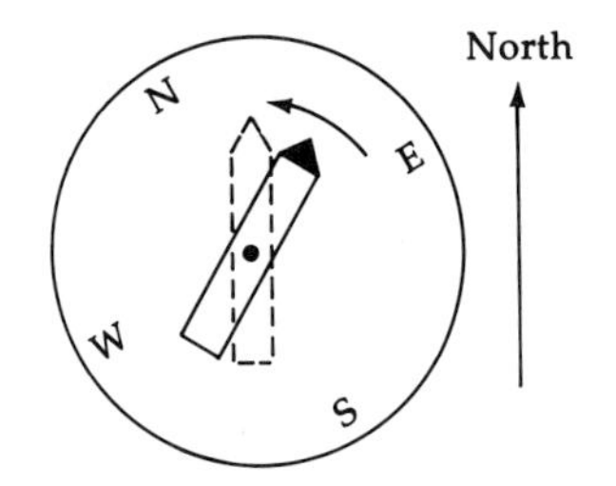

Figure 16-1. Compass needle swings around to point north.

The magnetic attraction or repulsion between two poles diminishes as the distance between them increases. In fact, the magnetic force depends on distance in the same way as the electrostatic and gravitational forces: it falls off with the square of the distance. Magnetism obeys an inverse-square law.

There is one important distinction between electric charges and magnetic poles. Although an isolated electric charge can exist in the absence of any other charges, as far as we know magnetic poles *always* occur as north-south pairs. No experiment to date has produced a magnetic "monopole." For example, suppose you have a bar magnet like the one in Figure 16–3. If you break the magnet in half, you don't get an isolated north pole and an isolated south pole. Rather, you get two magnets, each with a north and a south pole. You could continue breaking them into smaller and smaller pieces, and each time you would get two more magnets.

MAGNETIC DOMAINS

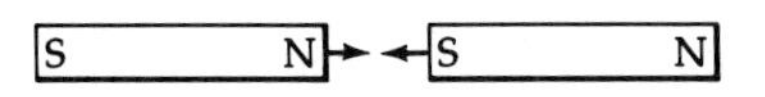

Figure 16-2. Opposite magnetic poles attract.

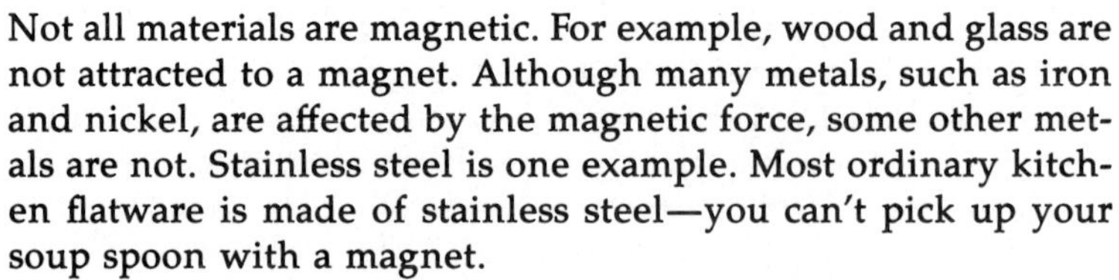

Not all materials are magnetic. For example, wood and glass are not attracted to a magnet. Although many metals, such as iron and nickel, are affected by the magnetic force, some other metals are not. Stainless steel is one example. Most ordinary kitchen flatware is made of stainless steel—you can't pick up your soup spoon with a magnet.

So the question arises, why are some materials affected by a magnet and others not? Why can iron be magnetized by an already-existing magnet, while pure aluminum cannot? Let us try to understand the answers to these questions.

We discussed breaking a bar magnet into several smaller magnets; you can imagine continuing this process until you are left with many thousands of submicroscopic bits. Each would be a magnet, having a north pole and a south pole. We might consider the original magnet to be made up of submicroscopic magnets. In most materials these tiny magnets are randomly oriented, so that the magnetic force of each individual is canceled by its neighbors. The overall effect is to neutralize the

Figure 16-3. Each broken piece is a magnet with N and S poles.

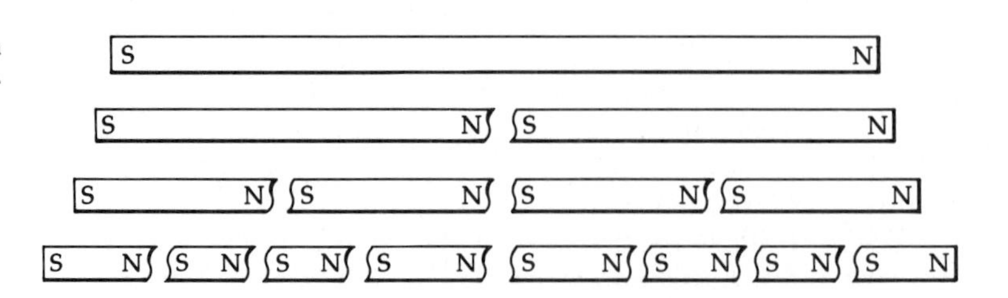

magnetic force throughout the material. This is illustrated in Figure 16–4, where the submicroscopic magnets are pointing from the south pole to the north pole. This randomization of the directions of the magnetic poles means that the material is nonmagnetic; no amount of stroking or coaxing by an external magnet will magnetize it.

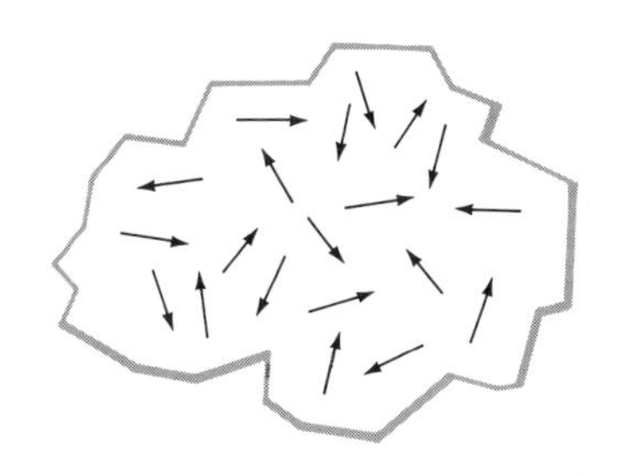

Figure 16-4. Most materials have the submicroscopic magnets oriented in a random fashion.

In a material such as iron, which *is* attracted to a magnet, the situation is quite different. Here, each submicroscopic magnet is aligned with some of its neighbors, the north poles for several adjacent tiny magnets pointing in the same direction. This means that pockets of concentrated magnetism will be scattered throughout the material, as shown in Figure 16–5. We call these regions over which magnetic forces are concentrated *magnetic domains.*

You can see in Figure 16–5 that the domains themselves are randomly oriented, with the result that the magnetic force of one domain neutralizes that of its neighbors. Although it will be attracted to a magnet, this material is not itself a magnet. However, we can make it into one by aligning the domains in a parallel arrangement by stroking with a permanent magnet, in much the same way that a breeze aligns a group of weather vanes. The magnetic strength of the material will be the sum of the strengths of all the domains. Figure 16–6 illustrates the aligned domains.

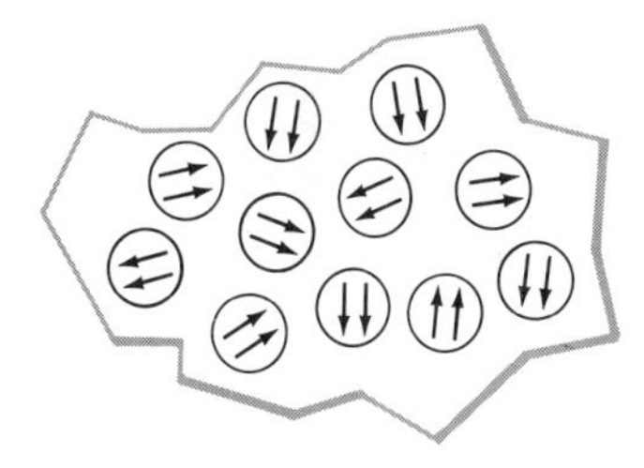

Figure 16-5. Randomly oriented domains in unmagnetized magnetic material.

It is fairly easy to align the domains, which typically are between 0.01 and 0.1 centimeter across. It is quite difficult to align the individual submicroscopic magnets. You can imagine that in magnetic materials such as iron, nature has already done the big alignment job by creating the domains. You add the final touch of magnetization when you bring the domains parallel.

This process can be taken the other way. That is, you can demagnetize a magnet by rerandomizing the domains. Striking a magnet with a hammer will decrease the magnetization. The magnetic repulsion in two magnets laid parallel, north pole to north pole and south pole to south pole, will gradually turn the domains away. Also, increasing the vibrations of the atoms by raising the temperature will disrupt the domains. The critical temperature above which an ordinary magnet shows no magnetic properties is called the *Curie temperature,* after the French physicist Pierre Curie.

Figure 16-6. Aligned domains in a magnetized magnetic material.

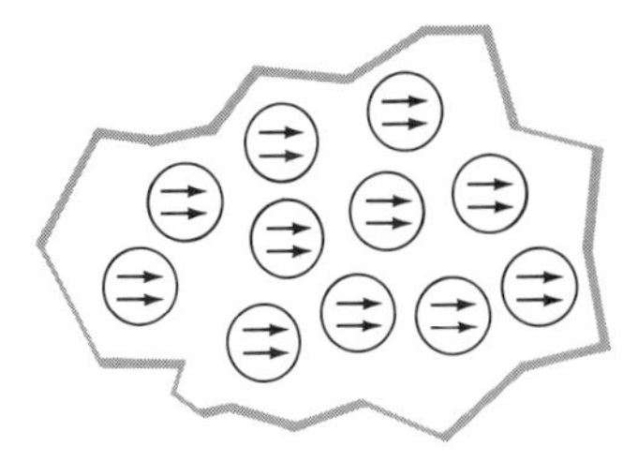

MAGNETIC FIELDS

We have become familiar with the notion of a field from our study of gravity and electrostatics. A magnet has associated with it a *magnetic field* that, as in the case of the other fields, represents a distortion of the surrounding space. Just as an object is affected by the gravitational field of another object, and an

Magnetic Fields in War and Peace

In this day and age, magnetism is familiar to consumers through such products as magnetic tape recorders, the magnetic memories of computers, and so forth. Magnetic fields themselves have been cleverly exploited over the years for both destructive and constructive purposes. Magnetic mines, for instance, have been used in time of war to destroy ships. The original devices employed in World War II were dropped to the bottom of the sea. When a ship passed over, a dip needle on the mine was caused to move a bit because the iron of the ship distorted the earth's magnetic field. This movement of the needle closed an electric circuit and the mine exploded underwater, sending up geysers that could snap a ship in two. Here is a physicist reminiscing about an early defense against such mines:

Another form of magnetic mine sweeper that I later saw was a relatively large ship carrying a strong magnet weighing many, many tons. This was a bar magnet going from the bow—way aft. This bar magnet, many feet in diameter, had coils wound around it which passed a current produced by big motor generators. The magnet was intended to produce magnetic fields ahead of the ship, so that as it steamed around a harbor it would explode magnetic mines before passing over them.

Magnetic fields are also tapped for benevolent purposes, as in the case of experimental magnetic pills, which may help cancer victims. The drugs used to kill cancer cells can be very dangerous to healthy cells, so doctors have to be careful about dosages—if too much is administered, a patient can die of side effects; if too little, death may come from the cancer itself. In hopes of solving this dilemma, molecular physicists at the National Magnet Laboratory are now testing a pill coated with ultrafine magnetic particles. These "magnetic micropills" are filled with anticancer drugs and injected into an artery feeding a malignant tumor. The portion of the body where the tumor is located is placed in a magnetic-field device that attracts the pill, which then saturates only that area with the drug. The tumor alone is affected; the dangerous drug does not go elsewhere via the bloodstream.

Thus, magnetic-field implements have been designed to threaten lives in war and to save lives from illness. The physics of magnetism has a wide scope indeed.

Sources: Francis Bitler, *Magnets: The Education of a Physicist.* Garden City, N.Y.: Doubleday & Company, Inc., 1959.

Caroline Rob, "Magnetic Pills," *Omni,* September 1982, p. 50.

electric charge is affected by the electric field of another charge, so a magnetic material responds to a magnetic field. We can illustrate the magnetic field by lines of force, shown in Figure 16-7. The lines indicate the direction of the magnetic force vector. The convention is that the lines run from the north pole to the south pole. As you can see from the figure, the shape of the magnetic field depends on the shape of the magnet.

You can observe the effects of magnetic field lines quite easily. Put a sheet of paper or a piece of glass over a bar magnet

Figure 16-7. Magnetic field lines.

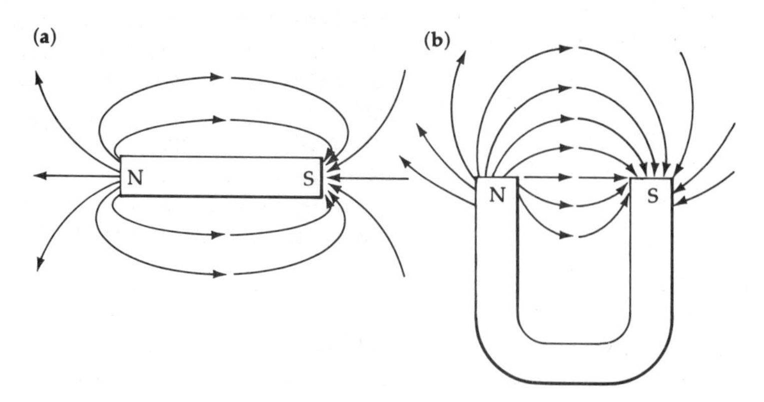

lying on the table. Iron filings sprinkled about on the paper or glass will be caught in the magnetic field. They will distribute themselves throughout the field, giving an indication of its shape. Figure 16–8 is a photograph of iron filings scattered through the field of a bar magnet.

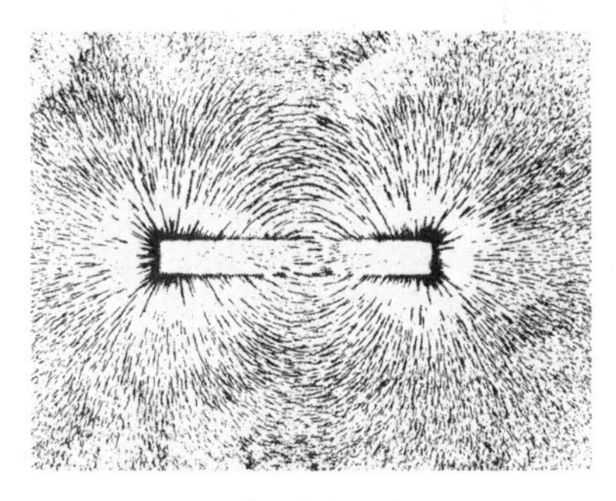

Figure 16–8. The field of a bar magnet.

Learning Checks

1. What are some similarities between the *electrostatic* force and the *magnetic* force?
2. What is an important distinction between electric charges and magnetic poles?
3. What are magnetic domains?
4. In terms of magnetic domains, what distinguishes an iron magnet from an ordinary bar of iron?
5. What is a magnetic field?

THE EARTH AS A MAGNET

As we said earlier, the compass was one of the earliest practical applications of magnetism. People naturally wondered what caused the compass needle always to point north. Many experiments over many years show that the earth itself has a magnetic field, although we still don't know exactly what causes it. The magnetic poles are located in the general vicinity of the earth's north and south geographic poles. Remember that opposite magnetic poles attract each other. So there is a south *magnetic* pole near the north geographic pole in the Arctic region, because this is the direction the north end of the compass needle points. The corresponding north *magnetic* pole is located in the Antarctic, in the region of the south geographic pole.

The magnetic and geographic poles of the earth are not in the same place. The earth's magnetic pole in the northern hemisphere is currently located just off Canada's Arctic shore, some 1200 miles from the north geographic pole. The magnetic pole in the southern hemisphere is on the shores of Antarctica, west of Ross Sea, about 1200 miles from the south geographic pole.

A superconducting magnet used in medical applications to detect heart disease, strokes, and cancerous tissue.

Because the sets of poles do not coincide, a compass in most parts of the world will not point due north. The angle by which the compass deviates from true north is called the *magnetic declination.* It varies from place to place on the surface of the earth. Figure 16–9 shows lines drawn through points of equal magnetic declination on a map of the United States. Only along the line of zero declination will the compass point directly toward the north geographic pole.

One feature of the earth's magnetic field that may be surprising is that it has shifted several times throughout history.

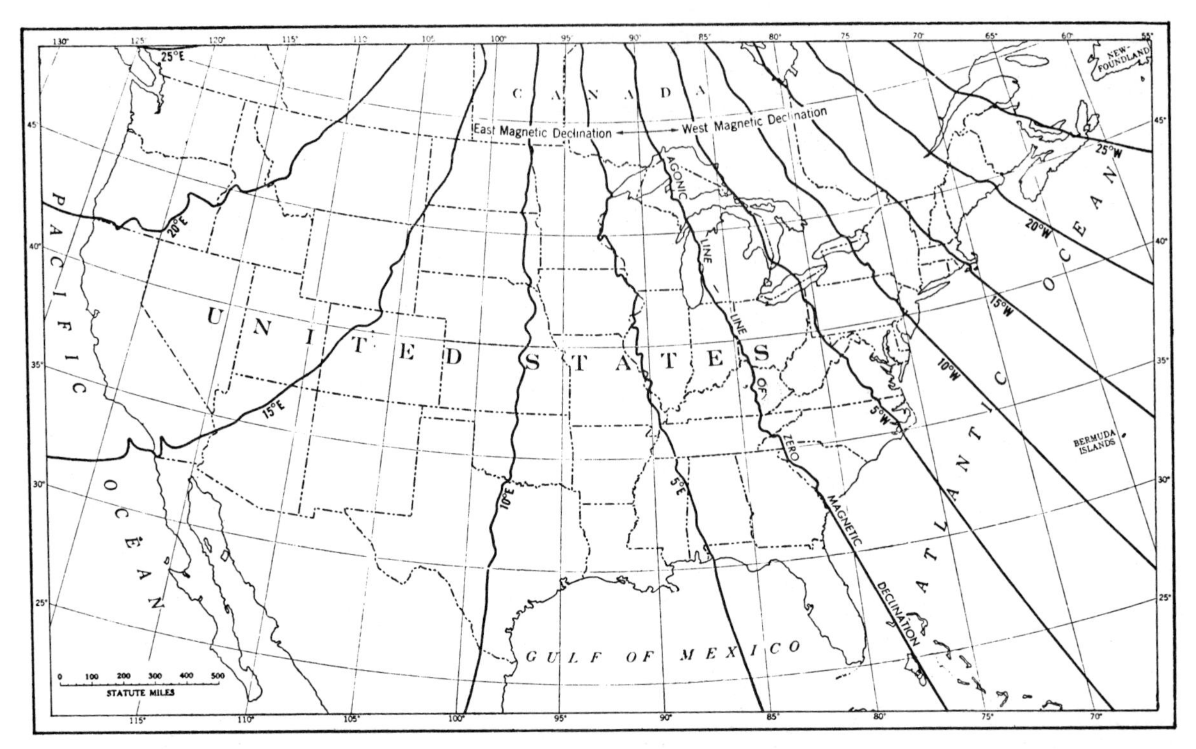

Figure 16-9. A map of the United States showing magnetic declination.

Geologists have observed that strata of rock containing magnetic materials—such as bits of iron—show different directions of magnetization from one layer to the next. For example, the magnetic materials in rock formed from volcanic lava are free to line up with the earth's field while the lava is still in the hot molten state. As the lava cools and hardens, the direction of this alignment becomes frozen. The magnetic particles are no longer able to reorient in response to another magnetic field. Because different layers of rock form at times when the earth's magnetic field has different directions, we are thus provided with a picture of how the field changes in time.

The story the rock tells is one of rather chaotic wandering by the earth's field. There seems to be no pattern to its changes, either in how often the field changes or by how much. Some examples: In 1580, when William Gilbert was performing some of his experiments, the magnetic declination in London was 11 degrees east of north. Currently, it is about 8 degrees west of north, and has been as great as 25 degrees west of north. Because the declination in London has shifted from *east* of north in Gilbert's time to *west* of north now, there must have been at least one time when the magnetic declination was zero. This happened in 1657; that year, a London compass needle would have pointed due north.

Learning Checks

1. What is meant by magnetic declination?
2. Why do we say that the *south* pole of the earth's magnetic field is near the *north* geographic pole?

ELECTROMAGNETISM

One of the great discoveries in all of history was that electricity and magnetism are related—two sides of the same coin, so to speak. Until early in the nineteenth century, the electrostatic force and the magnetic force were thought to be independent. Indeed, there was no reason to believe that they were connected at all. Hans Christian Oersted, a Danish physicist, happened on their interdependence quite by accident in 1819. He noticed that if he brought a compass up close to a wire connected to the terminals of a battery, the compass needle would swing around and stop in a position perpendicular to the wire. Playing around with this, he found that by switching the leads on the battery, thereby reversing the current, he could make the needle point in the opposite direction, again perpendicular to the wire. (See Figure 16-10.) When Oersted disconnected the wire from the battery, the needle again lined up with the earth's magnetic field, as if the wire were not there.

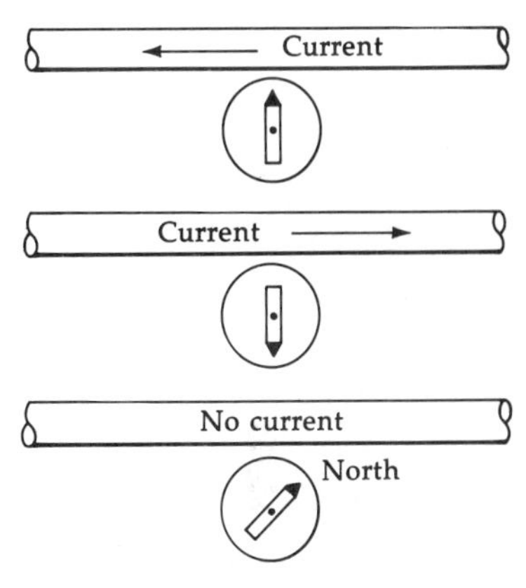

Figure 16-10. Alignment of the compass needle depends on whether the current flows and in which direction.

After Oersted announced his rather startling discovery, other physicists followed up with more experiments. A Frenchman, Dominique Arago, showed that unmagnetized iron filings were attracted by a wire carrying an electric current. Ampere, also in France, demonstrated that two electric currents behaved just like two magnets. If the currents were traveling in the same direction in two straight wires laid parallelly, the wires attracted each other. (See Figure 16-11.) Currents in opposite directions forced the wires apart. Clearly, then, electric current—that is, charge moving through the wire—has magnetic effects.

Figure 16-11. Interaction of current-carrying wires. (a) Currents flow in *opposite* directions – wires *repel* each other. (b) Currents flow in *same* direction – wires *attract* each other.

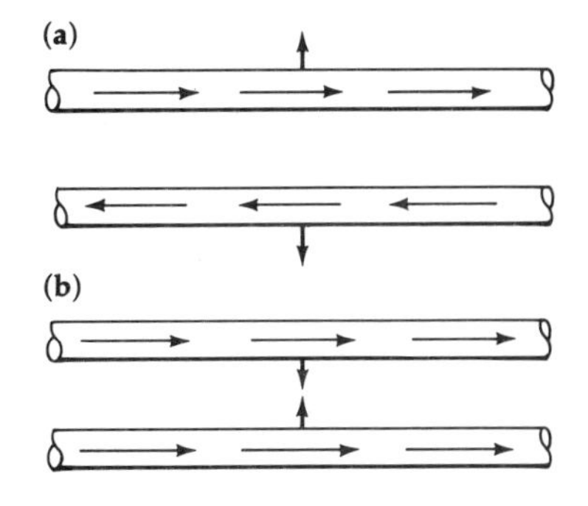

Many other experiments have served to produce this idea about magnetism: it occurs because of *moving* electric charges. The existence of an electric charge gives rise to an electrostatic field, as we have already seen. And if the charge moves, as in a conducting wire, the surrounding space is distorted in a second way, which we call the *magnetic field*. In Oersted's experiment, the permanent magnetic field in the compass needle interacted with the magnetic field created by the electric current, and the needle became aligned just as it would have with an ordinary magnet. In Ampere's experiment, the two magnetic fields in the wires interacted to attract or repel the wires, depending on the direction of the current.

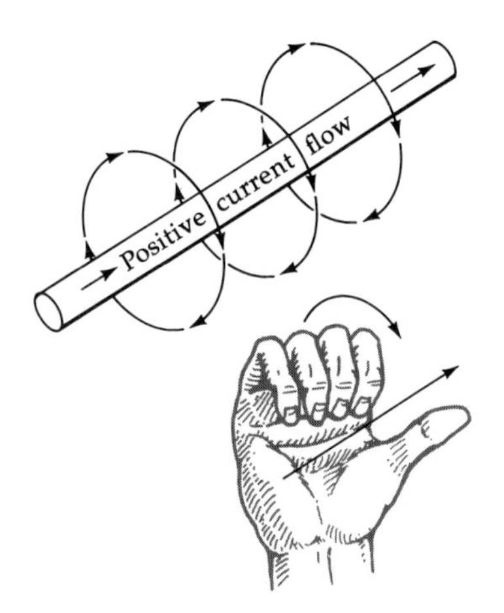

Figure 16-12. The righthand rule.

We can picture the field lines due to a current as concentric circles surrounding the wire, as in Figure 16–12. You can find the direction of the field like this: If you grasp the wire with your right hand so that your thumb points in the direction the current flows, your fingers curled around the wire point in the direction of the field lines. This shows why in Oersted's experiment the compass needle lined up perpendicularly to the wire. Obviously, if you reverse the current (that is, grasp the wire with your right thumb pointing the other way), you reverse the direction of the magnetic field. This also accounts for Ampere's observation with the two wires.

It is not difficult to visualize magnetic fields for current-carrying wires, but what about permanent magnets? What are the moving charges that cause the field in this case?

Answers to these questions involve the structure of matter at its most fundamental level. The electrons attached to each atom are, of course, charged particles. They have two kinds of motion (see Figure 16–13), which we can picture crudely as *orbital* motion, analogous to a planet going around the sun, and *spinning* motion, like a toy top rotating on its axis. Either of these constitutes a moving electric charge capable of generating a magnetic field. It turns out that the *electron spin* is the motion responsible for the magnetic fields of materials such as lodestone or iron.

As it happens, the electron can have only two orientations of its spinning motion. Again using our model of a spinning top, we can imagine that it rotates either clockwise or counterclockwise. Let us represent the clockwise motion by an arrow pointing up, like this $\uparrow$, and the counterclockwise spin by an arrow pointing down, or $\downarrow$. We can view each electron as a tiny magnet, whose field direction depends on the orientation of its spin.

Whether or not a given atom or molecule is magnetic depends on the combined effects of all the electrons. Most materials have their electrons paired off, one with spin up and the other with spin down; we can represent this situation by this symbol, $\uparrow\downarrow$. Since the electrons have opposite spins, the individual magnetic fields cancel each other, giving zero net magnetic field. So materials in which all electrons are paired are not magnetic. Many common substances—wood, glass, water—are like this.

On the other hand, magnetic materials, such as iron, are characterized by the fact that they have *unpaired electrons*, that

Figure 16-13. The electrons in (a) and (b) have opposite spins.

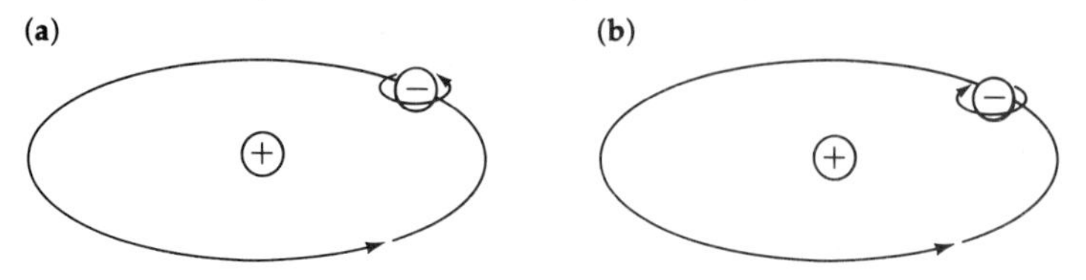

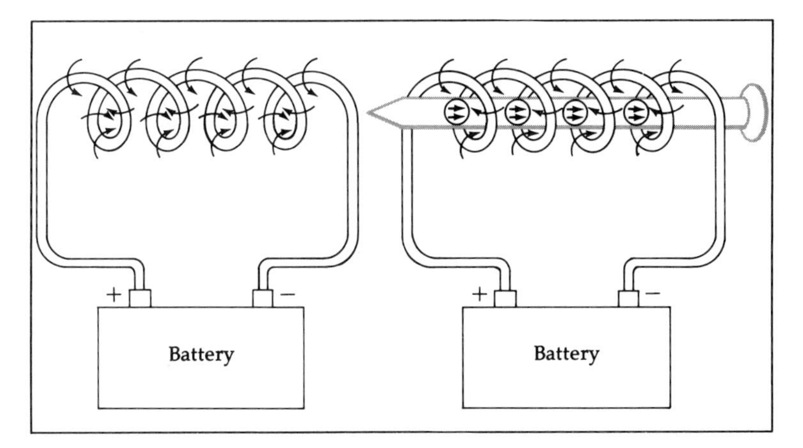

Figure 16-14. The field in the coil aligns the nail's domains.

is, electrons whose spinning-motion magnetism is *not* canceled. If an atom has two unpaired electrons we represent them as ↑↑, and such a material would be attracted to a magnet. Iron, for example, has four unpaired electrons per atom. It is these spinning electrons that give rise to the magnetic domains we discussed earlier in the chapter.

An *electromagnet* is a device in which an electric current is used to generate a magnetic field. You can easily make one, simply by connecting a wire to the terminals of a battery. The strength of the magnet can be increased by making a coil of several loops of the wire, since the magnetic field lines will reinforce each other inside the coil; that is, four turns of wire produce a field twice as strong as that with two turns, and so on. Inserting a nail into the coil further enhances the field strength, since now the magnetic domains of the iron are brought into alignment and contribute their magnetism also. (See Figure 16–14.) This magnet will pick up tacks, iron filings, and so on. When the wire is disconnected and the current no longer flows, the magnetic field drops to zero and the attracted materials fall away.

This kind of electromagnet has many uses in industry and in the everyday world. The giant magnets you see attached to cranes for moving large metallic objects are extensions of the nail-in-wire-coil idea: thousands of turns of wire wrapped around a large iron core. A simple application is in an ordinary doorbell, shown in Figure 16–15. The doorbell is designed so that the core of the electromagnet attracts the clapper when you complete the circuit by pushing a button. When the clapper is pulled to the magnet, this breaks the circuit, and the current stops flowing. The clapper is no longer held by the core, and a spring pushes it back to its original position where it strikes the bell. In this position, the clapper closes the circuit, thus reactivating the electromagnet, and this cycle is repeated over and over again, for as long as you push the button.

Figure 16-15. Common doorbell.

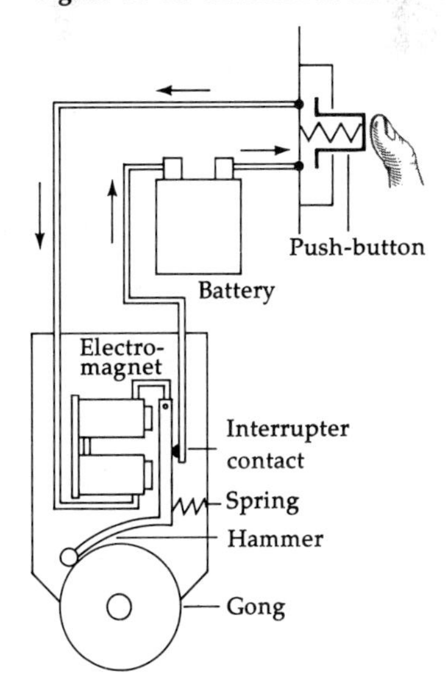

Learning Checks

1. What produces a magnetic field?
2. Explain Oersted's and Ampere's experiments.
3. Account for the magnetic properties of materials by the idea of paired and unpaired electrons.
4. Give two more applications of electromagnets.

Electric Motors and Generators: Induction

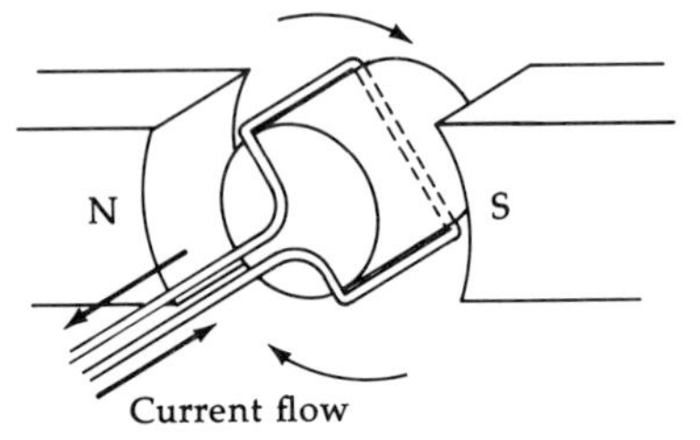

Figure 16-16. Simple electric motor.

Electric motors, so common in modern technology, are based on the principle that Oersted discovered. The purpose of a motor is to convert electrical energy to mechanical energy. A simplified diagram of an electric motor is shown in Figure 16-16. Here we have a loop of wire within the field of a magnet and connected to an external source of electricity. When current flows through the wire, the resulting magnetic field interacts with the field already present, producing a force on the loop that causes it to rotate. This rotary motion can be used, for example, to operate the turntable on a stereophonic sound system, the rotor of a sewing machine, and hundreds of other common devices.

We saw in Oersted's experiment that an electric current produces a magnetic field. But the reverse is also true; a magnetic field can be used to produce an electric current. Michael

New Focus on Electric Motors

In the United States, traditional electric motors, industrial and commercial, consume roughly two-thirds of all the electric energy generated. They use up a third more energy than automobiles and they consume the equivalent of 6 million barrels of oil (or 1.5 million tons of coal) per day. Pressed into service as pumps, compressors, blowers, and fans, the common industrial motors are only about 50 to 70 percent efficient when fully loaded.

Engineers are now zeroing in on this area to help with the energy crisis. They feel that innovative changes can be made in motors to diminish the amount of energy used. Designers have thus been attempting to improve the utilization of electricity inside motors and to reduce internal power losses. In these efforts, changes are made in the core steel of rotors, in the amount of copper in the windings, and in other functional aspects as well.

Some of the major design alterations now being implemented and tested for marketing have increased motor efficiency around 10 percent. This is good news for manufacturers, industrialists, consumers, and government officials, all of whom have been encouraging this trend.

Sources: D. V. Edson, "Electric Motors Save Energy, Cut Costs," *Design News*, August 6, 1979, pp. 46–48.

Mary-Sherman Willis, "Energy Conservation and the Electric Motor," *Science News*, July 21, 1979, pp. 54–55.

Faraday, an English physicist of the nineteenth century, discovered this in 1831.

It would be difficult to overemphasize the importance of Faraday's discovery. Mechanical energy and how to use it had been understood for a long time. But Faraday paved the way for converting mechanical energy—that is, energy of motion—into electrical energy. This was truly a revolutionary finding, for it changed forever the way in which energy is used by society. Figure 16-17 shows a representation of Faraday's experiment. The iron ring is wound with two coils of insulated wire, one of which is connected to a battery through a switch for opening and closing the circuit. The second coil is connected to a galvanometer, a device for measuring electric currents. The observation that Faraday made was this: When he depressed the switch to close the circuit, the galvanometer needle made a quick deflection and then returned to zero. When he released the switch, thereby shutting off the current from the battery, the galvanometer also deflected but this time in the *opposite* direction, and then quickly again returned to zero.

Let us try to interpret Faraday's results. (Refer to Figure 16-18.) When the switch is closed, current begins to flow in the first coil. It builds up over a short period of time and soon reaches a steady value. This current creates a magnetic field that matches the amount of current—it starts at zero and swells to a maximum strength corresponding to the steady current. The iron core enhances this magnetic field and directs it through the turns of the second coil of wire.

During the time the magnetic field is building up, it induces a current in the second coil, indicated by the deflection of the galvanometer needle. This induced current is short-lived, however—it disappears when the original current and magnetic field reach their steady-state values. For this reason, the galvanometer again flips back to a zero reading and stays there.

When the switch is released, we get another surge of current in the second coil, reversed in direction from the initial induced current. This time the current drops from its steady flow to zero, as does the accompanying magnetic field.

Here is the point: It is while the magnetic field is *changing*—either growing or dying out—that the current flows in the second coil. There is *no induced current* during the time that

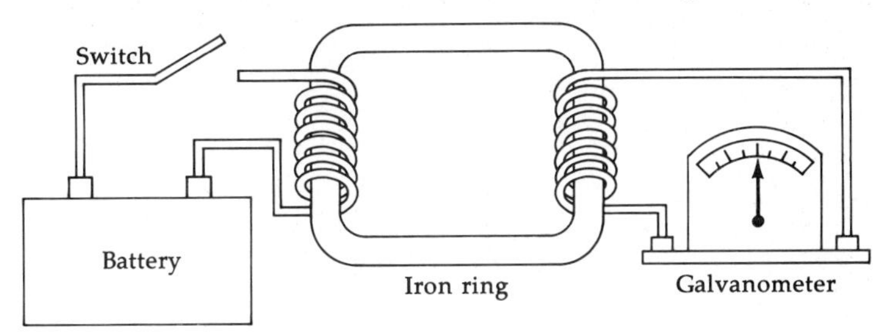

Figure 16-17. Faraday's induction apparatus.

Figure 16-18. Interpretation of Faraday's results.

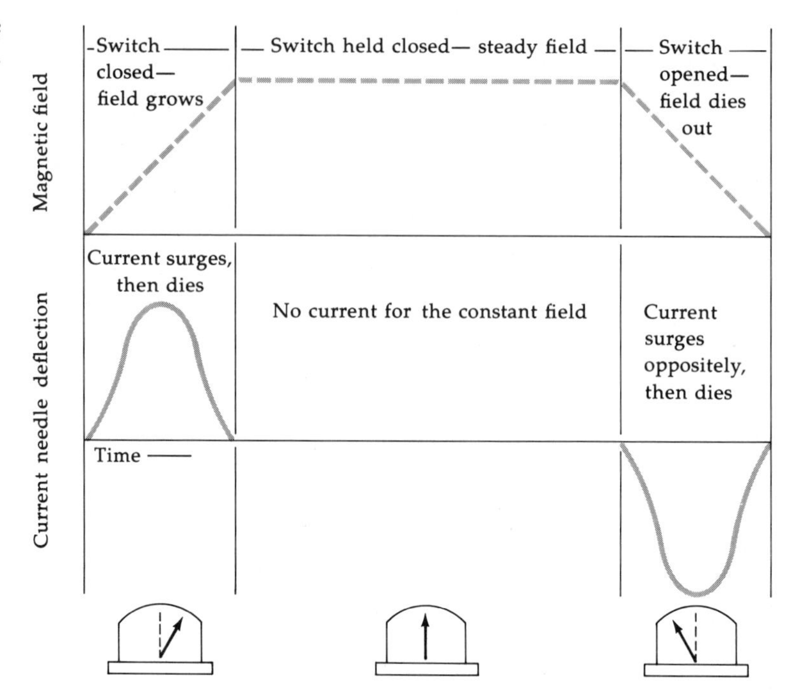

the field remains at a fixed strength, either at zero or at its maximum. So it is the *changing* magnetic field that induces the new current.

Faraday's idea was that the current is induced, not by the *existence* of the magnetic lines of the force, but by the *motion* of those lines across the wire. As the field changes due to the growing current in the first coil, the magnetic lines cut across the second coil, producing the current. This induced current dies out when the field stops growing, since the lines no longer cut the wire. When the current from the battery is shut off, the lines of the dying magnetic field again cut across the wire, inducing another brief current, this time in the opposite direction. Figure 16–19 illustrates the expanding and contracting lines of the changing field.

Note that whether or not the field lines cut across the wire depends on how the wire is oriented relative to the field. Thus,

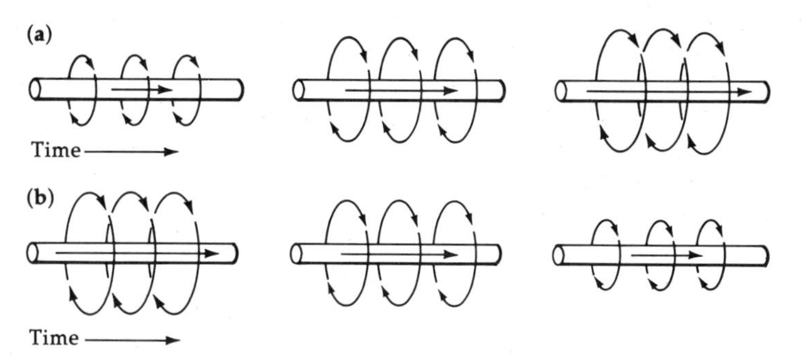

Figure 16-19. Changing field. (a) Field expands as current grows. (b) Field contracts as current dies out.

if the field lines and the wires are parallel, the lines do not intersect the wire and no current is produced. (See Figure 16-20.) The wire experiences the maximum change in the field when it is perpendicular to the field. And we get various amounts of change for situations between the parallel and perpendicular configurations.

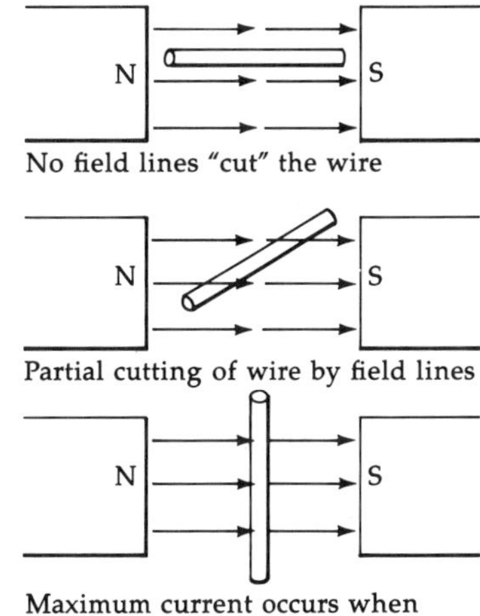

Figure 16-20. Induced current depends on orientation.

Now, this is an interesting discovery, but what practical value does it have? Isaac Asimov relates the story that Faraday was demonstrating his findings to a London audience that included William Ewart Gladstone, at the time a newly elected member of Parliament and later four times Prime Minister. After the lecture, Gladstone put this very question to Faraday, who replied, "Sir, in twenty years, you will be taxing it!"

The practicality of Faraday's discovery depends on the fact that it is only the *relative motion* of the wire and the magnetic field lines that induces the current. That is, it doesn't make any difference whether you hold the wire steady and move the magnetic field, or keep the field stationary and move the wire. Also, it doesn't matter whether the magnetic field comes from a permanent magnet or an electromagnet.

Perhaps the most important application of Faraday's discovery is the *generator,* which, as the name suggests, is used to generate electricity. Let us look at a simple generator to illustrate the principle, in Figure 16-21. Here we show a single loop of wire that, by some external mechanism, is made to move through the field of the magnet. As the wire cuts the field lines, a current is induced. It flows in one direction (relative to the field) in one section of the wire, and in the opposite direction in the other section. The induced current may be collected

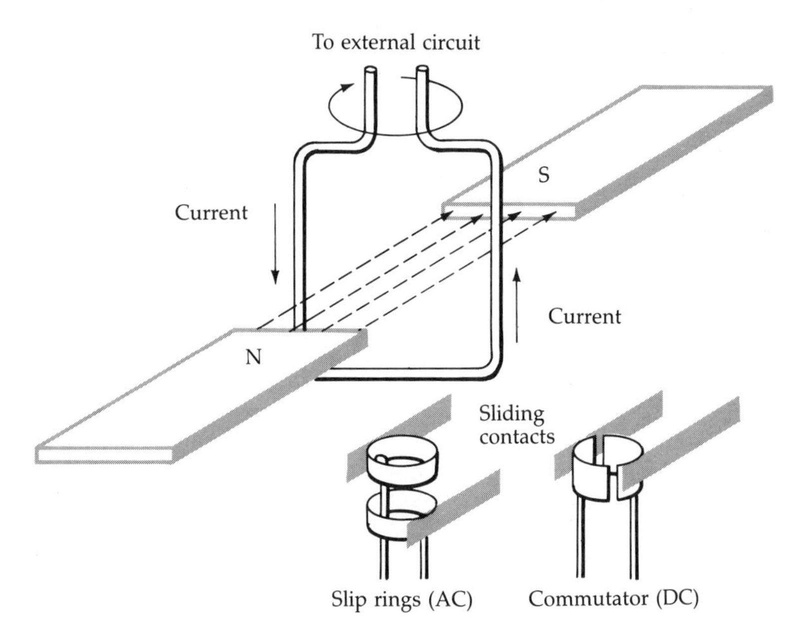

Figure 16-21. Simple generator.

as alternating current by use of the slip rings, or as a pulsating direct current by use of the two half-rings, known as a commutator.

This is the principle by which a generating station, such as a hydroelectric power plant, produces electricity. Water flowing over a dam turns a large wheel containing coils of wire that cut the field lines of a magnet surrounding the wheel. The current produced is carried by conducting cables to provide the electrical energy used in homes and offices. Generators for running bicycle lights are designed to apply the same idea on a smaller scale. Here the friction of the generator against the bicycle wheel supplies the motion of wires across lines of a permanent field.

One common application of the generator is in gasoline-powered lawnmowers. You start the engine by pulling a crank with a rope. The housing of the crank is a permanent magnet, and the current induced by turning the crank through the field of the magnet is carried to the sparkplug, where it creates a spark, igniting the gasoline.

Learning Checks

1. Explain Faraday's discovery.
2. If a magnetic field is moved so that the lines are always parallel to a conductor, will there be any induced current? Explain.
3. Give an example of a generator other than the ones mentioned in the text.
4. In Faraday's experiment, how is the strength of the field produced by the current related to the number of turns of the first wire?
5. What is the effect of doubling the number of turns in the second wire?

Transformers

Part of the reason that electricity is such a useful form of energy is that it can be conveniently delivered over many miles. You don't have to live close to a generating station in order to enjoy the fruits of electrical energy. However, if large amounts of current are transmitted over long distances, so much energy is lost as heat that hardly enough electricity is left to bother with. To get around this problem, a device called a *transformer* is used in the delivery of electricity.

Let us look again at the relationship among power, current, and voltage. It is this (see page 262):

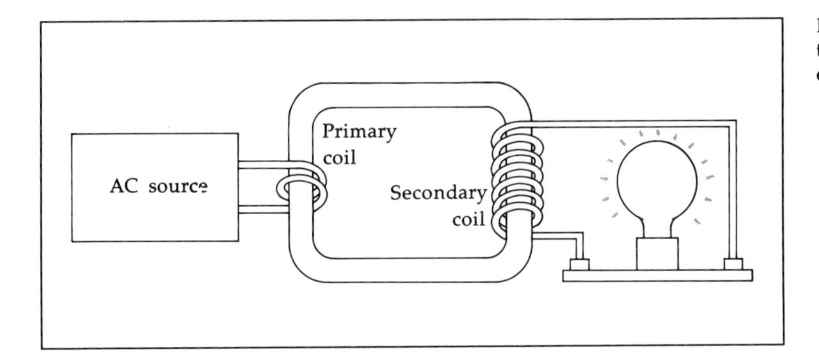

Figure 16-22. The changing field in the primary coil induces alternating current in the secondary coil.

power = voltage × current

That is, a watt is a volt times an ampere. This equation tells us that for a given amount of power, we can have various combinations of voltage and current. Suppose, for instance, that we buy 120 watts of power from the electric company. It could be 12,000 volts at 0.01 ampere, or it could be 0.12 volt at 1000 amperes; either combination is equal to the same power, 120 watts. So, for delivering a given amount of power (or energy) over long distances, we need a device that kicks up the voltage so that the current is reduced. Otherwise the current traveling through the lines will cause tremendous heating, which wastes the transmitted energy. By increasing the voltage, we can reduce the current and eliminate this heat-loss problem.

On the other hand, we need to get the voltage back down to reasonable levels when the electricity gets to the circuits in a building. The voltage coming from a city substation may be 2000 volts or so. Blasting your stereo with a jolt that size will leave you with nothing but a pile of charred rubble.

In short, then, what we need is a mechanism for manipulating voltages and currents up and down. The transformer does the trick. We show a simple one in Figure 16–22. As you can see, this is just the apparatus that Faraday used in his 1831 experiment. The primary and secondary coils are wound on an iron ring that provides maximum concentration of the changing magnetic field. This time, rather than using a direct current, as Faraday did, we use an *alternating* current. This means that the amount of current will always be either rising or falling (sixty times each second for a 60 cycle current), so the strength of the magnetic field through the iron ring is also either rising or falling. The lines of force continually expand outward and collapse inward. As they do, they cut back and forth across the secondary coil, inducing an alternating current in step with that in the primary coil.

The relative values of the current and voltage in the two

coils depend on the relative number of turns. For example, in the figure there are three times as many turns in the secondary coil. This means that the voltage in the secondary will be three times that in the primary; but because the power must remain the same, the induced current will be one-third that in the primary. This is a *step-up* transformer: the induced voltage is "stepped up" relative to the input voltage. If the current in the primary coil of a step-up transformer has a potential difference of 120 volts and the number of secondary turns is ten times that in the primary, then the secondary coil will have 1200 volts. Correspondingly, the current in the secondary will be one-tenth that in the primary.

A *step-down* transformer has fewer secondary turns, and therefore *reduces* the output voltage with a corresponding *increase* in current. Both types are used in electrical transmission. The step-up transformer kicks up the voltage coming from the generator, reducing the current for the long trip to its destination. There, a step-down transformer reduces the voltage and increases the current to the level needed to run the various appliances and other electrical devices.

Learning Checks

1. Explain the use of AC rather than DC in a transformer.
2. What is it that remains constant in a transformer (ignoring small losses as heat)? What important rule of physics applies here?
3. Why are transformers used for power transmission?
4. A certain transformer has its secondary and primary coils in a ratio of 5 to 1. Is it a step-up or a step-down transformer? If the input voltage is 1000 volts, what is the induced voltage?

SUMMARY

Magnetism is a phenomenon familiar in compass needles, natural materials such as lodestone, toy permanent magnets, and large industrial electromagnets.

Magnetic materials respond to *magnetic fields*, which are distortions of space similar to electric and gravitational fields. The earth has a magnetic field whose north and south poles roughly coincide with the geographic poles of the earth. We can picture matter as being made up of submicroscopic magnets that, in most materials, are randomly oriented, and their magnetic effects are canceled. In materials that are magnetic, however, these tiny magnets are aligned in groups, called *mag-*

netic domains, which in turn can be aligned to create a permanent magnet.

For many centuries electricity and magnetism seemed to be unrelated, because their effects are quite different. However, experiments by Oersted, Arago, Ampere, and others demonstrated that the two are actually manifestations of the same basic causes. Magnetism is produced by *moving electric charges.* This may involve the current in a wire, giving rise to electromagnetism; or it may be the spinning motion of the electron, which accounts for magnetic effects in such materials as iron.

We take advantage of the connection between electricity and magnetism in *electric motors, generators,* and *transformers.* An electric motor converts electricity into mechanical energy, whereas a generator converts the energy of mechanical motion into electrical energy. A transformer alters the voltage and current in a circuit.

Exercises

1. Since every atom in a block of wood is a tiny magnet, why isn't a block of wood magnetic?

2. A blacksmith can reduce the magnetic strength of an iron bar by heating it. Explain.

3. Will a magnet exert a force on an electron at rest? What about a moving electron? Explain.

4. A magnet will distort the image on the screen of an oscilloscope. What does this tell you about how the image on the screen is produced?

5. Suppose that in a wire perpendicular to this page a current flows toward the page. Is the magnetic field directed clockwise or counterclockwise? How do you decide?

6. Why will either pole of a magnet attract an unmagnetized piece of iron?

7. You can weaken the magnetic field of a magnet by dropping it on the hard floor. Explain.

8. How is a motor similar to a compass?

9. An electrician suspends a loop of wire between the poles of a permanent magnet so that the plane of the loop is parallel to the pole faces. Describe what happens if a direct current is sent through the coil. What if an alternating current is used instead?

10. A generator is rotated at a certain speed to produce 75 volts. What will the output voltage be if the rotation speed is doubled?

11. Figure 16–23 shows a "flying doughnut machine." The doughnut is an electrical conductor that flies up and off the core when you close the switch in the coil of the electromagnet. If the doughnut is cut as shown, it does not fly off. Explain these observations.

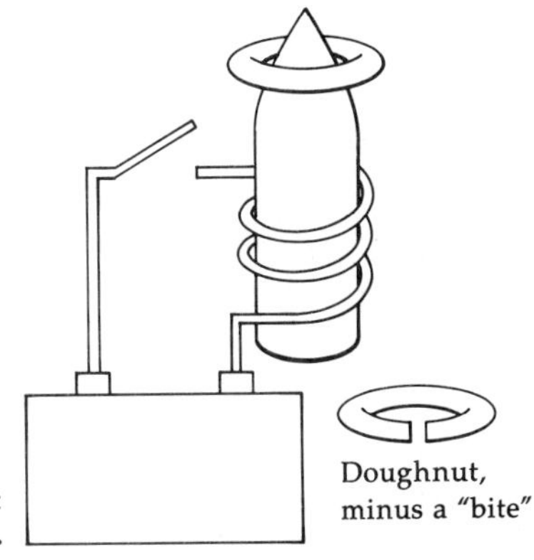

Figure 16-23. The flying doughnut machine.

12. The voltage input to a transformer is 1000 volts and the output is 10,000 volts. What is the ratio of turns in the primary to secondary coils? Is this a "step-up" or a "step-down" transformer?

13. A transformer has 40 turns in the primary coil and 160 turns in the secondary coil. Is this a step-up or step-down transformer? By what factor does it change the voltage?

14. A manufacturer rates its automobile batteries at 12 volts. One such battery is to supply energy to a device that operates at 2400 volts. There are 6000 turns in the secondary coil of the transformer. How many turns are in the primary coil?

17

WAVE MOTION AND VIBRATIONS

An ocean wave is a picturesque example of the properties of wave motion. Waves of light, however, allow us to see its color and movement, and sound waves signal its power as it rumbles and crashes on shore.

Wave motion is one of the most familiar kinds of motion in our world. Waves are all around us. In fact, most of the information we have about our surroundings comes to us by waves. We hear horns blow and children talk because of sound waves. We see colors and shapes and patterns because of light waves. There are water waves, such as those created when a pebble hits a still pond. You can see waves made by the wind as it blows across a field of wheat. When you strum a guitar, waves travel along the strings, which in turn produce the sound waves that you hear.

Nature provides a rich variety of wave phenomena that appear to our senses in vastly different ways. The rainbow you see in an oil patch on a rain-soaked street may seem to have nothing in common with the wake that trails a speed boat as it skims across a lake. But, in fact, we can interpret both effects on the basis of some simple properties of waves. The waves of the ocean surf that crash onto the shore are closely related to the sonic boom of a swiftly moving airplane. The sound-producing vibrations of a violin string are akin to the color-producing oscillations of electrons in dyestuffs. And so on.

Because wave motion is the origin of so many phenomena in our environment, and because an understanding of waves is basic to modern technology, it is important to learn some fundamental information about waves and their properties. We shall do so in this chapter. One of the most important instances of wave motion—sound—is taken up in the next chapter, followed by a chapter that focuses on musical sounds. We shall then be prepared to begin our study of light in Chapter 20.

VIBRATIONS AND WAVES

All kinds of wave motion have this feature in common: they result from *vibration,* the back-and-forth movement of some

Waves:
The Common Denominator

Many poets, musicians, and scientists have all, in their own way, connected the differing phenomena of color and sound. The famous French poet Arthur Rimbaud fused speech and colors in a poem entitled "Vowels":

A black E white I red U green
O blue—I'll tell
One day, you vowels, how you
came to be and whence.

The world-renowned cellist and composer Pablo Casals conceptually combined music and color:

Rainbow! All music is a rainbow!

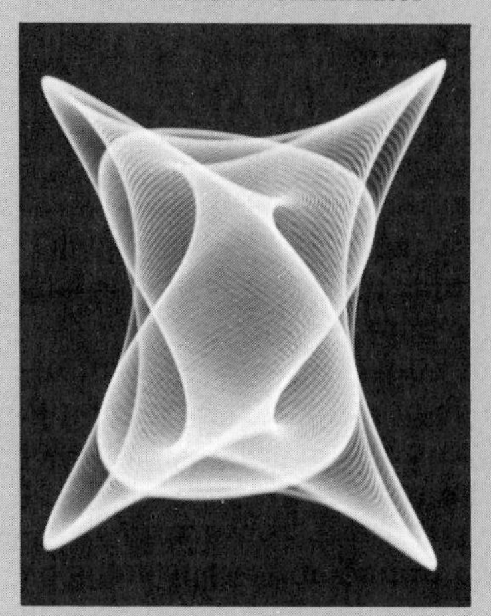

And the German physicist and anatomist Hermann von Helmholtz fancifully linked colors with notes of the diatonic scale to amuse himself:

G ultraviolet

.

.

E indigo blue

.

.

G yellow

Thus, the insights of physics both reflect and fuel the creative imagination.

An electronic waveform created on an oscilloscope.

object that is responsible for the propagation of the wave. It might be a violin string, or a cork on a fishing line bobbing up and down in the water, or a flag waving to and fro as a breeze blows across a school yard.

We can see the connection between a vibrating object and the wave it generates by considering a simple example. Suppose you hold the handle of a knife tightly against a table top and flip the blade end vertically with your finger. The blade vibrates back and forth so rapidly it appears as a blur. Gradually, the vibrational motion dies out, and the blade relaxes to its original, motionless position.

Now imagine that you tie one end of a light rope to the blade of the knife and the other end to the wall. Figure 17–1 shows a "slow motion" picture of what happens when you flip the blade. Pulling the blade up away from its rest position displaces the portion of the rope tied to the knife, which in turn pulls up the next bit of rope, although not quite so much. When you release the blade to let it head back the other way, successive portions of the rope displace their neighbors, which in turn pull on the next bit of rope, and so on. The effect is to send a disturbance, called a *wave pulse,* down the rope. Meanwhile the knife blade continues to vibrate, sending one wave pulse after another, alternately above and below the original position, or *equilibrium position,* of the rope. The result is a traveling wave that propagates along the rope for as long as the knife blade vibrates.

Figure 17-1. Vibration of the blade sends a wave pulse down the rope.

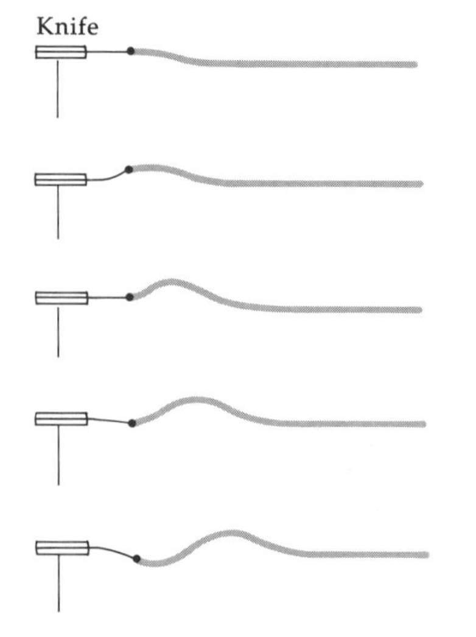

Figure 17–2 shows a portion of the wave, which we may use to illustrate some quantities characteristic of all waves. The *wavelength,* commonly labeled by the Greek letter lambda (λ), is the distance between successive identical points in the wave. It

might be the distance between two crests, or between two valleys. In essence, the wavelength is the smallest portion of the wave that contains all the features of the entire wave. The *amplitude* is the height of a crest or the depth of a trough; it is the maximum distance that the medium carrying the wave is displaced from equilibrium.

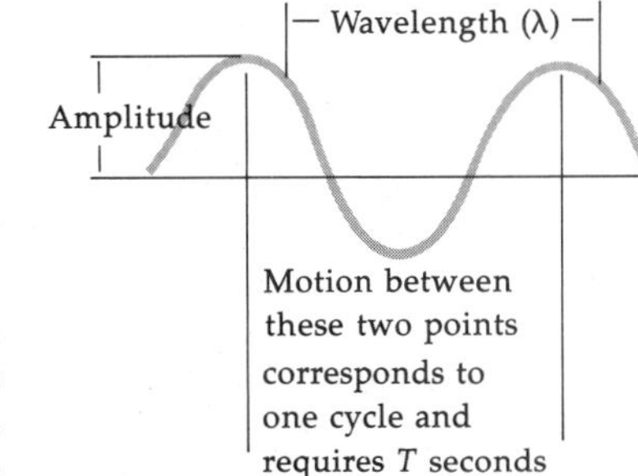

Figure 17-2. Wave definitions.

Because the wave travels along the rope, we are interested in how the wave behaves in time. The *period* (T) is the time required for the wave pulse to travel a distance of one wavelength. Another way of looking at the period is this: It is the time required for the vibrating object (the knife blade, in our illustration) to complete *one cycle* of its motion. By a *cycle* we mean a round trip: the blade moves from equilibrium to its highest position, then to its lowest, and finally back to equilibrium. One full cycle of vibration means one full wavelength of the traveling wave that is produced. The period is the time in which this cycle takes place.

Suppose that you stand beside the waving rope and count the number of wavelengths that pass by you each second. This number is called the *frequency* (f) of the wave. This also corresponds to the number of full vibrations, or cycles, that the vibrating knife goes through each second. So the frequency is measured in *cycles per second.* The modern name for cycles per second is *hertz* (abbreviated Hz), after the German physicist Heinrich Hertz (1857–1894), who pioneered in the study of transmitting electromagnetic waves. For example, if the knife blade vibrates back and forth ten times per second, you will see ten crests pass you each second, and the frequency of the wave will be 10 hertz.

It should be clear to you that there is a connection between the period of the wave and the frequency. If the frequency is high—that is, if many wavelengths pass by a particular point each second—then, of course, it takes a short time for the vibrating object to go through one cycle. We can be more specific. If the frequency f is 10 hertz (that is, 10 cycles per second), then each cycle requires one-tenth of a second, so $T = 1/10$ s. This shows that there is a reciprocal relationship between f and T, and we can write it as

$$T = \frac{1}{f}$$

If the period is small, the frequency will be high, and vice versa.

Finally, we can define the *wave speed,* which, as you might imagine, tells how fast a pulse of the wave travels. Suppose we have a wave whose wavelength λ is 3 meters and whose frequency f is 2 hertz. This means that the crests of this wave are 3 meters apart and you will count two of them going by you each

> It is interesting to note that we ourselves are vibratory phenomena:
>
> *After all, our hearts beat, our lungs oscillate, we shiver when we are cold, we sometimes snore, we can hear and speak because our eardrums and larynxes vibrate. The light waves which permit us to see entail vibration. We move by oscillating our legs. We cannot even say 'vibration' properly without the tip of the tongue oscillating. . . . Even the atoms of which we are constituted vibrate.*
>
> *R. E. D. Bishop,* Vibration.

More "Fore!" and Physics

We have referred earlier to the popular game of golf to illustrate some principles of physics. Now we do so again because one of the latest "musts" of the professional golfer is an excellent example of frequency at work: the precision-made, frequency-matched golf club.

Frequency-matching is a method for making the "flex" of clubs coincide accurately within a set. To do this, the grip end of the club is fixed in a vise and the head is then vibrated like a tuning fork. Sensing devices measure the number of times per second the vibrating club passes a given point (that is, its frequency).

The whole set is frequency-matched when the difference in number of vibrations per second between successive clubs is the same. For instance, if the frequency of oscillations per second in the 2-iron is 5.17 and 5.29 in the 3-iron (which yields a frequency difference of 0.12), then the frequency of the 4-iron should be 5.41 and the 5-iron 5.53, and so forth. Adjustments are done to put an entire set in synchronization.

Precision in the details of golf club manufacture, such as head weight, overall length, and so on has always been of concern to professional golfers. They feel that subtleties such as frequency-matching can make the difference between a perfect shot and one that is nearly perfect.

Source: "Up Front," *Sciquest*, September 1981, p. 3.

second. So, at the end of 1 second, the first crest that passed will be 6 meters away from you—that is, 6 meters' worth of wave will go by you each second. Therefore, the wave speed is 6 meters per second, which we get by multiplying the wavelength (3 meters) times the frequency (2 cycles per second). In general, then, we have

$$v = f\lambda$$

where v is the wave speed.

For any particular material, one characteristic is the speed of a wave moving through it. The wave speed depends on the physical properties of the material. For instance, the speed of sound waves moving through air will depend on the temperature. At a particular temperature, the speed will be a constant number, the same for any wave moving through the air, regardless of wavelength and frequency. Waves move across the water surface at a speed determined by the temperature and density of the water; but for any given set of conditions, the speed will have a fixed value.

The strings of a guitar differ in their thickness and on how tightly they are stretched. These parameters determine the speed that waves travel along the string when the guitar is strummed. If the string is quite massive, as in the case of the bass strings, which are thick, then the wave speed is less than it is for the higher-pitched ones. Also, as you stretch the strings more tightly with the guitar's thumb screws, you increase the speed of traveling waves.

Figure 17-3. Wavelength and frequency are inversely related.

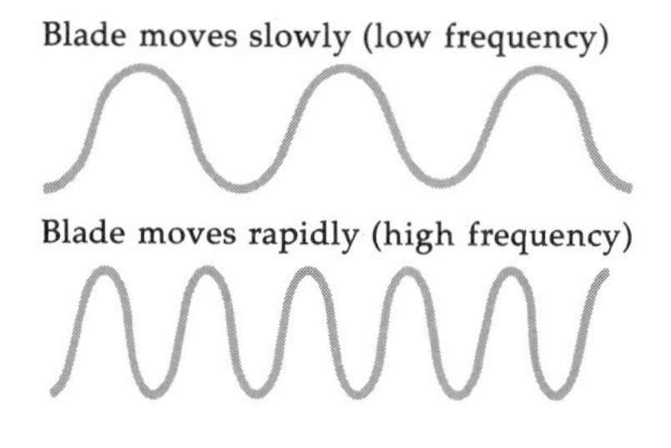

Because the properties of a substance fix the speed of waves passing through it, the frequency and wavelength are connected in a reciprocal way, as the preceding equation shows. You can see, for example, that if our knife blade moves slowly up and down, so that the frequency is quite small, the generated wave will have a large wavelength. (See Figure 17-3.) But if the blade flutters rapidly, as it does when you flip it, the frequency will be high and the corresponding wave will have its crests and valleys very close together. In other words, the wavelength will be small. In both cases, however, the product of f and λ will be the *same*—the speed with which the wave travels down the rope is a characteristic of the *rope*. It is a constant for any wave moving along that rope.

Learning Checks

1. Give the definitions of wavelength, frequency, and period of a wave.
2. What is meant by one cycle of a vibration?
3. How does the speed of a wave depend on the frequency and wavelength?
4. For a given material, what quantity is the same for any wave traveling in that material?
5. What happens to the speed of waves traveling on a guitar string if the string is tightened?

WAVE MOTION: PROPAGATION OF ENERGY

Line up a hundred closely spaced dominoes in some pattern, tip the first one over, and watch what happens. You see that the disturbance created by the first falling domino travels along the line formed by the remaining ones but that each individual domino stays at home, not moving away from where you put it. What catches your eye is the moving wave pulse, not the individual dominoes. In our knife-blade-and-rope illustration, focus on one small section of the rope—imagine painting a small portion red. Notice that as the wave travels *along* the rope, the red chunk merely moves up and down, oscillating about its equilibrium position as the wave travels by. You can produce waves on a still pool of water by dropping pebbles one by one onto the surface. The waves move outward in concentric circles. But if a cork or chip of wood floats on the water's surface as the wave passes by, it doesn't move along with them. Instead, like the red bit of rope, it just bobs up and down with the frequency of the passing waves, never getting very far from its original location.

The point of these illustrations is this: When a wave moves through a medium, the material of the medium does not move along with the wave. Rather, it vibrates to and fro about its equilibrium location. It is obvious that *something* travels along with the wave, and that something is *energy*. The energy that you give the first domino, when you tip it over, is delivered to the next, and the next, and so on. The passage of the energy is just the disturbance that you see. Wave propagation, then, is one way of transmitting energy. In fact, a good description of wave motion is the *transmission of energy between two points without the transfer of matter between the points.*

As an example, you have probably seen a demonstration where an opera soprano shatters a glass by singing a certain note. It is not any *material* that breaks the glass, but *energy,* in the form of a sound wave that travels through the air. The individual molecules of the air do not go very far.

The energy that a wave delivers depends on the wave's amplitude. For example, if you fasten one end of a rope to a wall and shake the other end up and down, you will send a series of crests and valleys traveling as a wave down the rope. If you shake the rope very hard, moving your hand through a wide sweep, the wave will have a correspondingly large amplitude. It takes more energy to do this than to barely wiggle the rope up and down, in which case the wave moving along the rope will have a small amplitude. The ocean waves that crash onto the beach deliver more energy if they are bigger—that is, if they have a larger amplitude.

Figure 17-4. Ripple tank diffraction experiment.

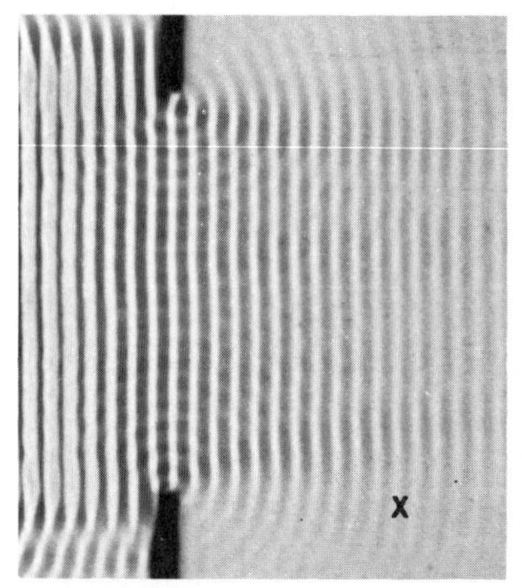

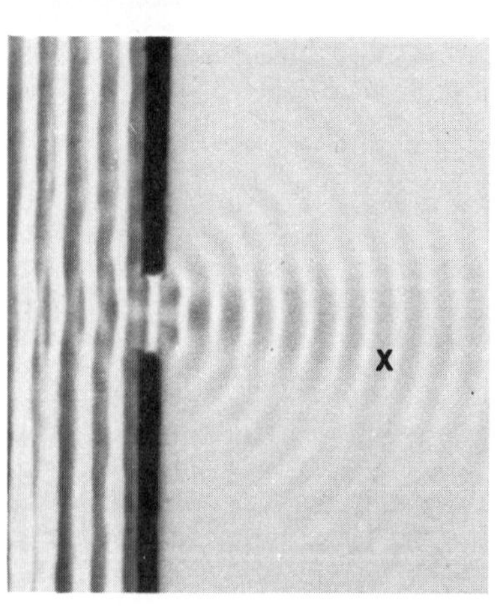

In general, we can say that the energy of a wave is proportional to the *square* of the amplitude

$$(\text{energy}) \propto (\text{amplitude})^2$$

This means that a 6-foot wave on the ocean has four times the energy of a 3-foot wave, even though its amplitude is only twice as large.

DIFFRACTION AND INTERFERENCE: THE TRADEMARKS OF WAVES

So far in our study of physics, the "things" we have encountered have been particles, or collections of particles. Particles are localized, which means that they are confined to a fairly small region. When a golf ball sits on the tee, there is no question as to where it is.

Now we come to waves, which are quite unlike particles. Waves are not localized. They are spread out, often over large distances. As you sit and read these words, you can say where the book is. It is on your desk, or in your lap. However, if a friend speaks to you from across the room, you can't really say where the sound wave "is." It is all over the place.

In this section we are going to look at two very important properties of waves that result from this "spread-out" characteristic. These properties, called *diffraction* and *interference,* provide a sharp contrast between waves and particles.

Diffraction

Figure 17–4 presents photographs of an experiment with water waves in a ripple tank, which is a shallow vessel containing water and equipped with a device for making waves. The photographs show a parallel series of alternating crests and troughs, called a *wave front,* that travels from the left toward a wooden barrier with an opening. In the top photo, where the opening is large compared to the wavelength, you can see that the principal effect of the barrier is merely to block off part of the wave front. The rest continues through the opening toward the righthand side of the tank, apparently unaware that the barrier is present. The barrier casts a sort of "shadow" in front of the advancing waves.

But now look at the bottom photo. The only change in the experiment is that here the opening is much smaller, about the size of the wavelength. However, the effect is much different now. You can see that the waves emerging to the right of the opening are circular rather than straight. The wave front has changed direction on passing through the small opening. This change of direction of the wave is called *diffraction,* and it is a distinctive feature of wave motion.

Imagine that a bug is floating on the surface of the water, at the positions marked X in the photographs. In both pictures the X's are in identical positions relative to the near edge of the opening. In the top photograph, the bug is in the shadow cast by the barrier and will hardly know the wave is present. The surface of the water remains smooth, so the wave doesn't affect the bug much. But in the bottom photograph, diffraction causes the wave to bend around the corners. In this situation the bug can't hide from the wave front. The bug will "feel" the ripples as the diffracted wave passes by.

Compare this situation with that for particles. Imagine replacing the water wave experiment with some children throwing stones at a fence. (See Figure 17–5.) The stones go straight through the fence without bending around the corners, no matter the size of the opening. In fact, making the hole smaller simply cuts down on the number of stones that make their way through. It changes nothing fundamental about the experiment. Diffraction doesn't occur for particles. It is a phenomenon peculiar to waves.

Let us look again at the top photograph in Figure 17–4. Even though most of the wave front proceeds through the opening undisturbed, you can see some diffraction at the edges of the barrier. Diffraction of one degree or another always takes place whenever a wave front encounters a barrier.

Diffraction on the Job

Diffraction has many down-to-earth applications in industrial settings. Some big aircraft, for example, use a perforated, translucent sheet to help suppress engine noise. For quality control checks on these sheets, a quick and simple method was needed to measure the diameters of the perforations, or holes. Diffraction patterns made by shining laser light through the sheets aided in speeding up measurements of holes 0.006 to 0.008 inch in diameter.

Source: B. J. Hogan, "Diffraction Pattern Speeds Hole Measurement," *Design News,* March 3, 1981, pp. 74–75.

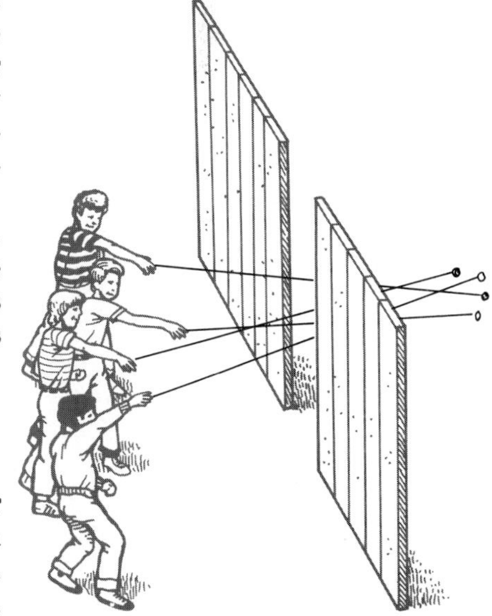
Figure 17-5. The stones are not diffracted.

Interference: The Principle of Superposition

When raindrops fall onto a puddle of water, and the circular waves move out from the various places the drops hit, each wave seems to pass through the others as if they weren't there. The motion of any single wave is uninfluenced by the rest.

However, the surface of the water *is* affected by all the waves, indicated by the variety of lines and patterns produced when the waves meet and overlap. This is an example of *interference*, another very important property of waves, and one that also makes their behavior quite distinct from that of particles.

Interference is one instance of a fundamental rule for waves, called the *principle of superposition*. In this sense superposition implies "laying one thing over another," and the principle applied to waves has quite a simple meaning. It is this: Whenever two or more waves travel through the same medium, the resulting effect at a given place is found by adding the amplitudes of the individual waves. Another way to look at superposition is to imagine the individual waves replaced by a single wave. The disturbance created by this new wave is simply the algebraic sum of the disturbances of its individual components.

Let us take the case of wave pulses traveling down a rope. Suppose you and a friend hold opposite ends of the rope, and each of you sends a pulse down the rope. We imagine that the two pulses are identical. Figure 17-6 shows what happens. When the two pulses arrive at the same location, their amplitudes add together to give a new amplitude that is the sum of the two. Note that the two pulses continue on their way, as if neither were aware of the other. In this example, where the resulting amplitude is larger than the amplitudes of the individual pulses, we have what is called *constructive interference*.

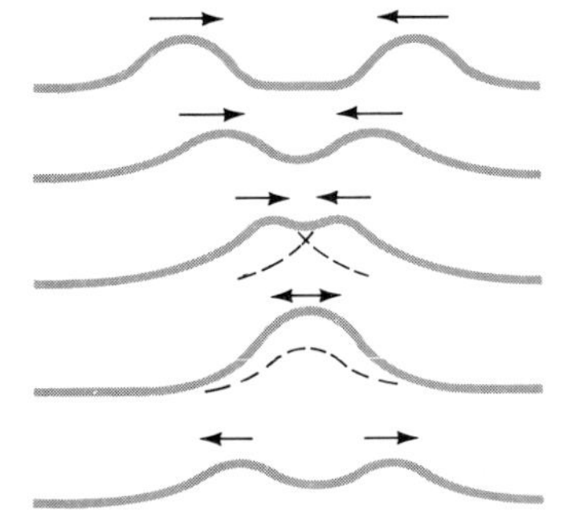

Figure 17-6. Constructive interference.

Figure 17-7 illustrates the opposite extreme, the superposition of a crest and a trough. This is called *destructive interference*. The two pulses cancel in their effect on the rope at the point where they meet. While one pulse tries to lift the rope, the other is trying to pull it down. The net result is that, at this particular point, the rope remains at equilibrium, as if no wave has passed that point at all.

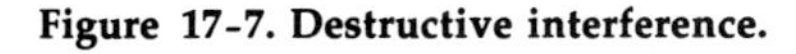

Figure 17-7. Destructive interference.

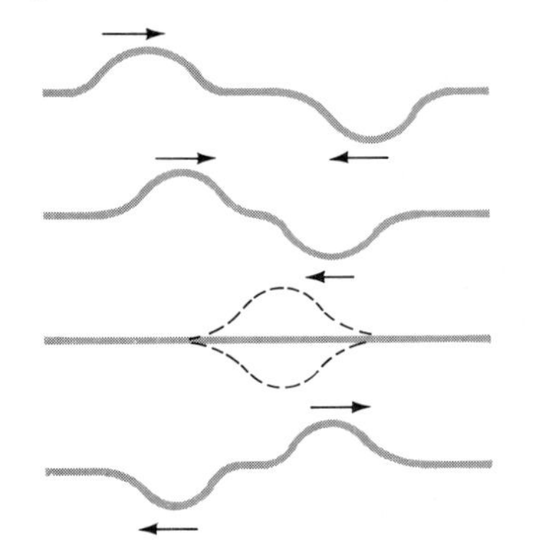

You should note carefully this feature of destructive interference. Even though *two* waves travel along the rope, it behaves (at the point where the waves meet) exactly as if nothing has happened; that is, it is possible for one wave plus one wave to equal zero wave. This can't happen with particles. It isn't possible for them to cancel out the way waves do.

Waves traveling on water provide a good means of studying interference. Figure 17-8 is a photograph of the pattern caused by two sets of interfering waves in a ripple tank. This time, the ripple tank is fitted with bobs or plungers that move vertically in and out of the water, generating waves like those made when you throw rocks into a pond. In this photograph, the bobs move with the same frequency, resulting in a symmetrical interference pattern.

When waves from the two sources meet, there are places where crests overlap crests and troughs overlap troughs, result-

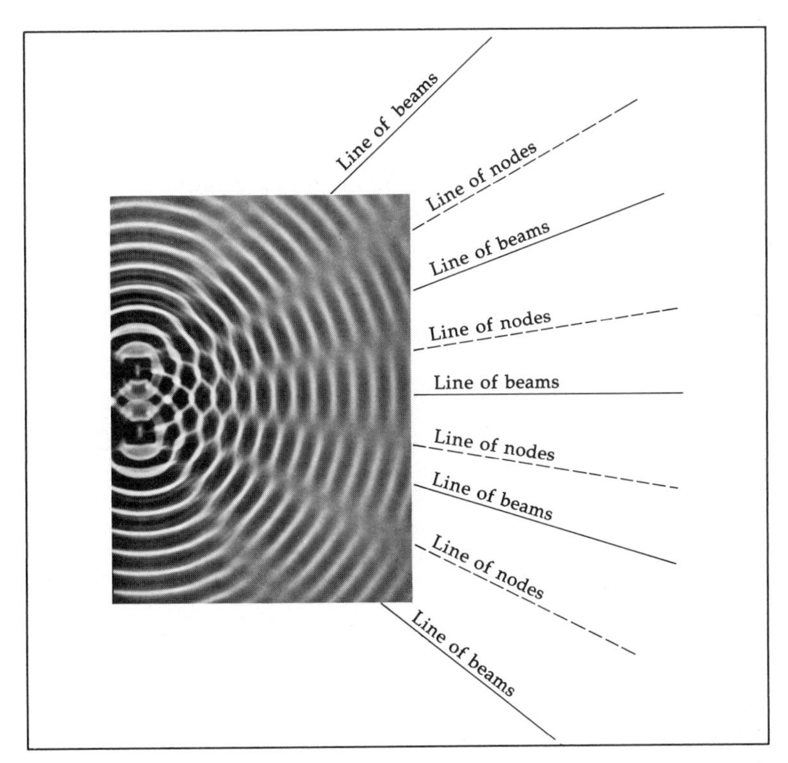

Figure 17-8. Interference for water waves.

ing in constructive interference. Here the waves reinforce each other. The symmetry of the arrangement, due to the equal frequencies of the outgoing waves, makes it appear that the disturbance is moving radially outward from the sources along straight lines, called *lines of beams.* Between the lines of beams, we see that a crest from one source combines with a trough from the other. The waves cancel, giving destructive interference. The result is a *node,* a point where the surface of the water remains unaffected, as if no wave had been there. Such cancellation in neighboring portions of the waves results in a *line of nodes.* You can see that the interference pattern consists of alternating lines of beams and nodes.

Figure 17–9 shows a ripple tank pattern of the combined effects of diffraction and interference. As the wave front moves from left to right, the two small openings diffract the waves. Notice that the resulting circular waves are identical to those that might be produced by vibrating bobs. These circular waves then interfere, producing the same nodes-and-beams pattern as in Figure 17–8.

We cannot overemphasize the importance of interference and diffraction as players in the story of waves. Together they form a trademark, a fingerprint revealing the presence of waves. As we continue our study, we will be able to understand many interesting and familiar aspects of sound and light as instances of diffraction and interference.

Figure 17-9. Diffraction and interference for water waves.

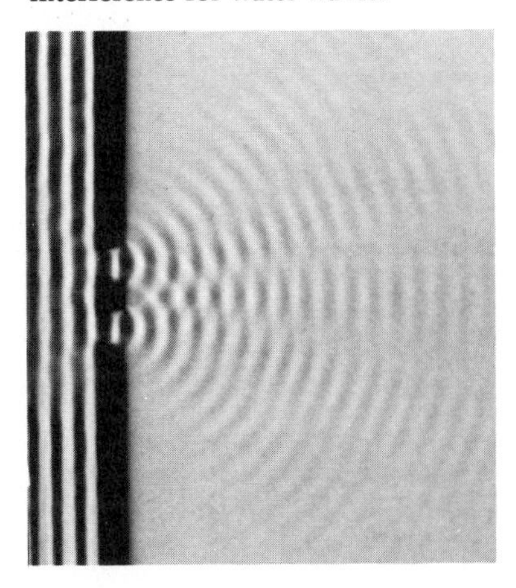

Television screens often display various interference patterns.

> **Learning Checks**
>
> 1. Distinguish between *interference* and *diffraction.*
> 2. Compare and contrast *constructive* and *destructive* interference.
> 3. How do diffraction effects depend on wavelength?
> 4. What is a *node?*

STANDING WAVES AND RESONANCE

We learned about interference of waves traveling along a rope by discussing what happens when you and a friend send pulses down a rope from opposite ends. The same effect can be achieved if you give your friend a rest and tie that end of the rope to the wall. The fixed end will reflect the waves that you send down the rope. These reflected waves will return toward you with the same frequency, and they will interfere with those you continue to make.

If you shake the rope up and down with some *arbitrary* frequency, the superposition of your waves and those reflected from the other end will give something of a mess, just a jumble of motion of the rope.

However, if you shake the rope up and down with any of several *particular* frequencies, a neat and symmetrical pattern will emerge. The rope will appear to vibrate in one, two, three or more segments. This phenomenon is called a *standing wave,* (see Figure 17-10), because it looks as if the waves are not traveling down the rope at all, but standing in one place. What happens is this: The waves you propagate at this special frequency match the reflected waves in just such a way that crests from one end and troughs from the other meet at *fixed* points along the rope. At these fixed locations, then, we have *destructive* interference—the waves coming from one end are exactly wiped out by those moving from the other direction, and the effect on the rope is the same as if no wave existed. Again these special points are called *nodes,* exactly analogous to the nodes we saw in the ripple tank experiment with water waves. The points where the maximum *constructive* interference takes place are called *antinodes,* which are analogous to points along the lines of beams in the ripple tank. Note that each antinode is located midway between two nodes.

Figure 17-10. Standing waves on a rope.

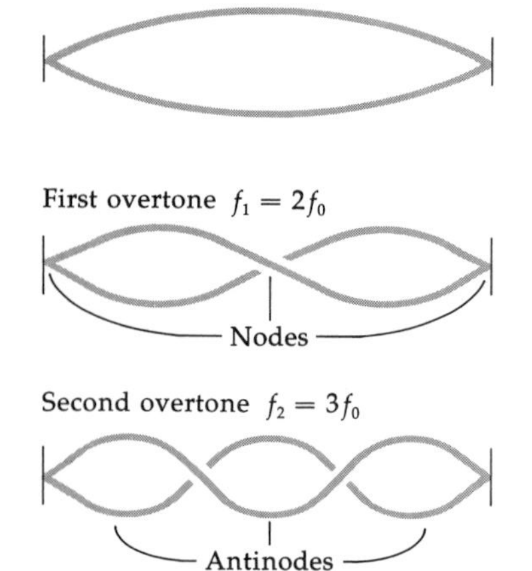

We said that standing waves will occur if you shake the rope at one of the "right" frequencies. What does this mean? Look again at Figure 17-10. Notice that the distance between two nodes is equal to *one-half the wavelength* of the wave comprising the standing wave pattern. So only those frequencies that allow this special situation—that is, this particular location of points of complete destructive interference—are the "right" ones. These particular frequencies in turn depend on the characteristics of the medium (the rope in our illustration). Recall that the product of frequency and wavelength gives the wave's speed, $v = f\lambda$, where v is determined by the physical properties of the rope, such as its tension and mass.

For a rope of a given length and characteristic wave speed, shaking it at one particular frequency will cause it to vibrate in a single segment. This frequency is called the *fundamental frequency,* or first harmonic, f_0, and the corresponding wavelength is twice the length of the rope. To vibrate the rope in two segments requires a frequency of twice the fundamental, called the *first overtone* (or second harmonic)

$$f_1 = 2f_0 \qquad \text{(first overtone)}$$

At this frequency, the wavelength is equal to the length of the rope. If the rope is shaken at a frequency three times the fundamental, it will vibrate in three segments

$$f_2 = 3f_0 \qquad \text{(second overtone)}$$

and the wavelength is two-thirds of the rope length. We could go on with this: A standing wave will emerge whenever the rope is shaken at any frequency that is a *whole number* multiple of the fundamental. Those multiples in between, such as $1.25f_0$, or $3.33f_0$, or $0.82f_0$, will not work.

The fundamental frequency f_0 and the higher harmonics f_1, f_2, f_3, and so on make up a set of *characteristic* or *natural frequencies.* All objects—not just strings or ropes—have their own natural frequencies. They depend on the physical characteristics of the object—its shape, the material of which it is made, which parts of it are restricted in their motion, and so on.

All kinds of objects may be made to exhibit standing waves. Figure 17–11 shows photographs taken from a movie of a marked rubber drumhead vibrating with several of its natural frequencies. Since the edge of the drumhead is clamped to the rim, and thus not allowed to move, the edge must be a node. Only those vibrations that have the edge as a node are allowed.

An interesting way of studying standing waves on surfaces is provided by *Chladni plates.* A metal plate clamped at some point is covered with salt or grains of sand. Bowing the plate with a violin bow causes the plate to vibrate at one of its natural frequencies, and standing waves are set up along the surface. The grains of salt bounce away from the antinodes and gather along the nodes, where there is little or no motion, thus revealing the pattern of the standing wave. Figure 17–12 illustrates some of the possible results. Different patterns are pro-

Figure 17-11. Standing waves on a drumhead.

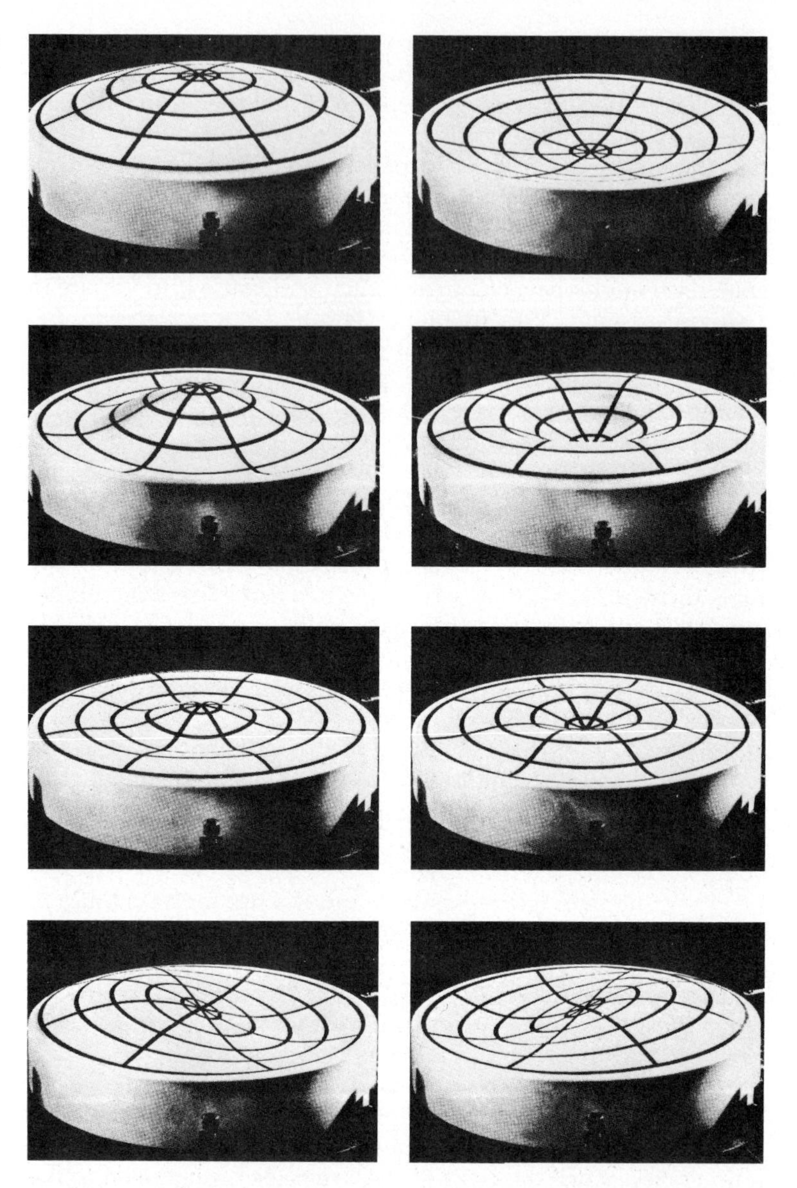

duced, depending on where the plate is clamped and bowed.

The pattern of the standing wave also depends on the shape of the Chladni plate. Figure 17-13 shows a violin-shaped brass plate, bowed at two different positions as indicated by the dark dots.

In addition to their connection with standing waves, the natural frequencies of vibrating objects are important in another, related aspect. If an object is made to vibrate because of some force that is applied with a frequency equal to one of these natural frequencies, the vibration will have a very large amplitude. The force doesn't need to be very *strong*. It is the

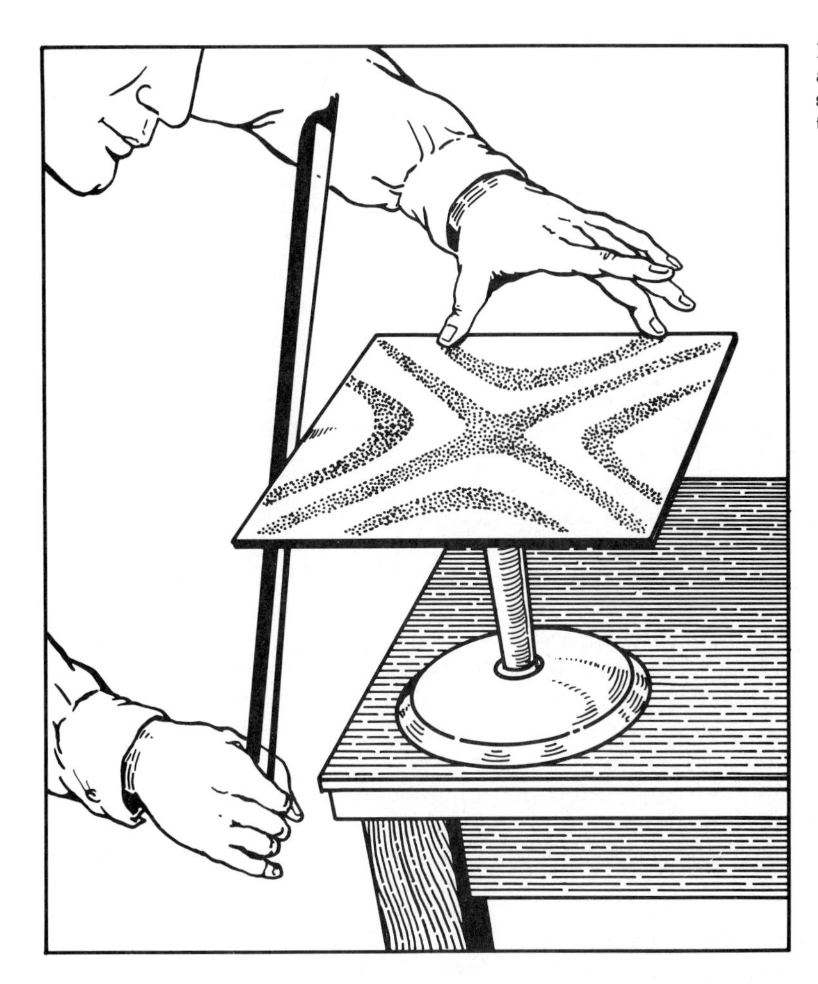

Figure 17-12. Using a violin bow and a heavy metal plate covered with sand to show how sound waves travel in a solid.

frequency with which you apply the force, rather than how great the force is, that causes the amplitude to build up. This phenomenon is called *resonance.* The natural frequencies of the object are sometimes referred to as *resonant frequencies.*

Let us take a common example. A child's swing on a playground is simply a pendulum, which has a single natural frequency that depends only on its length. Children first beginning to use such a swing quickly learn that they can go very high in the air by "pumping" the swing at just the right instants. This "pumping" is a force that is applied with the swing's natural frequency, giving a resonance condition. The amplitude builds up swiftly, even if the swing is not pushed very hard. It is the timing of the force, not its size, that is critical. The child learns to stop the swing by pumping at the "wrong" time. That is, when the pumping force is applied out of phase with the natural frequency, the swing's amplitude rapidly drops.

Resonance shows up in all kinds of wave motion. If you are

Figure 17-13. Violin-shaped Chladni plates, bowed at locations indicated by dots.

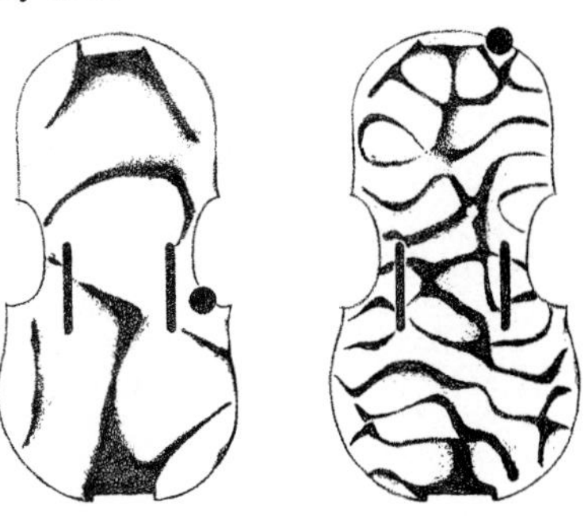

one of those people who sing in the shower, notice next time that some notes appear much louder than others, and you seem buried in the sound. When this happens, it means that you have sung a note corresponding to one of the natural frequencies of the "box" in which you are showering. Another example: Wet your finger and draw it around the rim of a wine glass. The glass will "sing." Notice that pitch depends on how much wine is in the glass.

If the tires on your automobile are out of balance, they will vibrate with a frequency that depends on the car's speed. You can feel the car "shimmy" whenever you reach a particular speed because the vibration of the out-of-balance tires matches one of the natural frequencies of the car. Again, this is resonance. Notice that the shimmy goes away if you get above or below this critical speed.

Perhaps the most dramatic example of the effects of resonance occurred in the fall of 1940 when resonance vibrations led to the collapse of the bridge across the Tacoma Narrows near Tacoma, Washington. On the morning of November 7, four months after the newly built bridge was opened for traffic, high winds caused the 2800 foot central span to vibrate in a twisting motion. The midline separating the two lanes of traffic was a node for this vibration. The roadbed oscillated up and down with an amplitude of several feet. This resonance condition persisted for about an hour until the bridge finally failed, with the central span breaking up and falling into the waters of the Puget Sound below. The film made by a news photographer on the scene is a remarkable illustration of resonance in the real world.

The first section of the span of the Narrows Bridge as it broke in two and fell 190 feet into the waters of Puget Sound. Note the car (right of center on the bridge). It belonged to a newspaper reporter who crawled 500 feet to safety along the tottering structure.

One simple way of demonstrating resonance is illustrated in Figure 17-14. Five pendulums are attached to a common horizontal rope. Pendulums A and D are matched, having the same length, as are C and E (with a different length from the A-D pair). Pendulum B is the odd ball in the crowd. What happens if we start one pendulum swinging, say pendulum A? Since its frequency matches that of D, the latter will soon begin to swing because of the energy that A has transferred along the horizontal rope. That is, A and D are in resonance. The other pendulums, mismatched in frequency, will also swing a little, but with not nearly so large an amplitude as D. Because of the frequency match of A and D, they readily share energy, and as time goes on this energy is transferred back and forth between them.

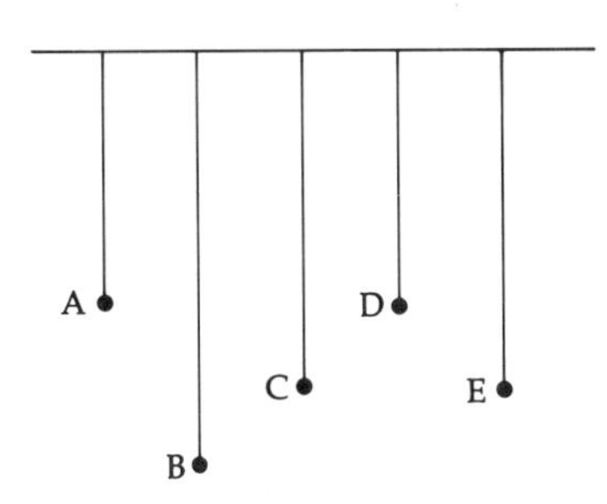

Figure 17-14. Pendulum arrangement for demonstrating resonance.

If pendulum C is set swinging, resonance will cause its partner, E, to swing also, with very little energy transferred to the others. Pendulum B, having no other pendulum with which to be in resonance, will have very little effect on its neighbors. It will neither deliver nor receive much of the energy that you put into the system by pushing one of the other pendulums.

Learning Checks

1. Explain what is meant by a *standing wave.*
2. If the fundamental frequency of some vibration is 25 hertz, what is the frequency of the third harmonic?
3. What is a *node?* Is the word "node" used in the same sense for a standing wave as for interference of water waves? Explain.
4. In resonance, how does the frequency with which an external force is applied compare with the natural frequency of the vibrating object?

SUMMARY

All waves result from the *vibration* of some object. You can send a *traveling wave,* or series of *wave pulses,* down a rope fixed at one end by shaking the free end up and down.

The *wavelength* of a wave is the distance between successive identical parts of the wave. The *amplitude* is the maximum displacement of the medium from its equilibrium, or rest, position. A *cycle* of the wave corresponds to one wavelength and results from a complete round trip of the vibrating object. The time it takes for one cycle is called the *period.*

Frequency is the number of wavelengths of a traveling wave that pass a fixed point each second. It is therefore also equal to the number of cycles the vibrating object goes through each second.

The period and the frequency are reciprocals of each other: $T = 1/f$ and $f = 1/T$. The *speed* of the wave tells how far a given portion of the wave travels in one second. It is related to the wavelength and the frequency: $v = \lambda f$.

Every material has its own special speed at which waves can travel through it or across it. This speed depends on what the material is made of, its temperature, and so on.

Energy is transmitted in a wave. The energy a wave carries is proportional to the *square* of the wave's amplitude.

Diffraction and *interference* are two very important trademarks of waves. In diffraction, the wave changes direction as it passes some obstruction. Diffraction effects are noticeable whenever the dimensions of the barrier are about the same as the wavelength of the wave. *Interference* means that the material in which two or more waves travel behaves as if the amplitudes of the waves were added together. *Constructive interference* refers to reinforcement: two waves produce the effect of one big wave. *Destructive interference* means cancellation: two waves nullify each other and the material behaves as if no wave were present.

Standing waves occur whenever waves of particular characteristic frequencies interfere in such a way that destructive and constructive interference take place at fixed locations. The frequencies at which standing waves exist are whole-number multiples of the lowest or fundamental frequency.

Resonance is marked by a very large amplitude of vibration and occurs when an object is made to vibrate by a force that is applied repetitively with the natural frequency of the vibrating object.

Exercises

1. What is the difference between the speed of a wave traveling down a rope and the speed of a small section of the rope?

2. Describe in your own words what is meant by the period of a wave and why it is the inverse of the frequency.

3. What role does superposition play in the production of standing waves?

4. When the cork on your fishing line bobs up and down in the water, how does the frequency of its motion compare with the number of waves that pass it each second?

5. For research purposes, an oceanographer places a plunger in the sea. When this plunger moves regularly up and down through the surface of the water, it produces waves of a particular wavelength. Increasing the plunger frequency causes the wavelength to decrease. Why is it not possible to change the frequency of the plunger without altering the wavelength of the water waves?

6. Explain how a child swinging on the playground is an example of resonance.

7. The wavelength of a certain wave is 10 meters and its frequency is 5 hertz. What is the speed of this wave?

8. In Exercise 7, if the wavelength is reduced to 2 meters, what will be the new frequency? Will the speed also be different?

9. Suppose that the fundamental frequency of a rope fixed at both ends is 9 hertz. If the rope is made to vibrate with two nodes between the ends, what is the frequency of this vibration? Which overtone is this?

10. If an ocean wave 2 meters tall has 50 units of energy, how much energy will a wave 10 meters tall have?

11. A water wave has a wavelength of 20 centimeters and a speed of 1 m/s. What is its frequency?

12. The fundamental frequency of a certain rope fixed at both ends is 5 hertz. What are the frequencies of the first three overtones? Is it possible to produce standing waves by shaking this rope at a frequency of 24 hertz? Explain.

13. The fundamental frequency of a rope tied between two posts is 3 hertz. If the rope is made to vibrate by an external force applied at 9 hertz, how many nodes will there be? Which overtone is this vibration?

14. In Exercise 13, how many segments make up the standing wave?

15. Sound waves travel through air at a speed of around 330 m/s. What will be the frequency of a sound wave whose wavelength is 11 meters?

16. Suppose that the cork on a fishing line bobs up and down on the surface of a pond to produce waves at the rate of two each second. What are the period and frequency of the waves?

17. In Exercise 16, do you have enough information to calculate the wavelength of these ripples? If not, what additional data do you need?

18. How does the principle of superposition apply to the production of standing waves?

19. What is the frequency of the minute hand of a clock? The second hand?

20. The alternating current in a household circuit changes polarity sixty times each second. Give the frequency and period of this current.

18

Sound

Vibrations of sound surround us whether we hear them or not. Some waves that strike our ears are indiscernible; others, unfortunately, nearly unavoidable.

Sound is an obvious part of our lives. Different sounds mean different things. They bring on varieties of feelings. They evoke memories of people or past events. We use sounds for communication, for pleasure, for setting moods, for identifying places and things. Traffic in the city, the crowd at a football game, the murmur of a mountain stream—each is marked by its own peculiar sound.

Our concern here is in understanding the physics of sounds—how they are made, what happens to them, their features and traits. What we learned about waves in Chapter 17 is a foundation on which we can build our understanding of sound, for sounds are waves. The connection between sound and wave motion has been known since the time of the early Greeks. Every sound you hear, from the quietest rustle of leaves to the booming of a cannon, from the throbbing beat of rock and roll to the melodic strains of Mozart, is a wave that strikes your ear.

SOUND WAVES

Like the other waves we have studied, sound waves are produced because some object vibrates. Hold your fingers lightly against your throat as you talk, and you will feel the vibrations. Strum a guitar and you can see the strings vibrating back and forth so rapidly that each string looks blurred. As long as the blurs are there, the string is vibrating and the sound persists. By clamping down on the strings to stop their vibrations, you cut off the sound.

We can learn much about the nature of sound waves by considering what happens when we strike a tuning fork. (See Figure 18–1.) The prong of the fork moves very rapidly right, left, right, left, in a periodic motion having a characteristic frequency. As it moves to the right, the prong shoves together air molecules immediately to its right, forming a small volume of compressed air. Because the pressure in this volume is greater

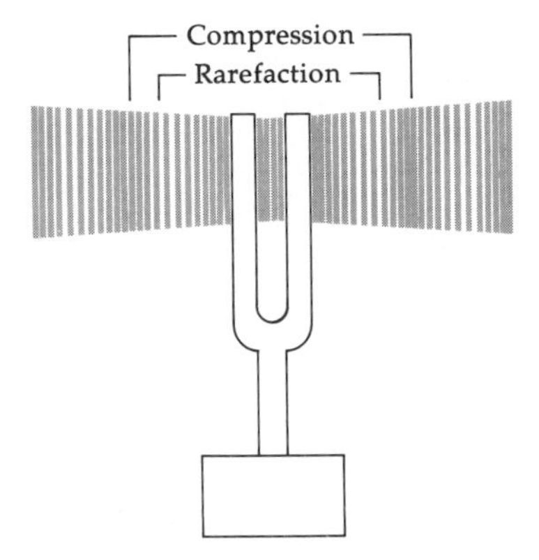

Figure 18-1. The vibrating fork produces regions of compression and rarefaction.

than that in the neighboring region of undisturbed air, the compressed molecules bounce apart, pushing against this normal region and compressing it. These compressed molecules in turn push against their neighbors, and so on, in a sort of domino effect. In this way, a volume of compression moves outward in all directions, an expanding sphere with the tuning fork, the source of the disturbance, at its center. This is exactly analogous to the ever-widening circle of the ripple formed when you drop a pebble on the surface of a pond.

In the meantime, the tuning fork prong has started its swing back to the left. This creates more room in the volume of air to the right. Here the air expands and there are now fewer air molecules in a given volume. We say that the air is *rarefied,* and we call this a *region of rarefaction.* Now the pressure is higher in the neighboring undisturbed air, which therefore pushes into the volume of expanded air and is itself rarefied in the process. In this way, a rarefied volume expands outward as a sphere trailing behind the region of compression.

As the prong of the tuning fork continues to oscillate from right to left and back again, alternating volumes of compression and rarefaction travel outward from their source for as

Theaters and Bones: The Science of Acoustics

If you are thinking of earning your living as a physicist, one field that might interest you is that of acoustical consulting. Acousticians are called upon to participate in designing performing arts theaters, opera houses, symphony halls, and auditoriums. They suggest how to get rid of unwanted echoes, dead spots where sounds from the stage cannot be heard, and background noises. People active in this area may be asked into communities to recommend noise control measures for urban or suburban streets, hired to perfect the acoustical properties of musical instruments, invited to consult in the design of hearing aids, or engaged to help perfect synthetic voices for computers.

Such physicists often work side by side with engineers, who have known for years that all materials give off their own characteristic sound waves when stressed to the breaking point. Engineers use these stress-wave emissions to measure the structural integrity of buildings, bridges, aircraft, and automobiles. If extra stress is applied to a beam with a minute crack, say, the acoustic emissions in this case will differ from those of a structurally sound beam.

One clever biomedical engineer with acoustical leanings applied this method to the human skeleton by sending ultrasonic pulses directly into bone and through it to an electronic listening device. Not only did x-rays bear out the damage discovered by the sound waves, but the "bone-listening" could detect hairline fractures too tiny to be identified by x-rays.

As a science, acoustics is in its infancy, with a bright future ahead.

Sources: Denise Grady, "Cyril Harris: A Man of Sound Advice," *Discover,* July 1982, pp. 36–42.
"Sound Finds Broken Bones," *Science Digest,* February 1982, p. 98.

Though visually impressive, the ceiling of the New York State Theater contained serious errors in acoustical design. An acoustical consultant and an architect plan a new ceiling.

long as the fork vibrates. Each cycle of the prong—that is, one round trip—corresponds to one compression-rarefaction pair. What we show in Figure 18–1 is a sliver of the concentric spheres formed by these alternating volumes.

Notice that for sound waves, the air molecules oscillate back and forth along the line of the outward-traveling disturbance. We call this a *longitudinal* wave (or sometimes *compressional* wave) to distinguish it from the waves on strings and water surfaces that we have considered before. These are called *transverse* waves, because the parts of the vibrating object move *across*, or perpendicular to, the direction the disturbance travels.

A common "Slinky" is useful for demonstrating both transverse and longitudinal waves. Figure 18–2 shows one hanging from the ceiling. By moving the free end back and forth parallel to the floor, we can produce a transverse wave, just like waves on a string. If we move the free end up and down, we generate a longitudinal wave. The alternating regions of gathered and separated coils along the Slinky are like the volumes of compression and rarefaction, respectively, in a sound wave traveling through air.

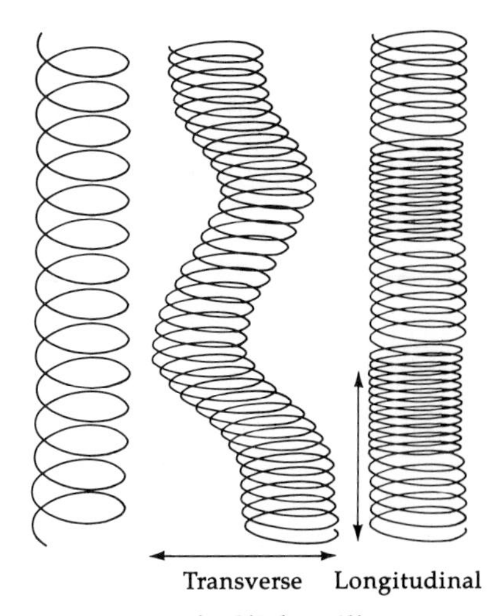

Figure 18-2. The Slinky will carry both transverse and longitudinal waves.

Perhaps it is easier to visualize the wave character of a transverse wave than a longitudinal one. After all, for waves traveling along a rope or across a water surface we can actually see the peaks and valleys. The wave pattern is obvious. These peaks and valleys don't show up in a longitudinal wave. However, we can *represent* a longitudinal wave by drawing a picture that looks just like a transverse wave, in which the wave crests correspond to the regions of compression and the troughs correspond to the rarefactions. This is shown in Figure 18–3. You can see, for example, that the wavelength is the distance between a point of maximum compression and the next neighboring maximum compression. The other terms, such as frequency and period, that we defined for transverse waves follow quite naturally when applied to longitudinal waves.

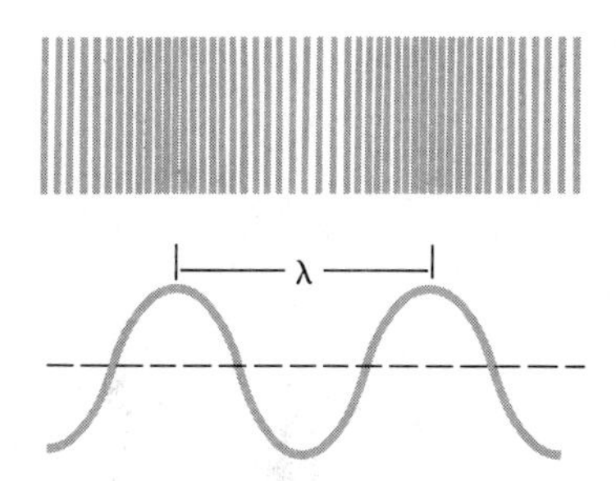

Figure 18-3. Representation of sound by a wave.

Remember, this picture is only a *representation* of the sound wave. We have drawn it merely to illustrate the wave properties of sound.

Just as in the case of the falling dominoes, or the waves on water, individual air molecules don't travel very far from home base when sound waves pass by. As an illustration, suppose someone in the next room rips open a package of limburger cheese. You will *hear* the sound of the package being opened long before you *smell* the strong odor of the cheese. Again, it is the energy the waves transmit, not the molecules themselves. The odor that finally reaches your nose gets there because of the random thermal motions of the molecules in the air; the molecules are not driven through the air by the sound wave.

Sound can travel through other materials besides air. Put your ear to the floor and you can hear someone walking across the room. Sound travels quite readily through water, as you can demonstrate to yourself by banging two rocks together when you are underwater in a swimming pool. The old cowboy-and-Indian movies always have an army scout who puts his ear to the ground and warns the cavalry patrol of a war party off in the distance. In fact, the phrase "keeping an ear to the ground" came into being because of this ability of the earth to carry sound.

Sound waves can move through any gas, liquid, or solid. In every case, it is the back-and-forth vibrations of particles about their equilibrium positions that make up the traveling longitudinal wave. The other side of the story is this: It is *impossible* for sound to travel in the *absence* of material particles, that is, in a vacuum. Something must exist to carry the disturbance away from the vibrating source, and in a vacuum that "something" is missing.

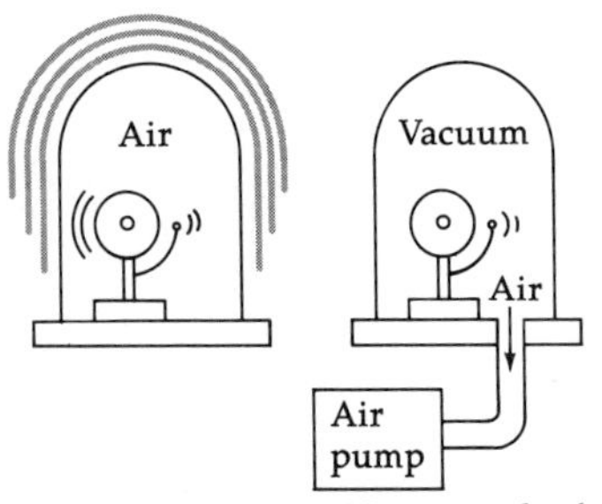

Figure 18-4. Sound must have something to carry it.

Figure 18–4 illustrates a nice demonstration of this feature about sound. Here we have a battery-operated doorbell that is set ringing inside a glass dome equipped with an air pump. You can hear the bell ringing as long as air remains in the dome, but as the pump removes more and more air, the loudness of the sound diminishes. Finally, when the dome is nearly evacuated, the ringing sound stops, even though the clapper continues to bang away.

The American Underwater Band experiments with making music in the medium of water. Affected by temperature, salinity, and density changes, sound waves behave differently in water and in air. (a) New instruments are adapted for use in the new environment. (b) Familiar instruments are tried to hear what they sound like submerged.

Learning Checks

1. Explain the terms *compression* and *rarefaction* as used in describing sound waves.
2. Distinguish between *transverse* and *longitudinal* waves.
3. In a wave traveling along a wire, the amplitude is the distance the individual segments of the wire move away from their normal, undisturbed positions. What corresponds to the amplitude of a sound wave?

The Speed of Sound

We saw earlier that the speed of a wave depends on the medium through which it travels. This is true of sound waves as well. The speed of sound depends on the kind of material, its structure, its temperature, and so on, and will not change for a particular medium under a given set of conditions. For air at 20° C, sound travels at about 340 m/s (or about 760 mi/h). This varies with temperature, from around 330 m/s on a cold winter's day to about 355 m/s in the heat of summer.

Sound travels faster in solids and liquids than in air. In general, the more dense the material, the greater will be the speed of sound through it. For example, in water at 20° C, sound travels 1450 m/s, or about four times its speed in air. Metals are quite good sound carriers. The speed of sound in iron at 20° C is more than 5000 m/s. By putting your ear to the railroad track you can hear a train coming long before you could by listening through the air.

As we shall see later, light travels extremely fast compared to sound. This is the reason you can often *see* an event off in the distance before you can *hear* the sound associated with it. For example, if at a track meet you stand at the finish line for the 100 meter run, you will see the puff of smoke from the starter's pistol a little before you hear the shot. At this distance, your seeing the smoke is virtually simultaneous with the firing of the pistol, since the speed of light is so very great. But the noticeable time delay in your receiving the crack of the pistol occurs because the sound travels much more slowly. Moving at 340 m/s, the sound takes about three-tenths of a second to cover the 100 meters between the starter's pistol and you. This time interval is not large, but it is quite obvious.

Physics and Surgery

Because intense ultrasonic waves can cause great pain and swift breakdown of body cell structure, they have been proposed for use as weapons. (Luckily, appropriate sonic projectors are inefficient and cumbersome, so this idea has been tabled for the time being.) But ultrasonics (the study of frequencies beyond the limit of human hearing) is also being applied in beneficial ways. One is in surgery, where a newly invented "scalpel" substitutes high-frequency sound for the cutting edge of metal.

How does this work? Cutting vibrations emanate from the tip of an instrument shaped like a pencil. The tip oscillates 23,000 times a second across a distance half a millimeter or less and squirts out water to cleanse the area being cut.

Over a thousand virtually bloodless operations have already been done with this procedure. This physics-inspired surgical cutting method should have a favorable psychological impact on preoperative patients worried about going under the "knife."

Source: "Sound-Wave Surgery," *Science Digest*, October 1981, p. 102.

For the same reason, there is a time delay between the lightning and the thunder of a distant storm, 1 second for every 340 meters between you and the lightning. Conversely, whenever you hear the clap of thunder almost simultaneously with the flash of lightning, you can be sure that the lightning struck very close by. A good rule of thumb is this: The sound of the thunder goes about 1 mile in each 5 seconds after the lightning flash.

Loudness: The Intensity of Sound

We learned in Chapter 17 that a wave carries energy as it moves outward from its vibrating source and that the amount of energy is related to the amplitude of the wave. As you might expect, the loudness of a sound depends on how much energy is carried by the sound wave.

The *loudness* of the sound is determined by the wave's *amplitude,* which in turn is governed by how vigorously the source vibrates. If a pianist gently strikes a piano key, the hammer taps the piano string softly, making it vibrate with a small amplitude. A quiet sound results. But if the pianist bangs down hard on the key, the same note sounds much louder, since now the string vibrates with a much larger amplitude.

The amplitude of a sound wave is a measure of the pressure produced by the vibrating object. Let us go back to our tuning fork. If we really clobber it, the prongs move back and forth in a wide sweep, compressing the air more violently than if we tap it lightly. A loud note differs from a soft note in that, for the former, the compressed volumes of air are more compressed and the rarefied volumes are more rarefied. It is this greater difference in the amount of compression that gives us the amplitude of the sound wave.

Obviously, compressing the air wave more requires more energy, so it makes sense that the energy and compression for a sound wave are related in exactly the same way that energy and amplitude are for an ocean wave.

Now let us consider more carefully what all this has to do with the loudness of the sound. Loudness is related to the wave's *intensity,* which is the amount of energy delivered during some length of time across a given area. It is energy per second per unit area; that is, the intensity of a wave is related to its *power.* Sound waves ordinarily carry very small amounts of power. Everyday conversational sounds involve around 1000 microwatts (a microwatt is one-millionth of a watt), and the ear is sensitive to much lower power as well. The faintest sound a normal human ear can hear has an intensity of about 10^{-12} watt per square meter (10^{-12} W/m^2). At the other extreme, the loudest sound most people can stand without pain has an intensity of about 1 W/m^2.

It turns out that the difference in loudness of two sounds depends on the *ratio* of their powers, not the difference of powers. For instance, a 6000 microwatt sound will be heard with the same relative loudness compared with a 3000 microwatt sound as a 2000 microwatt sound does in comparison with one at 1000 microwatts—even though the *difference* in power is 3000 microwatts in the first case and only 1000 microwatts in the second. But in *both* cases the louder sound has twice the power of the softer. This power ratio is what the ear detects.

Look at this in another way. Suppose you are listening to the radio at a sound level corresponding to 1000 microwatts, and you turn the volume up until the power is 2000 microwatts. By adding 1000 microwatts of power to the sound, you have increased the loudness by a certain amount. Now turn the volume up again to correspond to an additional 1000 microwatts, giving 3000 in all. The sound will be louder still, of course, but this time the change in loudness will not be as great as in the first case. The first ratio, 2000/1000, is larger than the second, 3000/2000.

So the loudness of sound is not directly related to the power itself. Rather, it changes with what is called the "logarithm" of the power. The logarithm of any number is the exponent to which 10 is raised to get that number. For example, the logarithm of 10 is 1, that of 100 (that is, 10^2) is 2, for 1000 (10^3) it is 3, and so on. If one sound carries, say, 2000 microwatts and another carries 20,000 microwatts, the power ratio of the sec-

Table 18-1
Loudness of Common Sounds

Source of Sound	Loudness, dB	Description
Rocket engine (nearby)	180	
Jet takeoff (nearby)	150	
Rock concert (2 meters from amplifier)	120	Threshold of pain
Subway train; siren at 30 meters	100	
Very noisy factory	90	Constant exposure can damage hearing
City traffic	70	
Normal conversation	60	
Library	45	Quiet
Soft whisper (5 meters)	30	Very quiet
Rustling leaves	20	
Normal breathing	10	Barely audible
	0	Threshold of hearing

ond to the first is 10, whose logarithm is 1. So the difference in noise level is called one *bel*, a unit named after Alexander Graham Bell (who, in addition to inventing the telephone, did other pioneering studies in the physics of sound). If one sound is 1000 times more powerful than another, it is 3 bels louder, and so on.

It is more convenient to use the decibel (one-tenth of a bel), abbreviated dB, as the unit of sound level. Each time the power is multiplied by 10, we add 10 dB to the sound level. If the threshold of hearing, at an intensity of 10^{-12} W/m^2, is assigned a value of 0 dB, then 10^{-11} W/m^2 would be 10 dB—the power is ten times that at threshold. Increasing the power by another factor of 10, to 10^{-10} W/m^2, raises the sound level to 20 dB. Table 18-1 (on the preceding page) gives the decibel figures for some common sounds.

Figure 18-5. Megaphone.

Controlling the Direction of Sound

The energy of a sound wave, and therefore its intensity, can be channeled in one direction. In this way the sound is made louder. A cheerleader at a basketball game uses a megaphone to direct the sound to a section of the crowd. (See Figure 18-5.) Speaking tubes on naval vessels allow seamen in various parts of the ship to communicate with each other. You can make such a tube by using a length of an old garden hose. (See Figure 18-6.) A listener at the opposite end can hear even the smallest whisper clearly and distinctly.

The idea behind these sound-direction devices is simple. When the sound wave is created, it spreads in all directions away from the vibrating source. This means that a listener some distance away will receive only a small fraction of the energy carried by the wave. (See Figure 18-7.) But if instead we direct the sound wave—with a megaphone, for example—then the listener receives almost all of the energy. The portion of the wave that was "wasted" in being spread out is now captured by the listener.

Pitch: The Frequency of Sound

The sounds you hear during the course of a day are all quite different, from the high squeal of a tire laying rubber to the deep swell of a bass fiddle. The property that distinguishes

Figure 18-6. Speaking tube.

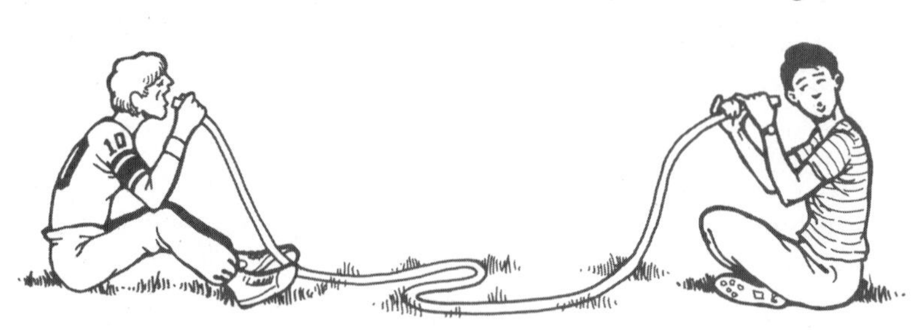

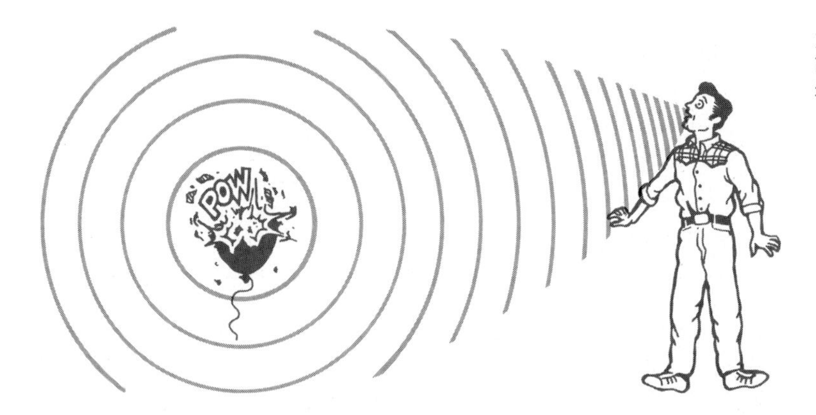

Figure 18-7. When the balloon bursts, the listener gets only a small fraction of the sound wave.

sounds in this way is called *pitch.* The frequency of the waves primarily determines the sound's pitch. High-pitched sounds, like tunes played on a piccolo, are due to high-frequency waves; the lower frequencies correspond to the low-pitched notes, such as those of the bass drum and tuba. The normal human ear can detect frequencies between about 20 hertz at the bass end of the spectrum to about 20,000 hertz on the high side. For other animals, the audible range may be quite different. Bats can hear frequencies up to 100,000 hertz. The audible range for dogs may go as high as 50,000 hertz. Blowing a high-frequency whistle designed for calling dogs produces a sound to which the human ear simply is not sensitive.

By permission of Johnny Hart and Field Enterprises, Inc.

Physics and Sociology

The World Soundscape Project is a Canadian-based research program that carries out pioneering investigations into the relationships of individuals and communities to their acoustic environments. It has prepared sound studies of many small villages. These investigations are often interdisciplinary, taking into account the architecture, psychology, history, and communication of the locale under study. Part of this effort involves identifying and cataloging the various pitches of the soundscape and the effects on residents. These appear as musical scores. Here is a representative sample of their findings:

In Europe the current is 50 Hertz giving a musical pitch of approximately G sharp. It is this fundamental that sounded from the cardboard factory. The other pitches resulted from the presence of strong harmonics. An additional feature of the [village] soundscape was the piping F sharp of the train whistles. The resulting aggregate of pitched sounds was a dominant ninth chord, quite in tune.

Further investigation of some malfunctioning street lamps gave us additional tones: C natural, C sharp and D; but the predominant character of the . . . soundscape was that 9th chord. Returning to the glass factory we were surprised to detect that its low rumble was centred on C sharp, giving the whole chord a quite classical harmonic resolution, just as the glassworks itself gives the town its fundamental character. . . . Of course, one does not hear these things standing still. One has to move about. As one does, the town plays its melodies.

Source: Bruce Davis, *Five Village Soundscapes.* Burnaby, British Columbia, Canada: World Soundscape Project, 1977.

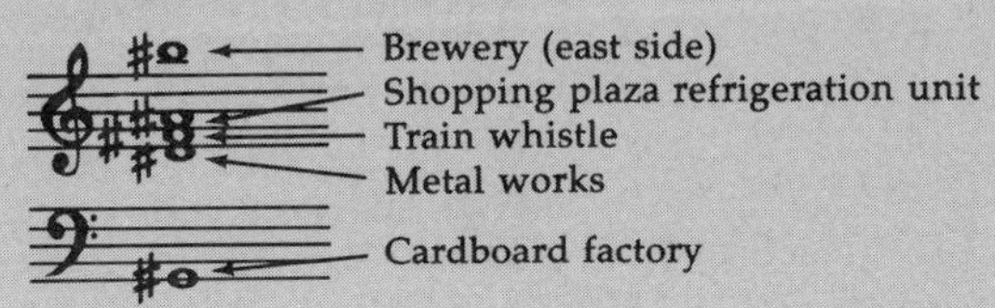

Because the frequency of a sound wave is the same as that of the vibrating source, the wide variety of pitches that we hear results from the various frequencies at which the objects in our environment can vibrate. Let us take the strings of the guitar as examples. We can combine what we know about wave speed and wavelength to understand how the guitar is able to provide its extended range of pitches.

In Figure 18–8 we show again some of the possible vibrations for a string clamped at both ends. When you pluck a guitar string, it vibrates in several of these modes at the same time, all of which correspond to standing waves. Any wave whose frequency is different from that of a standing wave will be rapidly wiped out because of interference as its reflected partners travel up and back along the string. Only the standing-wave frequencies survive.

Figure 18-8. Three possible vibrations of a guitar string.

Because all the strings on the guitar have the same length, they all have the same possible wavelengths for their characteristic vibrations. So, in order to get different frequencies, the strings are made in such a way that the *speed* of the waves is different on each string. Remember that the frequency of the wave is equal to the speed divided by the wavelength:

$$f = \frac{v}{\lambda}$$

This shows that for a fixed wavelength, such as we have for the guitar strings, the frequency varies directly with the speed of the wave on the string. High-speed waves will give high frequencies, low-speed waves low frequencies. On the guitar, each string has a different mass—the greater the mass, the less the wave speed, and the less the wave speed, the lower the frequency. Thus, the very thin wires give the high-pitched notes, while the thicker, more massive strings produce the lower frequency bass notes.

If this were all there were to it, the six string guitar could only produce six notes. But there is more. As everyone knows, you can change the pitch for a particular guitar string by tightening or loosening it. This is another way of altering the speed of the waves that travel up and down the string. Tightening the string increases the tension, or force, within the string. This makes it possible for waves to travel faster, and thus we get higher frequencies. This result is quite familiar: To raise the pitch for a guitar string, you simply tighten it.

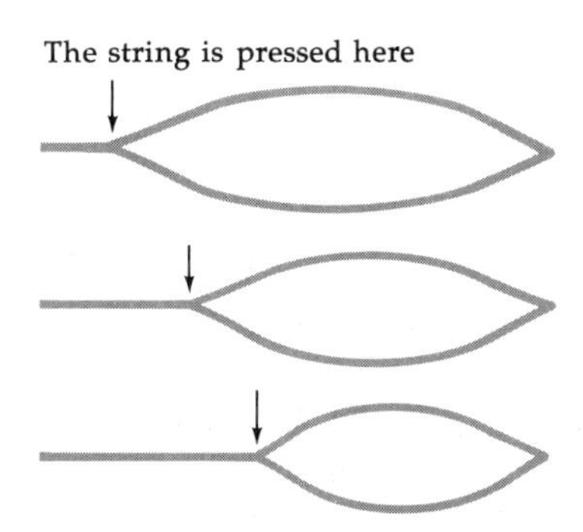

Figure 18-9. Pressing the string farther along the fingerboard shortens the wavelength and raises the frequency (only the fundamentals are shown).

Finally, you can also change the pitch by pressing the string against the frets at different positions on the fingerboard along the neck of the guitar. By doing so, you are effectively shortening the string—or at least shortening the portion free to vibrate. Pressing the string against the fret relocates the position of the clamped end. Because the string is now shorter, the wavelengths of the possible standing waves are also reduced. A smaller wavelength means a higher frequency—and that is what you observe: as you finger the string farther and farther away from the tuning screws, the pitch gets higher and higher. (See Figure 18–9.)

These various ways of obtaining different pitches make the guitar an extremely versatile instrument. By combining a few strings having several different masses, tensions, and effective lengths, the guitar is capable of producing quite a broad range of possible frequencies.

The siren provides a good illustration of the relationship between pitch and frequency. Figure 18–10 shows one way to make a siren. Air forced through the tube is allowed to pass through holes punched at regular intervals in a turning wheel. As the wheel turns faster, the holes pass more frequently in the path of the air, so the pitch of the sound rises. Slowing the wheel down lowers the pitch, because the frequency of the air puffs decreases.

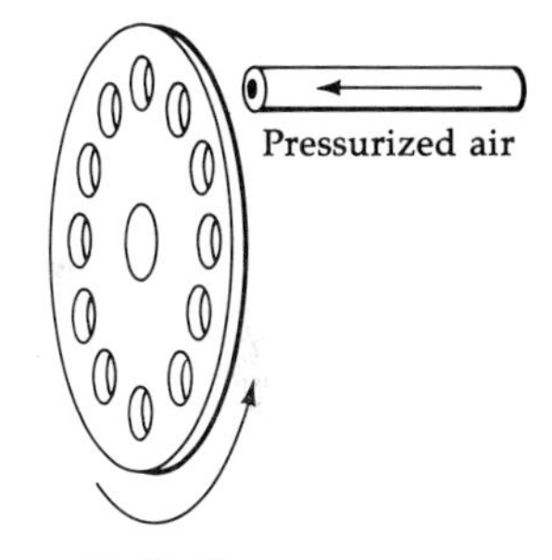

Figure 18-10. Siren.

Learning Checks

1. What is the speed of sound in air?
2. How is the speed of sound related to the frequency and wavelength?
3. What is the connection between the energy in a sound wave and the loudness of the sound?
4. Why does tightening the string on a guitar result in a higher pitch?

Diffraction of Sound

Up to this point, we have pictured sound waves as emanating from a vibrating object and traveling outward in all directions. As far as any listener is concerned, the sound has traveled along a straight line from the source. Indeed, if you are outdoors and a friend calls to you from across an open field, you can see him or her by turning toward the direction from which the sound has come. The sound of your friend's voice has moved directly to you in a straight line.

But now ponder another situation. As you sit and read these words, you are aware of various sounds—the television set playing in another room, children calling to each other in the yard, a dog barking up the street. Clearly, not all of these noises can reach you along linear paths from the sources. A straight line between you and the barking dog might encounter trees, automobiles, certainly a wall or two. If the sound traveled always and only in straight lines, we would only be able to hear sounds made by objects in our field of view. Our perception of the world would be very different indeed.

So, do sound waves follow straight lines or don't they? The answer to both questions is yes. In addition to moving directly from source to listener, sound waves can also bend around corners; that is, they can be *diffracted*, in much the same way as the water waves in a ripple tank that we discussed in Chapter 17. We have said that diffraction is one of the distinguishing features about waves in general, so it is not surprising that it is an important phenomenon for sound waves.

Remember the important criterion for diffraction: The wavelength must be approximately the same as the size of the object it encounters. If the object is much larger, it casts a "shadow" by effectively blocking a portion of the wave. If it is much smaller than the wavelength, the wave passes by almost unaffected. Only if the diffraction criterion is satisfied will we get waves bending around corners.

Let us look at the numbers, to help understand why diffraction of sound waves is so common. We can calculate a typical sound wavelength with our by-now-familiar relationship among wavelength, frequency, and speed. The convenient form is

$$\lambda = \frac{v}{f}$$

We have learned that the speed of sound is about 340 m/s. What about the frequency? Well, it turns out that ordinary sounds have frequencies on the order of a few hundred hertz. Middle C, for instance, is at 264 hertz. Putting these numbers together, we get

$$\lambda = \frac{340 \text{ m/s}}{264 \text{ Hz}} = 1.3 \text{ m}$$

that is, the wavelength for middle C is about 1.3 meters.

Thus, if a wave of about this length encounters a barrier with an opening roughly 1 meter wide—a doorway, for instance—it will be diffracted. The sound wave will bend around the edges of the door and spread out into all parts of the room.

Light, on the other hand, has wavelengths very much smaller than the dimensions of ordinary objects—only a tiny fraction of a centimeter. Our everyday experience, therefore, is that light waves travel in straight lines. Trees and people and bicycles cast shadows but diffract sound waves.

Figure 18-11 illustrates these points in showing how you can hear a radio in the next room without seeing it. The *sound* from the radio is diffracted throughout the room you are in. But the *light* that bounces off the radio is not diffracted. The light waves striking the doorway pass on through. The rest are blocked by the wall. So, unless you happen to be in a part of the room that is on a direct line to the radio, you won't be able to see it.

By making this same calculation for other frequencies, you can see that wavelengths of sound vary from around 17 meters for the low-frequency waves to about 0.02 meter at the high frequency end. It happens that the sizes of most familiar everyday objects are at least of this order of magnitude. So diffraction of sound is quite a common occurrence.

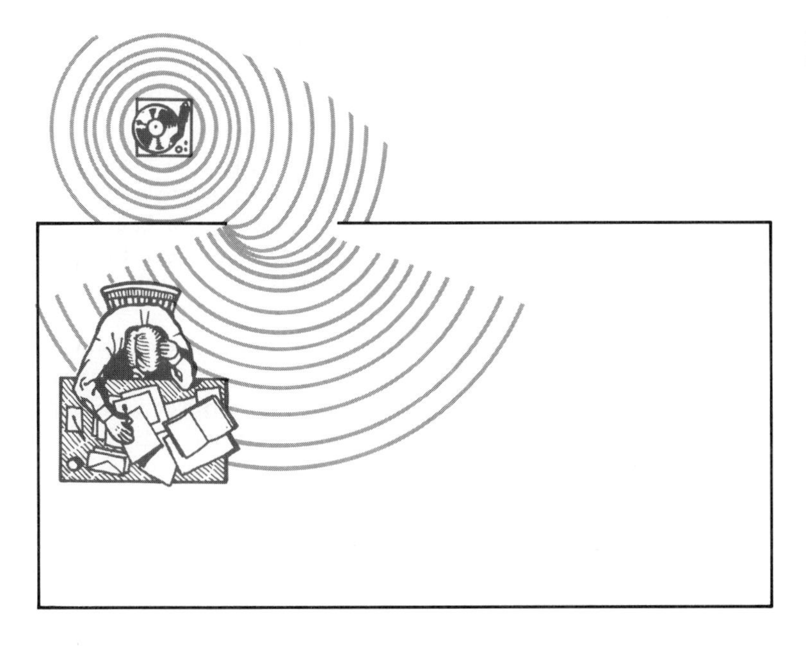

Figure 18-11. The sound is diffracted through the open door.

Beats:
The Interference of Sound Waves

Sound waves will interfere with one another much like other waves; that is, they also follow the principles of superposition. One of the most striking illustrations of sound interference occurs whenever two sound waves having nearly equal frequencies are superimposed. The resulting sound has a periodically changing loudness called a *beat*. The two waves combine to form a single composite wave of alternately rising and falling intensity, as the original waves first reinforce each other, then cancel, then reinforce again, and so on.

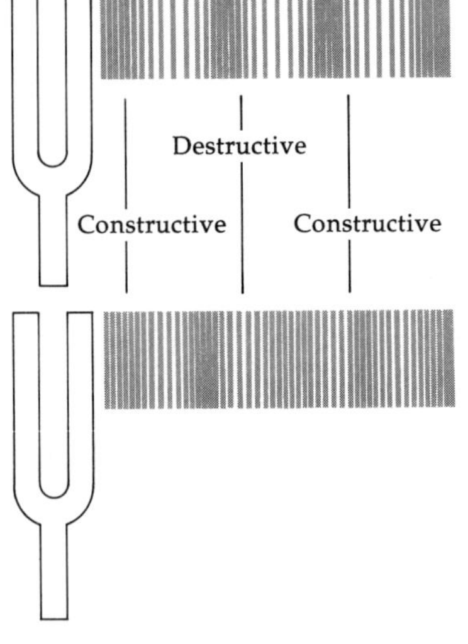

Figure 18-12. Beats from tuning forks.

Figure 18–12 illustrates how this happens. The two tuning forks, having nearly the same frequencies, will produce nearly the same patterns of compressions and rarefactions. However, they are slightly different. Because of this mismatch, occasionally two compression regions overlap, raising the sound level, while farther along a compression region overlaps a rarefaction, eliminating the sound. The result is a rising and falling of the intensity, a periodic swelling and dying out of the sound.

The frequency difference between the two sources determines the "beat frequency"—that is, the number of beats per second. Suppose that two tuning forks have frequencies of, say, 250 and 252 hertz, respectively. Let us imagine that they start out exactly together—crest matching crest, trough matching trough. Physicists say the waves are "in phase." As time goes on they gradually get out of phase. The second crest of the 252 hertz note appears a little sooner than the second crest of the 250 hertz note. The third crest appears still sooner and the fourth sooner yet.

After half a second, the 250 hertz note has completed 125 vibrations and the 252 hertz note has completed 126 vibrations; that is, the 125th crest of the former overlaps the 126th crest of the latter. So they are back in phase. However, the higher-frequency note has gained one complete crest, like a distance runner who laps an opponent going around a track. At this point the sound level is at a maximum—we get a beat. This matching of crests will also show up at the end of the full second and again for every half-second thereafter. Each second there will be two beats.

Following this line of reasoning, we obtain a very simple result: The frequency at which the beats occur is just equal to the *difference* in frequency between the two sources. Sounds differing by 5 hertz will give five beats per second, those 2 hertz apart will beat two times each second, and so on. As the frequencies for the sources are made more nearly the same, the time span of a single beat grows. For larger frequency differences, we get a more rapid, staccato beating pattern. The ear can distinguish up to about ten beats per second. Beyond that, the two notes are heard simply as separate sounds, with no beat effect apparent.

Musicians take advantage of beats to tune their instruments. Suppose an open string on a violin and a second string fingered to give the same note are plucked simultaneously. Beats will occur if the two are slightly off in frequency. Using the fingered string as a reference, the violinist can adjust the tension of the open string until the beats disappear. When this happens, the strings are in tune.

Learning Checks

1. Can sound be said to travel in straight lines, as light does? Explain.
2. Suppose the speed of sound in air were ten times its actual value. Would this cause you to change your answer to Question 1?
3. What would be the beat frequency for a middle C tuning fork sounded against one tuned to D above middle C, at 294 hertz? Would you be able to notice the beats?

MOVING SOURCES OF SOUND: THE DOPPLER EFFECT AND THE SONIC BOOM

Up to this point, we have assumed that both the listener and the source of the sound are stationary. In this section we are going to see what happens when they are moving relative to each other. This relative motion of the source and the listener can produce two important effects, the "Doppler effect" and the "sonic boom."

The Doppler Effect

When an automobile coming toward you blows its horn as it passes, the sound has a lower pitch as the automobile moves away than it had while approaching. You hear this as a quick drop in pitch at the moment the car passes by. The same effect shows up when you sit at a railroad crossing and listen to the whistle of a passing train drop in pitch. Or, if you are riding on the train and pass a crossing, the clanging bell drops in pitch as you go by. In each of these examples, the frequency that the listener hears depends on the relative motion of the listener and the source creating the sound.

This phenomenon is called the *Doppler effect,* named after a nineteenth-century Austrian, Christian Johann Doppler, the physicist who first studied and explained the effect in 1842. The story goes that Doppler persuaded members of his community orchestra to give a demonstration of this effect. They mounted

The Doppler Effect in Research

The Doppler effect is a powerful tool for many researchers. Scientists looking into the range of sounds made by dolphins, for instance, have used a Doppler probe, an instrument hand-held in various positions on the dolphin's head. Beams of ultrasound are projected into the animal's skull and are sometimes aimed to reflect from the moving surfaces of air spaces inside the head. The sound returned is modified in frequency as measured by Doppler shifts. Analysis of this and other data can determine what happens anatomically when dolphins utter a "buzz," "click," or other noise.

Weather scientists have developed a sophisticated tornado-warning system using an offspring of the Doppler radar employed by police to catch highway speeders. Doppler radar (which applies the Doppler effect principle to radio waves) uses shifts in the frequency of the electromagnetic echoes from a storm system to measure motions inside it. The research station at the National Severe Storms Center in Norman, Oklahoma, can make nearly a million such measurements in 1 minute. These data are computer-analyzed. The Oklahoma researchers have been able to spot tornadoes in their formative stages an average of 214 minutes before they strike.

Measuring Doppler shifts in the light waves from celestial objects has long aided astronomers. A shift to longer wavelengths (toward the red end of the light spectrum) tells an astronomer that the observed object is receding along the line of his or her sight. A shift toward the blue end (shorter wavelength) indicates the source is approaching. Astronomical observations of galaxies have been dominated by red shifts, an indication that the universe is expanding.

Sources: R. S. Mackay and H. M. Liaw, "Dolphin Vocalization Mechanisms," *Science*, May 8, 1981, pp. 676–677.

"A New Twist in Forecasting," *Time*, April 28, 1980, p. 67.

a railroad flatcar and played their instruments while the train moved along past the people of the town.

In the Doppler effect, the frequency of the waves as they are received is different from the frequency that the source produces. Figure 18-13 shows this for water waves. When the source and the observer are stationary with respect to one another, the frequency of the waves the observer receives is just equal to the frequency with which the plunger producing the waves dips in and out of the water. But now suppose that the source moves toward the observer. While the first wave is traveling across the water the source is moving, so that when the second wave is produced it is closer to the first wave than it would have been with the source stationary. The same is true of each succeeding wave. This results in the waves being "bunched up" as they reach the observer. So they are received at a greater frequency than they are produced, because the "bunching" has the effect of delivering more waves to the observer in a given period of time than when the source is fixed.

As the source moves *away* from the observer, each wave pulse is *farther away* from the previous one than it would have been with the source fixed. This has the effect of spreading the waves out, and fewer reach the observer in a given length of time. The frequency is reduced. If this is a sound wave, the pitch drops, just as you observed with the passing automobile horn.

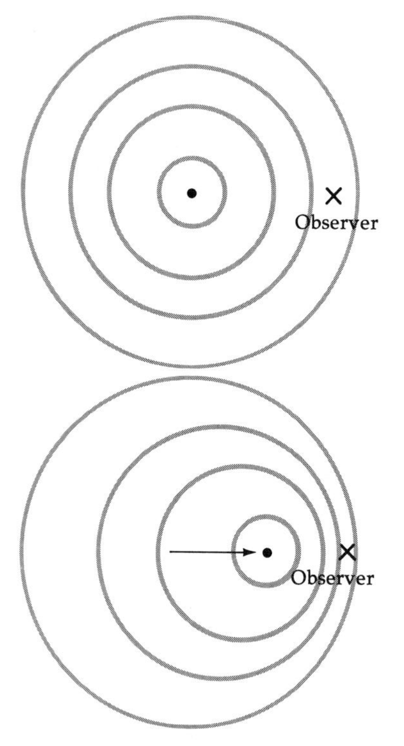

Figure 18-13. The observed frequency changes when the source moves.

Perhaps an analogy will be helpful in a further understanding of the Doppler effect. Imagine that two postal clerks sorting mail stand at opposite ends of a conveyor belt. The belt moves toward clerk A at some fixed speed—say, 5 m/s. If clerk

B begins to put boxes on the belt at a rate of say, 10 per minute, A will see that the boxes move toward him at 5 m/s with a frequency of 10 boxes per minute. But now if clerk B begins to walk toward A at 1 m/s, still loading boxes at the 10-per-minute rate, the boxes reaching A will be spaced closer together than before. They will therefore reach him at a frequency *greater* than 10 boxes per minute. On the other hand, if clerk B walks *away* from A, the boxes reaching A will be farther apart, so their frequency will be *less* than 10 boxes per minute. In all cases, of course, the speed the boxes travel is just that of the belt, 5 m/s.

When we apply this to sound waves, the speed of the conveyor belt represents the speed of sound in air, and the frequency at which clerk B puts boxes on the belt is analogous to the pitch of the sound produced by the source. Observing the "box frequency" is analogous to hearing the sound.

We can now understand how the pitch of the auto horn is shifted. The person riding in the car hears the true pitch of the horn. Someone standing at the side of the road hears the pitch Doppler-shifted up or down, as the car approaches or recedes.

Notice that there is no change in the *speed* of the wave when there is relative motion of source and observer. Remember that the wave's speed depends on the nature of the medium through which the wave travels. It is only the *frequency* that changes in the Doppler effect.

The Sonic Boom

Let us go one step further with this analysis. Suppose that our wave-producing plunger in the ripple tank moves across the water faster than the waves do. Not only does the plunger chase its own waves, it catches and passes them. This means that the source will be *beyond* the enlarging circle of the first wave before it produces the second, and beyond the second before it generates the third, and so on. Figure 18-14 illustrates the effect. You can see that the waves will overlap, constructively interfering with each other as they spread out across the water. This overlapping, in which the amplitudes of the waves add together, produces a *bow wave*, the familiar V-shaped wave that trails behind a speed boat as it skims across a lake.

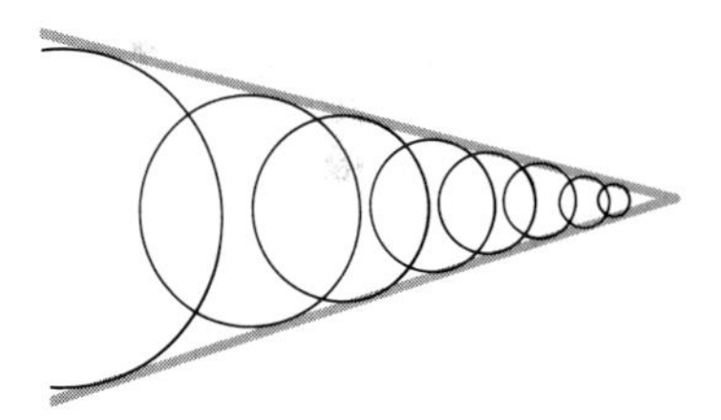

Figure 18-14. Bow wave for a source that moves faster than wave speed.

When a plane flying faster than the speed of sound passes overhead, the superposition of air waves, analogous to the speedboat's wake, is responsible for the *sonic boom*—an annoyingly loud crack that rattles windows and jangles nerves. Just as the speedboat produces a V-shaped bow wave on the water surface, so a supersonic aircraft makes a similar bow wave, or shock wave, in the air. Let us try to understand how this happens.

Physics and Ecology: Sea Lions and the Space Shuttle

The pilots of most supersonic jets, sensitive to the adverse effects of sonic booms on people and buildings, now travel at high speeds only while above water or uninhabited land—uninhabited by people, that is.

Biologists who work with island wildlife report that sonic booms upset seals and birds. Now, one locale—San Miguel Island off the southern California coast—is being exposed to a supersonic boom: the thundering noise made by America's space shuttle.

The United States Air Force carefully selected the launch and reentry path of the shuttle to avoid subjecting humans to the consequences of its explosive booms. In the process, though, major seabird rookeries and the breeding grounds of the California sea lion and several species of seals were exposed. Ecologists on site have stated that the sonic boom created by launch and reentry could stampede beached animals and cause trampling of sea lion pups. Compounding the problem is the fact that sea lion mothers and pups locate one another in crowds by their ability to recognize one another's voices. The social structure of this species would collapse if they suffered even temporary hearing loss from a sonic boom.

United States Park Service and Air Force scientists are conducting experiments and considering solutions, such as detouring the shuttle or scheduling launches only in nonbreeding seasons, which would curtail sonic-boom harm to the wildlife most vulnerable to it—the infants.

Source: David Zimmerman, "Shuttle Booms," *Science 80*, January/February 1980, pp. 80–81.

As we have seen, the speed of a wave depends on the properties of the material in which it travels. The speed of sound in air is determined by how fast the air molecules can rebound after one compression, become restored to normal, and undergo another compression. It is also this rebound rate that determines whether the air molecules can get out of the way of a plane speeding through them. So, as the plane's speed approaches that of a sound wave, it closes in on the limit of the molecules' ability to rebound and "clear a path." The air in the immediate vicinity of the plane remains compressed for as long as the plane maintains this speed. In the early days of flying, this "sound barrier" was thought to establish an upper speed limit for planes, for the volume of compressed air puts considerable strain on the structure of the aircraft. People believed that the plane would pull apart if it were to fly as fast as sound. Improvements in design enabled planes to withstand such strain, and the sound barrier was first broken on October 1, 1947.

When an aircraft flies at supersonic speeds, the regions of compressed air pile up together, in the same way that water waves superpose behind a speedboat. This creates a sharp dividing line between the strongly compressed regions and the normal surrounding air. It is this shock wave, trailing behind the plane in the shape of a cone, that produces the sharp crack of the sonic boom. As this cone sweeps through the air, it creates tremendous and sudden changes in the air pressure. Your ear registers these pressure changes as a sharp crack. A bullet whizzing close by your ear can produce the same sensation, on a smaller scale. The snap of a bullwhip is a miniature sonic boom.

As you can see from Figure 18–15, the angle of the shock-

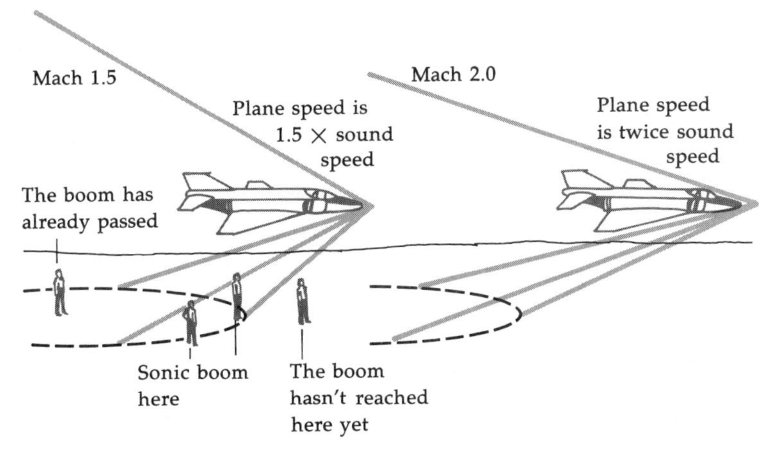

Figure 18-15. The sonic boom.

wave cone depends on the ratio of the plane's speed to the speed of sound. This ratio is called the *Mach number,* in honor of Ernst Mach. Mach was an Austrian physicist and philosopher who worked during the late nineteenth and early twentieth centuries. He was the first to investigate the theoretical consequences of motion at sonic and supersonic velocities. A Mach 1 airplane can equal the velocity of sound, a Mach 2 plane can fly twice this speed, and so on. For normal air at room temperature, Mach 1 is about 760 miles per hour.

Many people think the sonic boom is heard at the instant the plane exceeds sound speed. This isn't true. Since the shock wave is responsible for the boom, you will hear it when the shock-wave cone reaches you. Figure 18-15 shows that the cone cuts the earth's surface in a parabola. Several people standing at different places along the parabola would all hear the boom at the same time—perhaps several minutes or hours after the plane actually penetrated the sound barrier. Those ahead of the cone will hear the boom later, while those behind have already heard it.

Learning Checks

1. What is the Doppler effect?
2. Suppose that instead of a sound source moving toward you, you move toward the source. Will this produce a Doppler shift of the frequency you hear? Explain.
3. How is a bow wave on water related to the Doppler effect? How are they different?
4. What is meant by a *sonic boom?*

SUMMARY

There is an infinitely rich variety of sounds in our world—some that are harsh and shrill, others that are pleasant and soft, still others that are lively and stimulating. They come to us from myriad sources in an endless diversity of *pitch* and *loudness.* And yet we can gain a great deal of understanding about sound by mastering some fundamental information about waves. The study of sound provides a good example of what physics is all about: the understanding and interpretation of what seem to be widely disparate observations on the basis of some fundamental, simple ideas.

Sound is a longitudinal wave made up of alternating compressions and rarefactions of the material through which the sound energy passes. Sound may travel through a gas, liquid, or solid at a speed that depends on the make-up and conditions of the material. Its speed in air at room temperature is about 340 m/s; sound travels faster in liquids and fastest in solids, which are denser than air.

The *loudness* of a sound is expressed in *decibels,* a designation that makes use of the logarithmic relationship between loudness and power. The *pitch* of a sound is primarily determined by the frequency of the wave.

We hear sounds made by objects we cannot see because of the *diffraction* of sound waves by buildings, doors, and other ordinary objects. Diffraction of sound can occur because its wavelength is on the order of the dimensions of things in our environment. Like other waves, sound can undergo *interference.* The effect called *beats* results from interference of waves of slightly different frequencies.

If the source of a sound and the listener are moving with respect to each other, the listener will hear a pitch different from the one produced by the source. This dependence of frequency on relative motion is a general property of waves called the *Doppler effect.*

Whenever a wave generator moves through a medium faster than the speed of the wave itself, the waves pile up into a shock wave. In the case of sound, this produces the *sonic boom* created by supersonic aircraft and, on a reduced scale, by the tip of a bullwhip.

Exercises

1. A metal worker uses a hammer to strike a long solid metal rod with a motion parallel to the long axis of the rod. What kind of wave is produced? If the worker strikes it perpendicularly to the axis, what kind of wave occurs this time?

2. Explain why the bass strings of a piano are wrapped with wire.

3. Is it possible for a Mach 0.75 airplane to cause a sonic boom? Explain.

4. What part does superposition play in sonic booms?

5. Why does the pitch of a guitar string increase when you push it down against a fret?

6. The tremor of the ground from a distant explosion can be felt before the sound can be heard. Explain.

7. Astronomers describe the moon as a "silent planet." Explain.

8. Does the speed of sound depend on its frequency? If so, how would an orchestra sound?

9. Two identical tuning forks emit sounds of the same frequency. How might you produce beats between them?

10. A Mach 2 aircraft flies above your campus. Will everyone on campus hear the sonic boom at the same time? Explain.

11. In some concert halls, there are certain seats where it is difficult to hear the performers, and yet in adjacent seats it is easy to hear. Explain.

12. Sometimes it is difficult to distinguish between a sonic boom and an explosion of dynamite. Account for the similarities.

13. Explain the role of wave interference in the production of shock waves.

14. Police radar used in detecting speeding motorists is based on the Doppler effect. How might this work?

15. Particular sounds on a guitar, called *bell tones,* may result when you pluck a string that is lightly touched at its midpoint. The bell tone is a very pure tone one octave higher than the fundamental for that string. Explain.

16. Across a large parking lot you watch as a child dribbles a basketball. The ball strikes the asphalt once each second, and you hear a sound at precisely the instant the ball hits the pavement. The child then catches it, and you hear one more sound of ball on pavement. How far away are you from the child?

17. A supersonic transport plane takes off from Houston and heads north over Dallas and Denver. It moves through the sound barrier just as it passes over Dallas and maintains a supersonic speed. Is it true that the people in Denver will hear the sonic boom but those in Dallas won't? Explain.

18. A guitar string has a fundamental frequency of 264 hertz. With what frequency (fundamental) will it vibrate if you press it to a fret one-fourth the way down the fingerboard?

19. A violinist is tuning her violin. The out-of-tune string makes three beats per second with a string whose frequency is 440 hertz. What are the possible frequencies for the out-of-tune string?

20. In Exercise 19, you notice that the frequency of the beats *decreases* when you tighten the out-of-tune string. What was its frequency originally?

21. What is the speed of a Mach 1.5 aircraft?

22. Two tuning forks having frequencies of 440 and 436 hertz, respectively, are struck at the same time. What is the beat frequency? How long will each beat last?

23. Compare the intensity of a 50 decibel sound with one at 30 decibels.

24. Taking the speed of sound to be 340 m/s in air, determine the wavelength of a 340 hertz sound wave.

25. What is the frequency of a sound wave whose wavelength is 0.5 meter? Is a tone of this frequency in general audible to humans?

The physics of sound has its highest expression in music.

19

MUSIC

The physics of sound has its highest expression in music. All of the various facets of wave motion—frequency, loudness, resonance, standing waves—take on extra importance when we examine the physical aspects of music. In this chapter we will take a brief look at how music is a special application of what we have learned about sounds and waves.

MUSICAL SOUNDS

Music plays such an important role in this and other cultures that we can easily forget how closely related it is to other sounds that we hear. But of course, like all sounds, the music you hear comes from vibrating objects that generate waves. They might be waves from a resonating air column, such as an organ pipe, or from brass instruments such as the trombone and cornet. The waves might be produced by a stringed instrument, such as a harp or a viola. In the case of the woodwinds, a vibrating reed sets up air column vibrations that generate the music.

We have seen how vibrations on a string give sound waves of particular frequencies. If you pluck or bow a string between two thumbtacks, the sound you get is not very loud. Violins and guitars and other stringed instruments have a *sound box* to intensify the sound, making it loud enough to be heard across the expanse of a concert hall. The vibrating strings, coupled to the sound box, cause the box to vibrate as well as the air trapped inside. These all add their sounds together. The result is a musical tone louder than that due to the string alone.

Vibrating pipes or columns of air have standing waves analogous to those on a string. Because the air is confined by the walls of the pipe, it is different from the outside air in that its physical surroundings are different. The ends of the pipe are boundaries between these two "kinds" of air. At the open ends the air is free to vibrate, so any standing wave in the pipe will have an *antinode* at an open end. Close the pipe off at one end and you create a *node;* here, the air is no longer free to vibrate.

Mapping the sound box. To obtain a graphic representation of the acoustic intensity of a guitar sound box, researchers arrange a grid of microphones to pick up sounds, which are turned into a three-dimensional image by a computer. The inset is a velocity reconstruction of the back plate of the guitar showing the modal structure.

For a pipe of a given length, the standing waves that are possible depend on whether or not both ends are open. Look at the case of a pipe open at both ends, shown in Figure 19-1. Antinodes are located at the ends, and a node is halfway between them. The standing wave having the *fundamental* frequency will be the one for which these are the only nodes and antinodes present, as in part (a) of the figure. The wavelength of this wave is exactly twice the length (L) of the pipe; that is, the pipe contains one-half a wavelength. Part (b) shows the waveform for the first overtone. Note that there are still antinodes at both ends—this condition *must* be satisfied—but now there is also an antinode at the center with nodes on either side. For this standing wave, the wavelength is equal to the pipe length, and the frequency is therefore twice that of the fundamental. The other portions of the figure show the next two overtones. You can see that the wavelengths in these situa-

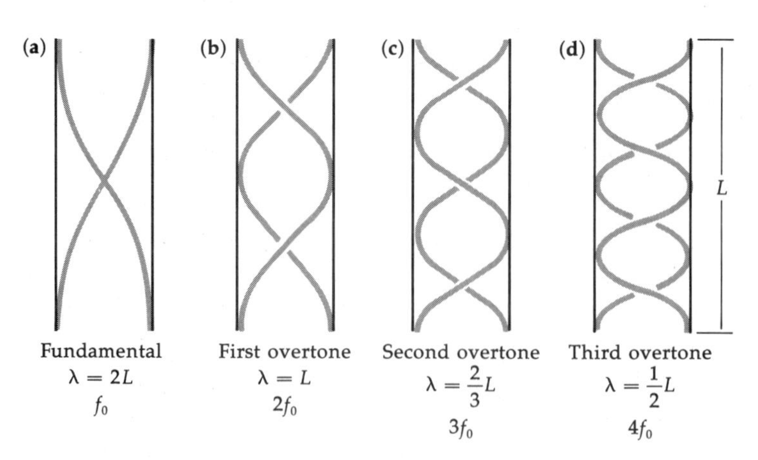

Figure 19-1. Standing waves in an open pipe.

tions are $2L/3$ and $L/2$, respectively, corresponding to frequencies equal to *three* and *four* times the fundamental.

Now consider the pipe closed at one end. The closed end must always be a node, and this creates a drastically different situation. Look at the illustration in Figure 19–2. In this case, the fundamental frequency will occur for a standing wave having an antinode at the open end and the nearest node at the closed end. This means that the wavelength will be *four* times the length of the pipe, since only one-quarter of a wavelength will fit in the pipe. Hence, the fundamental frequency here will be *half* that for the pipe open at both ends. So, merely by closing off one end of the open pipe, we can decrease its fundamental frequency by half.

You can easily demonstrate this to yourself. Take an ordinary drinking straw, open at both ends. Blow across one end and notice the pitch of the resulting sound. Now close off the bottom end with your finger and blow again. You will hear a pitch noticeably lower than before.

The rest of Figure 19–2 shows higher overtones for the pipe closed at one end. It is interesting to note that the next higher frequency is *three* times the fundamental, and the one after that is *five* times the fundamental. So by closing off the end of the pipe, not only do we lower the pitch of the sound, we also wipe out all the even harmonics—those frequencies that are some even multiple of the fundamental frequency.

Changing the length of the pipe also changes the pitch of its notes, for essentially the same reasons that you change the pitch of a guitar string by "shortening" it—that is, pressing it against a fret. For a shorter pipe, the wavelength of the standing wave for the fundamental vibration, as well as for the overtones, is smaller, resulting in a higher frequency and hence a higher pitch. For this reason the booming bass notes from a pipe organ come from the longest pipes. Other instruments are

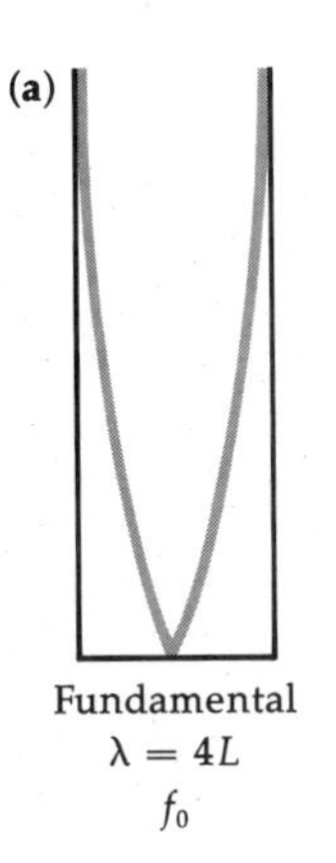

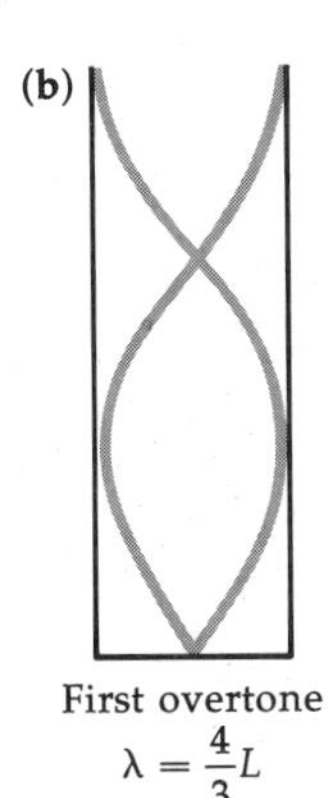

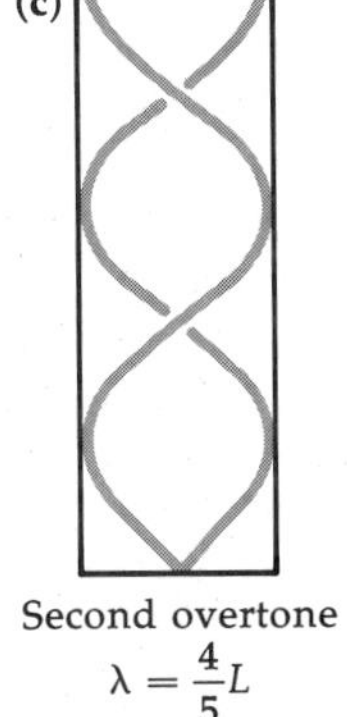

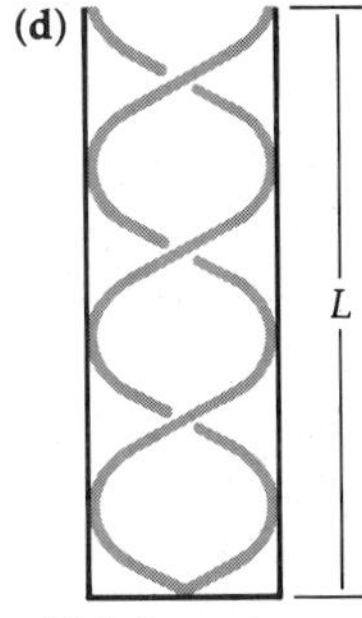

Figure 19-2. Standing waves in a closed pipe — the even harmonics are missing.

designed so that the musician can easily change the length of the vibrating air column. The slide of a trombone serves this purpose, as do the valves on a trumpet.

Learning Checks

1. What is the purpose of the sounding board of a piano?
2. What advantage is there in attaching a tuning fork to a sound box?
3. How is an organ pipe analogous to a violin string?
4. You blow across the top of an empty soft drink bottle and hear a certain tone. Now you pour some water into the bottle and blow again. Will the tone have a higher or lower pitch? Explain.
5. The wavelengths of the fundamental standing wave in a hollow pipe open at both ends is smaller than that in an identical pipe closed at one end. Explain how the smaller *wavelength* implies a higher *pitch.*

THE QUALITY OF A MUSICAL TONE

We have seen that a sound can be characterized by its *loudness,* related to the amplitude of the wave, and by its *pitch,* which corresponds to the frequency. For musical tones, there is a third distinguishing feature, called the *quality,* or *timbre.* Let us see what this means.

Suppose you are in a room containing a piano, a French horn, and a clarinet and someone plays the same note—middle C, say—on each instrument, one after the other, each with the same loudness. If only loudness and pitch mattered, the three notes would sound identical to you, because they all have the same frequency and amplitude. But they *aren't* all the same—they sound quite different from one another. You have no trouble picking out which tones come from which instrument. No one is going to confuse the sound of the piano with that of the French horn. Even if they are played at the same time, you are still able to discriminate among the three. Otherwise, there would be no point in having the different instruments in an orchestra. Each seems to have its own special flavor, or richness, some "feeling" that is its own particular trademark. It is this flavor of the tone that musicians call *quality.*

The quality of a tone comes about in this way: If you pluck a guitar string tuned to a particular frequency—let us again use middle C as our example—then the pitch you hear, of course, is middle C, meaning that the string vibrates at 264 hertz. But

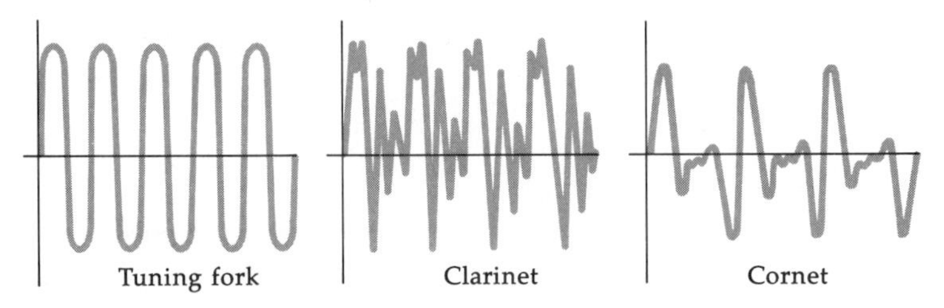

Figure 19-3. Same note, different instruments.

that's not all. We have seen that the guitar string is capable of vibrating at *many* resonant frequencies, not just the fundamental. The others correspond to the various standing waves available to it. In fact, when you pluck the string, it can vibrate at all these frequencies at once. That is, not only does it produce the fundamental frequency, it also puts out tones corresponding to the overtones as well. The fundamental frequency establishes the pitch, because its vibration has the largest amplitude, making it the loudest. But the overtones—528, 792, and 1056 hertz (2, 3, and 4 times the fundamental), and so on—also contribute to the sound. What you hear results from adding these various frequencies together. That is, the sound is a superposition of all the possible standing waves. So it is the mixing of the overtones with the fundamental that provides the special sound for each instrument.

The quality of the sound depends on two things: (1) which overtones combine with the fundamental and (2) their relative loudness. Figure 19-3 illustrates the idea. Here we show the waveforms for a tuning fork, a clarinet, and a cornet as they appear on an *oscilloscope*, which is an electronic device that converts sound into visible wave patterns displayed on a TV-type screen. The three instruments all played the same note to produce these patterns—440 hertz, which is A above middle C. As you can see, the wavelength (and hence the frequency) is the same for all three, but each wave has its own distinct shape. Your ear and brain can readily distinguish among the wave shapes, interpreting them as distinct sounds.

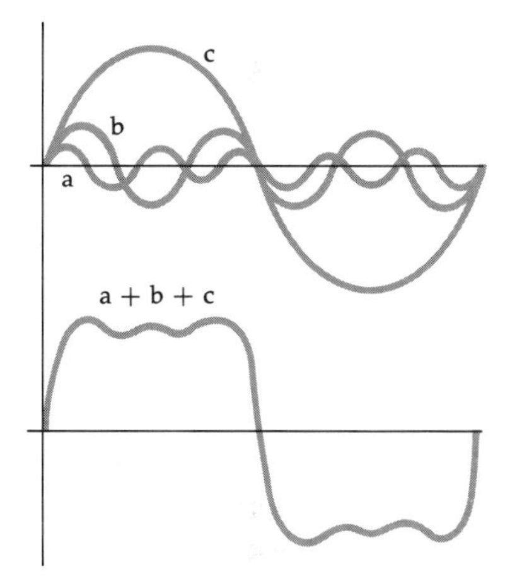

Figure 19-4. Individual waves and their composite.

Figure 19-4 illustrates how such rather complex wave patterns can be constructed for a simple case. Notice that the final wave results from superposing a fundamental wave and different overtones with varying levels of loudness. Here again we have an instance where the principle of superposition is important for understanding wave phenomena.

Figure 19-5 shows what we would get if we "separated" the composite oscilloscope tracings for the tuning fork, clarinet, and cornet into their fundamental and overtone components. The graphs give the particular overtones, and relative intensities for each, that mix to make the characteristic sound. You can see that the tuning fork vibrates with only the fundamental frequency of 440 hertz, but the clarinet and cornet each have their own unique set of overtones mixed in.

Figure 19-5. Breakdown of the waves in Figure 19-4.

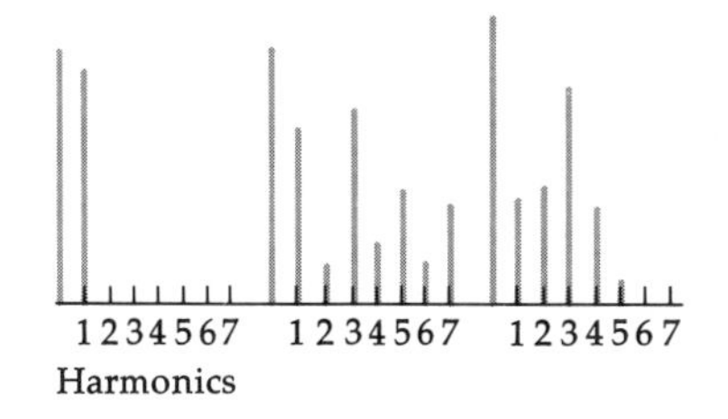

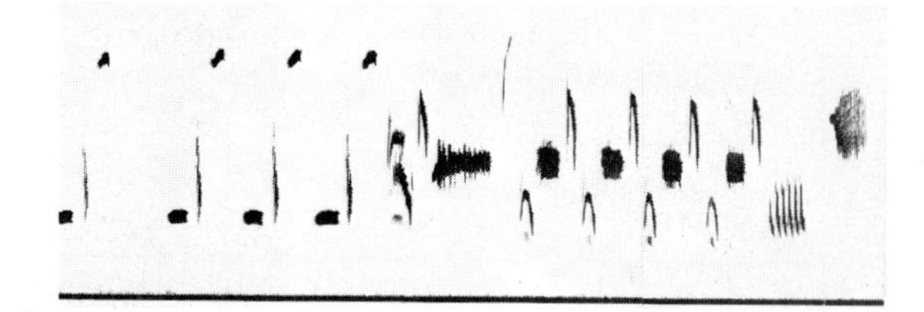

a

b

c

The human ear detects and responds to myriad notes of the natural world. (a) Oscilloscope representation of a song sparrow. (b) Spectrogram of humpback whale songs. (c) Musicians of the Winter Band take to the open sea to exchange music with the whales.

Understanding what is behind the quality of a musical tone helps us appreciate what a marvelous "detector" the human ear is. It can take the various notes played by the different instruments in an orchestra, sort them out, and enable us to hear the beautiful music that results. We can hear the soft notes and the loud ones; we can even concentrate on one or two sections of the orchestra if we wish. The next time you listen to your favorite group, be reminded how remarkable it is that you can listen as you do.

High-fidelity sound equipment must be capable of reproducing the overtones in order to deliver faithfully the tones of the original instruments. For example, manufacturers of high-quality speakers often advertise reproducibility of frequencies up to 50,000 hertz. Even though the human ear is sensitive only up to about 20,000 hertz, the speaker must be able to respond to much higher frequencies in order to include a sufficient number of overtones in the final mix. Otherwise, that jazz group just won't sound quite right.

MUSICAL SCALES

People have noticed connections between music and mathematics for many hundreds of years. Pythagoras, the Greek mathematician and philosopher of the sixth century B.C., was apparently the first to notice that pleasant sounds resulted from vibrating sources that were related to each other by ratios of small whole numbers. For instance, he found that similar strings having length ratios of 3/2 or 4/3 gave tones that seemed to go well together.

We can use our knowledge about sound waves to try to understand what Pythagoras discovered. For example, the note A above middle C has a frequency of 440 hertz. When sounded against a note of half that frequency, 220 hertz, it produces a tone that is pleasing to the ear. Recall that the beat frequency for two notes is the difference in their frequencies. So in this case the beat frequency, $440 - 220 = 220$ Hz, just matches the pitch of the lower note. The two notes, in a frequency ratio of 2/1, seem to melt into each other. In fact, they sound so much alike that we call both by the same name, A.

We can make a musical scale by introducing into the interval between the low A and the high A other notes whose frequencies bear some orderly relationship to one another. In the diatonic major scale, there are six such notes, labeled B, C, D, E, F, and G. Including both A's, we have eight notes, called an *octave* (from the Latin word for "eighth"). Figure 19-6 shows such an octave and the frequencies that match each note.

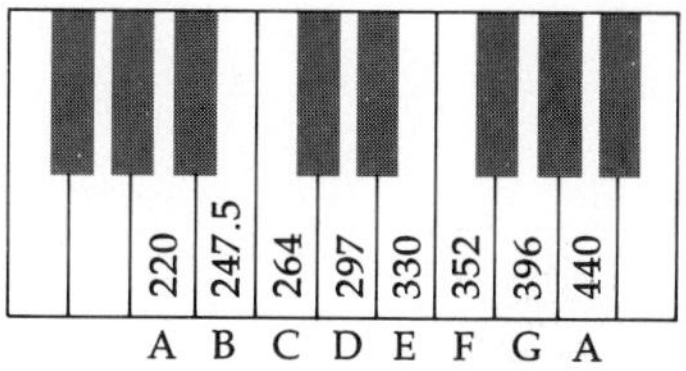

Let us look at the ratios of some of these notes. For example, the frequencies for F and C have a ratio of 352/264, which we can reduce to 4/3. The ratio for G to E is 6/5. The triad of notes C, E, and G have their frequencies in the ratio 4/5/6. This combination is called a *major chord.*

In all these, the beats reproduce the notes and their overtones to give pleasant sounds, blendings of tones that seem to the ear somehow to go well together. In the example of F and C, the beat frequency is $352 - 264 = 88$ Hz. Notice that 88 hertz is 352/4 as well as 264/3. This means that middle C, at 264 hertz, is the third harmonic of the beat frequency, while F, at 352 hertz, is its fourth harmonic. So, in a sense, F and C can be considered members of the same "family" of tones. The ear likes to hear them together.

220 247.5 264 297 330 352 396 440

A B C D E F G A

Figure 19-6. Frequencies of A-major scale.

SUMMARY

Standing waves are very important in the creation of music by strings, pipes, and other instruments.

In addition to *loudness* and *pitch,* musical tones are also characterized by their *quality,* or *timbre.* Each instrument mixes together the *fundamental* and various *overtone* frequencies to give its own unique sound.

Musical scales are groups of notes differing from each other by frequencies whose ratios reduce to fractions of small whole numbers.

Exercises

1. Give the physical properties of a sound wave that correspond to the human sensations of pitch, loudness, and timbre.

2. How do we distinguish between a note played on a piccolo and the same note sung by a soprano?

3. You are watching a jazz group tune up. Two strings of a bass fiddle are tuned to the same frequency. The bass player plucks one of them, lets it vibrate for a few seconds, and then places her finger on it to stop its vibration—and you notice that the other string has begun to vibrate even though it hasn't been touched. Explain.

4. A closed organ pipe has a fundamental frequency of 300 hertz. What are the frequencies of the first three overtones?

5. Answer Exercise 4 for an open pipe, also at 300 hertz.

6. Would a plucked violin string vibrate for a longer or shorter time if the violin had no sound box? Explain.

7. An open organ pipe has a fundamental frequency of 300 hertz. Another pipe, which is closed, has the same first overtone as the open pipe has. Do the pipes have the same length? If not, which is longer? Explain your answers.

8. A closed organ pipe is 4 meters long. Give the wavelength and frequency of the fundamental and first two overtones.

9. Do Exercise 8 for an open pipe of the same length.

20

LIGHT: ITS PROPERTIES AND BEHAVIOR

The lighthouse beacon is one of the older ways we use light for communication. With technology, the use of light has increased to include neon signs, telephone switchboards, runway lights, and computer terminal screens.

Light is among our earliest and most common experiences. Babies begin to respond to their visual surroundings almost as soon as they first open their eyes. Everything we see affects us partly because our eyes intercept the light and pass it along to the brain for interpretation. Our enjoyment of the rich variety of colors, brilliances, and shapes in our environment begins with light. We are bathed in it, all the time. It furnishes much of our information about the world and each other. A few minutes in a darkened room will convince you of the important role light plays in your life.

Think for a moment of as many things as you can that involve light—the blue of the sky, the colors of a rainbow, the infinite starry nighttime sky. These are things you can see. But there are many other such phenomena that you *cannot* see. For example, the signal that comes through the air to a television set is invisible, but it is the same kind of thing and involves the same kind of physical processes as visible light. As we shall see, the part of nature that we call light is only a very small portion of a more general phenomenon.

Particles or waves?—the question "What is light?" is one that has puzzled physicists and philosophers of science for centuries. In this chapter and the next, we are going to be concerned primarily with how light behaves under a variety of circumstances. Answers to questions about the *nature* of light depend on what sort of experiment we do to observe its *properties.* What we shall find is this: Light apparently has a dual essence. Under some circumstances it behaves like the water waves and sound waves we have already studied. In other situations, the effects of light seem to indicate that it is much like a steady stream of BB's fired from a rifle. Both of these interpretations are necessary, it turns out, in order that we have the complete picture of light as we now understand it.

> *We all* know *what light is, but it is not easy to* tell *what it is.*
>
> *Samuel Johnson, according to Boswell*

Our experience tells us that light travels in straight lines. Although you may not realize it, you believe very strongly in this straight-line-propagation idea. We navigate ships and survey land in accordance with this belief. In his book on optics, published in 1704, Isaac Newton described light as a stream of particles. This would account for our observations that light follows straight-line paths. Newton's authority was so compelling that his *corpuscular theory*—that light is composed of corpuscles, or particles—gained a strong foothold and was championed by Newton's successors for more than a century.

Christian Huygens, a prominent physicist of the seventeenth century, proposed a competing theory. He felt that light was made of waves. As we shall see later in the chapter, Thomas Young did an experiment in the early nineteenth century that showed an interference pattern for light, exactly analogous to the interference of sound and water waves. This confirmed Huygens's notion of the existence of light waves.

Work on another front, also in the nineteenth century, gave strong indication of the source of light. In Scotland, James Clerk Maxwell theorized that charges and currents on the one hand, and electric and magnetic fields on the other, are connected by a wavelike disturbance traveling through space at just the speed known for light. Scientists concerned about the nature of light found a compelling link between Maxwell's equations and Young's light-wave experiments. The view that light is an example of Maxwell's electromagnetic waves gradually took a firm hold.

But the successful theory of light as a wave ran into a snag early in the twentieth century. In Chapter 22 we will look at some experiments that indicated that light is not a wave but a "particle," although not a particle in the same sense that Newton suggested. The resulting search for a new theory eventually led to the development of *quantum mechanics,* the present-day explanation of nature's structure at the size level of electrons and atoms. We are going to look at the quantum description of matter in Chapter 23. As we shall learn, there is a certain symmetry in nature: matter also shows the same sort of wave-particle duality evident for light.

THE ELECTROMAGNETIC SPECTRUM

The light waves to which our eyes respond make up a small part of a sea of similar waves called the *electromagnetic spectrum.* Figure 20–1 illustrates this spectrum, arranged so that the various regions are shown according to their wavelengths. The *visible* portion is what you see every day. It runs from red light in the long wavelengths to violet light in the shorter wavelengths. On either side of this comparatively tiny strip are waves similar in nature to visible light. However, our eyes are not sensitive to them. At the long-wavelength end are infrared

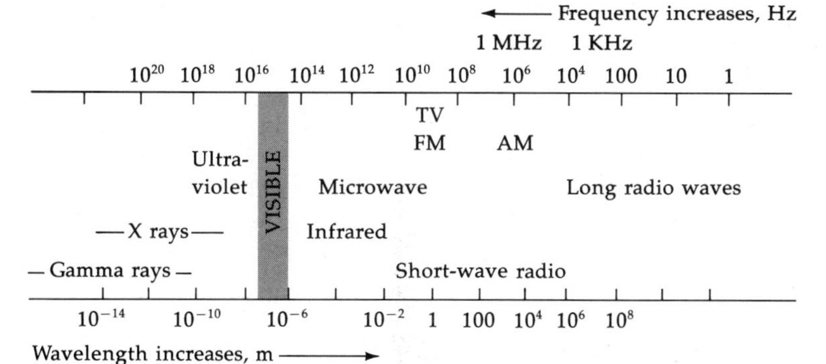

Figure 20-1. The electromagnetic spectrum.

(literally, "below red") waves, microwaves, and radio waves. Extending past the blue at the short-wavelength end are ultraviolet ("beyond violet") waves, X rays, and gamma rays. You can see the great differences in the wavelengths throughout the spectrum. They range from several meters for radio waves to billionths of a centimeter for gamma rays.

Visible light occupies that band of the spectrum where the wavelengths range from about 7.6×10^{-7} m at the red end to around 3.8×10^{-7} m for the violet. A convenient unit for wavelengths in the neighborhood of the visible region is the *angstrom* (Å), named in honor of the nineteenth-century Swedish astronomer Anders Jonas Angstrom. One angstrom is equal to 10^{-10} m. Written out as a decimal number, this is 0.0000000001 meter. So red light has a wavelength of about 7600 angstroms and violet light, 3800 angstroms.

There is no sharp cutoff for any of the colors. They blend into each other as we move from red toward the shorter wavelengths, through the oranges, on into the yellows and greens, and finally to violet.

Let me emphasize that the entire electromagnetic spectrum is made up of the *same kind of waves*. They differ only in their wavelengths and frequencies. They all travel at the same speed in a given medium, which we have seen is $c = 3 \times 10^8$ m/s, or 300,000,000 m/s, in a vacuum. And, as Maxwell's theory predicts, they all derive from the same physical occurrence: the acceleration of charged particles. Let's have a closer look.

Figure 20-2. The electric field (E) and the magnetic field (B) are perpendicular to each other and to the direction of motion.

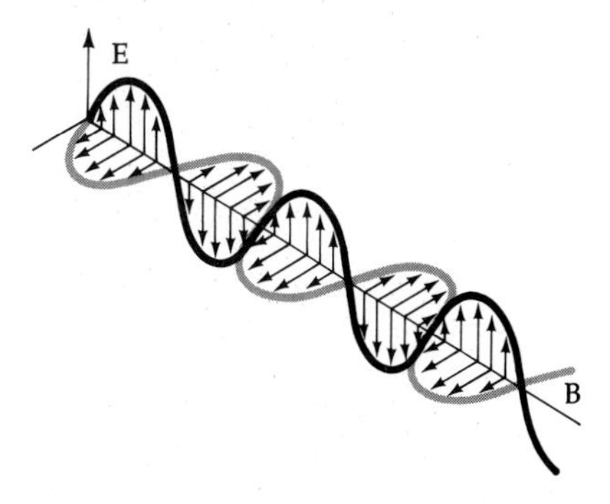

Recall from Chapter 17 that mechanical waves, such as water waves and sound waves, result from the back-and-forth motion of some object, such as a tuning fork or a floating chip of wood. Also, a wave carries energy as it emanates from the source. These attributes hold true as well for an electromagnetic wave, in which case the object generating the wave is a particle carrying an electric charge, such as an electron.

As the electron moves to and fro (accelerates), the electric and magnetic fields, which are changing in time, carry away energy. (See Figure 20-2.) We say that the energy *radiates* away

What Do Oil Slicks, Farming, Moths, and the Military Have in Common? In a Word: Infrared

Infrared is a spectral region that has inspired a wealth of ideas. Infrared scanners mounted on research jets, for example, can map oil spills on the ocean and determine how thick they are, what kind they are, and where they are headed. The infrared "eye" used in this process senses heat over an ocean area several miles wide and is able to detect oil on the surface because of the differences in temperature between the oil and the water. The United States Coast Guard has used this method to place booms and dispersants in the proper locations to protect coastal regions threatened by these traveling slicks.

The United States Army has capitalized on infrared radiation to protect its soldiers in battle. Far-infrared devices can detect the warmth given off by the body of an enemy hiding behind a tree, can locate buried land mines night or day, and can penetrate thick smoke. These and other enhanced-detection instruments have altered the nature of guerrilla warfare by providing conventional forces with the night vision necessary to combat guerrilla units operating under the cover of darkness.

An infrared "gun" will be used, not by soldiers, but by farmers—to manage crops more efficiently. A technique developed by a physicist and soil scientists of the United States Department of Agriculture is based on relating a crop's temperature to that of the surrounding air to determine when irrigation is required. When soil moisture is depleted below a certain level, plants become "stressed," and their temperature rises. When a crop's temperature is more than that of the air around it for a specified time, irrigation is indicated. The infrared "gun," easily carried in a holster, can thus assess crop needs so that farmers won't irrigate when it isn't necessary.

Humans do not have a monopoly on infrared use. The flight behavior of moths is affected by various kinds of infrared signals, and these insects may well utilize infrared radiation in their mating-communication scheme. The debate over whether moths in fact receive infrared radiation continues, although one researcher noted that in the presence of infrared light a moth's antenna vibrates, its abdomen curves toward the light source, and its proboscis uncoils. Of what importance is such information? The same as in the other examples we have cited: human control—of insects, crops, the consequences of certain disasters, and the dark night.

Sources: H. S. Hsiao, *The Attraction of Moths to Light.* San Francisco: San Francisco Press, Inc., 1972.

"Sensitive Detectors Trace Ocean Oil 'Pancakes,' " *Design News,* October 10, 1979, p. 18.

"Tech Front Lines," *Popular Mechanics,* January 1981, pp. 104–106.

"Infrared 'Ray Gun' Improves Farm Crop Management," *Industrial Research,* September 1979, pp. 90–91.

from the accelerating charge; hence, we often refer to these waves as *electromagnetic radiation.* The frequency of this radiating wave will be the same as that of the vibrating electron, and its wavelength will depend on the speed with which light can travel in the particular material. If the electron's frequency corresponds to a wavelength between 3800 and 7600 angstroms, we will see the light. If the electron frequency is outside this range, we won't see the resulting radiation. Our eyes are not sensitive to any electromagnetic wave that is longer than red or

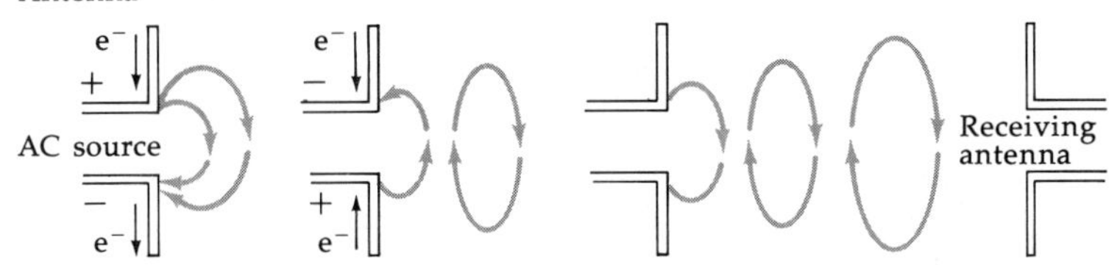

Figure 20-3. A radiating antenna.

shorter than violet. We may still be able to detect it, however. Infrared-sensitive film, for example, allows us to record pictures caused by waves somewhat longer than the 7600 angstrom red light at the upper end of the visible spectrum.

Figure 20-3 shows a transmitting antenna, a common example of a source of electromagnetic radiation. This is a simplified drawing of an antenna like that for a TV station or the CB radio in an automobile. The ends of the antenna are attached to opposite terminals of an AC voltage source. Because the current periodically alternates, each terminal switches back and forth between positive and negative several times every second. The charges flowing along the antenna respond to this sign switch by changing the direction in which they move. In this way the antenna provides a steady supply of accelerating electric charges.

Recall two things from our study of electricity and magnetism: An electric current produces a magnetic field, and a changing magnetic field gives rise to an electric field. This is exactly what we have with the antenna. The alternating current produces a magnetic field. Because it is a changing field, it in turn creates an electric field. As this process is repeated over and over again, an electromagnetic wave (see Figure 20-2), consisting of these changing electric and magnetic fields, travels outward from the antenna.

Look at an ordinary incandescent light bulb. It provides a common example of visible light produced by accelerating electric charges. When the current is turned on, electrical energy surges through the bulb's filament, a tiny wire that forms the heart of the bulb. The electrons in the wire jostle to and fro as a result of their electrical energy. This jostling means that the electrons are accelerating, and they oscillate within the proper range of frequencies to give light waves visible to the eye. Thus, we see the light.

Usually the light we see is not just one color but a mixture of colors. That is, we ordinarily do not see light of one particular frequency but of a combination of frequencies. *White light* is the term given to a mixture of all the frequencies—all the colors of light—in the visible spectrum. A common household light bulb produces white light, indicating that the electrons in the filament vibrate at all possible frequencies within the visible range. Light from the sun as we see it is almost white (the earth's atmosphere has an effect on our perception of the color that we shall discuss later).

We remember that the speed of a wave is constant for a given medium. The same equation given in Chapter 17 relating the speed (v), the wavelength (λ), and the frequency (f) can now be applied to electromagnetic waves. It is common to label the light speed c:

$$c = f\lambda$$

For example, suppose your favorite radio station is located at 1000 kilohertz (kHz) on the AM dial. This tells you that the station's broadcasting antenna moves electric charges back and forth 1,000,000 times each second, generating a radio wave having this same frequency. We can calculate the wavelength by writing

$$\lambda = \frac{c}{f} = \frac{3 \times 10^8 \text{ m/s}}{10^6 \text{ s}^{-1}}$$

$$\lambda = 300 \text{ m}$$

So the radio wave is 300 meters long. Compared with radiation in the visible region, this is a very long wavelength and a very low frequency. Orange light of 6000 angstroms, for example, has a wavelength of $\lambda = 0.0000006$ meter, so its frequency is

$$f = \frac{c}{\lambda} = \frac{3 \times 10^8 \text{ m/s}}{6 \times 10^{-7} \text{ m}} = 5 \times 10^{14} \text{ Hz}$$

or, written out, $f = 500{,}000{,}000{,}000{,}000$ hertz. As we progress farther toward the blue and into the ultraviolet, wavelengths continue to get smaller and frequencies higher.

Because water waves, sound, and other mechanical waves require some material to do the "waving," it is natural to suppose that light waves should also need matter for their existence. In fact, this was one of the reasons early physicists thought the ether to be so essential. We saw in Chapter 6 that the ether was "necessary" to provide a fixed frame of reference against which to measure motion. Another reason was the conviction that some material substance was required to transmit light waves, much as a rope transmits a wave when you shake it. Indeed, the name *luminiferous* ("light-carrying") *ether* has its origin in this supposed function. However, the Michelson-Morley experiment demonstrated that the ether does not exist. Unlike mechanical waves, electromagnetic waves can travel through a vacuum. Ample evidence for this is the light that reaches us from the sun after having traveled through millions of kilometers of empty space. We now understand that as the radiation travels along, it is the electric and magnetic fields that undulate, not some material substance.

Learning Checks

1. Where does the name "electromagnetic spectrum" come from?
2. How are electromagnetic waves similar to sound waves? How are they different?
3. Compare light waves to ultraviolet waves: How are they alike? How are they different?
4. What must vibrate in order for an electromagnetic wave to be created?
5. A certain electron moves at constant velocity. Does it produce electromagnetic radiation? Explain.
6. How far apart are the crests of an electromagnetic wave that you see as blue light? What is the vibration frequency of the electric charge that produces this wave?
7. Are radio waves a form of light? Explain.
8. How fast does a TV signal travel through a vacuum?

THE WAVE NATURE OF LIGHT

We have discussed light as waves coming from an accelerating electric charge. But how do we know that light travels in waves? That is, what makes us believe that electromagnetic radiation has wavelike properties? Some of the most convincing evidence comes from experiments in which light waves undergo *diffraction* and *interference.*

Diffraction

Consider diffraction first. If a small shaft of light, such as from a pinhole in a window shade, strikes a very narrow slit, the light passing through is spread out. Had the light merely gone straight through, it would simply produce an image of the slit. But, in fact, we don't see this at all. What we see is a broad band, as if the light had somehow been smeared out. It has been *diffracted,* just like water waves passing by a barrier or sound waves going through a doorway and fanning out into the next room. Rather than traveling in straight lines, as we are led to believe light does, in this case it bends around the corner.

We don't observe the diffraction of light very often. Under ordinary circumstances, it is an extremely small effect. This is because the wavelengths of visible light are almost unimaginably small compared to the sizes of objects in our everyday world. Recall from Chapter 17 that waves going through a slit are diffracted only if the slit size and the wavelength are of roughly the same magnitude. The sun shining through a window shows essentially no diffraction, because the window

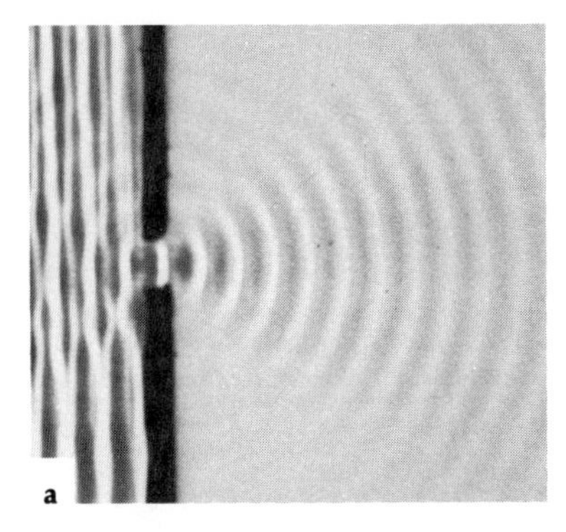

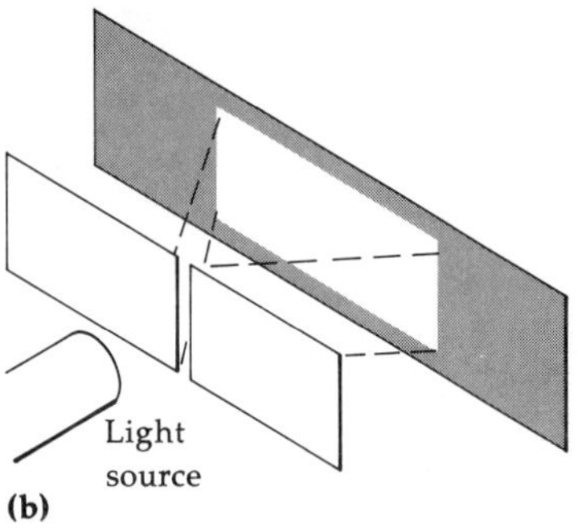

Figure 20-4. Diffraction of waves. (a) Diffraction of water. (b) Diffraction of light.

opening is many millions of times larger than the wavelengths of the sun's radiation. For us to observe diffraction effects, the diffracting slit must be tiny, on the order of a few millionths of a meter. Otherwise, the light goes straight through. (See Figure 20–4.)

One of the reasons for believing light travels in straight lines is that it seems to produce sharp shadows. Well, it does and it doesn't. That is, it all depends on what we mean by "sharp" and on what sort of effects we are looking for. Objects of ordinary size cast shadows by blocking the path of the incoming light. This is analogous to the "shadow" caused by barriers in the path of water waves. (See Figure 20–5.) Shadows are sharp in the sense that they reveal the shape of the object. However, if you look closely at a shadow you can see that it is fuzzy at the edges. The exact position of the line indicating where the light's path has been cut off is not very clear. The light, while producing a sharp enough shadow to indicate the shape of the object, nevertheless bends around the edges enough to leak some light into the shaded region. Hence the shadow is slightly blurred.

To summarize, then, we can say that diffraction effects for light, like those for water waves, are more pronounced with an opening whose size is on the order of the wavelength of the light. Figure 20–6 is an illustration. We can conclude from all this that while it is generally true that light travels in straight lines (like a stream of particles), it can also be diffracted by very narrow slits and at the edges of objects (like a wave).

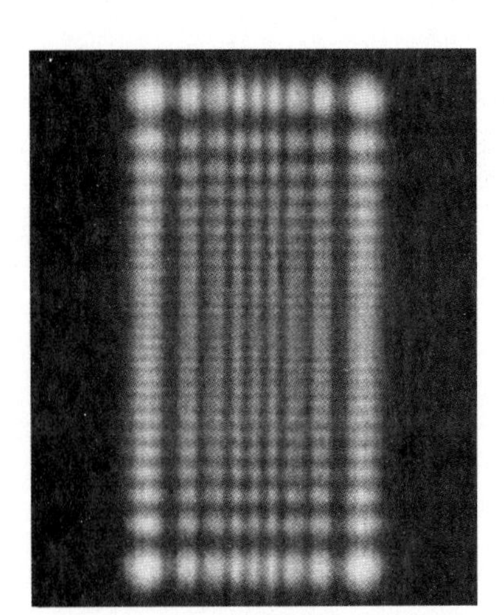

Figure 20-5. Diffraction by a rectangular aperture.

Interference

Perhaps the most compelling reason for thinking of light as a wave comes from a simple experiment first performed by the British physicist Thomas Young in 1801. Young caused a thin pencil of light to fall on two slits, very narrow and very closely spaced. When he displayed the light passing through the slits on a screen, it showed an *interference pattern.* (See Figure 20–7.) That is, instead of seeing two overlapping bright regions corresponding to light from the two slits piling up, Young observed several alternating bright and dark bands. This pattern is just

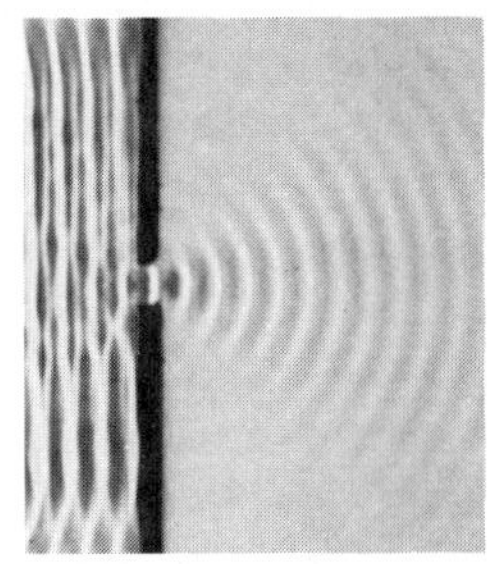

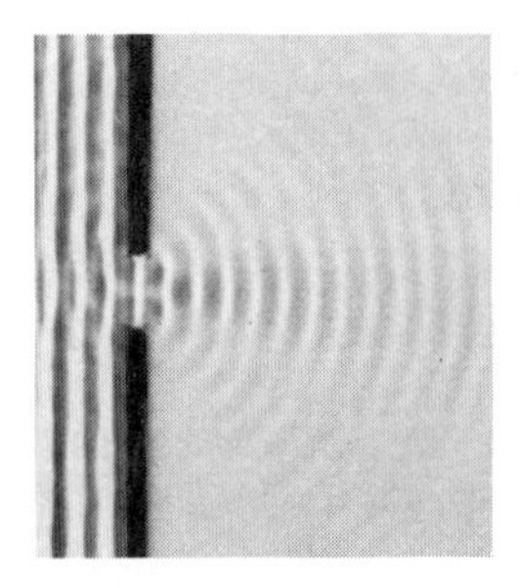

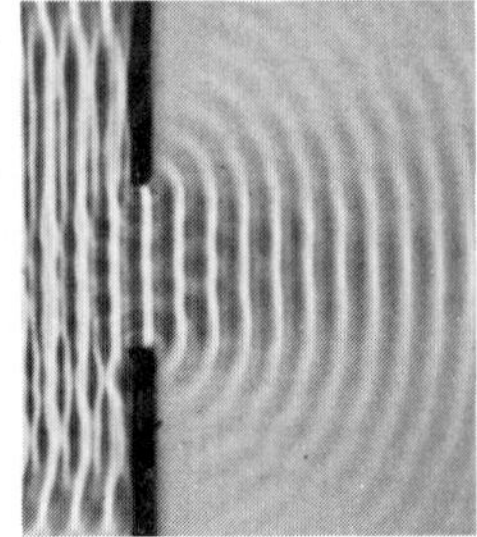

Figure 20-6. Light can be diffracted like a wave.

like the one produced by interfering water waves in a ripple tank. (See Figure 20–8.) In fact, a detailed analysis of the Young experiment accounts for the interference pattern on the basis of the waves from the two slits alternately canceling and reinforcing. The cancellation producing the dark regions occurs when the waves are out of phase, crest meeting trough. Bright bands result from reinforcement that takes place when the waves are *in phase,* crests matching crests and troughs matching troughs.

Interference also occurs in the smeared-out diffraction band from a single narrow slit. If light waves coming from two different parts of the slit overlap out of phase, they will cancel and produce a dark band. You can see this kind of interference, using no equipment other than your fingers. Look at a light source through the slit between two fingers and gradually close this slit down until light just passes through. You will be able to see the alternating dark and bright bands.

The particle theory of light completely fails to account for these interference effects. If light were a stream of particles passing through two slits, we could expect the greatest intensity of the light to show at the two points along straight-line paths from the source. (See Figure 20–9.) Very little overlapping would be expected. There would be no alternation of dark and bright bands. In fact, just the opposite takes place. The greatest intensity occurs in the region *between* the slits, called the *central maximum.* From the corpuscular theory, we would expect essentially no light in this region, since no particles would land there after passing through the slits.

In our earlier discussion of wave motion, we emphasized the fact that these two phenomena, diffraction and interference, are very important characteristics of waves, serving as a sort of trademark. This is no less true in the case of light. Diffraction and interference are the key features in establishing the wave behavior of electromagnetic radiation.

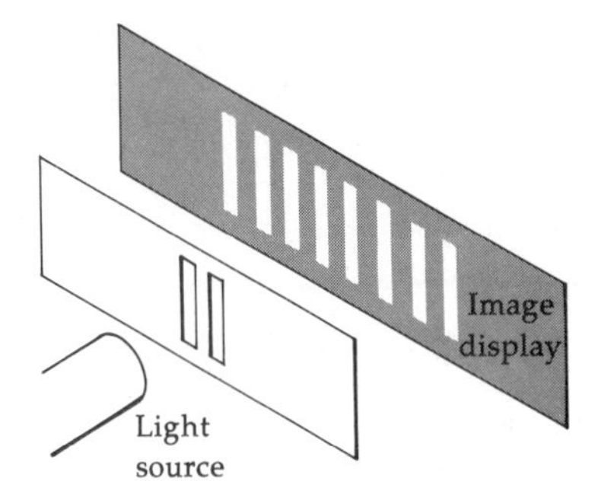

Figure 20-7. Two-slit interference pattern.

Figure 20-8. Interference of water waves.

Learning Checks

1. Can you give a definite answer to the question, "Does light travel in straight lines?" Explain.
2. How can interference effects occur for light passing through a single slit?
3. How do the interference and diffraction of light favor the wave theory over the particle theory?
4. Explain why, when people are talking outside your room and off to one side of the open door, you can hear them but not see them.
5. Explain the observation that shadows have fuzzy edges.

Figure 20-9. The two-slit experiment for particles.

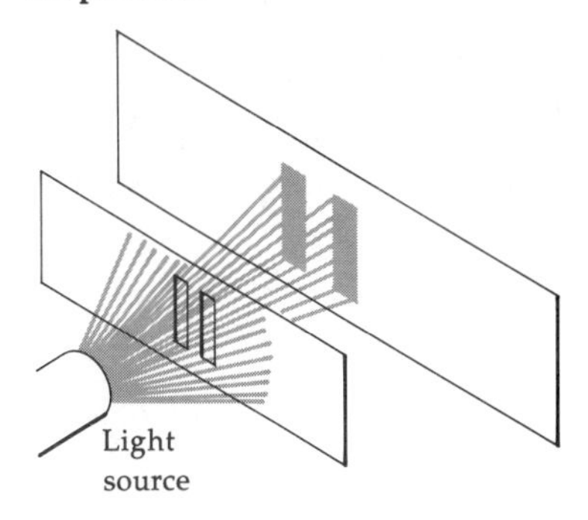

REFLECTION AND REFRACTION

In our discussion so far, we have learned that diffraction and interference are manifestations of the wave nature of light. We turn now to two other phenomena that occur when light interacts with matter and that depend less strongly on the wave nature of light. These are *reflection* and *refraction.* Both involve the path followed by the light as it encounters a material object.

Reflection

As you sit and read this book, glance about at the various things in your view. Except for sources of light, such as the sun or an electric bulb, the objects you see around you are visible because light reflects off them. Most of the light we see is reflected light. The moon, for example, has no source of light of its own. We see the moon because of the sunlight that bounces off it and travels to earth.

We can distinguish two kinds of reflection: *specular* and *diffuse.* Specular reflection, taken from the Latin word for mirror, is reflection from a glassy surface like a mirror or a highly polished metal plate. It produces clear images, such as the reproduction of your face when you look in a mirror. Diffuse reflection, on the other hand, does not produce such images. Diffuse reflection of light from the wall of your room enables you to see the wall, but you can't see your image if you stand in front of it.

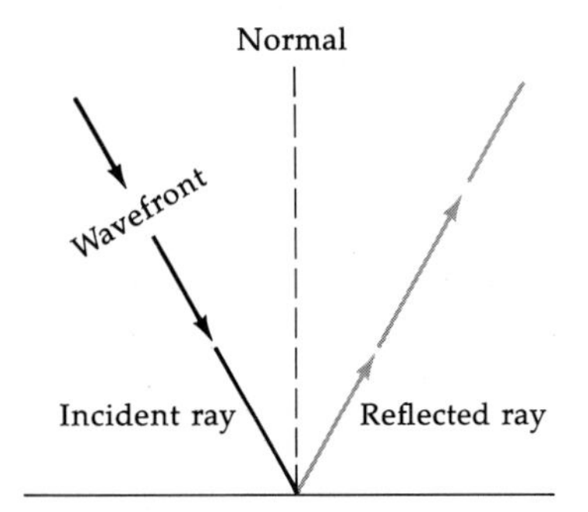

Figure 20-10. The law of reflection.

In either case, the light bouncing off a surface obeys a simple rule, called the *law of reflection.* This law is illustrated in Figure 20-10. Here a beam of light strikes a smooth surface, such as a mirror. We can represent the leading edge, or *wave front,* of the waves making up the light beam by a line perpendicular to it called the *light ray.* The ray and a line perpendicular to the reflecting surface, the *normal,* define an angle that we call the *angle of incidence.* The angle that the reflected light makes as it leaves the surface is labeled the *angle of reflection.* Simple observation tells us this:

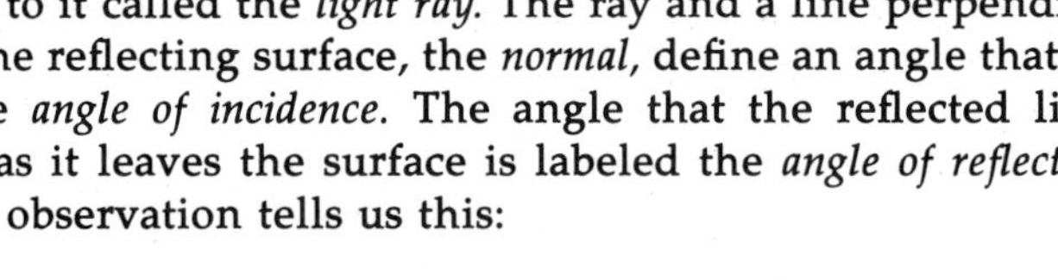

$$\text{angle of incidence} = \text{angle of reflection}$$

In this situation, light behaves very like a particle. Thus, light moving in along the normal strikes the surface and doubles back along the same track. This is what a billiard ball does after bouncing straight off a cushion. If the ball strikes the cushion at any angle other than 90 degrees, it rebounds at the same angle on the other side of the normal. So also does the beam of light.

Suppose light coming from an object reflects off some surface. If the reflecting surface is a plane mirror, the object has its image in a location that seems to be behind the mirror. We can see why this is so from Figure 20–11. The triangle in front of the mirror reflects an infinite number of rays from every point on its surface. Some of the rays strike the mirror and rebound according to the law of reflection. Let us follow a particular ray, one that eventually winds up in your eye. It leaves the triangle, follows the path indicated by the arrows, reflects off the mirror surface, and enters your eye. Because we believe so strongly that light travels in straight lines, it appears that the light has originated from an image *behind* the mirror and traveled along the dashed-line path. The same will hold true for all the other rays originating from the triangle. They also appear to come from behind the mirror. Because the angles of incidence and reflection are equal, the image appears to be as far *behind* the mirror as the object is in front. In actual fact, of course, no light goes behind the mirror at all. It only looks this way because of the zigzag path that the light follows.

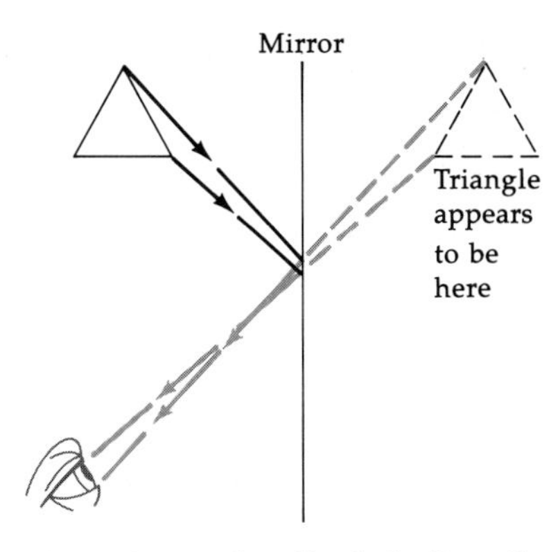

Figure 20-11. The triangle looks as if it were behind the mirror.

For diffuse reflection, the situation is somewhat different. Suppose we consider several parallel rays striking two surfaces, one smooth and the other rough. (See Figure 20–12.) We get specular reflection from the smooth surface; the reflected rays remain parallel as they rebound. The angles of incidence and reflection are the same for all rays. From the rough surface, however, the reflected rays are strewn about in all directions. Each individual ray still obeys the law of reflection as it rebounds from a tiny segment of the surface. But because the surface is rough, each segment presents to the incoming rays a normal line having a different direction, so the reflected rays seem to go off chaotically. No single normal line represents the entire surface, as it does with the mirror or polished metal. So, even though the *incident* rays may be parallel, they will not remain parallel upon reflection; because of the irregular surface, they won't all have the same angles of incidence. One ray might strike the surface at an angle of, say, 10 degrees, while a neighboring ray, landing on a different part of the surface, strikes at an angle of 15 degrees. Your eye perceives no image of an object behind the surface as it does in the case of specular reflection.

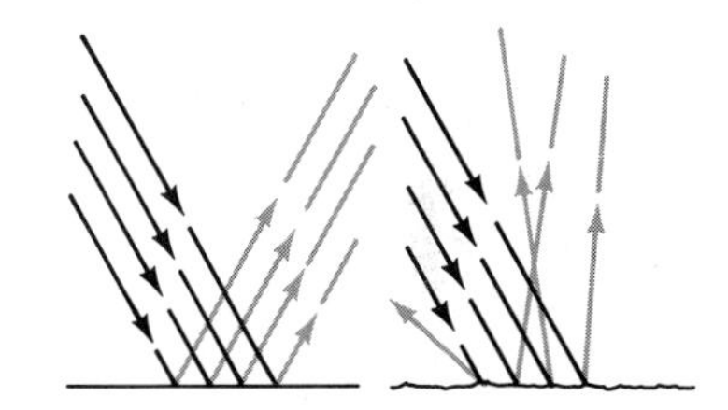

Figure 20-12. The type of reflection depends on the surface.

Learning Checks

1. Give several examples of the two kinds of reflection. Which kind is more common?
2. What is a *light ray?*
3. For diffuse reflection, the reflected rays are not parallel even though they originate from parallel incident rays each obeying the law of reflection. Explain.

Refraction in the Atmosphere

Atmospheric refraction is a concern of astronomers, geodesists, meteorologists, and those who investigate the sightings of lake monsters. Astronomers are concerned because refraction can distort celestial images. This is why the sites of astronomic observatories have to be selected with care (such as away from the "heat islands" created by cities). Geodesists are scientists who study the shape and dimensions of the earth, its gravity, and its magnetism. Geodesists must be aware of refraction because its presence may mean that corrections will have to be introduced into their measurements and calculations. Meteorologists take an interest in the varying appearances of refraction in extreme weather conditions. And the researcher of the lake monster phenomenon looks to atmospheric refraction as a possible explanation for the reports of reputable eyewitnesses. The people who see lake "monsters" may well be viewing the distorted (and therefore unrecognizable) image of a familiar creature or landmark.

Sources: W. H. Lehn, "Atmospheric Refraction and Lake Monsters," *Science*, July 13, 1979, pp. 183–185.

E. Tengstrom and G. Teleki (eds.), *Refractional Influences in Astronomy and Geodesy*. Boston: D. Reidel Publishing Co., 1979.

Refraction

If you stick a pencil halfway into a glass of water, the pencil appears bent. In fact, viewed from certain angles, it appears to be broken in two. The part above the water's surface seems to be separated from the submerged part. This is illustrated in Figure 20–13. What is happening here?

Again, we tend to interpret what we see in terms of our belief that light travels in straight lines. But here is where the speed of light gets into the act. Light travels at *different speeds* in water and in air. Upon passing from one to the other, it does *not* travel in a straight line; instead, because of the speed difference, the light ray changes direction at the boundary between the two. Bending of light rays because of different speeds in different media is called *refraction.* We say that the light is *refracted* at the interface between the two materials.

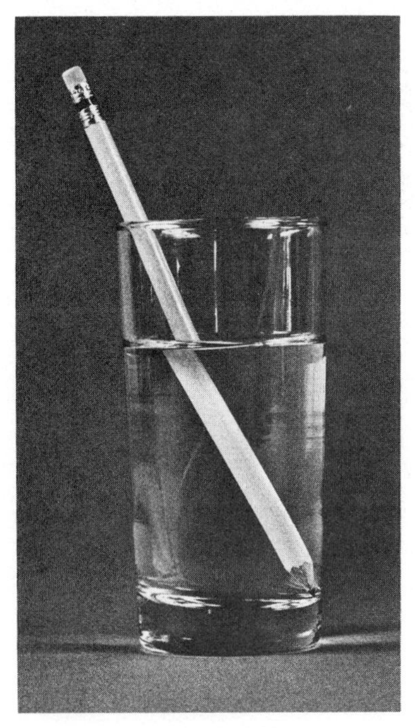

Figure 20-13. Pencil in a glass of water illustrates refraction.

We have learned that the speed of light in a vacuum is about 300,000,000 m/s. When traveling through transparent material such as air or glass or water, the electromagnetic waves interact to some extent with the matter. This slows them down. In water, for example, the speed of light is about 75 percent of its speed in a vacuum, or around 225,000,000 m/s. This is still incredibly fast by our everyday standards, but it is different enough from the speed in a vacuum to produce the pencil-in-the-water effect. Air slows down the light only slightly, by about 0.03 percent, so we can take the speed of light in air to be the same as in a vacuum. In some glassy materials, the light speed is cut nearly in half. Diamond slows the light down to an unusually slow speed of only about 120,000,000 m/s, or 40 percent of its speed in a vacuum.

Let us see how these speed variations result in refraction. Figure 20–14 shows a ray of light traveling through air and striking the surface of water at some oblique angle with the normal, the angle of incidence. As it crosses the air-water interface it slows down and consequently is bent *toward the normal.*

The path the light ray takes through the water makes a *smaller* angle with the normal, this time called the *angle of refraction.*

We can see why this bending takes place by considering the waves themselves rather than just the light ray. Figure 20–15 shows the light waves, where the crests are represented as parallel lines, as they begin to strike the surface of the water. As one end of the wave front starts to penetrate the water, it slows down and travels a shorter distance in a certain length of time than the other end, which is still traveling at the higher speed through air. By the time the next portion of the front enters the water, it has gained ground on the first. Each succeeding portion of the wave front gains on the one before it and loses ground to the one following, so that the entire wave front has swung around more toward the normal.

We can draw an analogy by replacing our wave front with a line of soldiers marching abreast from a paved surface onto a muddy field. (See Figure 20–16.) Those who enter the field first are slowed down by the mud. They march at the same cadence as before, but now their steps are naturally shorter. The soldiers in the next rank, marching at the faster speed on the pavement, will get closer to those in the mud; the overall effect will be to change the direction of march toward the normal to the interface of pavement and field.

Now consider the situation in which the light goes from the water into the air—for example, you sit beside a swimming pool and look at an object under water. The light travels through the water from the object, across the interface, and through the

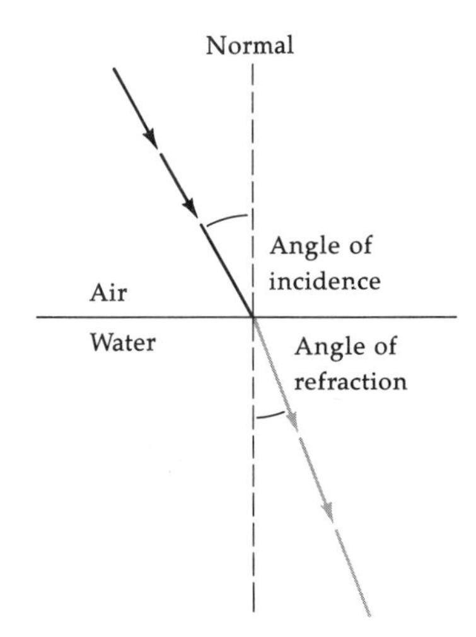

Figure 20-14. The light ray bends toward the normal as it crosses into the water.

Figure 20-15. Wave fronts slow down as they enter the water.

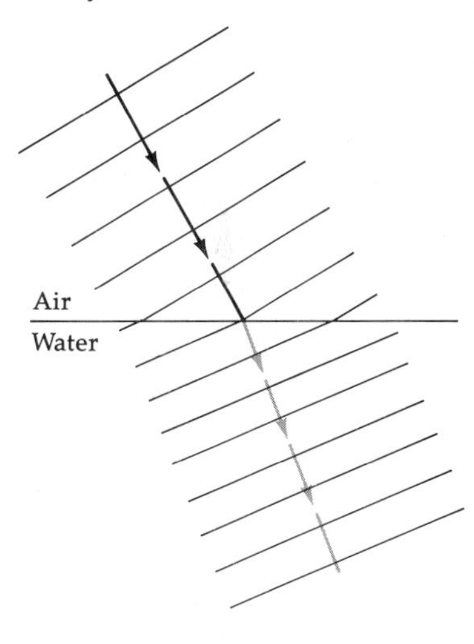

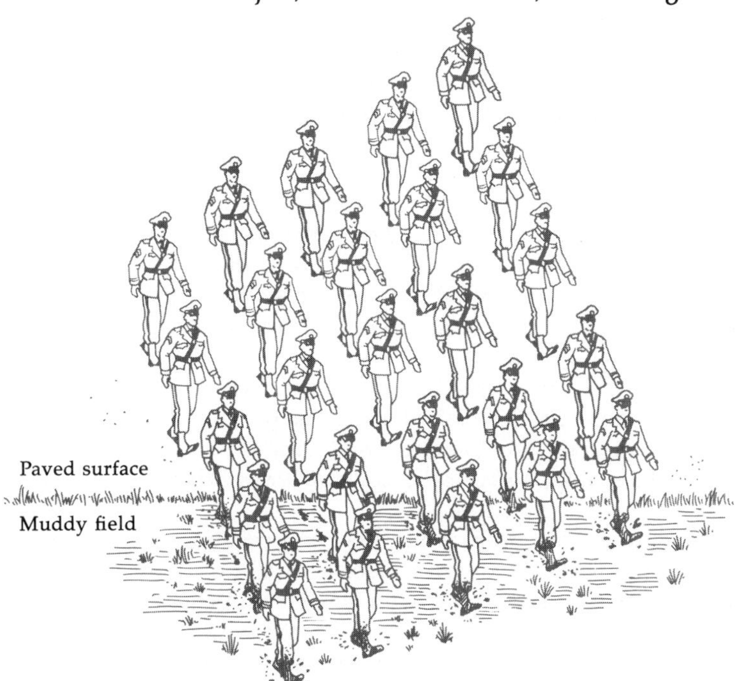

Figure 20-16. The soldiers slow down as they enter the muddy field.

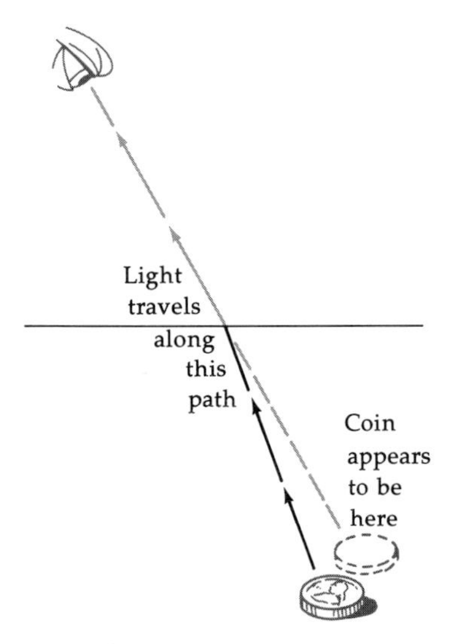

Figure 20-17. Don't believe everything you see.

air to your eye. The roles of the two angles are now reversed. What used to be the angle of incidence becomes the angle of refraction, and vice versa, but the light still travels along the same path as before. This time it speeds up as it crosses between the two materials and bends *away* from the normal, making an angle of refraction *larger* than the incidence angle. An easy way to keep this in your mind is to remember that, in the medium where the light has the greater speed, it will make a larger angle with the normal line.

What does all this have to do with our "broken" pencil? Figure 20–17 illustrates the situation when you look at a coin under water from above the surface. The light from the coin is refracted at the surface and moves to your eye along the path indicated by the arrows. However, it appears from your point of view to have traveled along the dashed-line path (since you believe in straight-line travel), and so the coin seems to be in a location different from where it actually is. For the same reason, the submerged half of the pencil is not where it appears to be; it is "out of sync" with the top half, the light from which is unrefracted on its way to your eye, and the pencil looks broken. In this case, "seeing is deceiving." This sort of thing is a real problem, say, for a sportsman on land trying to harpoon a submerged fish. The fish isn't where it appears to be.

Physicists use the fact that light travels at different speeds in different materials as a way of characterizing substances. It is convenient to use a number, called the *index of refraction*, or *refractive index*, for this purpose. The index of refraction for a material is defined as the speed of light in a vacuum divided by its speed in the material. That is,

$$n = \frac{c}{v}$$

where n is the index of refraction for the material in which light travels with the speed v. For example, since light speed in water is 75 percent of its vacuum speed, then the index of refraction of water is 1/0.75, or 1.33. In general, the value for n for any material is greater than 1, since v will always be less than c.

Other examples are glasses, which have indices of refraction between 1.5 and 2.0, depending on the specific type of glass. Flint glass, for instance, has an n value of 1.65. For diamond, it is 2.42; thus, light travels slower through diamond than through the glasses.

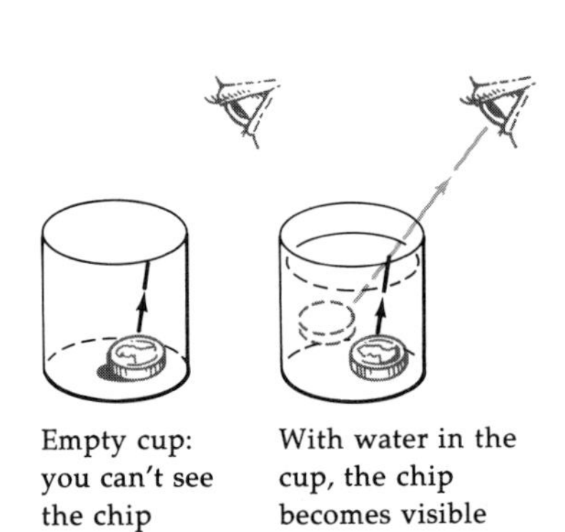

Figure 20-18. Refraction demonstration.

Figure 20–18 shows a simple demonstration you can do to illustrate refraction of light. Place a small object in the bottom of a cup and position your eyes so that the object is just out of your view. If you now fill the cup with water, the object will gradually become visible, even though neither it nor your eye

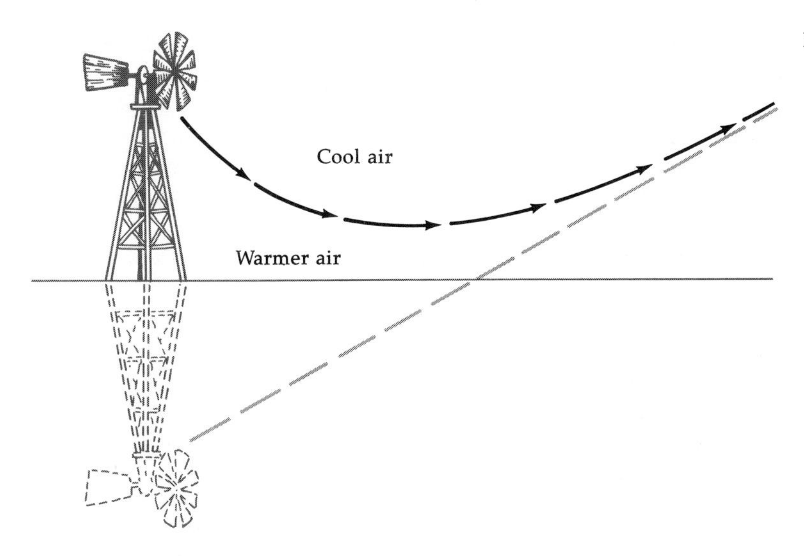

Figure 20-19. A mirage.

has moved. Refraction causes the light to take a different route. My 10-year-old puts it this way: "It's like when you're taking a bath, and you reach for your toe and miss it!"

Refraction is responsible for some atmospheric effects you have probably seen many times. *Mirages* are caused by refraction of light at the interface between hot and cool air, as illustrated in Figure 20–19. Light travels slightly faster in the less dense warm air. On a hot summer day the air near the asphalt highway will be warmer than air at a slightly higher level, so light from the sky or roadside scenery will be refracted along the path shown in the figure. The eye sees the light as if it were coming along the straight dashed line, giving the appearance of light reflected from the road surface. Of course, it isn't reflected at all, but it certainly looks like it. Because of our experience, we are conditioned to attribute this to reflection from a wet surface. The brain tends to interpret the scene in the same way as the one where trees lining the shore of a lake are mirrored on its surface. Hence, we are led to believe that water is on the road, only to see it disappear as we approach. Old western movies on the Late Show would hardly be complete with-

Seeing was not only deceiving but treacherous to settlers crossing California's Mojave Desert. This mirage is appropriately called Silver Dry Lake.

out some ancient thirsty desert rat being fooled in just the same way.

Sometimes the situation is reversed and the cooler air is near the ground. Distant objects that are actually beyond the horizon can appear suspended in the sky. Ranchers in West Texas tell of sometimes being able to see a neighboring town early in the morning, only to have it disappear later in the day. This is called *looming*. The light rising through the cool air near the surface is refracted back toward the ground by the warm upper air heated by the early morning sun. Later, as the ground also heats up, the temperature of the air is more nearly uniform, and the light is no longer refracted. Looming may also account for reports of unidentified flying objects. A distant automobile's headlights that reach a viewer through a long, gentle refractive curve seem to be speeding circles of light in the sky.

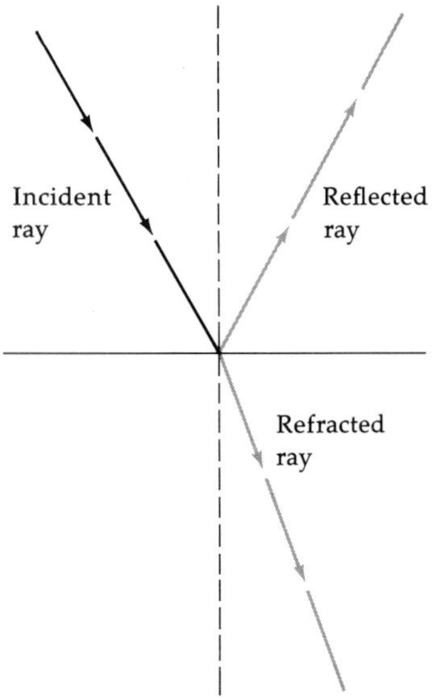

Learning Checks

1. What is meant by the *index of refraction?*
2. How is the refraction of light related to its speed in different media?
3. Why does a fish under water appear closer to the surface than it actually is, as viewed from above the water?
4. When light goes from air to water, its wavelength changes. Does its frequency also change?

Figure 20-20. The incident ray is reflected and refracted.

Total Internal Reflection

When light strikes the surface of a transparent substance such as water or glass, part is reflected and part travels through the substance as refracted light. (See Figure 20-20.) These two parts of the original beam obey the laws of reflection and refraction.

Now suppose we consider light moving from water into air, striking the boundary at an oblique angle as shown in Figure 20-21. Part of the light is reflected back into the water, while the refracted portion is bent away from the normal as it enters the air. As we have already learned, if we gradually increase the angle of incidence, the angle of refraction also gets larger and larger. If we keep this up, we will eventually get to a point where the refracted light just skims along the water's surface. This means that the angle of refraction is 90 degrees. That angle of incidence that gives a 90 degree angle of refraction is called the *critical angle*. When the light is incident at an angle larger than the critical angle, none of it is refracted. It is *all*

Figure 20-21. Total internal reflection.

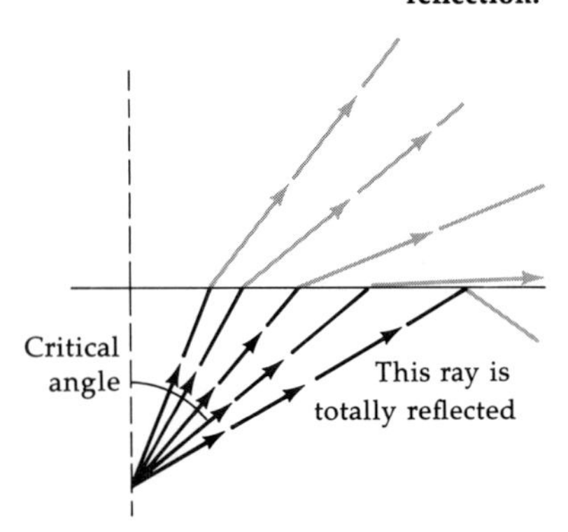

reflected back into the water. This is called *total internal reflection.*

Total internal reflection can produce some strange effects, such as the one shown in Figure 20–22. The swimmer seated at the edge of the pool appears to have two pairs of legs, one that we can see from under water along a direct line, the other appearing by total internal reflection to be dangling above the water's surface.

Figure 20-22. Strange effects are produced by total internal reflection.

Total internal reflection can occur only when the path of light passing between two materials is such that the angles of refraction are always *larger* than the angles of incidence. This happens when the light passes from a material in which it has a lower speed into one in which it moves faster. Total internal reflection cannot take place, for example, when the light goes from air into water. It can only happen when the light goes the other way.

The critical angle depends on the indices of refraction for the two materials. In all our examples, we assume that the light goes from a transparent material into air. When the transparent material is water, the critical angle is about 48 degrees. For flint glass ($n = 1.65$), it is approximately 37 degrees. If you are under water and look up along a line making an angle of just less than 48 degrees with the normal, you will see the horizon. Light just barely grazing the water's surface is refracted into the water along a path corresponding to the critical angle. Peering along a line at an angle of greater than 48 degrees, you will be able to see the bottom of the pool.

The larger the index of refraction for a substance, the smaller the critical angle. This means that the light must be traveling along a more nearly vertical path to escape from a material with a high refractive index.

For example, we have seen that diamond has an unusually high n value, 2.42. Its critical angle, 24.4 degrees,is correspondingly small. Once a light ray gets inside a diamond it has a relatively difficult time getting out. Being continually reflected internally, it is effectively trapped. Diamonds have their characteristic sparkle for this reason. Light that gets in bounces around for a while until it happens to approach a face at an angle smaller than the critical angle. It can then escape, perhaps coming to the viewer's eyes from an unexpected direction. This is illustrated in Figure 20–23.

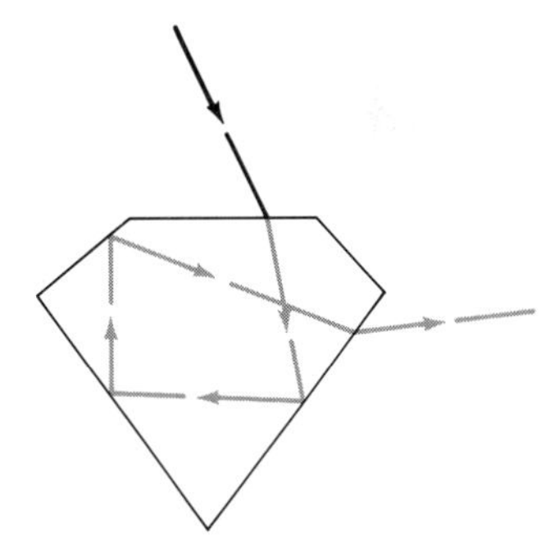

Figure 20-23. Light entering the diamond is momentarily trapped by total internal reflection.

"Light pipes" are a practical application of total internal reflection. Bundles of thin glass fibers are coated with material having an index of refraction between that of air and the glass. A light ray moving almost parallel to the axis of one of the fibers is trapped, for it always strikes the fiber from the inside at an angle greater than the critical angle. It is thus confined to the pipe until it can escape from the other end, no matter what the shape of the flexible pipe. Surgeons can use such a light

A Living Light Pipe

Botanists have discovered that plants have light-piping capability. When a pinpoint laser was beamed into one end of a curved segment of tissue from corn, from oats, and from mung beans, the light traveled about an inch through to the other end.

"It's all based on a simple phenomenon—total internal reflection," the woman who made the discovery explains. "Light entering a plant-cell compartment at the proper angle is reflected from one side to the other, moving through the entire length of the cell in a zigzag pattern, the same way it travels through optic fibers."

The plants are only about 1 percent as efficient at this job as optical fibers are. But the finding has exciting implications. "What it boils down to," the discoverer concludes, "is that plants have a complex visual system, perhaps as intricate as our own."

Source: "Plants See by Fiber Optics," *Science Digest,* January 1983, p. 19.

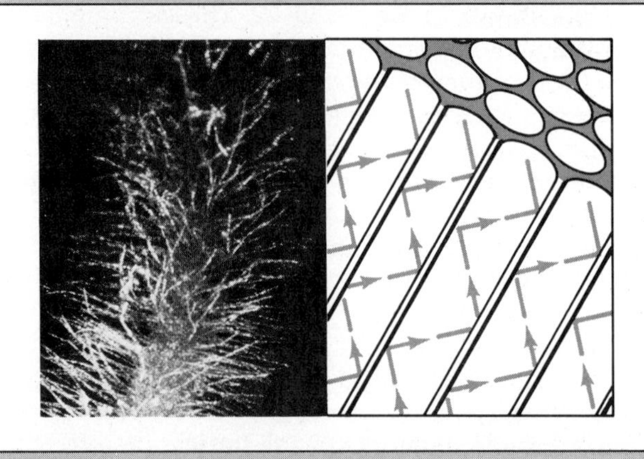

The light beamed into this segment of a curved corn root is reflected internally, zigzagging through cellular compartments just as if it were traveling through multifiber optic bundles.

pipe to illuminate hard-to-see places, such as the inside of your stomach.

Learning Checks

1. What is meant by the *critical angle* for total internal reflection?
2. Can total internal reflection take place for a beam of light going from air into water? What about for light going from flint glass into water? Explain.
3. Can a fish look *up* and see the bottom of the pool? Explain.
4. For a fish under water, what is the angle between the two horizons limiting its view?

LENSES AND OPTICAL INSTRUMENTS

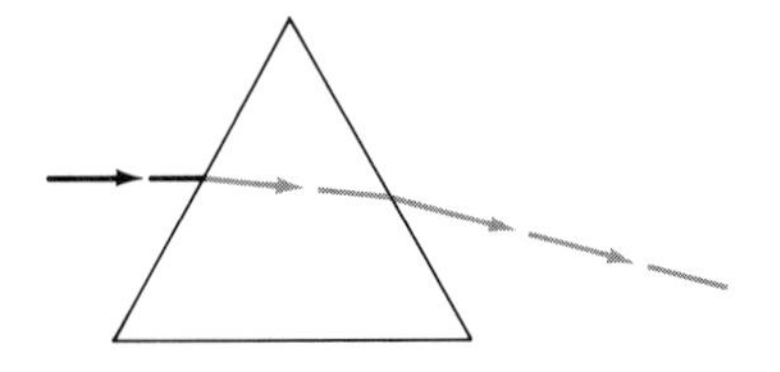

Figure 20-24. Refraction by a prism.

The optical instruments so familiar in our everyday world—cameras, telescopes, microscopes, magnifying glasses, binoculars—are designed to perform a variety of tasks. However, they all have one feature in common. Each takes advantage of the refraction of light to carry out its mission—making small objects look large, or faraway objects appear close up, and so on. The important component in each of these instruments, and the feature that causes refraction of light, is the *lens.* First we will learn how a lens works. Then we will be in a position to understand how lenses can be used in various combinations as aids to the human eye.

As a prelude to learning how a lens works, let us see what happens when light goes through a prism. (See Figure 20–24.) A prism is made of glass or some other transparent refracting material. The one in the figure has a triangular cross-section. When a beam of light strikes the left face, it is bent downward toward the normal, as we expect, and follows the indicated path to the right face. When it emerges into the air, the beam is now bent *away* from the normal, because its speed is greater in air. But because the two faces of the prism are fixed at an angle relative to one another, this second bending is *also* downward, and the light follows the path shown. The effect of the prism is to bend the light twice in the same direction.

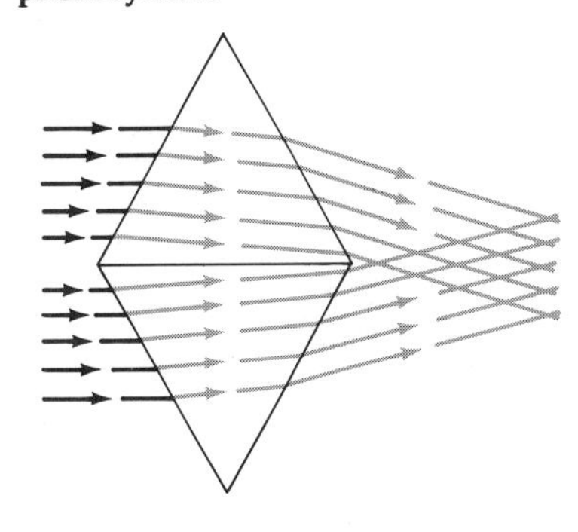

Figure 20-25. Focusing by a double-prism system.

Figure 20–25 shows what happens when we put two identical prisms together, base to base, and shine a shaft of light parallel to the common base. The rays striking the upper prism are bent downward, and those striking the lower prism are bent upward. This means that the light emerges to the right in two beams that cross each other. The rays of light in the upper beam are all parallel, because they were bent the same amount by the prism. The same holds true for the rays in the lower beam, bent upward. These two beams cross over a broad front.

The rays within each of the half-beams are all parallel because the normals to every point on each face of the two prisms are parallel. But suppose instead of a straight-edged double prism we make one whose faces are smoothed out into spherical segments. This is illustrated in Figure 20–26. Now the normals are no longer parallel. Light coming in near the top will have a large angle of incidence and so will be refracted the most. Similarly, the light entering near the bottom will be bent upward the most. As we move toward the middle, the angles of incidence for parallel rays become smaller and smaller. A ray going directly through the middle will not be refracted at all. It will emerge along the same path by which it struck the glass.

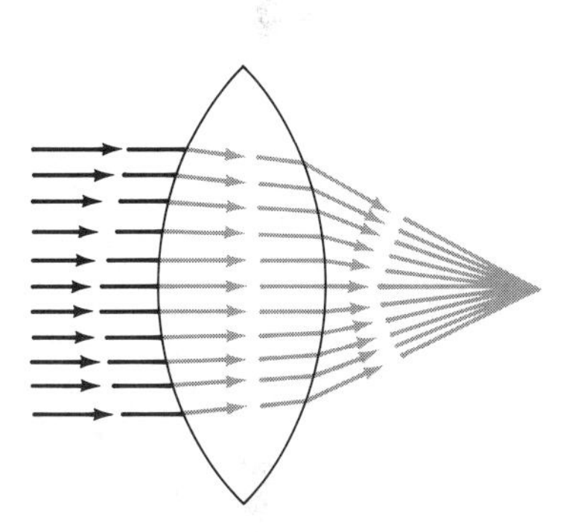

Figure 20-26. Converging lens.

You can see that with this new arrangement the emerging rays within a half-beam are no longer parallel. Instead of intersecting over a broad front, the two half-beams tend to converge to a point. A smoothed-out double prism of this type has the shape of a lentil seed, and so it is called a *lens* (from the Latin word for "lentil"). The name has come to mean any refracting material with a curved surface.

The kind of lens we have described is called a *converging* lens, because it bends the light rays so that they converge more or less to a point. We can make a *diverging* lens by switching our two prisms, joining them apex to apex, as shown in Figure 20–27. Now the two half-beams are bent *away* from one another—they

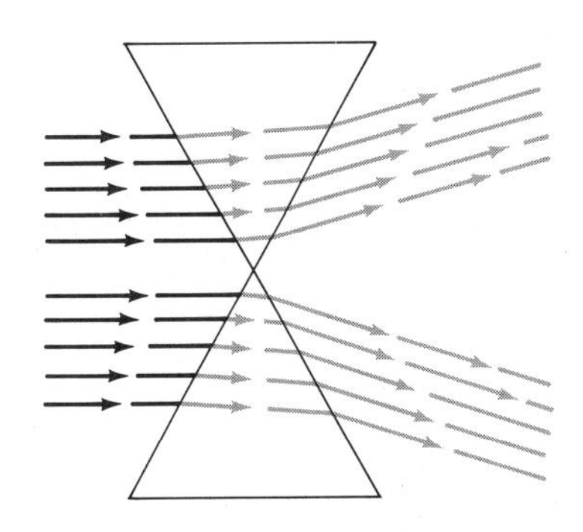

Figure 20-27. Divergence by a double-prism system.

Figure 20-28. A diverging lens.

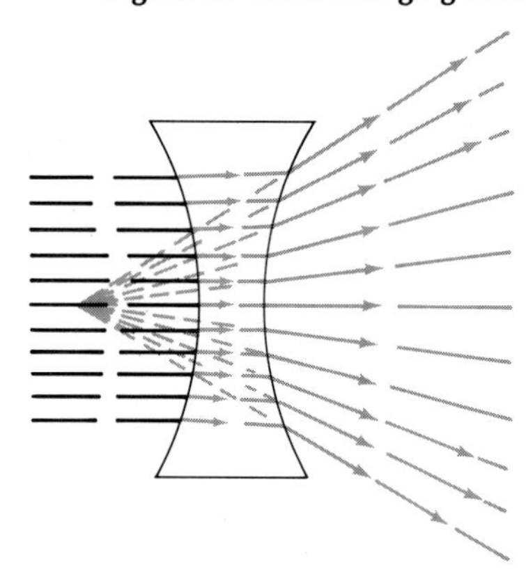

diverge. Figure 20–28 illustrates the smoothed-out version of the diverging lens. Here, viewed from the right side, the refracted rays appear to diverge from a point.

Lenses usually have spherical surfaces. If we look at a lens in cross-section, we can imagine that each face is the arc of a circle. (See Figure 20–29.) The *center of curvature* of the lens is the center of this circle. Notice that the center of curvature lies on a line drawn through the center of the lens, called the *principal axis,* or *lens axis.* For a converging lens, the *focal point* is the point to which all parallel rays are focused. For a diverging lens, it is the point from which the divergent rays seem to come. The *focal plane* is the plane perpendicular to the lens axis and running through the focal point. Incident parallel rays that are not parallel to the lens axis will be focused somewhere in the focal plane, not at the focal point. The *focal length* is the distance from the middle of the lens to the focal point. Notice that since there are two lens surfaces, there are two focal points and two focal planes.

Image Forming

The main function of a lens is to form an image of an object. A telescope makes an enlarged image of a distant planet. Microscopes do the same for small things. The kind of image formed—that is, whether it is larger or smaller than the object and whether it is upside down or rightside up—depends on what kind of lens is used and where it is located. We can see this by drawing a *ray diagram.* Figure 20–30 illustrates the idea for an object positioned beyond the focal point. We can locate the

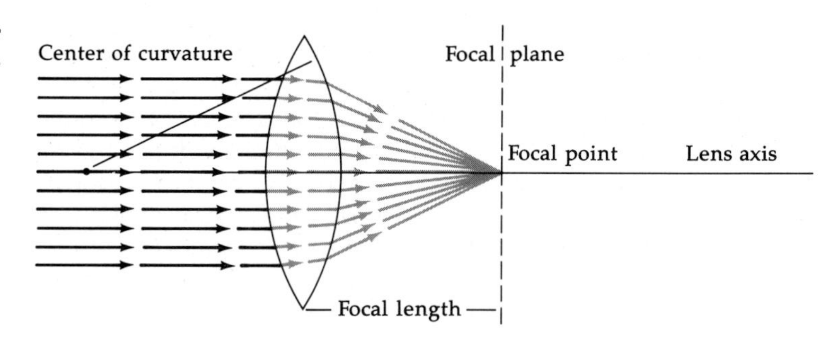

Figure 20-29. Nomenclature for a lens.

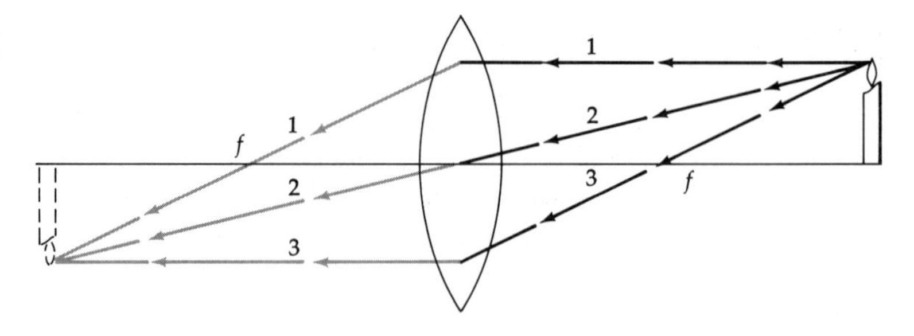

Figure 20-30. The object is beyond the focal point.

image by drawing only three of the infinite number of rays leaving the object and striking the lens. Ray number 1 approaches parallel to the lens axis, so we know it is bent to go through the focal point. Ray number 2 goes through the thickest portion of the lens; it hardly changes direction at all and goes straight through. Ray number 3 travels from the object through the first focal point. When it strikes the lens, refraction takes it parallel to the principal axis on the other side. The place where these three rays intersect is the location of the image of the particular part of the object from which they came. We could make this same argument for every other point on the object, so the entire image is located as shown in the figure.

The image in this case is a *real image,* so called because light from the object actually goes to the image location. If we were to put a screen at this point, we would see the image on the screen. This is the sort of arrangement used with a slide projector. Note that the image here is inverted, which is why you put slides in the projector upside down.

When we put the object closer to a converging lens than the focal point is, we get a different result. Figure 20–31 illustrates a common magnifying glass. The ray diagram shows that the light appears to be coming along the dashed lines. This means that the two rays look as if they come from the top of the image. Hence, the image is larger than the object. The lens serves to enlarge the angle through which we can see. For this reason, the image is magnified.

Notice that in Figure 20–31 the light rays do not actually pass through the image location. We could not project this image on a movie screen. Such an image is called a *virtual image.* Plane mirrors produce virtual images: no light goes behind the mirror to the image position. The name is perhaps a little misleading. A virtual image is a perfectly good one, and you can see it quite nicely. But it is to be distinguished from a real image, for which the light rays really do appear at the image location.

A *microscope* takes advantage of both these image-forming situations to create an enlarged image of very small objects.

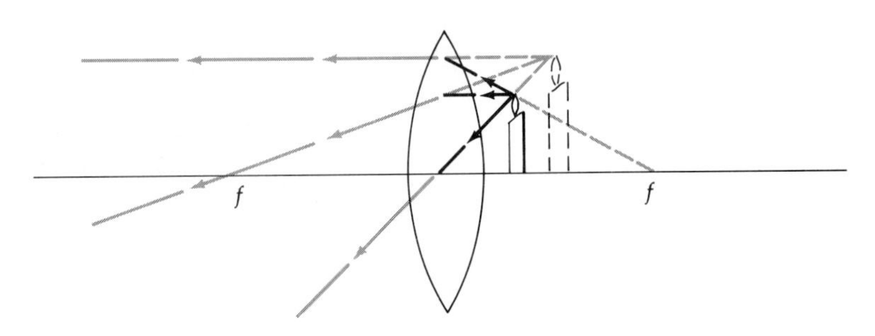

Figure 20-31. The object is between the focal point and the lens.

Figure 20-32. Microscope.

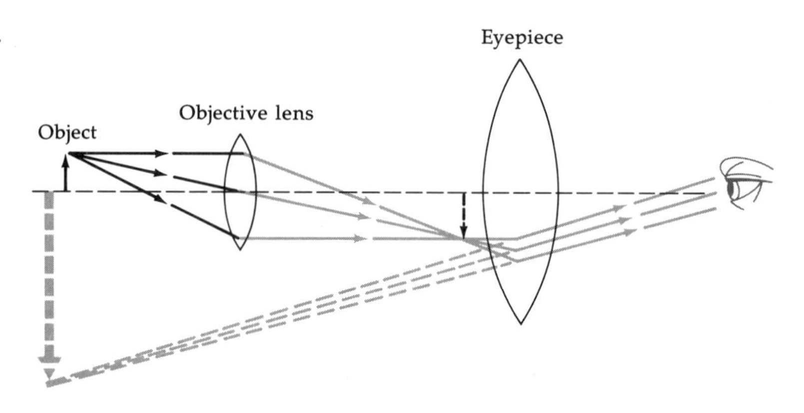

The *objective lens* (see Figure 20–32) has an extremely short focal length. This allows the object to be conveniently located beyond the focal point. As we have seen, this forms a real, inverted, enlarged image. This image acts as the object for the second lens, called the *eyepiece*, which is simply a magnifying glass. The image for the eyepiece is virtual and enlarged. It is the one you see when you view a tiny object through a microscope. The combination of lenses enables us to achieve a much greater magnification than would be possible with a magnifying glass alone.

A major departure from traditional telescope design, the multiple mirror telescope at the Mt. Hopkins Observatory in Arizona boasts a compound eye of six primary mirrors (one is shown uncovered here), each with a diameter of 72 inches. This arrangement creates light-gathering power the equivalent of a conventional telescope with an aperture of 176 inches. Light captured by each mirror is brought to a common focus and is maintained as a single image by a complex electronic control system.

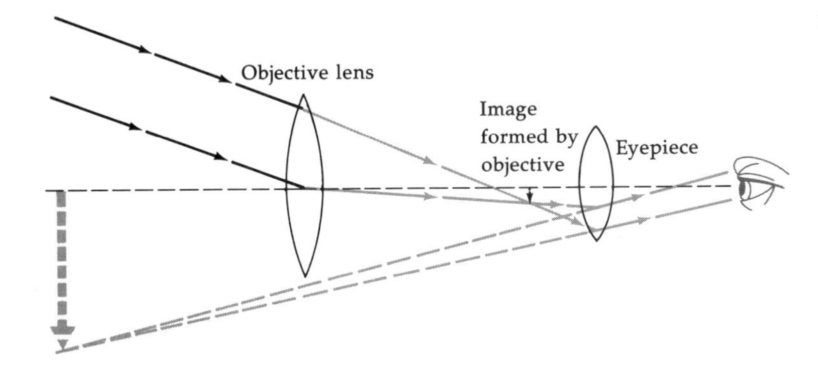

Figure 20-33. Telescope.

The principle by which a *telescope* operates is the same as that for the microscope, except that the roles of the objective lens and the eyepiece are reversed. (See Figure 20–33.) The objective lens has a large focal length. Because the object, such as a star or planet, is extremely far away, its image formed by the objective lens is quite near the focal point. This image serves as the object for the eyepiece lens. The image of the eyepiece is the one you see. It is magnified because it subtends a larger angle than does the distant object.

As with the microscope, the final telescope image is upside down. This is no problem for viewing astronomical bodies—you don't really care whether you see Saturn upside down or rightside up. But for terrestrial telescopes and binoculars (which are just twin telescopes), upside down images can be a real problem. Terrestrial telescopes get around this by using a third lens to re-invert the image. Binoculars (Figure 20–34) usually have prisms that invert the image by total internal reflection. (See Figure 20–35.) Prisms are used instead of mirrors because they reflect nearly all the incident light. Mirrors, on the other hand, absorb some of the light that strikes them, and hence their images are not as bright. You can see that the arrangement in Figure 20–35 gives the light a longer path than for a single trip through the tube. This allows the manufacturer to use objective lenses with large focal lengths.

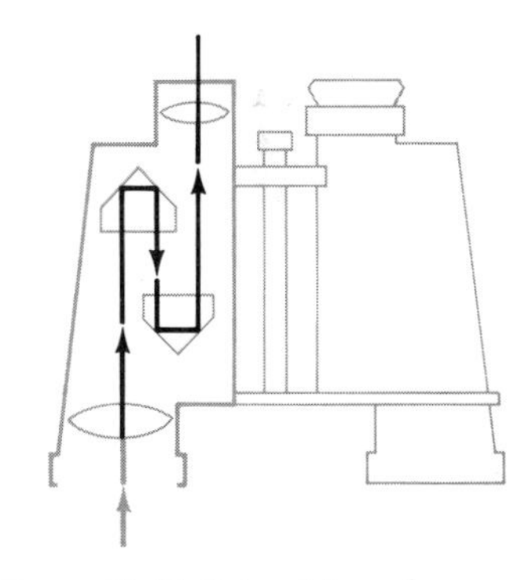

Figure 20-34. Prism binoculars.

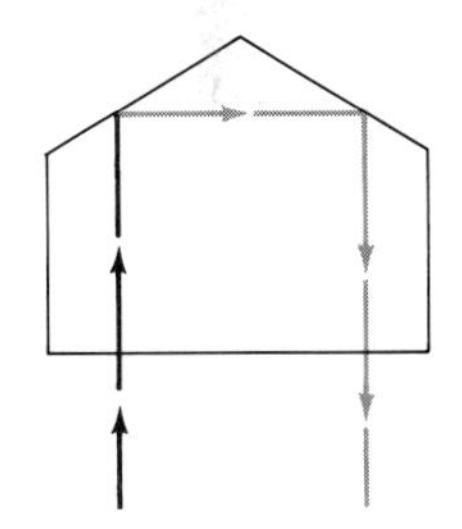

Figure 20-35. Total internal reflection in a prism.

Learning Checks

1. Account for the names *converging* and *diverging* lens.
2. Distinguish between a *real* and a *virtual* image.

The Camera

The word *camera* comes from the Latin phrase *camera obscura*, which means "dark room"—a description of the forerunner

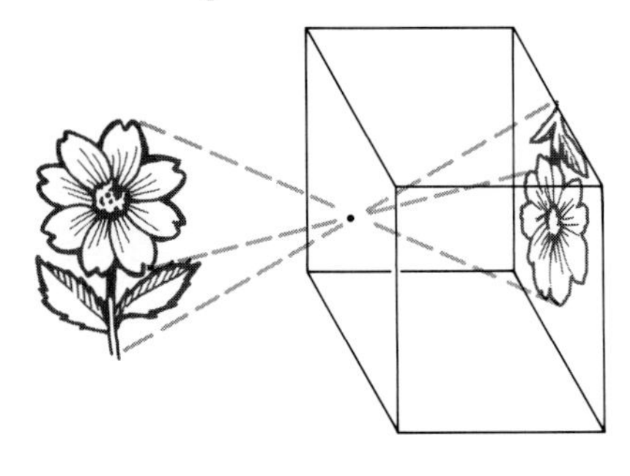

Figure 20-36. Pinhole camera.

of the modern camera that finds so many uses in our lives. In the fifteenth century, Leonardo da Vinci was among those who experimented with image formation in dark rooms. A single hole let in light from an object to form an image on the opposite wall. A portable version of the camera obscura is what we now call a *pinhole* camera. (See Figure 20–36.)

The question of focusing does not arise with a pinhole camera because the rays forming the image are straight-line. They undergo no refraction on their journey from the object. If the hole is small enough, the image will be quite sharp regardless of the distance from the opening to the opposite wall. The difficulty is that, with such a small hole, only a tiny shaft of light can enter the camera. You get a rather dim image. Making the hole larger admits more light, of course, but this blurs the image because of overlapping subimages from different portions of the hole. In a nutshell, then, the choice with a pinhole camera is between sharp dim images or fuzzy bright ones.

With the modern camera, we solve this dilemma by enlarging the opening and inserting a converging lens to gather the rays into a sharp image. Now, of course, we have introduced the problem of focusing, because the image will be sharp only in one plane. Where this plane is located depends on the distance between the object and the lens, as illustrated in Figure 20–37. For objects a great distance away, the image will form at the focal plane. For objects closer to the lens, the image will be located somewhere behind the focal plane. All but the simplest cameras allow for focusing—that is, positioning the lens at various distances from the film for photographing near and distant objects. In old-fashioned cameras, the lens is mounted on an accordionlike extension. This has been replaced in the modern versions by a screw attachment that can be rotated to move the lens forward or backward.

Figure 20–38 shows a schematic of a camera. It is a light-tight box containing a lens, a shutter, a diaphragm, a film holder-changer, and a viewfinder for locating the scene to be photographed. The shutter and diaphragm together control the exposure. The shutter regulates the length of time light is al-

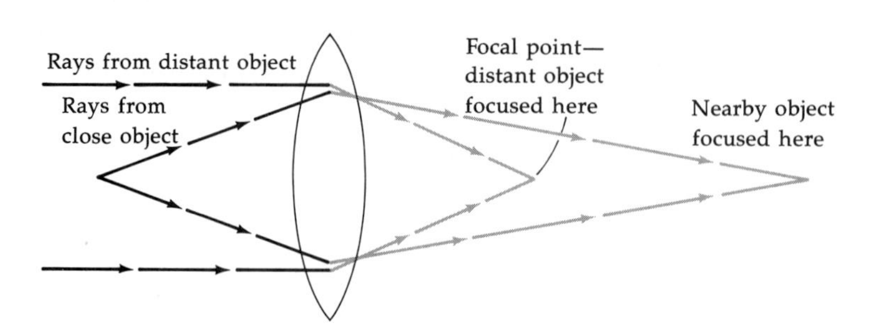

Figure 20-37. Position of focus depends on object distance.

lowed to stream into the camera, and the diaphragm fixes the size of the opening through which the camera admits light.

One type of shutter, called the *leaf shutter,* consists of a series of blades or leaves fitted just behind the lens. The shutter opens by a mechanism that swings the leaves simultaneously outward to uncover the lens opening. After a preset length of time the leaves close again, blocking out the light. Typical leaf-shutter exposure times range from 1 second to 1/500 second.

The *focal-plane shutter,* as the name implies, is mounted directly in front of the focal plane. It is a series of blinds that move laterally across the face of the film. The blinds are arranged to form a traveling slit whose width establishes the exposure time. Focal-plane shutters normally operate with exposure times between 1 second and 1/1000 second.

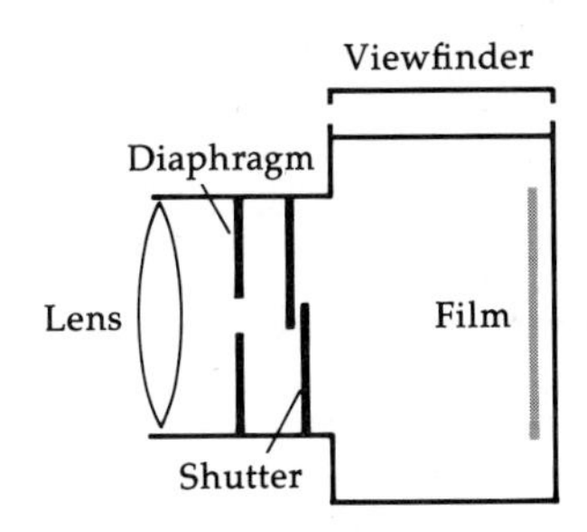

Figure 20-38. Camera schematic.

The diaphragm, or aperture, is the shutter's partner in controlling exposure. An opening with an adjustable diameter, the diaphragm determines the effective size of the lens and limits the amount of light it admits to the camera. On a sunny day or for a brightly lit scene, the diaphragm opening may be small. For dim lighting it must be wider, allowing more light to bathe the film.

The brightest images require lots of light squeezed together. This implies a large lens diameter and a short focal length. Modern cameras denote these two values with the so-called *f*-stop number, which is the ratio of the focal length of the lens to the diameter of the diaphragm:

$$f = \frac{\text{focal length}}{\text{diaphragm diameter}}$$

A small *f*-stop indicates that the maximum opening at the lens can be large, showing how dim the light can be for good pictures. For example, a "50 mm $f/2$ lens" has a focal length of 50 millimeters and the diaphragm can be opened to a maximum diameter of 25 millimeters.

Camera manufacturers ordinarily mark the stop setting by a series of *f*-numbers on a scale, such as 1.4, 2, 2.8, 4, 5.6, 8, 11, 16. As the *f*-stop numbers increase, the diameter decreases. The amount of light the camera admits varies with the area of the diaphragm opening, which in turn is proportional to the square of the diameter. For example, since the $f/5.6$ opening has half the diameter of the $f/2.8$, it admits only one-fourth as much light. The numbers on the scale are chosen so that each setting lets in twice as much light as the next higher *f*-stop. If the exposure time is 1/100 second for an *f*-stop of 4, then for the same exposure at $f/2.8$ the shutter must be set to remain open for half as long, or 1/200 second.

Figure 20-39. The human eye.

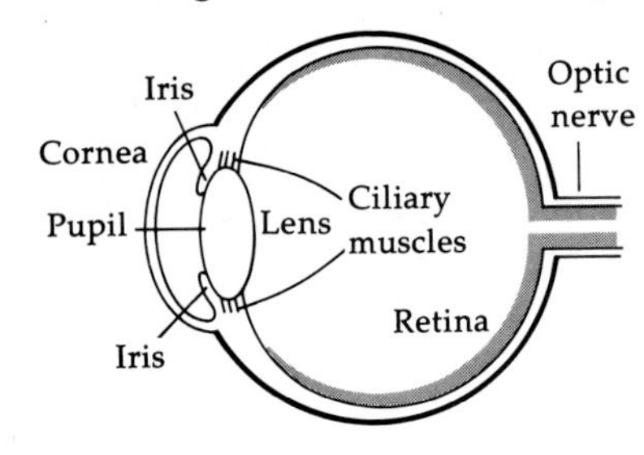

The Human Eye

The *eye* is, without question, the most important optical instrument. Figure 20–39 is a drawing that gives the main features of the human eye. The muscles that hold and maneuver the eye are fastened to the sclera, a tough outer membrane. The foremost part is the *cornea*, a transparent section that serves as a converging lens. The *iris* functions like the diaphragm of a camera, automatically enlarging and contracting to allow the proper amount of light through the *pupil* into the eye's interior. The pupil and iris make up the colored portion of the eye.

Located behind the iris is a second converging lens, called simply the *lens.* Teamed with the cornea, it makes the eye a double-lens system. The cornea and lens focus incoming light onto the *retina,* the light-sensitive inner coating on the rear of the eyeball. It is a mosaic of two kinds of cells, the *cones* and the *rods.* The cones respond to the light by sending color messages to the brain, whereas the rods are more sensitive to dim light but do not distinguish between colors.

We have seen that the point of focus for a given lens depends on the location of the object. In a normal, relaxed eye the focal length of the cornea-lens combination is such that images for objects from several meters to an infinite distance away are in focus at the retina. But for objects much closer—30 centimeters, say—a process called *accommodation* takes place. Small sets of muscles, the *ciliary* muscles, change the shape of the lens, squeezing and thickening to make it more converging. This shortens the focal length. The light rays are focused on the retina rather than behind it, where they would converge without accommodation. The closer the object, the more squeezing the ciliary muscles must do. In moving an object closer to our eyes, we eventually reach the so-called *near-point,* where accommodation reaches its limit. The muscles cannot change the lens shape enough to focus on an object closer to the eye than this point.

You may have noticed that as people get older they have to hold a book or newspaper farther and farther away from their eyes. The common complaint is that "my eyesight is fine, but my arms aren't long enough." The ciliary muscles weaken and the lens becomes more rigid with advancing age. This causes the near-point to recede gradually. It varies from about 10 centimeters in a young child to more than 40 centimeters in an elderly person. The formal name for this condition is *presbyopia,* from the Greek meaning "old man's vision."

Two quite common defects of the eye are *nearsightedness* and *farsightedness.* A nearsighted person's eyeballs are either deeper than the focal length of the double lens or the lens is curved too much. Each of these situations results in distant objects being focused in front of the retina. (See Figure 20–40.)

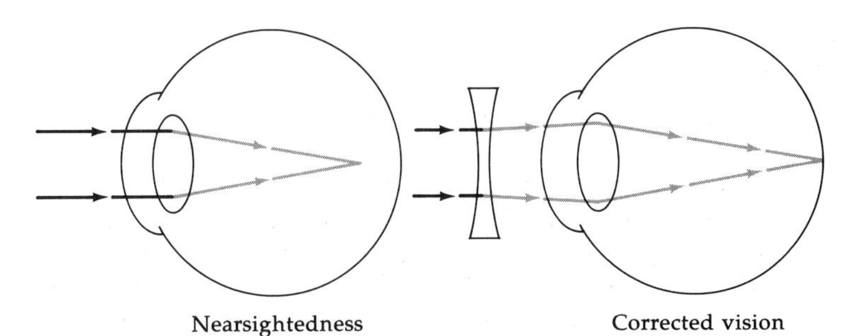

Figure 20-40. Nearsightedness.

Near objects are naturally brought to focus farther back on the retina. Nearsighted people know that if they squint their eyes they can focus more easily on distant objects. Squinting reduces the size of the opening through which the light comes, analogous to the pinhole camera arrangement where a clear image does not depend on depth. The technical term for nearsightedness is *myopia,* from the Greek for "shut vision," referring to this continual squinting.

Farsightedness, however, results from the opposite condition. Either the eyeball is too shallow or the lens is too flat. Here the focusing takes place behind the retina, spawning the name *hyperopia,* "vision beyond." With accommodation, distant objects are readily brought into focus, but accommodation reaches its limit too far away. Near objects are not seen clearly. (See Figure 20–41.)

Both conditions can be corrected by additional lenses in front of the cornea—the familiar eyeglasses or contact lenses. Since myopia is caused by too much convergence, a diverging lens does the trick. Farsighted persons need more convergence, so a converging lens corrects this defect.

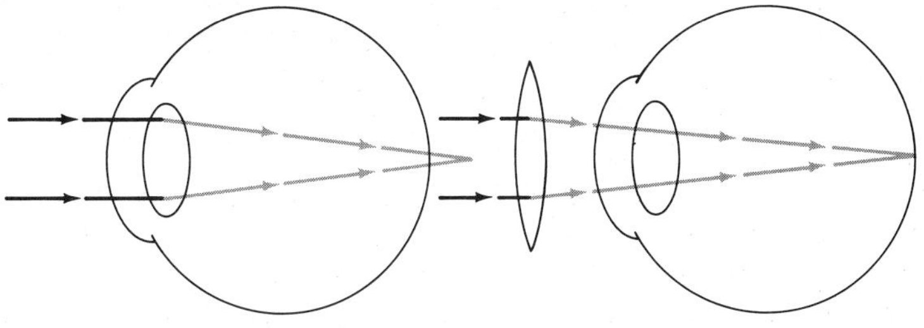

Farsightedness

Corrected vision

Figure 20-41. Farsightedness.

Learning Checks

1. List the components of a camera and their counterparts in the eye.
2. Why is focusing not a problem with a pinhole camera?

SUMMARY

Light is made of *electromagnetic waves.* These waves are produced whenever a charged particle accelerates. Visible light is only a small portion of the electromagnetic spectrum, with wavelengths ranging between about 7600 angstroms (red) and 3800 angstroms (blue).

Light may or may not appear to travel in straight lines, depending on the situation. *Diffraction* and *interference* effects demonstrate the wave nature of electromagnetic radiation.

When light strikes different surfaces, we can get either *specular* or *diffuse reflection. Refraction* of light takes place when it crosses from one material to another in which its speeds are different. The *index of refraction,* which is the ratio of light speed in a vacuum to that in a material, determines how much the light is bent when passing through the material.

Optical instruments such as telescopes and microscopes make use of the refracting properties of *lenses,* which cause beams of light either to converge or to diverge from a point. Lenses can be combined in different ways for various imaging purposes.

The *camera* and the *eye* are similar optical instruments whose function is to record an image, either on film or on the brain. Lenses are used to correct such eyesight deficiencies as *myopia* (nearsightedness) and *hyperopia* (farsightedness).

Exercises

1. What is a shadow?
2. What kind of lens is used in a simple magnifying glass?
3. Does the particle theory of light account for interference effects? Explain.
4. Explain how refraction produces a mirage.
5. Describe a simple way to determine the focal length of a lens.
6. Explain how an interference pattern can result from light incident on a single slit.
7. Among the colors of the rainbow are green, violet, red, blue, orange, and yellow. Arrange these colors in order of wavelength from short to long.
8. Sometimes we speak of light "rays" and sometimes light "waves." Explain.
9. When a fish looks up through the water surface at an object in the air, will the object appear to be its normal size and distance above the water? Use a diagram to explain.
10. Suppose a photographer takes a picture that is blurred because the camera was not focused properly. Explain.
11. Draw a ray diagram to show how looming can occur.
12. Stars seem to twinkle when seen from earth but not when seen by astronauts beyond the earth. Explain.
13. For a freeway commuter caught in traffic on a hot summer day, the cars ahead seem to shimmer. Explain.
14. An interior decorator places two plane mirrors on adjacent walls in a boy's room so that they come together at a corner. The boy places a baseball near the corner between the mirrors. Sketch the images that would be formed.
15. If you are under water and look up at the surface at an angle of 43 degrees, you can see an object on the horizon. What will you see by looking at an angle of, say, 56 degrees?
16. The theory of light that uses ray diagrams is called *geometric optics.* Explain why this is sometimes referred to as a "short-wavelength approximation."
17. A typical FM radio station broadcasts at a frequency of 100 megahertz (mega = 1,000,000) and a typical AM station is at 1000 kilohertz (kilo = 1000). What are the wavelengths of the signals from these two stations?

18. Approximately how wide a slit must be used to diffract red light ($\lambda = 7000$ Å)?

19. A tailor places his 6-foot-tall customer in front of an ordinary plane mirror for a fitting. What is the minimum length the mirror can be in order that the customer can see his entire image? Explain. (Hint: draw a ray diagram.)

20. In Exercise 19, does it matter how far away the customer stands? Explain.

21. Radio amateurs are allowed to communicate on the "10 meter band." What is the frequency of radio waves whose wavelength is 10 meters?

22. The speed of light in water is three-fourths that in air, and in window glass it is two-thirds that in air. What wavelengths will light (whose frequency is 5.0×10^{14} Hz) have in air, glass, and water?

23. You want to measure the speed of light in air by shining a flashlight on a mirror in the distance and viewing the reflected light. How far away must the mirror be in order for the light to take 1 second to reach it and return?

24. Suppose you want to photograph yourself in a mirror. For how many feet should you set the camera focus if you are 8 feet from the front of the mirror?

25. Alderite and cryptex are two hypothetical materials having different light speeds. When light travels between them, it makes an angle of 25 degrees with the normal in alderite and 17 degrees with the normal in cryptex. In which material is the speed of light greater? Which has the greater index of refraction?

26. Arrange these waves in order of increasing frequency: gamma radiation, violet light, radio waves, yellow light, infrared radiation.

27. Which has the larger critical angle, benzene ($n = 1.498$) or rock salt ($n = 1.54$), assuming that air is on the other side?

28. A lens is described as being "75 mm $f/3$." What is the focal length of the lens? What is the maximum diameter of the diaphragm opening?

29. A camera has an exposure time of 1/100 second for an f-stop of 2.8. For the same exposure at an f-stop of 4, how long must the shutter remain open?

Light appears to scatter as it radiates through the branches of a redwood tree in early morning fog.

21

LIGHT: SCATTERING, POLARIZATION, AND COLORS

The effects we have studied so far—diffraction, interference, reflection, and refraction—all take place when light interacts with material objects. Now we are going to consider this topic more closely. To do so, we need to understand a few things about matter at its most fundamental level. We will be helped in this endeavor by learning about the quantum theory in a later chapter, but it is adequate for now to take a less sophisticated approach.

INTERACTION OF LIGHT WITH MATTER

Size Effects

In our discussions of the diffraction of water waves (Chapter 17), we noticed that the effects depend to some extent on the size of the diffracting object compared to the wavelength of the water wave. A small stick poking through the surface of a pond hardly disturbs the wave motion on the water at all, but a large boulder casts a definite "shadow." (See Figure 21-1.) A slit or opening whose width is on the order of the wavelength diffracts a plane water wave into a circular one. On the other hand, with a large slit diffraction occurs only at the edges. The main portion of the wave passes through undisturbed.

Similar considerations apply to electromagnetic waves. What happens when light strikes an object depends on how the object's size compares with the wavelength of the light. Reflection and refraction occur when light encounters objects that are much larger than wavelengths in the visible part of the spectrum. In fact, the size differences are so great that we can ignore light's wave properties. This is why we can represent the wave front by a light ray indicating the direction the wave is

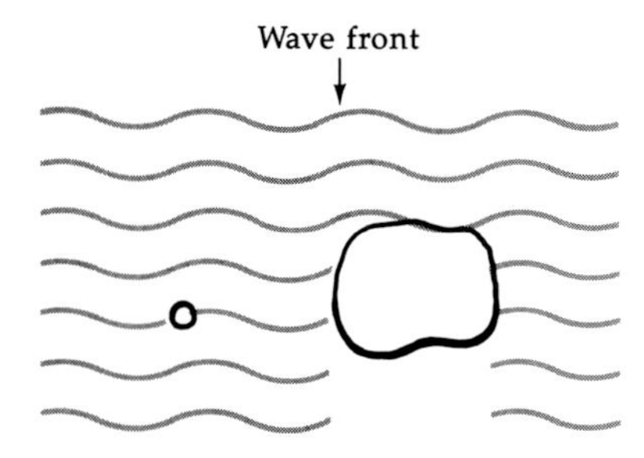

Figure 21-1. Object size is important in diffraction effects.

traveling. Diffraction of light, like diffraction of water waves and sound waves, is governed by relative sizes. Large objects show diffraction effects only at the edges, producing the definite but fuzzy-edged shadows we discussed earlier (Chapter 20).

Many practical applications can be traced to the relative sizes of objects interacting with electromagnetic waves. For example, whether you can see an object depends on how big it is in relation to visible wavelengths. If it is smaller than the wavelength of light, then—much like the stick poking through the water surface—it will not reflect or diffract the light. That is why we can't see single atoms. The distance across a typical atom is about 1 angstrom. Compare this with the shortest wavelengths of visible light, about 3800 angstroms in the blue region.

Reflecting surfaces will produce either specular or diffuse reflection, depending on the wavelength of the incident radiation. Figure 21-2 shows the large reflecting telescope at the observatory near Arecibo, Puerto Rico. The giant bowl-shaped dish is made of perforated metal. Compared with wavelengths of light in the visible region, the holes in the dish and the rough features of the metal are quite large. Diffuse reflection takes place; you can easily see the holes as you stand near the dish. However, the long-wavelength radio waves that reach the telescope from deep space do not "see" the holes or other rough features, so small by comparison. For this radiation, the dish serves as a highly polished mirror, producing specular reflection that focuses the radio waves to the long antenna suspended high above. The Arecibo telescope, then, is an example of an object that gives both diffuse and specular reflection. For long-wavelength radiation, the surface is smooth enough for the mirrorlike specular reflection. For shorter wavelengths, it is rough and the reflection is diffuse.

Figure 21-2. Arecibo Observatory.

The reception you get on your radio or television set is influenced by diffraction of electromagnetic waves by buildings, trees, mountains, and other obstacles. You may have noticed that the quality of TV and FM reception is affected much more by the direction of the antenna than is the reception on your AM radio. Radio waves in the AM range are much longer than FM or TV waves: several hundred or thousand meters for AM versus a few meters for FM and television. This means that the AM signal is diffracted around large objects quite readily and spreads out in all directions. The shorter FM and TV waves are more directional—that is, they travel in more nearly straight lines.

Learning Checks

1. How can a surface give rise to both *diffuse* and *specular* reflection?
2. Will blue light or red light be more effective for viewing small objects? Explain.

Scattering of Light

We have seen that reflection, refraction, and diffraction generally occur when radiation strikes relatively large objects. One explanation for this is that the individual atoms and molecules are very close together compared with the wavelength of the light. On the other hand, if the radiation encounters matter in which the atoms and molecules are widely spaced, then the wave properties of light come into play. *Scattering* is the general term used when light interacts with matter under these conditions.

Here, too, we have size effects. Suppose you shine a flashlight against a wall. You can see the spot of light on the wall, and you can see light at the glowing filament in the bulb. But if the room air is clean, you can't see the shaft of light between the wall and the flashlight. Only when there is some particulate matter, such as dust or smoke, in the path of the light are you able to see it. The smoke reflects or scatters some light sideways to your eyes, and the beam becomes visible. You can often see the sunlight streaming in through the windows if there are dust particles in the air to serve as scattering centers.

To understand how atoms and molecules scatter light waves, recall our discussion in Chapter 17 about *resonance*. You remember that an object such as a tuning fork begins to vibrate when it is "struck" by a sound wave at or near its own natural frequency. The better the match between the frequency of the incoming wave and the natural frequency of the tuning fork, the greater the amplitude of the vibration. The tuning fork will

Physics and Dentistry: The Scattering of Light Off Teeth

Researchers have found that a smooth tooth differs from a decayed tooth in the way it scatters light. When excited by a high-frequency blue beam, healthy teeth emit a yellow luminosity; teeth with cavities glow closer to the red range of the spectrum. Such changes can be seen either with the naked eye or with the aid of a light meter.

Thus, the technique shows some promise as a substitute for dental X rays, and thought is now being given to translating the idea into operational diagnostic equipment. The illustration shows a possible method, which conveys a sense of how smoothly facets of two diverse disciplines have been blended.

Source: "Light Spots Cavities," *Science Digest*, November 1982, p. 90.

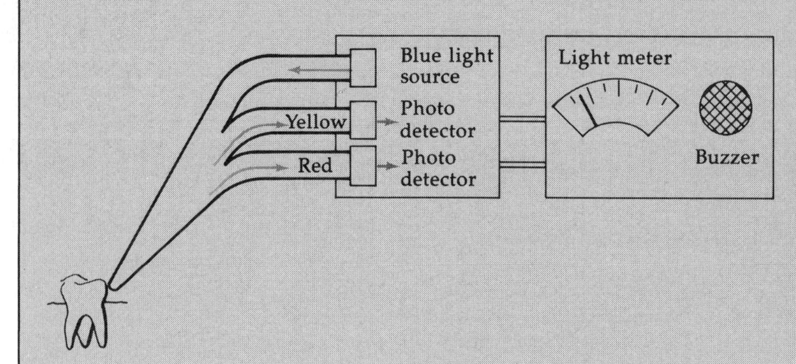

A blue light is projected onto a tooth through a stylus. The tooth will glow red if a cavity is present, and yellow if there is no decay. The reflected light is sent back to a photodetector, which triggers a buzzer in a light meter when decay is found.

produce its own sound wave as a result of the resonance. If hit by a sound wave whose frequency is quite different from its own, the fork will vibrate with a much smaller amplitude. The greater portion of the energy carried by the incoming wave in this case will simply be absorbed as heat.

When electromagnetic waves interact with matter, a similar thing takes place. Figure 21-3 shows what happens. The individual atoms and molecules contain charged particles—electrons and protons—that can act as tiny vibrators, waiting to be set in motion. An electromagnetic wave will do the trick. Remember that it is the electric and magnetic fields that make up the wave, and they interact with the fields of the charges. We can imagine that an electron is connected to the atom by a tiny spring whose stiffness gives the electron some particular frequency of vibration. If the frequency of the electromagnetic wave matches this natural frequency, we will have resonance. The electron will begin to vibrate at this frequency with a fairly large amplitude, in turn producing a wave that we will see as light if the frequency falls within the visible range. The term *scattering* arises because this light is sent out in all directions, not just the direction from which the incident wave came.

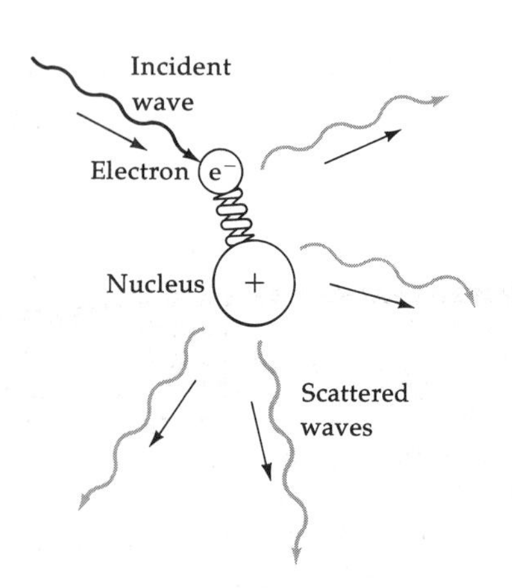

Figure 21-3. Crude model for light scattering by atoms.

Blue Skies and Red Sunsets

Nearly every day you see a beautiful product of light scattering. The blue color of the sky results from scattering of sunlight by the atmosphere. Let us try to understand this phenomenon.

Our atmosphere is made up mostly of nitrogen and oxygen molecules. The electrons of these molecules have resonances toward the blue-violet end of the spectrum. The oxygen and nitrogen nuclei are much heavier than the electrons and vibrate more slowly, much like the gentle swaying of a Cadillac on a rough road compared with the rapid bumping of a Volkswagen. These vibrations are matched in frequency by light in the red and infrared regions. So one might expect that the atmosphere would scatter both blue light and red light. However, compared with electrons, the nuclei vibrate with much smaller amplitudes and thus don't contribute as much to the scattering process. The net effect is that, for the majority of the visible wavelengths, the atmosphere is quite transparent. The air won't pay much attention to the longer wavelength colors—the reds, oranges, yellows—present in the white light coming from the sun. These waves of mismatched frequencies are absorbed as heat.

But the blue light is special. Its frequencies are matched to those of the vibrating electrons. Effectively, the air picks out the blue light, absorbs it in much the same way that a tuning fork absorbs sound of its own frequency, and re-radiates it. This emitted light is also blue, of course. It is scattered in all directions, giving our sky that beautiful blue color we enjoy so often.

Were there no atmosphere, the sky would look black rather than blue. Photographs taken by the Apollo astronauts on the moon show how black the sky looks without any air to scatter the light. Seen from space, the earth looks like a big blue marble, again because our atmosphere scatters light at the blue end of the spectrum.

Small amounts of visible light are scattered due to vibrations of electrons hit by light at frequencies that do not correspond to resonances. Here again, the blue dominates. It turns out that the amount of light scattered decreases in proportion to the fourth power of the wavelength. For example, red light at 7400 angstroms has twice the wavelength of blue light at 3700 angstroms, so the red is scattered only $(1/2)^4$, or 1/16 as much as the blue light.

In any event, the amount of light scattered by individual molecules is extremely small. Atmospheric scattering is only appreciable if the light travels through air many kilometers thick. Green light, for example, goes about 150 kilometers through the atmosphere before its intensity is cut in half.

We can see the effect of distance on scattering of sunlight by comparing how the sun appears at noon and at sundown. When the sun is directly overhead, the scattered blue light is subtracted from the sun's originally white light. What is left is the whitish yellow color we observe. When the sun begins to set in the late afternoon, its rays skim the surface of the earth,

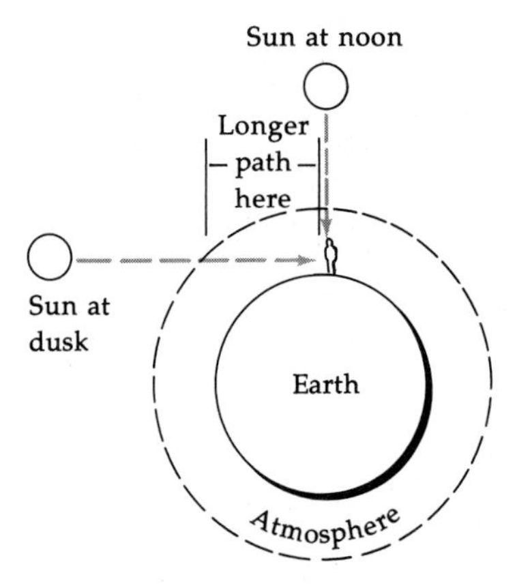

Figure 21-4. Light from the setting sun travels through a larger fraction of atmosphere than at noon.

traveling through a thickness of air many times greater than at noontime. (See Figure 21–4.) As a result, there is more scattering of the longer wavelength light, such as the green and yellow, leaving the sun with its beautiful red-orange sunset colors. Dust in the atmosphere provides additional scattering. The southwestern part of the United States is blessed with dazzling and spectacular sunsets for this reason.

There are other ways that scattering influences what we see. We learned in Chapter 20 that diffraction of light around the edges of objects causes shadows to be fuzzy. Scattering is also a culprit. It is scattered light, including light reflected from visible surfaces, that helps prevent a shadow from being perfectly black. On most days, enough scattered light comes through a window for you to see your surroundings even without artificial light or the sun's direct rays. During a solar eclipse, the earth becomes noticeably darker, but scattered light keeps away the total darkness you would have if the sun were switched off. A final example: Scattered sunlight drowns out the light of stars during the daytime.

Learning Checks

1. Where does the term *scattered light* come from?
2. How does the scattering of light depend on its wavelength?
3. Why are the resonances due to *nuclear* motions at longer wavelengths than those from *electron* oscillations?
4. Would the stars be visible to you if you were standing on the bright side of the moon? Explain.
5. Explain how the air is essentially transparent to most visible light.

POLARIZATION

When you look through Polaroid sunglasses at a pool of water, you can see the bottom better because the sunglasses reduce the glare from the water's surface. Wearing Polaroid sunglasses while driving in bright sunshine is safer for a similar reason—the glasses remove most of the glare of the highway. We say that light that passes through this special kind of glass is *polarized.*

Polarization is an effect that depends on the wave properties of light. Suppose you could pick out the waves in a beam of light. You would see the electric field vector oscillating perpendicular to the direction in which the light is moving. It might be up and down, left and right, or any of a variety of directions. The light from an incandescent bulb, for example, has its electric vectors oscillating in all directions because the vibrating electrons that generate the electromagnetic waves move about at random. Such light is said to be unpolarized. Any direction perpendicular to the line of travel is likely to contain one of the electric field vectors.

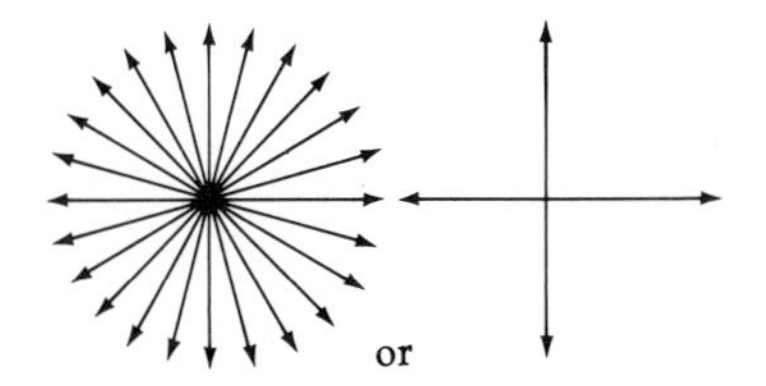

Figure 21-5. Equivalent representations of unpolarized light.

Figure 21–5 illustrates unpolarized light. The arrows in the lefthand part of the figure represent the electric vectors for light moving into the page. Since each vector can have one up-and-down component and one left-and-right component, we can simplify the picture. We need only two vectors—one up-down, the other left-right—to represent all the possible directions of the electric vector for unpolarized light. This is shown at the right in the figure.

Now, suppose our light beam runs into some object that blocks out one of these waves, say the left-right one. The up-down wave continues on its way. Only light waves whose electric vectors oscillate up and down are present in the beam. This light is now *polarized*, and we can represent it by a single vector.

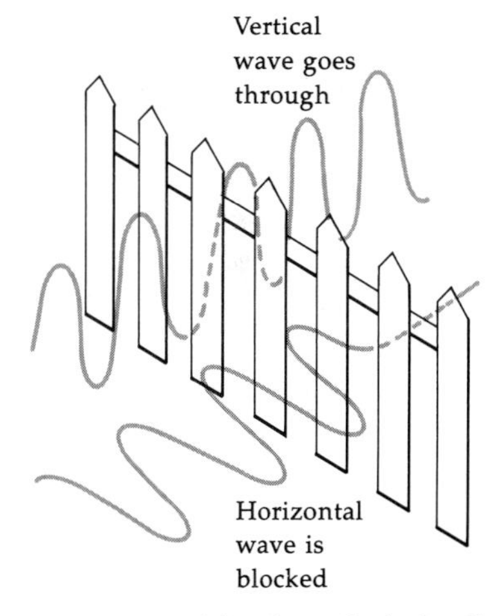

Figure 21-6. Picket fence "polarizes" waves on a rope.

Perhaps an analogy will help illustrate polarization of light. Figure 21–6 shows a rope passing through the gap between slats of a picket fence. A vertical wave, made by shaking one end of the rope up and down, passes cleanly through the fence. But a horizontal wave, which could result from shaking the rope from side to side, cannot pass through the fence. The pickets block it out.

Sometimes the term *plane polarized* is used for light that has been sent through such a blocking material. This is because the only electric vectors that survive are all confined to a plane, the up-down one in our analogy. The direction defined by this plane is the *polarization axis.*

Light passing through any material interacts with the atoms and molecules along the way. We can construct substances that will polarize light if we arrange the molecules in a particular way—that is, by aligning the "picket-fence gaps," so to speak.

One result of doing this is a polarizing filter, made of certain long-chain molecules called polyvinyl alcohol. Initially these molecules are randomly oriented. Unpolarized light passing through the material remains unpolarized. When sheets of this material are heated and stretched, the molecular chains line up parallel to one another and define a polarization axis. Only the electric field vector of the incident unpolarized light that oscillates along this axis can pass through. The emerging beam of light is polarized.

Figure 21-7. Crossed picket fences.

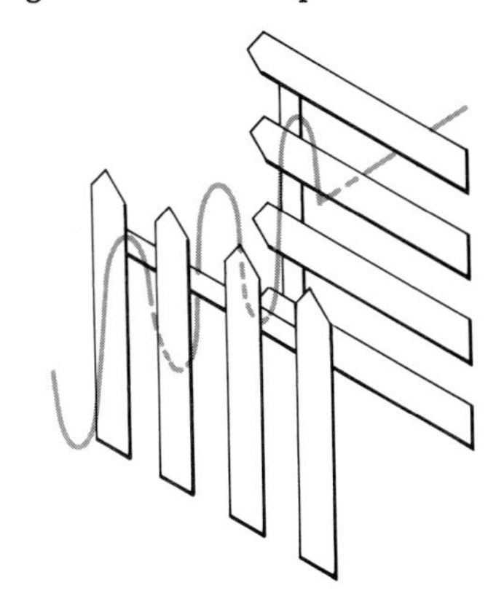

If a second sheet of the filter is placed behind the first with the two polarization axes at right angles, all the light is blocked. Suppose the first filter polarizes the light in the up-down direction. Then the axis of the second filter runs left and right, blocking the polarized ray coming from the first. We illustrate this with our picket fence analogy in Figure 21–7. You can demonstrate it for yourself by looking through the

The Guiding Light

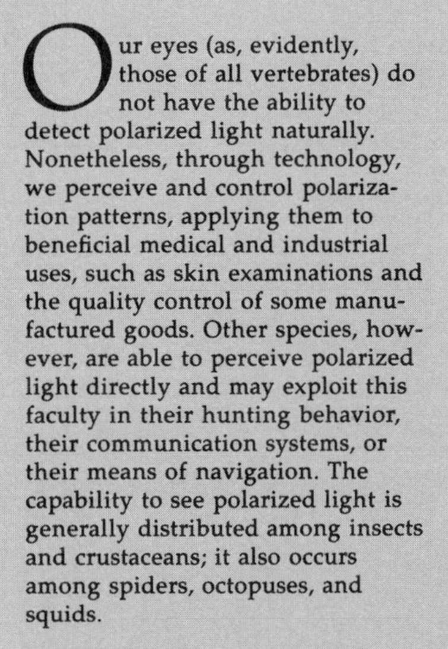

Our eyes (as, evidently, those of all vertebrates) do not have the ability to detect polarized light naturally. Nonetheless, through technology, we perceive and control polarization patterns, applying them to beneficial medical and industrial uses, such as skin examinations and the quality control of some manufactured goods. Other species, however, are able to perceive polarized light directly and may exploit this faculty in their hunting behavior, their communication systems, or their means of navigation. The capability to see polarized light is generally distributed among insects and crustaceans; it also occurs among spiders, octopuses, and squids.

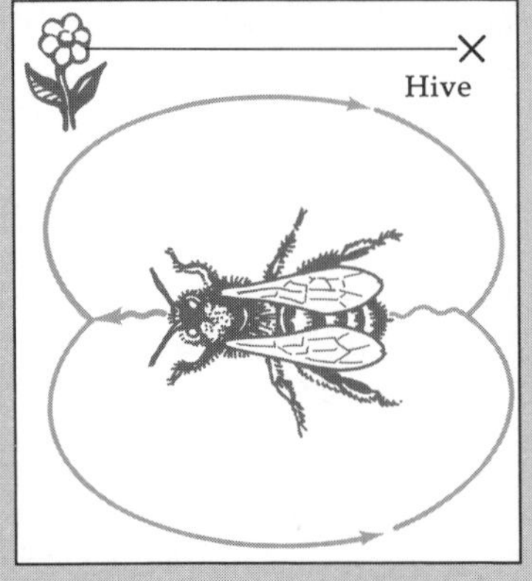

An insect use of polarization. The honeybee does a "waggle dance" to communicate the location of an outside food source to its hivemates. The plane of vibration of the polarized light of the sky that the bee was exposed to while foraging determines the direction of the dance.

Polarized Light and Sea Life
Oceanographers have determined that the major cause of submarine polarization is the scattering of light by microscopic particles in the water. Skylight enters the sea from above, and the average direction of the illumination is roughly vertical. The scattered light, moving horizontally toward an underwater observer, is partially polarized, with the electric vibration horizontal. One water flea swims in a direction perpendicular to the electric vibration direction. When placed in water free of particles (and hence with underwater illumination virtually unpolarized), this flea stops favoring movement in one direction. When polarization is restored (that is, suspended particles added), the flea resumes the perpendicular travel. The horseshoe crab detects the polarization of submarine light with ease and uses it for navigation.

Most sea organisms tend to swim perpendicular to the direction of vibration; a certain percentage swim parallel to it; and some, depending on the hour of the day, orient to it in yet other configurations. For all, though, polarization of light works as an underwater compass.

Polarized Light and Insects
Upon entering our atmosphere, a wave of sunlight will frequently collide with molecules of gas or tiny particles of dust. During such collisions, some planes of vibration are eliminated and the wave will finally vibrate in only one plane. The light has thus been polarized.

There are many insects that are guided by this light. By rotating polarized glass over the honeycombs of bees, the famous biologist Karl von Frisch demonstrated that honeybees, via their "dances" in the hive, use polarized skylight to convey the direction of a food source. When polarized light waves shift in the sky, or when they are blocked from view (by clouds, say), a number of insects will become incompetent fliers or will linger on a plant, going nowhere. Dragonflies, for instance, noted for their effortless flight and their hawklike hunting skills, will, in the absence of polarized light, stop traveling through unfamiliar territory and will change their hunting style from endless, active roaming to a kind of sedentary ambushing.

Thus, in the air and under the water, the phenomenon of polarization greatly influences the behavioral patterns of those species with eyes constructed to detect it.

Sources: W. A. Shurclift and S. S. Ballard, *Polarized Light.* Princeton, N.J.: D. van Nostrand Company, Inc., 1964.

Karl von Frisch, *The Dance Language and Orientation of Bees* (L. E. Chadwick, trans.). Cambridge, Mass.: Harvard University Press, 1967.

W. G. Wellington, "A Special Light to Steer By," *Natural History,* December 1974, pp. 47–52.

lenses of two pairs of Polaroid sunglasses and rotating one of them. Gradually the light becomes dimmer, until very little light passes through. At this point, the polarization axes of the two lenses are perpendicular. In fact, this is a good way to make sure that the glasses you buy are actually polarized. If either of the lenses is not made of polarizing material, the amount of light passing through will not change as you rotate them relative to each other.

Polarization provides conclusive evidence that light is made of *transverse waves,* like water waves. Our illustration of the wave pulses on a rope passing through the picket fence would not represent polarization if light were a longitudinal wave, like sound. Longitudinal waves cannot be polarized.

Polarized light is often present in nature. Much of the light reflected from water or other shiny surfaces is polarized along a horizontal axis. Figure 21-8 shows a representation of how this happens. When the unpolarized light strikes the water, the electric field vector perpendicular to the surface passes on into the water, while the horizontal vector glances off and rebounds at an angle governed by the law of reflection. This is like a rock that skips off a pond if you throw it with its flat side parallel to the water surface. The reflected light is thus polarized along a left-right axis. Polaroid sunglasses, on the other hand, have their polarization directed along a vertical axis, effectively blocking out this reflected light. They help reduce the glare from surfaces. When you are wearing polarized sunglasses, you can see this effect by rotating your head from side to side. Notice the increase in the glare as you make the polarization axis more nearly horizontal.

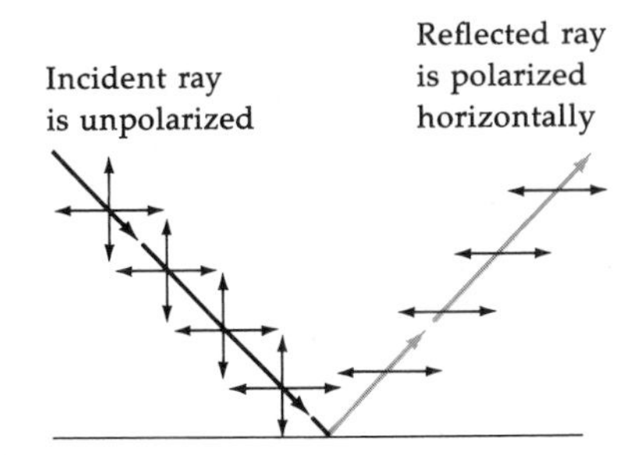

Figure 21-8. Polarization by reflection.

How much light is polarized by reflection depends on the refractive index of the material causing the reflection. We have seen that the light incident on water or glass or some other refracting material is partly reflected from the surface and partly refracted on through. It turns out that the reflected light will be *completely* polarized if the reflected ray is perpendicular to the refracted ray. This is illustrated in Figure 21-9, where the incoming ray is incident at just the correct angle to make the reflected and refracted rays at right angles to each other. This angle is called the *polarizing angle*, labeled Θ_p. With a little trigonometry, it can be shown that the tangent of this angle equals the ratio of the indices of refraction. That is,

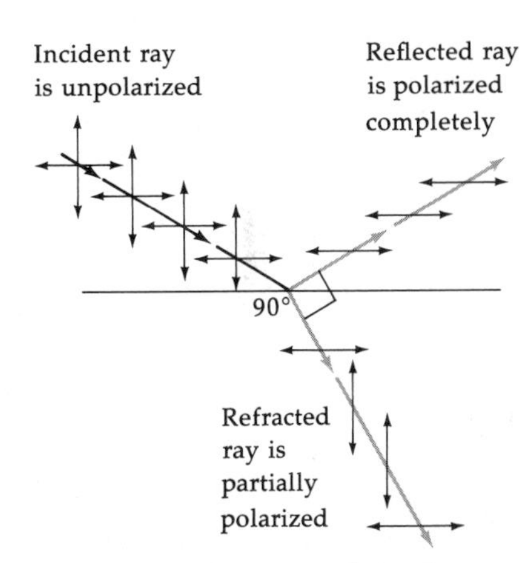

Figure 21-9. Complete polarization of reflected ray.

$$\tan \Theta_p = \frac{n_2}{n_1}$$

where n_2 is the index of refraction for the bottom material and n_1 that for the top. If the original beam passes through the air where $n_1 = 1$, then we have simply $\tan \Theta_p = n_2$. So, the polarization angle is fixed by the refractive index of the water. Light incident at any other angle will be only partially polarized. This is why only some of the light reflected to your eyes is cut out by the Polaroid sunglasses, since not all of the light will be incident at the polarizing angle.

Chemists use polarized light to study certain materials that rotate the polarization plane of light. The nineteenth-century French physicist Jean Baptiste Biot discovered that quartz crystals have this property. Various chemical compounds in solution will also twist polarized light. For instance, if you shine light from a polarizing filter on a solution of ordinary table sugar, the polarization axis of the emerging light will be rotated to the right, or clockwise. Grape sugar has the same effect and for this reason is often called *dextrose,* from the Latin *dexter,* meaning "right." In general, substances that rotate polarized light to the right are *dextrorotatory,* and those that rotate it to the left (counterclockwise) are *levorotatory,* from the Latin word for left, *laevus.*

You can determine the amount of rotation by placing a second polarizing filter between the sugar solution and your eye and rotating the filter until the light reaches maximum brightness. (See Figure 21-10.) The axes of the two filters will no longer be parallel, as they would be without the sugar solution in between. The angle between them tells how much the sugar has rotated the light.

Chemicals that rotate plane-polarized light are said to be *optically active.* In 1848, many years after Biot's discovery, the famous French chemist Louis Pasteur found that optical activity results from asymmetry in the structure of individual molecules. Organic molecules containing at least one asymmetric carbon atom—that is, a carbon atom attached to four *different* atoms or groups of atoms—can be optically active. These molecules can exist in two forms that are mirror images of each other. They are kin in the same way that your right hand is to your left. The two forms are chemically identical—they have the same chemical formula, undergo the same reactions, and so on—but they rotate polarized light in opposite directions because of the different spatial arrangement of atoms at an asymmetric carbon site.

Many biologically important chemicals are optically active. A good example is amino acids, the building blocks of proteins. Making these chemicals in the laboratory gives equal mixtures of both the righthanded and lefthanded forms. However, almost invariably they appear in nature with only one type of handedness. Quite often the body will accept only one form and simply excrete the mirror image. Or it may accept both forms but react differently to each. Two examples: (1) The levo-

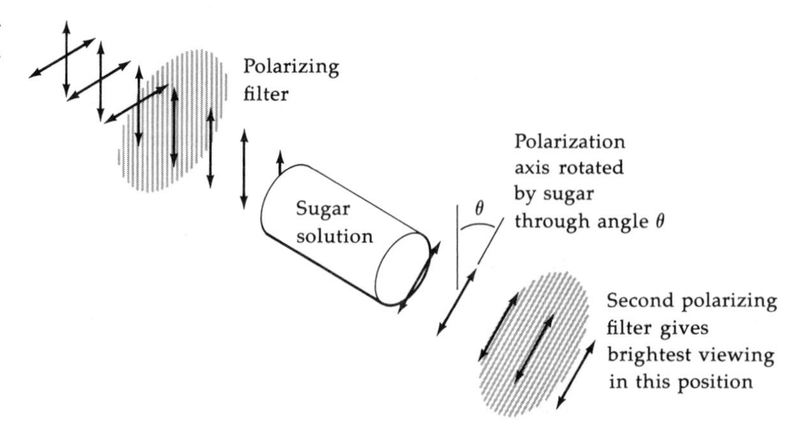

Figure 21-10. Optical activity of sugar.

rotatory form of nicotine is the one naturally occurring in tobacco; its dextrorotatory partner, which must be made artificially, is much less toxic. And (2) levoadrenaline is twelve times stronger than its mirror image in constricting blood vessels.

Learning Checks

1. What is *plane-polarized* light?
2. What is meant by the *polarization axis?*
3. Which direction is the polarization axis for sunglasses? Why?
4. What is the criterion for complete polarization from a reflecting surface?
5. What is meant by *optical activity?*

COLORS

We learned earlier in this chapter that a relatively small part of the electromagnetic spectrum is visible light, a band of colors ranging from red at long wavelengths to violet at shorter wavelengths. We can demonstrate this color spectrum by taking advantage of the refraction of light by glass and other transparent materials. Figure 21–11 shows a *prism,* a piece of glass with a triangular cross-section. As we have already learned, a light beam striking this prism will be refracted, taking a new direction as it travels through the glass at a slower speed. But the beam of light that emerges from the prism doesn't have the same appearance as the white light that went in. It is broken up into its component wavelengths, spread out to display the various colors of the visible spectrum.

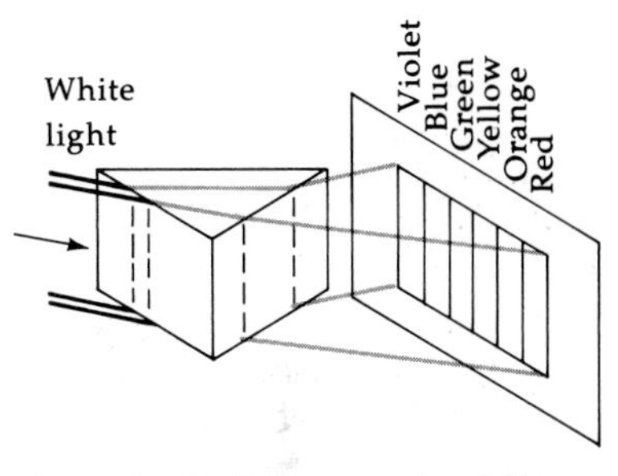

Figure 21-11. Prism spreads white light into rainbow of colors.

The Rainbow

You can see the same effect with a mirror and a pan of water. Lay the mirror face up on the bottom of the pan. Position the pan so that sunlight or the beam of a flashlight strikes the mirror. You will see the rainbow colors in the light reflected from the mirror. (See Figure 21–12.)

Figure 21-12. The water has the effect of a prism.

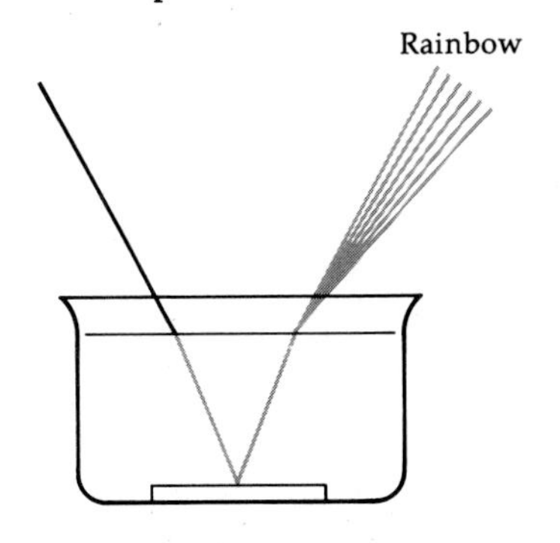

How do glass and water break up the white light into its different colors? We can understand what is happening from our knowledge of resonance.

Transparent materials like glass and water are made up of atoms and molecules whose electrons have their strongest resonances at frequencies corresponding to violet and blue light. As we have already seen, this is responsible for the blue color of the sky when the atmosphere scatters white sunlight. The longer wavelengths interact with these substances to a much smaller extent. So, as the white light moves through the glass prism or the water, the violet portion is strongly absorbed and

Watercolor of Kachina dancers, with a rainbow in the background, by José Bartolo Leute.

What is this prism
arcing the sky
when rain and sun have
duelled?

Rosemary Gilbert

re-emitted. This happens to the yellow light also, but to a lesser degree, and to the red light least of all. As a result, light of different wavelengths will travel at *different speeds* through these materials. The processes of absorption and emission take time, which is why the shaft of light slows down in the first place. The red light hardly pays any attention to the transparent material and moves on through, slowing down very little. At the other end of the spectrum, the violet light participates in the absorption-emission process many times over because of the strong match between its frequencies and those of the resonant electrons. As a result, the violet light takes longer to work its way through the glass than the yellow, which in turn takes longer than the red. The violet light's path thus changes the most.

The shape of the prism causes the light to bend twice. As it enters the glass it bends *toward* the normal and upon leaving, it bends *away*. The combination of the two nonparallel faces angles the light both times *away* from its original path, increasing the overall refraction effect. This spreads the white light into its component wavelengths—a spectrum of colors identical to the rainbow.

The rainbow that you see in the sky results from a combination of refraction and reflection of sunlight by droplets of water in the atmosphere. The spray of a lawn sprinkler produces a rainbow for the same reasons. In both cases the sun will be behind you and the water droplets in front. Let's see how this happens.

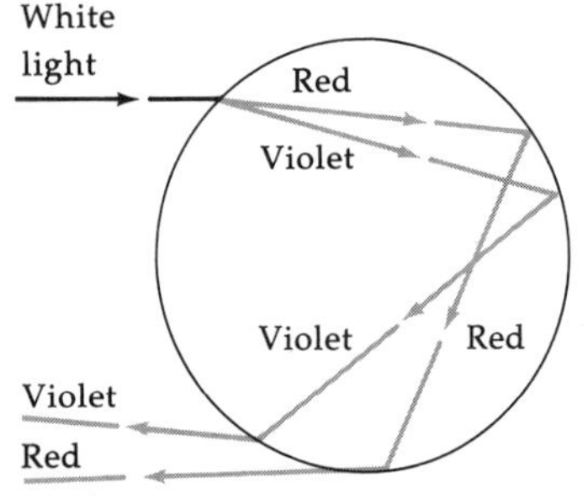

Figure 21-13. Reflection-refraction for a single drop of water.

Figure 21–13 shows a single drop of water with a ray of sunlight striking it from the left. Some of this light is reflected at the surface, and we don't really care about this part. The important part for us is the ray that is refracted as it travels on through the drop. Just as in the case of the prism, red light in this refracted ray is bent the least, violet light the most, on their way through the drop. The refracted ray strikes the back side of the drop and again is partly refracted and partly reflected. This time we want to follow the *reflected* ray. It travels back through the water droplet, strikes the edge again, and a *refracted* ray leaves the drop to travel toward your eye. As you can see from the diagram, the spreading of the white light by the refraction process means that light of only one color will reach your eye from any given drop. For example, if you see red light from a certain drop, you must look a little lower in the sky to see yellow, and lower yet to see blue. The tremendous number of water droplets within your field of view allow you to see all of the rainbow's colors.

The rainbow usually visible is arc-shaped because the part you see is a segment of a circle. Sometimes you can see a full-circle rainbow from an airplane. The shape is circular because the water droplets are spherical.

Learning Checks

1. Explain why blue light is refracted more by a prism than red light is.
2. Explain why the triangular shape of the prism helps spread colors out.
3. Do we see a complete rainbow from a single drop of water? Explain.
4. Which color of the rainbow is higher in the sky, blue or red? Explain.

Diffraction Gratings

It is possible to break up white light into its component colors in other ways besides passing it through transparent materials such as glass or water. One way is to take advantage of the diffraction of light as it passes through a very small hole or narrow opening. We have seen that when a pencil of light passes through a narrow slit, it is diffracted and spreads out into a broad band. Interference of light waves coming from different parts of the slit produces dark and light regions on a screen. This separation is more pronounced when the light passes through a *diffraction grating*. Instead of having one or two slits, a grating has many slits extremely close together.

One way of making a grating is to etch parallel lines on a sheet of glass, thereby creating alternate transparent and opaque regions. A typical laboratory grating may have 2000 lines per centimeter, giving each slit a width of 0.0005 centimeter.

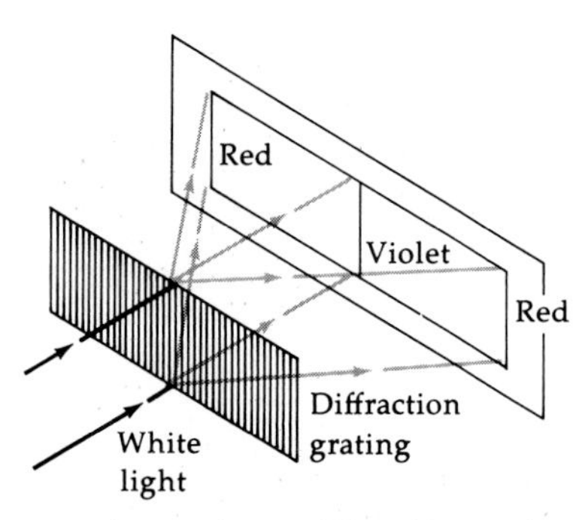

Figure 21-14. Diffraction grating spreads out the light.

A diffraction grating makes white light into a rainbow because different wavelengths are diffracted by different amounts. Recall that for a given slit width, the amount of diffraction decreases as the wavelength decreases. For example, very short-wavelength water waves are more likely to pass undisturbed through a ripple-tank barrier than are those of longer wavelengths. The same is true of light. The short-wavelength blue and violet colors can pass through relatively "unaware" of the grating, while the longer wavelengths toward the red end are diffracted more. Because of this difference in the extent of diffraction, the white light spreads out in a display of its component colors. (See Figure 21–14.)

Notice that the rainbow from a grating has violet in the center and red at both ends. This is different from a refraction-reflection rainbow such as we see in the sky, where the red light is refracted the *least*.

Preferential Absorption

The way light interacts with matter accounts for the colors of the various objects we see every day. You have learned that transparent materials such as water and glass have strong resonance in the ultraviolet and infrared but hardly interact at all with the intermediate visible wavelengths. Most molecules are like this.

However, there is a large class of molecules that *do* have electrons whose resonance frequencies lie in the visible region. They absorb and re-emit the reds, greens, oranges, and other colors that we see about us. These molecules are called the *dyestuffs*. They are long-chain organic molecules in which the electrons are more or less free to move. The dyestuffs give your clothes their bright colors. Their electrons are not so tightly bound as the electrons in, say, water molecules. As a result, they vibrate at lower frequencies than the electrons in transparent materials, much as a loose guitar string vibrates with a lower frequency compared to a tightly tuned string. It is as if the electrons in the dyestuffs were attached with floppier springs, not stiff ones like those in oxygen or nitrogen or water.

Consider a red dress, for example. It contains a dyestuff whose electrons have natural frequencies corresponding to red light. When white light from the sun or an electric bulb strikes the dress, these electrons pick out the red light, vibrate in resonance with it, and re-emit, or scatter, it, giving the dress its red color. The net effect is that everything is absorbed and held *except* red. The other colors—blue, green, yellow, and so on—are also absorbed but they produce no resonance, much as in the case of the mismatched tuning forks. This absorbed light is transformed into heat.

The process just described is sometimes called *preferential absorption:* the red dress appears red because it preferentially absorbs and re-emits light of that color. A black object has resonances for *all* the visible wavelengths, so it absorbs all the colors. None are re-emitted, and essentially all of the absorbed radiation is transformed into heat. White objects, on the other hand, absorb almost none of the incident light. This accounts for the differences in temperature of white and black objects left out in the sun. The black will be much hotter, because it absorbs essentially all the incident radiation and re-emits very little, retaining it as heat.

Now suppose that an "orange" shirt—that is, one that appears orange in white light—is illuminated with blue light. The shirt no longer appears orange, but black, because there are no wavelengths of light available for which it has strong resonances. The incident light is all absorbed, leaving the shirt with an appearance that we see as black. The world would look very different if the sun gave us some color other than the one we see.

Microwave cooking is based on preferential absorption. The central element of a common microwave oven puts out radiation whose frequency matches the frequency with which water molecules can rotate, or tumble end over end. That is, the radiation and the water in the food are in resonance. This means that the water molecules readily absorb the microwaves and begin rotating. As they rotate, they re-emit microwaves, which are scattered to other parts of the food, involving still more water molecules. The tumbling water molecules bump and rub against their neighbors. Heat generated by the resulting friction cooks the food. You may have noticed that the air inside a microwave oven does not get hot as the food cooks. This is because the air molecules are relatively transparent to the microwaves, which are deliberately confined to a narrow range of frequencies so as to be in resonance only with water and other food molecules having structural similarities with water.

Learning Checks

1. Account for the fact that blue light is *refracted* more than red, but red is *diffracted* more than blue.
2. Explain preferential absorption.
3. Explain the visibility of the dyestuffs.

Color Addition and Subtraction

We showed in Figure 21-11 how a prism disperses white light into a spectrum of rainbow colors. Now suppose we pass this rainbow of colors through a second prism that is upside down with respect to the first. As shown in Figure 21-15, we get white light back again. That is, when the various wavelengths

Figure 21-15. The second prism recombines the colors, forming white light.

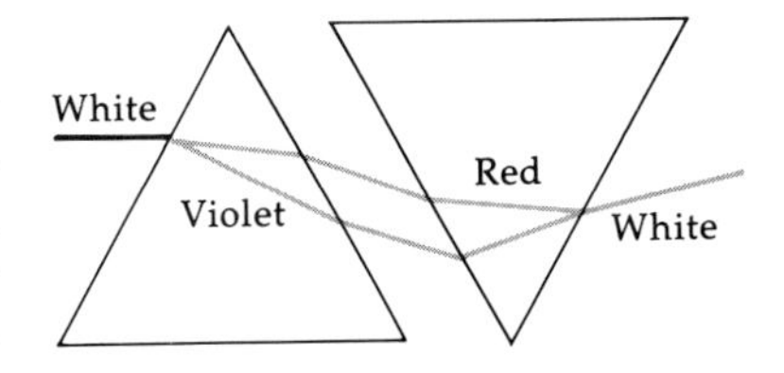

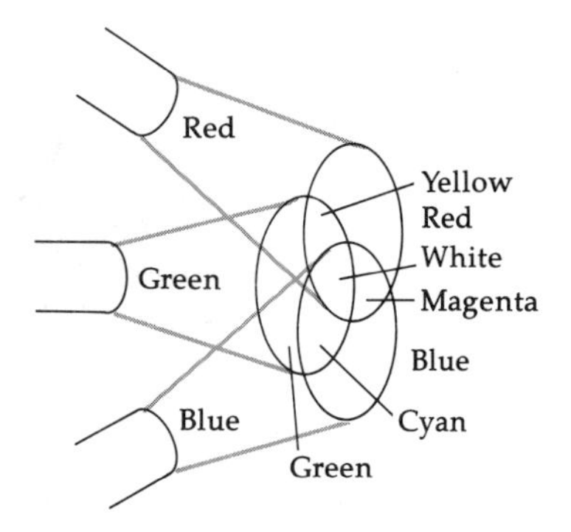

Figure 21-16. Mixing of lights.

of light making up the rainbow colors are recombined, they add together to again make white light.

In fact, we can get white light by adding lights from separate sources as long as we mix the colors in their rainbow proportions. It turns out that *any* color can be obtained by various mixtures of *red, green,* and *blue,* the colors of the long-, intermediate-, and short-wavelength regions of the spectrum. These are called the *additive primaries,* because the lights of various wavelengths are added together to give some final color. You can convince yourself of this by doing what Figure 21-16 suggests. Place red, green, and blue filters over the lenses of three flashlights, one filter on each. Shine the three lights on a white wall. White light will appear in the center, where all three primary colors overlap. Here the wavelengths for the various portions of the visible spectrum are added together before you see them. Additional colors are seen where only two of the beams mix together.

> *How important it is to know how to mix on the palette those colors which have no name and yet are the real foundation of everything.*
>
> Vincent Van Gogh

Now, suppose we mix together red, green, and blue paints. What happens in this case? We certainly don't get white paint. Instead, we get a dirty blackish-brown color. In this case, the final color results from the preferential absorption of each color in the mixture, so what you see results from the *subtraction* of colors rather than their addition, as in the case of colored lights. The red paint, remember, absorbs all the wavelengths except the red and a little bit of orange. The green paint absorbs the red, orange, and violet. It emits mostly green along with a little yellow and blue, the colors on either side of green in the spectrum. A similar absorption-emission occurs for the blue. So the final color from any mixture is the one left over after all the absorption has taken place.

Figure 21-17 shows color subtraction for white light passing successively through blue and yellow filters. The blue filter subtracts the long-wavelength light and allows only the violet, blue, and green to pass, since these are the colors corresponding to the resonance frequencies of its electrons. Of these three colors, only the green can pass through the yellow filter, for it transmits only green, yellow, and orange. The resulting light is green, the only one to survive the trip through both filters. Every small child knows that a blue crayon and a yellow

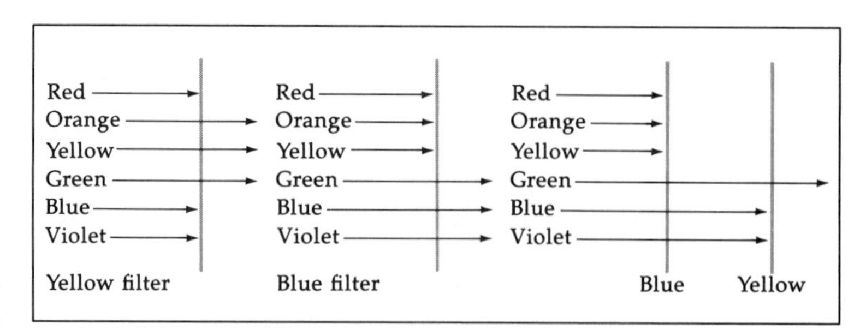

Figure 21-17. Color subtraction by filtering.

Physics and Art

"So strongly do these color phenomena appeal to me," the Nobel prizewinning physicist Albert A. Michelson declared in 1903, "that I venture to predict that in the not very distant future there may be a color art analogous to the art of sound—a color music, in which the performer can play the colors of the spectrum in any succession or combination . . . It seems to me that we have here at least as great a possibility of rendering all the fancies, moods, and emotions of the human mind as in the older art."

Beginning with the fantastic light and sound concerts of the 1960s, the art of mobile color has become all that Michelson could have hoped for, and more. Electronic instruments have been modified to turn sound into color, and some musicians have started to become artists whose media are the shifting "chords" of color sequences. These visual patterns may be controlled or free flowing; they can evoke a range of emotions and moods. They have been presented as entertainment and have been used as psychotherapeutic aids. Physicists and artists continually strive to create ever more advanced hardware to push this art to its limits.

Sources: T. D. James, *The Art of Light and Color.* New York: Van Nostrand Reinhold Company, 1972.

Albert A. Michelson, *Light Waves and Their Uses.* University of California Press, 1902.

Seated before an elaborate console, this mobile color artist can manipulate projectors, lamps, screens, and masks to achieve his color "chords."

Cadmium yellow lt.
Cadmium yellow dp.
Cadmium red lt.
Yellow ochre
Raw sienna
Light red
Burnt sienna
Burnt umber
White
Alizarin crimson
French ultramarine
Cobalt blue
Viridian
Ivory black

The typical palette of an advanced oil painter. Warm colors are placed along the top, cool colors along the side. Mixing color pigments is a subtraction process.

crayon mixed together on a white sheet of paper will give green, and the reason is exactly the same.

In summary, then, we see that mixing *lights* of various colors corresponds to *adding* together parts of the visible spectrum, whereas mixing *pigments* of different colors is a subtraction process.

Color subtraction accounts for the rainbow that can be seen in soap films or in a patch of gasoline on a wet pavement. Here the subtraction takes place not because of absorption, as in the case of paints or dyes, but because light waves from the different surfaces of the film interfere with each other.

Look at Figure 21–18 for an illustration of what happens in the case of gasoline on water. When white light illuminates the gasoline, part is reflected from the top surface, and part passes on through and reflects from the water surface. At some angle of view, the difference in path lengths for these parts will equal some odd number of half-wavelengths for a particular color. That is, it might be one-half, or three-halves, or five-halves the wavelength of yellow light. So the waves reflected from the two surfaces will cancel because of destructive interference. The yellow will be subtracted from the white light, and you will see blue.

Figure 21-18. Rainbow from an oil slick on a wet surface.

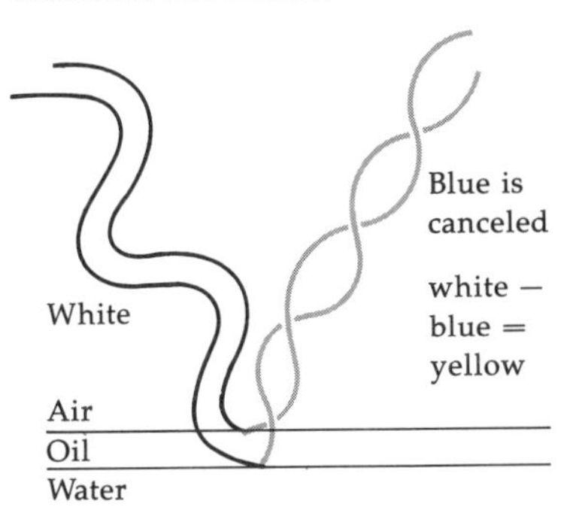

For another part of the surface, seen from a different angle, the paths traveled by the two waves will be such that some other color is subtracted from the white light. Over a wide enough field of view, this same subtraction process will allow you to see all the colors of the rainbow.

Learning Checks

1. Distinguish between mixing colored *lights* and mixing colored *paints*.
2. Look at Figure 21–17 and describe the result if the light passed first through the yellow filter, then through the blue.
3. What happens to the blue light that shines on an orange shirt?

SUMMARY

Light scattering takes place when light interacts with matter in such a way that the separation between individual atoms and molecules is large compared to the wavelength of the light. If the frequency of the light matches the natural frequencies of the electrons in the matter, the resulting resonance generates a new light wave of the same frequency. This wave moves out in all directions. Our sky looks blue because of such scattered blue light from atmospheric molecules.

Polarized light is light whose electric field is confined to one direction. It is useful in the study of certain chemicals that are *optically active.* They can rotate the polarization axis of light reflected from a surface such as water. The light will be completely polarized if the reflected and refracted rays are perpendicular to each other.

White light is produced by mixing together light of all wavelengths in the visible spectrum. Any color can be made by some combination of red, green, and blue lights—the *additive primaries.* Colors of objects are caused by *preferential absorption,* the absorption and re-emission of visible light by dyestuffs containing electrons whose resonance frequencies correspond to colors in the visible spectrum.

Exercises

1. How would you expect the color of the sky to appear at sea level compared with very high altitudes? Explain.
2. If light is scattered by the atmosphere, how is it possible to see mountains many kilometers away?
3. Explain the observation that snow-capped mountains look bluish from long distances.
4. The pages of this book are "white." What do we mean when we say this?
5. Can sound waves be polarized? Explain.
6. Two Polaroid sheets with their axes at 90 degrees block out almost all the light incident on them. What would be the effect of a third Polaroid sheet inserted between them with its axis oriented at 45 degrees?

7. Explain how light reflected from a metallic surface is polarized.

8. Explain why the wave coming from a radio antenna is polarized.

9. Suppose you are wearing a red shirt. Which does it absorb more effectively: long-wavelength or short-wavelength light?

10. Explain how scattering gives the deep clear ocean its blue color.

11. A red rose with green leaves is illuminated with red light. Describe what you see.

12. In Exercise 11, which become warmer: the red petals or the green leaves?

13. Suppose the sun emitted green light rather than white. Which color garment should you wear on a hot day? What about on a cold day?

14. What color will a white sheet appear to be in red light? In blue light? Explain.

15. Suppose you look at the reflection of a white lamp from a very thick piece of glass. You will see a white image and a colored one. Explain.

16. A single sheet of polarizing material transmits (ideally) 50 percent of incident unpolarized light. What percentage will two such sheets transmit if their polarization axes are parallel? Perpendicular?

17. Passing light through a flat piece of window glass does *not* break it down into its spectral colors. Explain.

18. Clothing stores often use colored lighting to achieve various effects in marketing their products. Suppose a store owner asks you to explain the physical difference between red light and blue light. How should you respond?

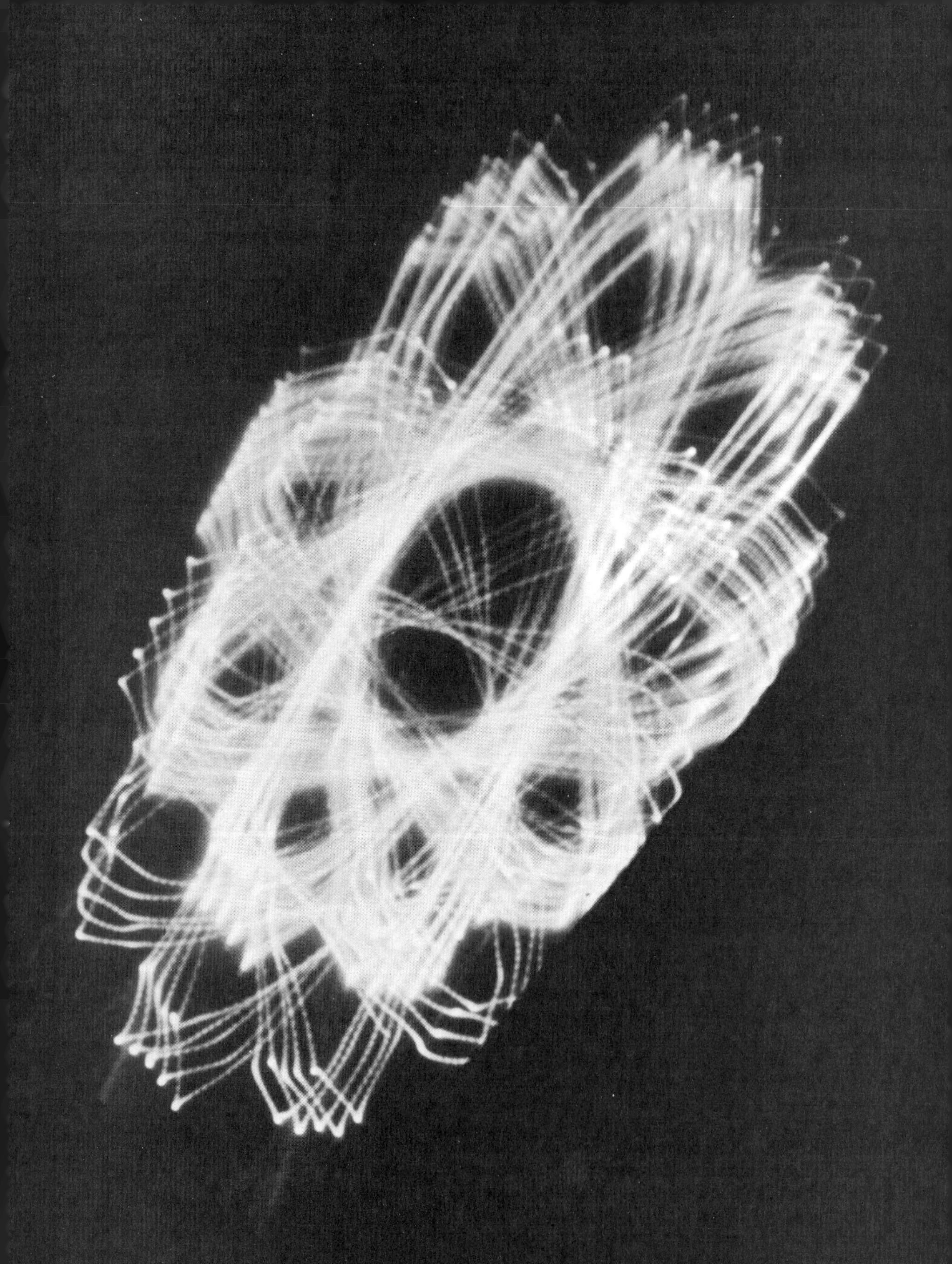

Using lasers to alter the motions of electrons, physicists continue to probe the inner structures of matter.

22

THE BEGINNINGS OF MODERN PHYSICS

By the end of the nineteenth century the physicist's picture of the universe was thought to be essentially complete. Matter was composed of particles that obeyed Newton's laws of motion. As we have seen, the conservation principles taken together with Newton's laws were quite successful in describing the motion of everything from planets and galaxies to falling rocks and rolling wheels. Heat energy and its effects on matter were well understood in terms of thermodynamics. Light was characterized as electromagnetic waves produced by accelerating electric charges. It traveled through space at a speed of 300 million meters per second, in agreement with Maxwell's equations.

So it all looked quite tidy. In fact, physics was so thorough in describing the universe that around 1880, so the story goes, the head of Harvard's physics department, John Trowbridge, was gently advising some students not to choose physics as a career because little of importance was left to be done.

As things turned out, however, the situation was not all that rosy. New experimental techniques enabled scientists to study matter and radiation, and their interaction, on a smaller and smaller scale. J. J. Thomson discovered the electron in 1897, and one of his students, Ernest Rutherford, developed a successful model of the atom in about 1910. When physicists tried to interpret experiments involving these objects—that is, bodies at a size level smaller than they had ever dealt with before—they ran into a blank wall. Evidence gradually accumulated that led to an unnerving conclusion: the laws of "classical physics," which are the rules we have been studying, simply do not apply when things are so very small, down at the level of the atom and the electron. There, it is a different game, played by different rules. And, naturally, scientists began to try to learn the new rules. All the new rules are not known yet, and the quest for them goes on.

> *If we ask, for instance, whether the position of the electron remains the same, we must say "no"; if we ask whether the electron's position changes with time, we must say "no"; if we ask whether the electron is at rest, we must say "no"; if we ask whether it is in motion, we must say "no."*
>
> *Robert Oppenheimer*
> Science and the Common Understanding

Two great revolutions in physics have occurred in this century, revolutions that not only have changed the physicist's picture of the world but have altered science itself, from a philosophical standpoint. Einstein's theory of relativity, which we have already studied, is one. We now turn our attention to the other, called *quantum mechanics* or the *quantum theory.* Together relativity and quantum mechanics form what is often called "modern physics." They are "modern" not only by being fairly recent but also in the sense of being really different, a radical departure from pre-twentieth-century ideas. Relativity, we have seen, changed traditional notions about space and time. Quantum mechanics, we shall learn, presents new and startling ideas about how the universe is made and operates.

In this chapter we are going to look at some of those experiments that were so troubling because they violated the classical laws. Then we will be ready in the next chapter to look at the new quantum theory to which they gave birth.

BLACKBODY RADIATION

One of the genuine pleasures of life is to sit in front of a blazing fire on a cold winter's night. Part of our enjoyment comes from the many colors in the flames and glowing coals—a mixture of reds and yellows, perhaps some blue, an intense yellow or white as the fire gets very hot.

We can see similar colors with an electric heater. When it is turned on, the heating element doesn't change its appearance initially, but as the temperature continues to rise we begin to notice some color. We see a dull red at first, then a brighter red, and finally some yellow mixed in.

As we have already learned, these colors are light waves, electromagnetic radiation emitted by accelerating charged particles in the glowing objects, jiggling about in response to the energy they receive. Why do we see these colors? And what makes the colors change as the temperature increases? Classical physics is not equipped to answer questions about such common, everyday experiences. The search for answers led to the new quantum theory.

All heated material bodies put out radiation of different wavelengths. Two things happen as the temperature rises: the *amount* of emitted energy increases, and the *frequency* of the waves also increases. In 1879 Josef Stefan, an Austrian physicist, showed that the amount of energy emitted as radiation is proportional to the fourth power of the absolute temperature. That is,

$$E = kT^4$$

where k is a proportionality constant. For example, if an object initially at room temperature, 300 K, is heated to 600 K, it radiates sixteen times as much energy at the higher temperature. The temperature has increased by a factor of 2, and $2^4 = 16$.

What about the color, or frequency, of the radiation? For an ordinary steam radiator at about 373 K, the frequency of the emitted "light" is in the infrared, and you can't see it. A radiator is invisible in a darkened room. But your skin feels the infrared waves as heat, which of course is the reason for the radiator in the first place. As an object gets hotter, it puts out energy of different colors, so to speak. At around 1000 K the radiation is visible as a dull red, like that coming from the eye of an electric range. More of the higher-frequency yellow light becomes mixed in as the temperature rises. Near 2000 K the brightly glowing object is still emitting infrared radiation, while the dominant color is reddish yellow. The filament of a light bulb at around 2300 K puts out white light, a mixture of most of the frequencies in the visible spectrum. You may hear something described as "white hot"—not a bad term, for it refers to an object hot enough to emit white light. Higher temperatures correspond to even higher frequencies.

For any given temperature, a hot object emits light of several frequencies, not just the *peak frequency,* which is the dominant one. During the years around the turn of the century, physicists tried to get an accurate picture of how the radiation is distributed among the various frequencies by studying the emission from a so-called *blackbody.*

Remember from our study of light that a dark object absorbs most of the light that strikes it, reflecting very little. A *blackbody* is the idealized extreme, absorbing *all* the radiation that strikes it and reflecting none. But even though it doesn't *reflect* the radiation, it does *emit* some of what it absorbs. We feel this as heat. Dark seat covers in a car that has been sitting in the sun for a long time feel hot because they are emitting some of the radiation they have absorbed. A black object still looks black because it is too cool to emit visible radiation. If its temperature increases enough, it will glow—the coil of an electric range and the glowing charcoal in a hot barbecue pit are good examples.

The question the physicists wanted to answer, then, was this: When a blackbody gives off radiation energy, how is this energy distributed among the various frequencies in the electromagnetic spectrum? To answer this question, they had to come up with a blackbody on which to do their experiments.

Although in practice there is no such thing as a perfectly black body, the German physicist Wilhelm Wien in 1895 concocted a way of achieving something quite close. Figure 22–1 shows a small hole poked in the wall of a cavity, like a keyhole into a large room. If we shine light into the cavity, it bounces around, being absorbed and diffusely reflected by the rough

Figure 22-1. The hole in the cavity serves as a "blackbody," absorbing light but reflecting little.

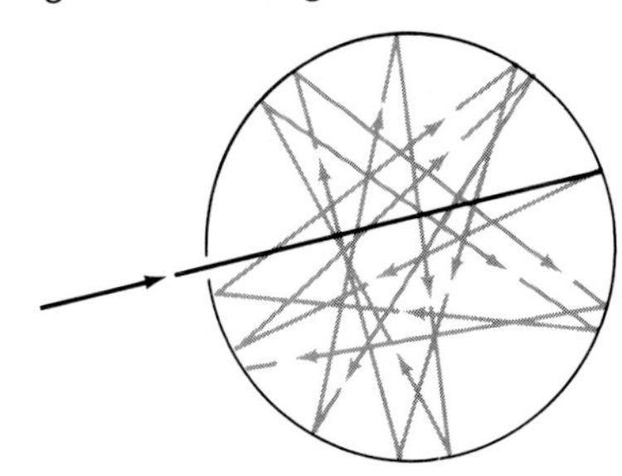

Blackbody Radiation and Cosmology

A revolutionary discovery was made in astronomy in 1965: A cosmic background radiation (CBR) was found to permeate the entire universe. When first measured, the CBR seemed to fit the spectrum of a blackbody emitting radiation at about 3 degrees Kelvin. This discovery sparked excitement because this was just the kind of radiation that had been predicted to be left over from the "big bang." Astronomers thus pointed to the finding as clear evidence for a primordial explosion. They cited the CBR as the last "echo" of this event.

This radiation, with its blackbody character, was believed to be a sort of "universal fundament" that moved smoothly and evenly, expanding along with the universe. So for the health of the big bang theory (which postulates an even distribution of matter and energy in all directions from the moment of the initial explosion), it is necessary that the blackbody be uniform—that no irregularities appear in it; and, indeed, early measurements detected none. However, new measurements made with more sophisticated implements were carried out in the late seventies. Signals from a certain sector of the universe indicated deviations in the blackbody radiation, thus altering the picture of the universe derived from the big bang assumption. What these new data mean is being debated, with a variety of explanations being advanced for it, and alternate models of the universe being constructed from it.

Plans are afoot to launch a Cosmic Background Explorer, a series of space probes to take the most refined blackbody measurements yet. The flights already have congressional approval. The future will tell what this mission yields. Meanwhile, cosmologists have time to speculate on what physical processes might have happened to "degrade" the CBR.

Sources: "Cosmic Blackbody: Confirmation and Questions," *Science News*, April 21, 1979, p. 260.

"New Evidence Supports Superclusters," *Industrial Research and Development*, February 1980, p. 57.

walls. Eventually almost all the light is absorbed. Very little survives to be reflected back out the small hole. So the tiny opening behaves like a blackbody: it absorbs light without reflecting any. For a given cavity temperature, the light emitted from the hole will have the frequency distribution of black body radiation appropriate for that temperature.

Figure 22–2 illustrates what Wien found in his studies. Here we have plotted the intensity (or brightness) of the blackbody emission in terms of its frequency. Each curve represents a different temperature for the cavity. For each temperature the most intense radiation is in the neighborhood of the peak frequency, and the curve tails off on either side. An object at 1000 K, for example, gives off most of its radiation in the infrared and red, and very little at low (microwave) or high (ultraviolet) frequencies. So this object looks red.

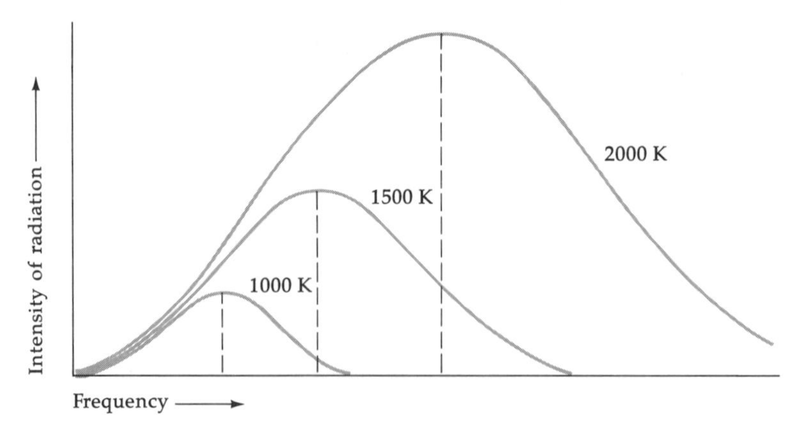

Figure 22-2. How the brightness of the emission from a blackbody varies with frequency for different temperatures.

For higher temperatures, the peak frequency shifts to higher frequencies. We see this as a color change in our electric heater; for example, the heating element turns from red to reddish yellow as it gets hotter. Wien found a very simple connection: The peak frequency is directly proportional to the absolute temperature:

$$f_{\text{peak}} = wT$$

where w is a proportionality constant.

With Wien's frequency distribution results in hand, physicists faced the problem of interpreting them. In England, Sir James Jeans and Lord Rayleigh (John William Strutt) noticed that Wien's curves resembled those describing the distribution of velocities for a gas at various temperatures. Recall that the temperature of a gas is proportional to the average kinetic energy of its molecules (Chapter 13). At equilibrium, the temperature is uniform throughout the gas, so on the average each atom has the same energy. This is called the *equipartition theorem*—on the average, the total energy of a system is divided equally among all the particles.

For any specific particle, however, there is some probability that its energy will be either greater or smaller than the average. Because the kinetic energy depends on the velocity, these deviations from the average energy give a distribution of velocities. Figure 22-3 shows this for various temperatures. Most particles will have the average velocity for any particular temperature, but some will be moving faster and some slower.

Look at Figures 22-2 and 22-3. They differ in details, but notice this similarity: in both figures, the peaks of the curves are farther to the right for higher temperatures. For the gas molecules, higher temperatures mean greater average velocities, and for the blackbody emission, higher temperatures mean higher peak frequencies. This similarity prompted Rayleigh and Jeans to apply the equipartition theorem to the thermal radiation problem. That is, just as the total energy of the gas is

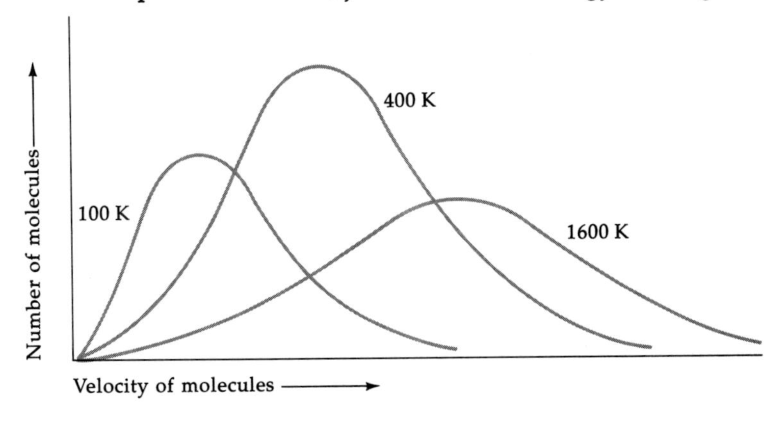

Figure 22-3. How the velocity of gas molecules is distributed at different temperatures.

Figure 22-4. Failure of Rayleigh-Jeans prediction.

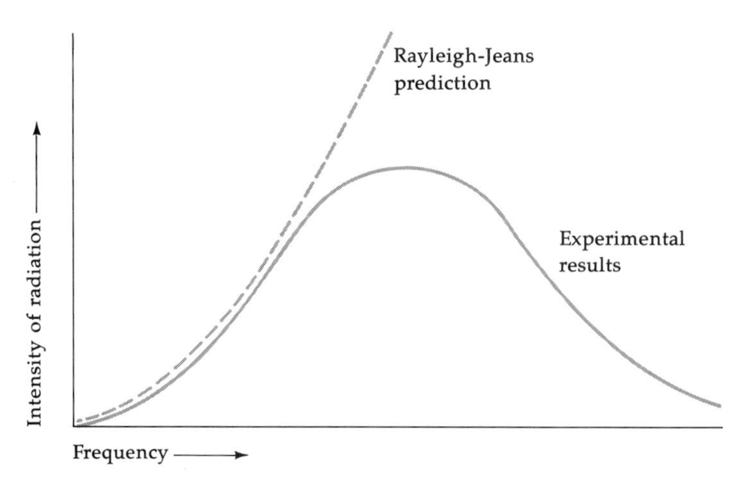

equally distributed among the molecules, they assumed that the total radiation energy of the blackbody is equally distributed among the possible frequencies and that all frequencies have the same probability of emitting radiation.

Figure 22–4 shows the results of this assumption, superimposed on the experimental curve. As you can see, it follows the experiment closely at low frequencies. But it fails miserably at ultraviolet and higher frequencies. Instead of turning over and tailing off, the Rayleigh-Jeans curve keeps on rising as the frequency gets larger.

How can we account for this "ultraviolet catastrophe"? When Rayleigh and Jeans applied equipartition to the blackbody problem, they forgot one important thing, one that made all the difference. Whereas the number of molecules in an enclosure is always *finite* (no matter how large), the number of radiation frequencies is always *infinite.* Distributing a given amount of energy among a large but *finite* number of molecules having various velocities is one thing. Distributing energy among an *infinite* number of oscillators having various frequencies is vastly different. The distinction is this: Because the number of frequencies is infinite, there are always more high frequencies than there are low ones. In going from low to high, you never run out of frequencies, although you *do* run out of velocities, because there is a finite number of molecules.

Figure 22-5. The infinite ladder.

Perhaps we can understand this by thinking about the rungs on an infinitely tall ladder. (See Figure 22–5.) They are numbered 1, 2, 3, 4, and so on to infinity. No matter which rung you stand on, there are *always* more above you than below you. If you have to paint the ladder using a gallon of paint, equally distributing the paint among the rungs, it should be obvious that most will go to the higher-numbered rungs, simply because there are more of them.

So it is with the infinite number of frequencies. No matter which one you pick out, there are always more to the high-frequency side than to the low-frequency side. This means that if each frequency were given the same amount of energy (just as the rungs were given the same amount of paint), the higher ones would get the lion's share of the total, simply because they outnumber the lower ones. Put another way, if a given chunk of radiation energy could choose any frequency it liked, it would most probably pick one of the more plentiful high ones.

Think of what would happen if Rayleigh and Jeans were correct. For one thing, it would be fatal to sit in front of a roaring fire. Instead of glowing with their pleasant red and orange colors, the hot coals would quickly dump their heat energy over into the high-frequency end of the spectrum and out would come a steady stream of deadly X rays and gamma rays!

The German physicist Max Planck is the hero of our tale, for it was he who solved this dilemma. An unlikely hero he is, too. Of all those who contributed to the early development of quantum mechanics, Planck was most heavily steeped in classical physics. Yet he was able to solve the blackbody problem by making an assumption that was ridiculous according to classical thermodynamics, in which he was expert. The only thing it had going for it was that it worked.

Planck made his proposal to the German Physical Society during Christmas week in 1899. What he suggested was this: It is wrong to assume that all the radiating oscillators have the same energy regardless of their frequency. Instead, each oscillator radiates an amount of energy that is *directly proportional* to its frequency. In symbols,

$$E = hf$$

where E is the energy radiated by an oscillator whose frequency is f. The proportionality constant h has come to be known as *Planck's constant*. With this assumption, and by selecting the value $h = 6.63 \times 10^{-34}$ J · s, Planck succeeded in calculating a distribution curve that closely fits Wien's experiment at all frequencies.

Learning Checks

1. What two things increase as the temperature of an object increases?
2. How is the total emitted energy of an object related to the temperature?
3. In a darkened room, will an object at room temperature (300 K) be visible? Explain.
4. What is meant by the *peak frequency* of a glowing object?
5. What does it mean to say that an object is "white hot"?

ENERGY QUANTA

The essence of Planck's idea is that the energy of electromagnetic radiation is not continuously distributed. Instead, it appears in packets, or bundles, the energy of each bundle being proportional to the frequency of the radiation. Because the size of the energy packets is of crucial importance, Planck called them *quanta* (singular, *quantum*) from the Latin word for "how much." Thus it is that light has particle aspects; the quanta are packets of energy, in much the same way that material particles are packets of matter. These light particles later came to be known as *photons.* (Note the *ons* ending, in analogy with material particles: electr*ons*, prot*ons*, and so on.) Light quanta, or photons, in the microwave region are small, because the waves have long wavelengths and low frequencies. Ultraviolet light is high-frequency radiation, so the photons in this region are large packets of energy compared with infrared and microwave quanta. X rays and gamma rays, being at even higher frequencies, are enormously large packets of energy.

Some simple calculations will illustrate the idea. Suppose we consider light with a wavelength of 4000 angstroms, which is in the violet part of the spectrum. This is 4×10^{-7} meter. Dividing this into the speed of light gives us the frequency

$$f = \frac{3 \times 10^{8}\ \mathrm{m/s}}{4 \times 10^{-7}\ \mathrm{m}} = 0.75 \times 10^{15}\ \mathrm{Hz}$$

The energy of a photon having this wavelength and frequency is obtained by multiplying the frequency by Planck's constant h:

$$E = (6.63 \times 10^{-34}\ \mathrm{J \cdot s}) \times (0.75 \times 10^{15}\ \mathrm{Hz})$$
$$E = 5 \times 10^{-19}\ \mathrm{J}$$

So this violet photon has an energy of 5×10^{-19} joule.

Compare this with an infrared photon, say, having a wavelength of 8000 angstroms, or 8×10^{-7} meter. Its frequency is

$$f = \frac{3 \times 10^{8}\ \mathrm{m/s}}{8 \times 10^{-7}\ \mathrm{m}} = 0.375 \times 10^{15}\ \mathrm{Hz}$$

Notice that this is only *half* the frequency of the violet photon having *half* the wavelength. Its energy is then

$$E = (6.63 \times 10^{-34}\ \mathrm{J \cdot s}) \times (0.375 \times 10^{15}\ \mathrm{Hz})$$
$$E = 2.5 \times 10^{-19}\ \mathrm{J}$$

This energy is only half that of the violet photon.

These calculations illustrate that the quanta are energy packets of various sizes, depending on the frequency of the radiation.

Let us see how Planck's quantum concept accounts for Wien's experiments. For a body at some temperature, there is a given amount of energy to be distributed among the various oscillators, each of which will then emit radiation of a particular frequency. Red light, for example, has only about half the frequency of blue light, and so needs only half as much energy. Each oscillator requires a full quantum of energy before it can radiate, and there is a greater possibility that red light will be emitted before enough energy can accumulate to satisfy the demands of a quantum of blue light. The higher the frequency, the smaller the chances of an oscillator absorbing enough energy to radiate before the lower frequencies are satisfied. It is this probability feature, brought on by the fact that the light is radiated in quanta, that cancels the ultraviolet catastrophe.

For example, let us go back to the red-hot coals of the fireplace. At this temperature, all the available energy is absorbed and re-emitted by the lower frequencies in the infrared and red regions. There is not enough energy to activate the frequencies in the blue, ultraviolet, X ray, and other regions. They are left out in the cold, so to speak. For still lower temperatures, even the red light does not get into the act.

Raising the temperature enlarges the pool of energy available for absorption, increasing the probability of satisfying the large appetites of the high-frequency quanta. This accounts for the shift of the peak of the radiation curve to higher frequencies.

In a nutshell, Planck's hypothesis is this: Each quantum of radiation has energy proportional to its frequency. The low-frequency quanta have a high probability of radiating; for those of high frequency the probability is small.

Learning Checks

1. What is a *blackbody?* Why do we say that it is the perfect emitter of radiation?
2. What is the "ultraviolet catastrophe"?
3. What is Planck's quantum idea?
4. Will an oscillator whose frequency is 1000 hertz absorb energy corresponding to a frequency of 100 hertz? Explain.
5. How does the quantum idea eliminate the ultraviolet catastrophe?

THE PHOTOELECTRIC EFFECT

Max Planck had decidedly mixed feelings about his energy quanta. He believed he had hit upon a fundamental and impor-

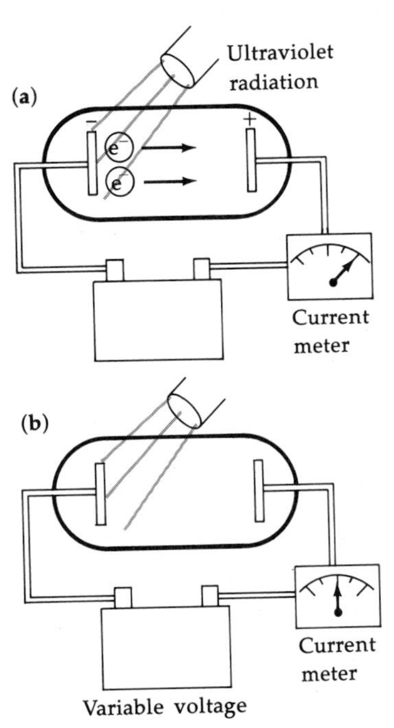

Figure 22-6. The photoelectric effect. (a) Electrons ejected from the negative plate travel to the positive plate and produce the current recorded on the meter. (b) When the receiver plate is made more and more negative by the variable voltage, it eventually repels the electrons so strongly that they do not flow. The current meter reads zero.

tant concept, but he regretted making such a sharp break with classical physics. His theory was not widely known for several years. Many people who were familiar with it considered the notion something of a lucky mathematical trick that happened to explain thermal radiation.

Then Albert Einstein in 1905 (the same year he published his work on special relativity) used Planck's quantum concept to explain a puzzling phenomenon known as the *photoelectric effect*. A beam of light, usually blue or ultraviolet, when shining on a metal produces a stream of electrons from the surface. These *photoelectrons*, as they are called, make up an electric current. We take advantage of this principle in light meters, "electric eyes," and motion picture sound tracks.

Figure 22–6 shows a simple arrangement for studying the photoelectric effect. Ultraviolet radiation aimed at the negatively charged zinc plates frees electrons, which flow toward the positive plate, producing an electric current. If the second plate is now *negatively* charged, the electrons will be repelled. This reduces the current. By increasing the voltage until the current stops, it is possible to determine the energies of the electrons ejected by the ultraviolet light. The faster the electrons, the greater this stopping voltage.

Careful experiments involving the photoelectric effect showed the following facts:

1. The higher the frequency of the incident light, the faster the ejected electrons move away from the metal. That is, the electrons' kinetic energy increases as the radiation frequency gets larger.
2. The photoelectrons' kinetic energy depends *only* on the frequency and is independent of how bright the light is (that is, its intensity).
3. Raising the intensity of the light increases the *number* of ejected electrons.
4. Each particular metal has a characteristic "threshold" frequency. If the frequency of the light is less than this particular frequency, no electrons escape, no matter how bright the light. For example, an extremely feeble beam of ultraviolet light ejects a stream of electrons from zinc; but red light, no matter how bright, has no effect.

The classical picture of light as a wave could not account for these results. For instance, the energy of a wave depends on its amplitude—at the beach you get knocked over by a big wave, not a little one. One would predict that since the intensity also increases with the amplitude, the more intense light will carry more energy. We would expect, then, that an electron bound to the metal would be set free if struck by sufficiently intense light. But this doesn't happen. A bright beam of red light doesn't bother the electrons at all.

Furthermore, if you believe that light is a wave you would predict that the kinetic energy of ejected electrons will increase as the light gets brighter. That is like making a baseball go faster by throwing it harder. But, in fact, the energy of the photoelectrons depends only on the *frequency* and has nothing to do with the intensity. A dim ray of blue light will produce electrons having greater kinetic energy than green light, no matter how intense. The blue light has greater frequency than the green. It became clear that the kinetic energy of the electron was somehow related to the frequency. The connection between energy and frequency was a complete surprise. Nothing in classical physics suggested it.

Einstein knew of Planck's theory, which was by now 5 years old. He carried it one step further and solved the photoelectric mystery. Planck believed the quanta to be involved only in the *exchange* of energy between matter and radiation, suggesting that perhaps they were somehow due to the internal structure of matter. Einstein, on the other hand, proposed that the light *itself* is made up of photons. A beam of light is simply a stream of these energy bundles traveling at the characteristic speed. When the light strikes the metal, one of its electrons absorbs a whole photon, taking on the full packet of energy. If the photon delivers enough energy to the electron, it will tear loose from the metal. In Einstein's view, the electron either absorbs the full quantum of the photon's energy or does nothing at all. There is no such thing as a fraction of a quantum, any more than there is a fraction of an electron. If there are only a few photons in the light beam, then only a few electrons can be ejected. The more photons there are available, the more electrons will be able to tear loose from the metal.

For light of frequency equal to f, the photons each have energy hf. If the frequency is high enough, part of the photon energy will go toward freeing the electrons from the metal. The remainder furnishes the kinetic energy for the electron as it flies away. We can write this in an equation:

$$KE = hf - hf_0 = hf - W$$

where W is called the *work function.* It is the amount of energy the electron needs to escape from the metal. The quantity f_0 is the *threshold frequency*—that is, the lowest frequency of light that will free an electron. Then we have

$$W = hf_0$$

So

$$hf = KE + hf_0$$

Photoelectricity

The three most important effects or manifestations of photoelectricity are commonly designated the *photoemissive*, the *photoconductive*, and the *photovoltaic*. Photoemissive cells are used in the reading of film soundtracks in movie projectors and in converting optical signals to electrical signals in television cameras. Photoconductive cells are found in switch-on relays for street lighting and in instruments that measure low temperature heat radiation. Photovoltaic cells are widely used in the exposure meters of photographers and act as "electric" eye detectors for the operation of relays.

The threshold frequency f_0 will be a constant for a given metal and, in general, different from one metal to another. In any case, if the light beam has photons whose frequency is smaller than f_0, then this light will not release any electrons for that particular metal.

This quantum theory explains all the facts about the photoelectric effect that were so puzzling. For example, the kinetic energy gets larger as the frequency rises. We can see this from the last equation: Since hf_0 doesn't change, then a bigger value of hf means a bigger value of KE. What about the intensity? Well, according to Einstein, a very intense beam of light has many photons, and dim light has fewer. Because each electron absorbs a single photon, the theory explains how the number of electrons increases with the intensity of light. Cranking up the intensity increases the supply of photons and thus provides more photoelectrons, raising the photoelectric current. We can also explain why red light will not work for most metals but blue light will. A red quantum is too small—its frequency is below the threshold, so it doesn't supply the electron with enough energy to elbow its way to the surface and break loose.

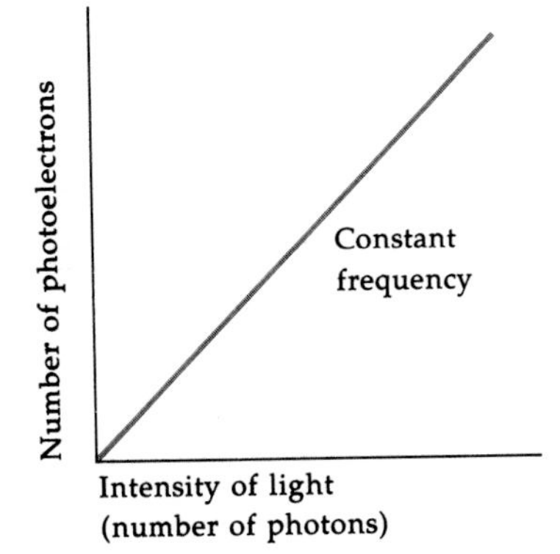

Figure 22-7. Photon growth with increase in light intensity.

What about the possibility of an electron escaping by absorbing more than one photon of low-frequency light? For example, two photons of red light at 7200 angstroms carry the same energy as a single blue photon at 3600 angstroms, so presumably an electron could be ejected by absorbing the two red photons. This doesn't happen. It is much more likely that the electron will reradiate one photon before absorbing the second. If one quantum won't do the job, then the electron stays put.

Einstein's elegant explanation thrust the idea of radiation quanta into the foreground of physics. The particle picture of light, which was the favorite scheme of Isaac Newton, reemerged in this new form. The photon became even more of a reality in the minds of physicists when the eminent American physicist Robert Millikan in 1916 confirmed Einstein's predictions with a series of careful studies on the photoelectric effect. Figures 22–7 and 22–8 show the results of his work.

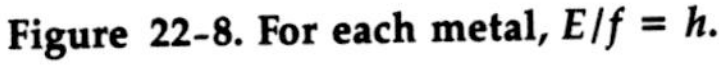

Figure 22-8. For each metal, $E/f = h$.

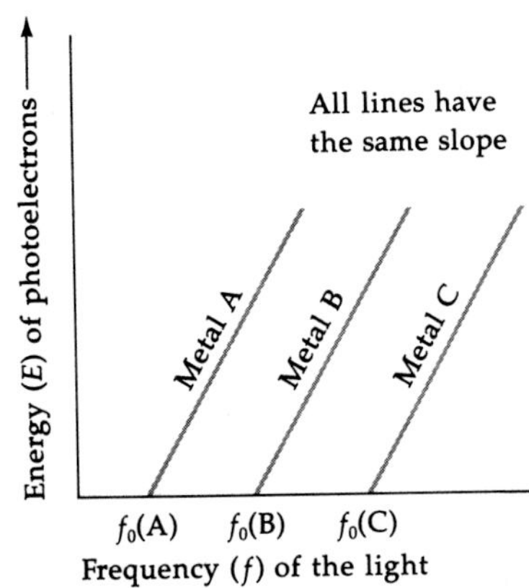

Figure 22–7 shows that the number of photoelectrons increases linearly as the light intensity—that is, the number of photons—grows. Figure 22–8 shows that Einstein's equation for kinetic energy is correct. Each straight line gives the data for a particular metal and illustrates how the kinetic energy varies with the frequency of the light. Note that each metal has a different threshold frequency. However, the slopes are all the same—and numerically equal to the number that Planck got for the constant h in the blackbody problem. Millikan's independent confirmation of Planck's constant was an important step in establishing the quantum as a real part of the physical world. It has been with us ever since.

Learning Checks

1. Explain the term *photoelectron.*
2. How is Einstein's idea about light quanta an extension of Planck's?
3. Explain how the particle theory of light succeeds, and the wave theory fails, in accounting for the photoelectric effect.
4. Explain the term *threshold frequency.*
5. Account for the fact that the lines in Figure 22–8 touch the horizontal axis at different places.
6. How might the photoelectric effect be useful in a light meter for photography?
7. Which has greater energy, a blue photon or a red photon? Explain.

THE SPECTRUM OF HYDROGEN

The introduction of energy quanta gave rise to a new idea in physics, that of *quantization.* Something is said to be *quantized* if it is characterized by discrete, whole packages rather than by a smeared-out, continuous distribution. Your television dial is quantized. You can watch channel 3 or 4 or 11, but trying to tune in channel 5.5 or 6.3721 makes no sense. The radio, on the other hand, can be tuned to receive any of a band of frequencies; the radio dial is continuous. The rungs of a ladder are quantized. (See Figure 22–9.) American money is quantized: what you have in your bank account is some whole-number multiple of the cent.

Figure 22-9. The rungs of a stepladder are quantized.

Planck's interpretation of the blackbody problem involves quantization of energy as it changes from a mechanical form in matter to a radiant form in light. Einstein showed that the light *itself* is quantized as photons. Symmetry suggests, therefore, that something about the structure of matter should be quantized, also. That this turns out to be true solved another of those experimental puzzles floating around early in this century.

The problem involves what is called the *spectrum* of an atom. Whenever an atom is given energy (or "excited") in some way, such as by an electric arc or collision with other particles, it releases this extra energy as electromagnetic radiation. However, instead of appearing in a continuous distribution of frequencies, this radiation gives rise to a spectrum of discrete lines. For the simplest atom, hydrogen, Figure 22–10 shows the part of the line spectrum that is in the visible region. The line spectra of hydrogen and other atoms could not be explained by the classical theory, which predicted a continuous spectrum like the rainbow you get by shining white light through a glass prism.

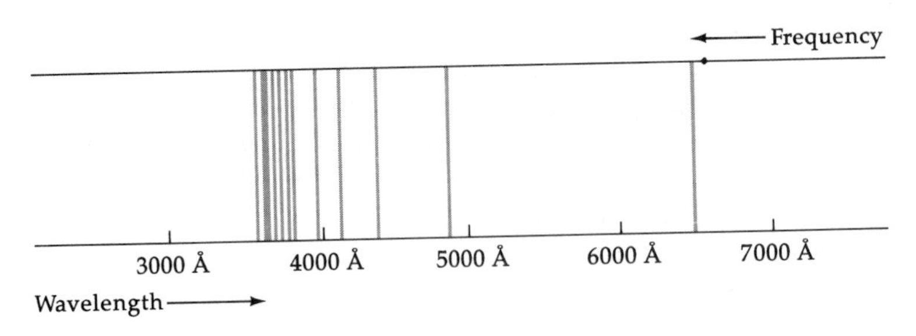

Figure 22-10. The spectrum of hydrogen (visible region).

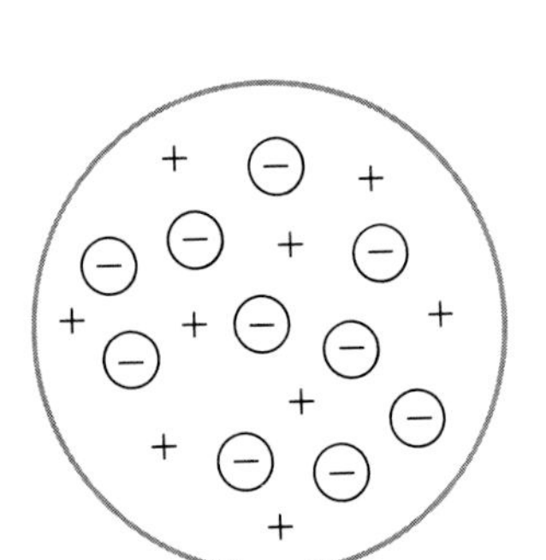

Figure 22-11. J. J. Thomson's "plum pudding" atom.

When J. J. Thomson discovered the electron in 1897 and showed that it was a part of the atom, he felt that it was necessary to develop some theoretical model for the structure of the atom that would account for the line spectra. Thomson's idea is sometimes called the "plum pudding" model. (See Figure 22-11.) He imagined that the negatively charged electrons were dispersed throughout a uniform sea of positive charge, much like the plums in a pudding (or like raisins in raisin bread, if you're not a plum pudding fan). When the atom becomes excited, its electrons should begin to vibrate about their equilibrium positions, like strings in a piano, and emit radiation having a set of discrete characteristic frequencies, corresponding to those in the observed line spectrum. Thomson hoped that he could use the methods of classical mechanics to calculate the distribution of the electrons in the atom and predict the frequencies in the spectrum. When Thomson and his students did these calculations, however, the plum pudding model proved to be a failure. The calculated line frequencies were not even close to those that they observed in the spectrum.

About that time, a young Danish physicist named Niels Bohr came to England to work with Thomson, who was director of the Cavendish Laboratory at Cambridge University. Bohr had earned his Ph.D. at the University of Copenhagen in 1911. He felt that a completely new model of the atom was necessary to account for the atomic line spectra. He favored one that would use some of the ideas of the new quantum theory. It seemed reasonable that if light were quantized, as Einstein had shown in the photoelectric effect, then the mechanical energy in the atoms emitting the light should also have discrete values.

Thomson was rather fond of his own model of the atom. As you might imagine, he was cool toward having young Bohr tamper with it. Sensing that "you can't fight City Hall," Bohr soon went to Manchester to work with the New Zealander Ernest Rutherford, a former student of Thomson's who later succeeded him as director of the Cavendish.

Rutherford was in the midst of his historic experiments on the internal structure of atoms. He bombarded thin metal foil with so-called *alpha* (α) *particles,* which were later found to be positively charged helium nuclei. He noticed that while most of them went through the foil undisturbed, a certain fraction

The Spectrum on the Job Site

An instrument for recording the spectrum of any given element is called a *spectrometer.* This instrument has been a successful research aid to innumerable branches of science and has proven itself invaluable for practical use in a great number of jobs. Customized versions can be found at many workplaces. In the effort to help reduce atmospheric pollution, for example, a spectrometer is used to analyze the exhaust gases of autos. A portable spectrometer used in the mining industry can measure small concentrations of tin within 30 seconds. In the rubber industry, a spectrometer can tell scientists within minutes how many metric tons of natural rubber can be extracted from an entire crop of the guayule shrub.

The field of spectrographic analysis has advanced steadily over the years, and the most recent refinements can be attributed to a new result and tool of spectroscopy—the laser, which we discuss in Chapter 24.

(a) Goodyear technician analyzing molecules in a pill-sized sample of guayule shrub (dark area in the glass tube) is able to estimate the amount of rubber in a single shrub or in an entire desert crop. (b) Measuring exhaust emissions is part of the program to control pollution.

Figure 22-12. Rutherford's scattering experiment.

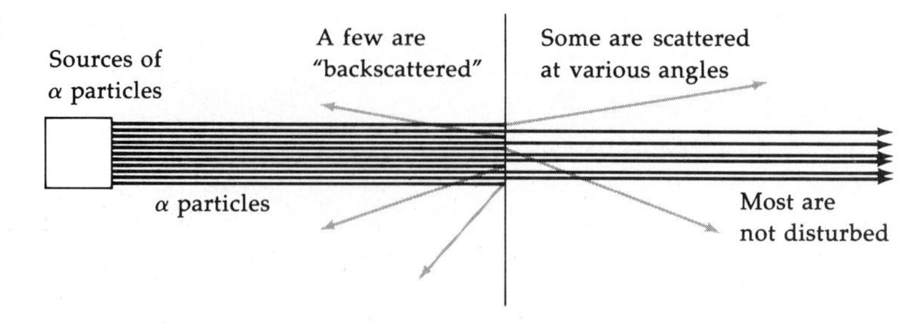

were deflected and scattered about at various angles. (See Figure 22-12.) Some α particles were even backscattered—that is, sent back almost along the original path they had followed from the source.

These results were completely incompatible with Thomson's model of the atom. Rutherford reasoned that the heavy α particles would not rebound off the smeared-out positive charge proposed by Thomson, any more than a speeding cannonball would bounce off a sheet of tissue paper. Instead, he felt that the positive charge of the target atom had to be concentrated in a small hard core, or nucleus. Only then could the electrostatic repulsion be great enough to deflect the α particles from a straight-line path.

Figure 22-13. Rutherford's nuclear atom.

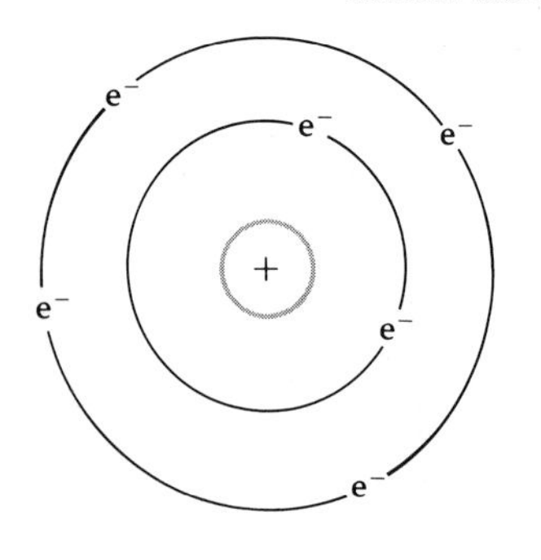

Thus was discovered the *nuclear atom* (Figure 22-13), with its positive charge and most of its mass confined to a superdense nucleus. The negatively charged electrons orbit the nucleus in circular or elliptical paths, much like planets going around the sun. Because of this model's resemblance to a miniature solar system, the electrons are often called *planetary electrons.*

Although the nuclear atom completely accounts for Rutherford's scattering experiments, there is one problem: such an atom cannot exist, at least not for long. The planetary electrons are accelerating, for they follow curved paths. Like all accelerating charged particles, they give out energy in the form of electromagnetic radiation; or they should, according to what we know about electromagnetism. If an electron's energy is to be conserved, the radiation must come from somewhere. The only source is its electrostatic potential energy, which the electron would have to give up by moving closer and closer to the nucleus. So an atom would put out a continuous flood of electromagnetic energy as its planetary electrons plunged toward the nucleus along ever-tightening spiral paths. (See Figure 22-14.) In a very short time, the electrons would crash into the nucleus, and the atom would cease to exist. So says the classical theory.

Figure 22-14. The orbiting electron must radiate its energy and crash into the nucleus.

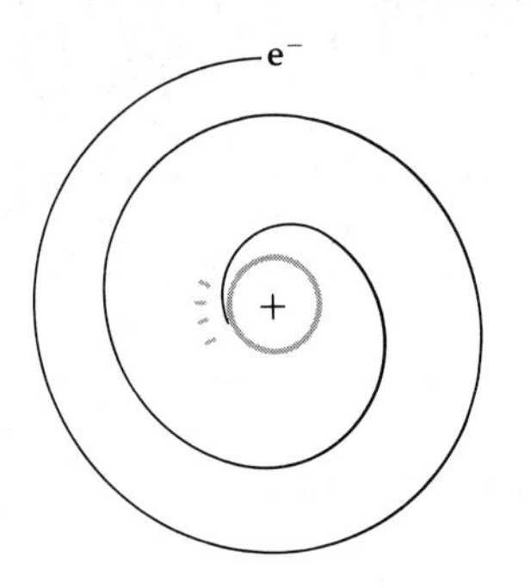

Now this is a severe shortcoming. After all, the very fact that you are sitting there reading this book demonstrates that atoms exist; they don't undergo any such spontaneous annihilation. This was the dilemma facing Niels Bohr when he came to Manchester: how to reconcile the Rutherford atom with the known results of classical electrodynamics.

Because the spectrum of hydrogen is the simplest, he chose to attack it first. Some earlier work Johann Balmer had done in 1885 aided Bohr in his search. Balmer, a Swiss schoolteacher, had calculated the frequencies of some of the lines in the hydrogen spectrum. He was struck by the regularities of the positions of these lines. Tinkering around with the numbers a bit, Balmer came up with a simple equation for calculating the frequencies of the lines in the visible part of the spectrum. This group of lines is now called the *Balmer series* in his honor. The equation is

$$f_m = 3.289 \times 10^{15}\left(\frac{1}{2^2} - \frac{1}{m^2}\right)\text{Hz} \qquad \text{(Balmer series)}$$

where f_m is the frequency of a line labeled m, which can take on only *integral* values starting at 3. That is, $m = 3, 4, 5, 6, \ldots$. This means that by substituting 3 for m, Balmer could calculate the frequency of a particular line. Using $m = 4$, he got a number that matched the frequency of the next line—and so on for the other integers in place of m.

Other series of lines discovered in different regions of the spectrum were found to match similar expressions. The *Lyman series* (discovered by Theodore Lyman) is in the ultraviolet region. Its lines have frequencies given by

$$f_m = 3.289 \times 10^{15}\left(\frac{1}{1^2} - \frac{1}{m^2}\right)\text{Hz} \qquad \text{(Lyman series)}$$

This is like the Balmer equation, except that $1/2^2$ $(= 1/4)$ has been replaced by $1/1^2$ $(= 1)$ so that m can start at 2. Thus, $m = 2, 3, 4, 5, \ldots$. Similarly, the *Paschen series* (after Friedrich Paschen) falls in the infrared region. The equation is

$$f_m = 3.289 \times 10^{15}\left(\frac{1}{3^2} - \frac{1}{m^2}\right)\text{Hz} \qquad \text{(Paschen series)}$$

in which $m = 4, 5, 6, \ldots$.

Note the pattern here: these equations can be written as

$$f = 3.289 \times 10^{15}\left(\frac{1}{n^2} - \frac{1}{m^2}\right)\text{Hz}$$

where n is fixed for a particular series and m takes on integral values beginning with $n + 1$. Each value of n corresponds to a different series of lines (see Figure 22-15): $n = 1$ gives the Lyman series, $n = 2$ the Balmer series, $n = 3$ the Paschen series, and so on for others that we have not mentioned.

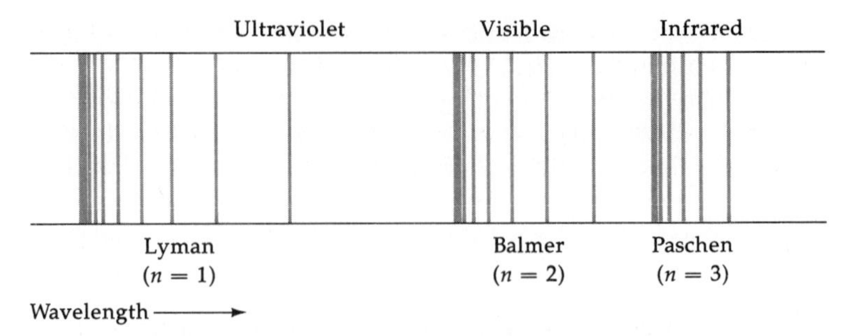

Figure 22-15. Series of lines in the hydrogen spectrum.

The German physicist Walter Ritz had also made an interesting discovery many years earlier, during Bohr's boyhood, that helped unravel the line spectrum mystery. What he found was this: If you add or subtract the frequencies of any two lines in the spectrum, you get the frequency of another line. This *Ritz combination principle*, as it is called, and Balmer's equations were simply curious bits of numerology—no one had devised any reasonable explanation for them.

Rutherford's experiments had indicated that the central-nucleus idea was correct, so Bohr decided to deal with the problems the single electron presented. Because the lines in the spectrum implied that the atom released energy in discrete amounts, Bohr felt that the energy states available to the electron were also discrete. In other words, he believed the mechanical energy of the electron in the atom to be *quantized.*

It was on this point that Bohr broke with classical physics. There was nothing in all the rich lore of mechanics that remotely suggested that an object should have its energy quantized. The swing of a pendulum, the flight of a baseball, the heat from a fireplace—for all these the energy always appears to be continuous, capable of taking on any value. But Bohr was undaunted. Much like Max Planck, he was searching for an explanation that worked, regardless of whether it was part of the dogma of physics.

Bohr made a series of assumptions that retained Rutherford's nucleus of concentrated positive charge but also accounted for the empirical findings of Ritz, Balmer, and the others. Here are his assumptions:

1. The radius of the electron's orbit cannot have arbitrary values, but is restricted to those values that make the angular momentum a whole-number multiple of $h/2\pi$.
2. An electron does not radiate energy while it remains in one of these orbits.
3. An electron may jump from one orbit to another. If it jumps to an orbit *nearer* the nucleus, it *emits* energy in the form of a photon whose frequency corresponds to the energy difference of the two orbits. It may also *absorb* a photon and jump to an orbit of *higher* energy farther from the nucleus.

Let us see how these assumptions account for the appearance of the hydrogen spectrum. The first one specifies that the electron can occupy only certain orbits. We show in Supplement 9 that the energy of a particular orbit is given in terms of a constant R (called the Rydberg constant) as

$$E_n = \frac{-R}{n^2} \qquad (n = 1, 2, 3, 4, \ldots)$$

where $n = 1$ is for the lowest orbit (the so-called *ground state*), $n = 2$ is for the next one farther out, and so forth. (The negative sign appears because we take the zero of energy to be for the electron pulled completely away from the nucleus.) According to the third assumption, the electron may jump from orbit number 4, say, to orbit number 2, giving out a photon whose energy, hf, is the difference between these two orbits. That is

$$hf = E_4 - E_2$$

$$hf = -R\left(\frac{1}{4^2}\right) - \left[-R\left(\frac{1}{2^2}\right)\right]$$

$$= R\left(\frac{1}{2^2} - \frac{1}{4^2}\right)$$

So the frequency f would be

$$f = \frac{R}{h}\left(\frac{1}{2^2} - \frac{1}{4^2}\right)$$

Notice that this is just the equation for the $m = 4$ line in the Balmer series if $R/h = 3.289 \times 10^{15}$ Hz. In fact, Bohr was able to show that he could account for all the lines in the spectrum by this procedure.

We can illustrate these ideas by an *energy-level diagram*, as in Figure 22-16. Each horizontal line labeled E_1, E_2, and so on,

Revelation

In *The Search,* novelist and physicist C. P. Snow describes how a college physics student feels when he first discovers Bohr's atom:

For the first time I saw a medley of haphazard facts fall into line and order. All the jumbles and recipes and hotchpotch of the inorganic chemistry of my boyhood seemed to fit themselves into a scheme before my eyes—as though one were standing beside a jungle and it suddenly transformed itself into a Dutch garden. "But it's true," I said to myself. "It's very beautiful. And it's true."

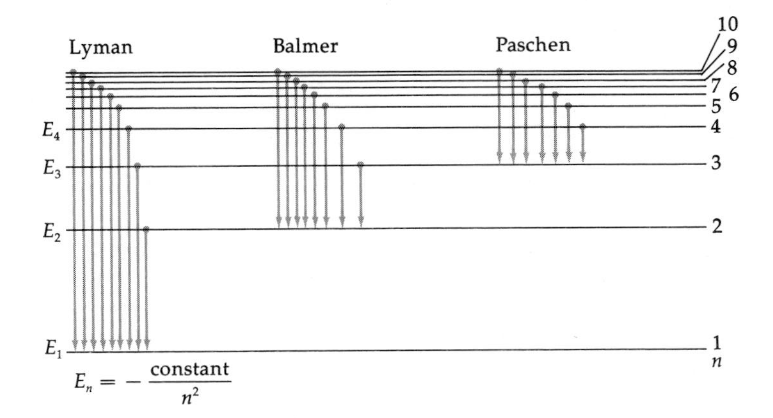

Figure 22-16. Energy-level diagram for hydrogen.

corresponds to the allowed hydrogen orbits numbered $n = 1$, $n = 2$, and so on. The vertical arrows represent *transitions* between these levels—that is, "jumps" of the electron from one orbit to another. Because each transition results in a photon given off, each arrow stands for a line in the hydrogen spectrum. Notice that a *single* line in the spectrum involves *two* energy levels—a starting and an ending orbit for the electron. As you can see, the lines in the Lyman series arise because of transitions from the higher levels down to E_1. These arrows are fairly long because the energy gaps are large, so these lines appear in the high-frequency ultraviolet region. The Balmer series' arrows, for which the electron jumps from the higher orbits down to orbit number 2, are shorter, representing the lower-frequency photons in the visible region. E_3 is the energy level for orbit number 3, the terminating level for transitions in the infrared Paschen series.

Figure 22-17. The Ritz combination principle.

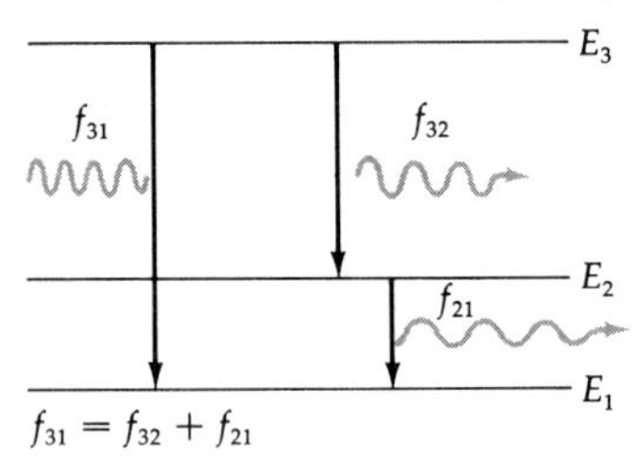

Bohr's scheme also accounts for the Ritz combination principle, as you can see from Figure 22-17. If the electron makes a transition from E_3 down to E_1, it will give out a photon whose frequency (which we will call f_{31}) is given by

$$f_{31} = \frac{(E_3 - E_1)}{h}$$

The electron could also make this trip in two steps, giving up a photon in each—from E_3 to E_2 and then from E_2 to E_1. These two transitions give two photons having frequencies f_{32} and f_{21}. Obviously, from the picture, the frequencies of these two photons added together give the frequency of the single photon for the transition in a single jump. That is,

$$f_{31} = f_{32} + f_{21}$$

which illustrates Ritz's discovery.

In order for an electron to get to a higher energy level—that is, to an orbit farther from the nucleus—it has to absorb enough energy to bridge the gap. This could be in the form of a photon of exactly the correct energy. If the frequency of the photon corresponds to some intermediate energy, the electron doesn't absorb it, because it can only occupy the allowed orbits and cannot absorb part of a photon.

Bohr's ideas proved to be an important first step in understanding the structure of atoms and their spectra. The Bohr atom is strictly true only for hydrogenlike (that is, one-electron) atoms. Later modifications to his theory were necessary to account for the properties of other, more complicated atoms.

Bohr himself offered no rationale for the quantization of

the angular momentum of atomic orbits. As we shall see in the next chapter, a French nobleman named Louis de Broglie provided a theoretical interpretation some years later. The importance of Bohr's contribution was that the ice had been broken. The notion that the classical physics of Newton and Maxwell was not correct for atomic phenomena began to grow.

Learning Checks

1. What is the Ritz combination principle? How does Bohr's theory account for it?
2. Distinguish between Thomson's "plum pudding" atom and Rutherford's nuclear atom.
3. How does the nuclear atom violate the theory of electromagnetic radiation?
4. What is an energy-level diagram? What do the levels represent?
5. What is the connection between lines in the spectrum and an energy-level diagram for an atom?

SUMMARY

Around the turn of the century, some experiments involving light and matter gave results inconsistent with Newton's laws of motion and Maxwell's electromagnetic theory. For the distribution of *thermal radiation from a blackbody,* classical physics predicted that the total energy would be equally distributed among all the possible frequencies. This led to the "ultraviolet catastrophe"—complete disagreement with the experiments at high frequencies. Planck resolved the question by assuming that the energy is emitted in *quanta,* or packets of energy, each proportional to the radiation frequency, $E = hf$.

The *photoelectric effect* is the production of a stream of electrons by irradiation of a metal surface with blue or ultraviolet light. The photoelectron energies are proportional to the frequency and independent of the intensity of the light, which is just the opposite of the classical prediction. Einstein provided the solution by assuming that the light is made up of *photons,* each having energy equal to hf. An electron may absorb only one complete photon, using its energy to escape from the metal and fly away at a speed that increases with the light's frequency.

Line spectra produced when light is emitted from *atoms* could not be explained by the classical theory, which predicted continuous radiation. Also, Rutherford demonstrated that the atom consists of a dense core of positive charge around which circulate orbiting electrons. Using the information from line spectra and from Rutherford's work, Bohr suggested that the electron orbiting the nucleus is restricted to particular orbits and that the atom emits or absorbs energy only when the electrons jump from one orbit to another. This quantizes the emission and absorption of the radiation by requiring that the mechanical energy within the atom be quantized.

Exercises

1. Explain how a blackbody is both a perfect *absorber* and a perfect *emitter* of radiation.

2. Explain the fact that for an infinite number of radiation frequencies there are more high frequencies than low ones.

3. How do Rutherford's experimental results show that Thomson's "plum pudding" atom cannot be a correct model?

4. Explain Wien's hole-in-a-cavity as a good approximation of a blackbody.

5. What can you say about the relative temperatures of reddish, bluish, and whitish-yellow stars?

6. If all objects radiate energy, why can we not see them in the dark?

7. How does Planck's theory explain the observation that the peak frequency increases as the temperature rises?

8. Arrange the following in the order of increasing energy: X rays, microwaves, infrared waves, yellow light, red light, violet light, gamma rays.

9. If we double the temperature of an object, what happens to the wavelength at which its radiation peaks?

10. If we double the frequency of a photon, what happens to the energy?

11. If we double the wavelength of a photon, what happens to the energy?

12. What determines the kinetic energy of electrons in the photoelectric effect?

13. What determines how many electrons are ejected in the photoelectric effect?

14. Suppose that ultraviolet radiation of a particular frequency strikes two metals, A and B. If the work function for metal A is larger than that for B, from which will the photoelectrons have the greater speed? Explain.

15. Explain how Einstein's theory accounts for the observation that the energy of emitted photoelectrons is independent of the intensity of the incident radiation.

16. The general equation for the frequencies of lines in the Balmer, Paschen, and Lyman series is

$$f = 3.289 \times 10^{15}\left(\frac{1}{n^2} - \frac{1}{m^2}\right)\,\text{Hz}$$

Explain why m must always be a larger integer than n.

17. How can the spectrum of hydrogen contain many lines when a hydrogen atom contains only one electron?

18. What do the lines in the spectrum of an atom represent?

19. Is energy conserved when an atom emits a photon? How about when it absorbs a photon? Explain both answers.

20. Suppose a certain metal has a work function of 3×10^{-19} joule. What is the lowest frequency of light that will produce photoelectrons for this metal? Also give the wavelength and color of this light.

21. If you double the temperature of a blackbody, by what factor does the total amount of energy given off increase?

22. One blackbody is at 100 K and another of the same size is at 1000 K. Which one has the higher peak frequency?

23. In Exercise 22, which gives off more energy at a wavelength of 7 meters? At 7000 angstroms?

24. Use the equation in Exercise 16 to calculate the frequency of the photon emitted when the electron in hydrogen falls from level 3 down to level 2. What is the color of this photon? What is its wavelength?

25. When a hydrogen atom goes from its ground state (level 1) to its first excited state (2), it must absorb a photon of frequency 2.5×10^{15} hertz. What is the energy separation between these two states?

26. Refer to Exercise 25. Is it possible for a hydrogen atom to absorb a photon whose frequency is less than 2.5×10^{15} hertz? Explain.

27. Compare the energy of a single photon whose wavelength is 7000 angstroms (a red photon) with that of a photon at 3500 angstroms (in the ultraviolet region).

28. Figure 22-18 shows an energy-level diagram of five levels for some arbitrary atom. Assuming that transitions are possible between any two levels, how many lines will appear in the spectrum of this atom?

E_5

E_4

E_3

E_2

E_1

Figure 22-18. Energy-level diagram.

29. Show how the Ritz combination principle applies in Exercise 28.

30. Refer again to Exercise 28. Suppose that the spectral line with the shortest wavelength is in the ultraviolet region. To which transition does this correspond?

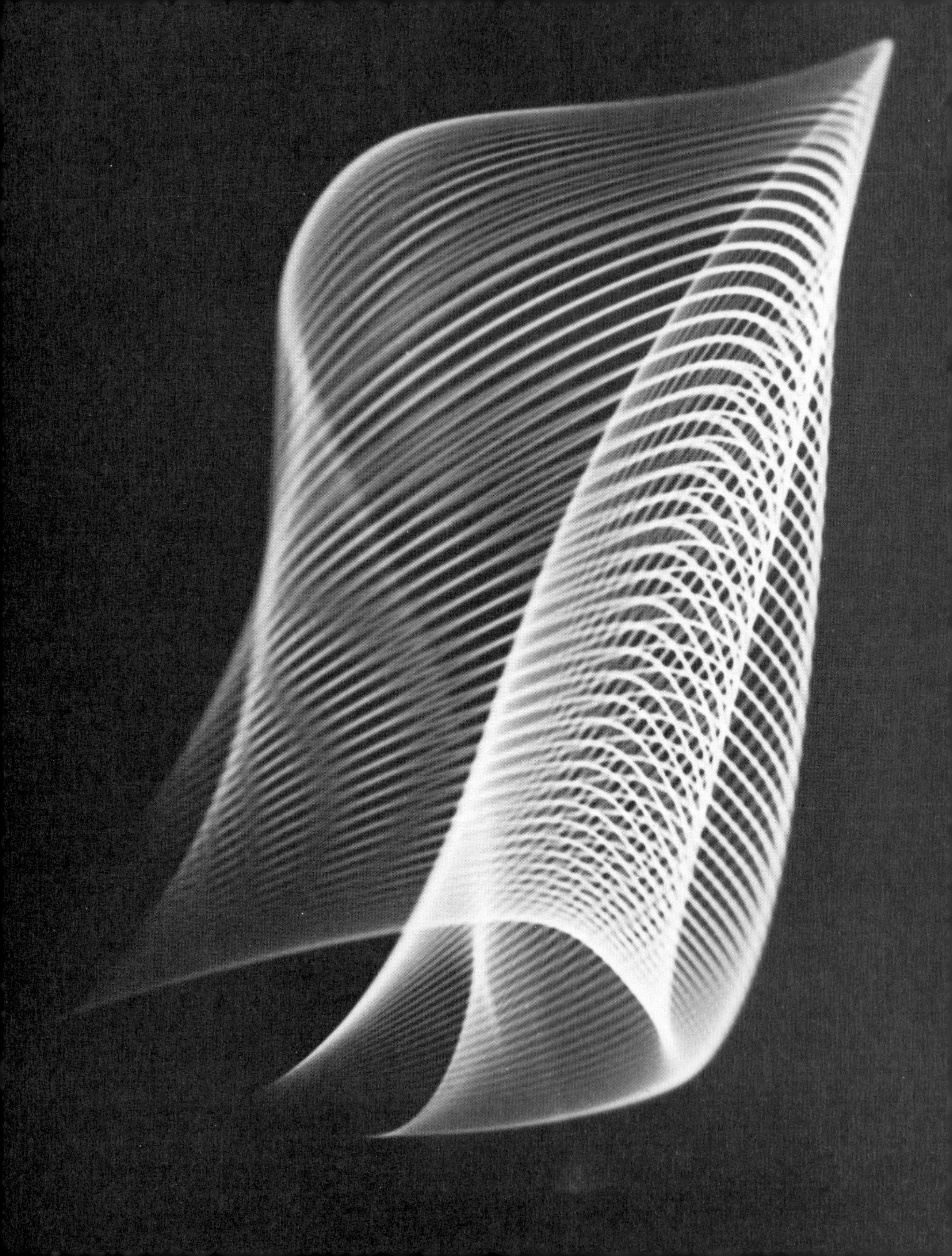

23

QUANTUM MECHANICS

Electronic wave forms and the computer combine to form visual music, called "oscillons" by the photographer.

We saw in Chapter 22 how experiments such as the photoelectric effect and the line spectrum of hydrogen forced physicists to conclude, reluctantly, that classical physics was inadequate to deal with nature at the submicroscopic level. Recall from our study in Chapter 8 that Einstein developed the special theory of relativity for a similar reason—that objects traveling at near light speed don't behave the way the classical theory predicts. So it was that quantum mechanics was born, out of a need to understand the laws governing light and electrons and interactions between them.

THE IDEA OF THE QUANTUM

The new concept on which this theory is based is quantization. That things like angular momentum and energy are quantized is strange and difficult to imagine, simply because the effects never show up in our everyday experience. Consider a simple pendulum, for example. (See Figure 23–1.) You know that if you pull the pendulum bob off to one side and let it go, it swings back and forth between the turning points. It gains kinetic energy at the expense of potential energy as it moves toward equilibrium, then overshoots and slows down as it trades kinetic energy for potential energy on its way to the opposite turning point. As far as we can tell, this interchange of kinetic and potential energy is a continuous process, not a jerky one. As the pendulum loses energy because of friction at its pivot, it slows down gently and smoothly, not in a series of discontinuous jumps. The pendulum apparently can have any energy state whatsoever. It is not confined to a collection of quantized states like those allowed the electron in Bohr's hydrogen atom.

The obvious question is why should there be a difference? After all, the pendulum is made up of electrons and other such particles whose energy states *are* quantized. Why don't we see

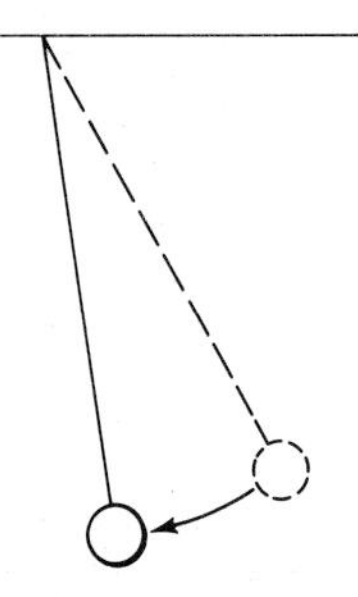

Figure 23-1. The pendulum's loss of potential energy is apparently continuous.

the pendulum's energy given up in quanta? Why shouldn't its energy be quantized like that of the electron?

We can understand this apparent contradiction by looking at the magnitudes of the numbers involved. The energy states allowed a periodic system with frequency f are given in terms of Planck's constant ($h = 6.63 \times 10^{-34}$ J · s) as

$$E_n = nhf$$

where n can be any integer 1, 2, 3, 4, . . . Now an ordinary pendulum might typically make one full cycle of its swing in 1 second. Its frequency is then 1 cycle per second (1 Hz), so

$$E_n = n(6.63 \times 10^{-34}\ \text{J} \cdot \text{s})(1\ \text{s}^{-1})$$
$$= n \cdot 6.63 \times 10^{-34}\ \text{J}$$

> *A quantum leap, no matter how infinitesimal, always makes a sharp break with the past. It is*
>
> - *the discontinuous jump of an electron from one orbit to another with the particle mysteriously leaving no trace of its path.*
> - *the instantaneous collapse of a wave of probabilities into a single real event.*
> - *the link between two entirely separate locations, events, or ideas, that magical moment when the previously inexplicable is suddenly explained, and a radical new theory is born.*
>
> *Fred Allen Wolf*
> "Taking the Quantum Leap"
> *Science Digest*
> December 1981, p. 88.

From this equation, E_1 is at 6.63×10^{-34} joule, and E_2 is twice this amount, at 13.26×10^{-34} joule. The essence of quantization is that there can be *no* values of energy between these two. E_3 is at 19.89×10^{-34} joule, and so on.

As you can see, the energy levels for this pendulum are separated by only 6.63×10^{-34} joule. This is a fantastically small number. One joule is about the energy needed to lift a baseball to a height of a few centimeters. Dividing this by the number 1 followed by 34 zeros means taking about a billionth of a billionth of a billionth of a billionth of a joule. Such an unimaginably small spacing between energy levels is simply not perceptible. The states are crammed so close together they *appear* to be continuous. The classical laws are more than adequate here because we can ignore the quantization.

We can look at this in another way by examining the sort of values of n required by this expression. If the pendulum bob has a mass of 2 kilograms and we start it swinging from a height of 1 meter, then its potential energy relative to the bottom of its path is

$$E = mgy = (2\ \text{kg})(9.8\ \text{m/s}^2)(1\ \text{m})$$
$$\cong 20\ \text{J}$$

Now, 20 joules divided by hf, the size of one quantum, gives us a value for n:

$$n = \frac{20\ \text{J}}{(6.63 \times 10^{-34}\ \text{J} \cdot \text{s})(1\ \text{Hz})}$$
$$n = 3 \times 10^{34}$$

This value of n, 3 followed by 34 zeros, is the number of energy levels between zero and 20 joules. It is an incredibly large number—much larger than the number of seconds that have ticked by since the beginning of the universe.

Such closely packed levels are essentially continuous by any measurement anybody can make. You can picture this as in Figure 23-2. For a given amount of energy, a few levels are well separated. But for a large number of levels spanning the same energy, the spacing between them is so tiny the lines drawn seem to run together, as if painted with a broad brush. The unimaginably small "jumps" that the pendulum makes as it loses potential or gains kinetic energy appear not as jumps at all. They make a continuous flow, much as the individual frames of a movie film produce continuous motion when run through a projector.

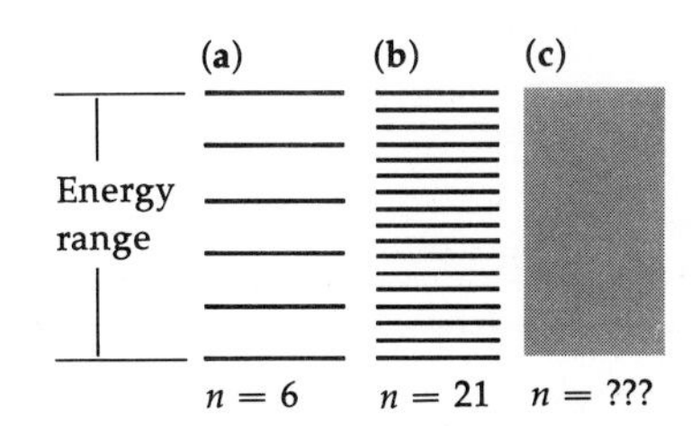

Figure 23-2. Energy level separation. (a) A few widely spaced levels. (b) More levels, but the energy is still discrete. (c) The large number of levels makes the energy seem continuous.

This is rather like approximating an inclined ramp by a staircase of individual steps. (See Figure 23-3.) If we start out with only a few steps separated by a rather large gap, each step is clearly identifiable. A ball rolling down the stairs decreases its potential energy in spurts, and we can watch this as it bumps toward the bottom. But as we put more and more steps closer and closer together, eventually we can't distinguish one "step" from the next. The ball rolls gently and gradually down this smooth ramp.

As another example, let us compare an electron in its orbit with a rock whirled in a circle on a string. Remember that one of Bohr's postulates for the hydrogen atom is that the electron's angular momentum is quantized, restricted to integer multiples of Planck's constant divided by 2π.* That is

$$L = \frac{nh}{2\pi} \qquad (n = 1, 2, 3, \ldots)$$

This means that the difference in angular momentum from one allowed orbit to the next is $h/2\pi$, or about 10^{-34} J · s. Let us see how this number compares with the angular momentum values for the electron and the rock.

Recall from Chapter 7 that the angular momentum for something moving in a circular path is the mass times the speed

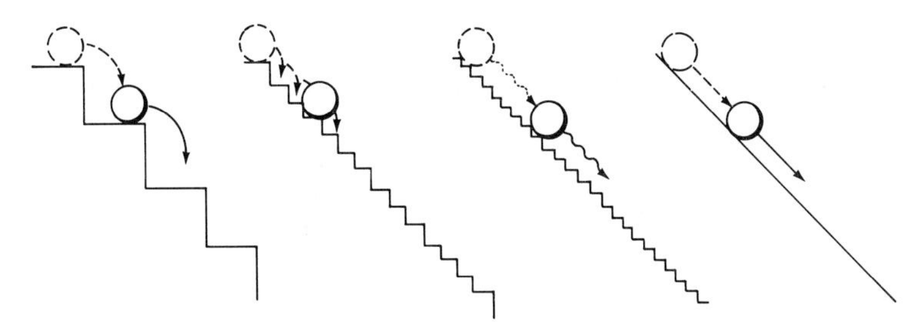

Figure 23-3. As more and more steps come closer together, eventually one "step" cannot be distinguished from the rest.

*Notice that angular momentum has units kg · (m/s) · m = kg · m^2/s = J · s, which are also the units of h.

times the radius of the circle: $L = mvr$. The electron has a mass of about 10^{-30} kilogram and its orbit's radius is about 10^{-11} meter. So if we assume a reasonable speed of, say, 10^6 m/s, then mvr is

$$10^{-30}\ \text{kg} \times 10^{-11}\ \text{m} \times 10^{6}\ \text{m/s} = 10^{-35}\ \text{J} \cdot \text{s}$$

In going to the next orbit, the angular momentum change of 10^{-34} J · s is of roughly this same magnitude. Here, then, the quantization is a very noticeable effect. The *separation* in angular momentum values is about the same size as their values in the individual allowed orbits.

On the other hand, look at the situation for the rock on the string. Let us suppose that the mass is 1 kilogram and we whirl it on the end of a 20 centimeter string at a speed of, say, 1 m/s. Since the angular momentum is mvr, we have

$$L = mvr = (1\ \text{kg})(1\ \text{m/s})(0.2\ \text{m}) = 0.2\ \text{kg} \cdot \text{m}^2/\text{s} = 0.2\ \text{J} \cdot \text{s}$$

In the next "allowed orbit" farther out, the rock's angular momentum should be greater than this by 1.05×10^{-34} J · s, according to the quantum mechanical rule. But 0.2 is more than a billion-billion-billion times larger than 10^{-34}. Adding such an incredibly small number makes a difference far too small to be detected by any measuring device, existing or imagined. You would get a much larger effect by adding the money in your pocket to the national debt of the United States.

Again, the allowed radii of the whirling rock are so closely spaced as to be continuous, so you can put the rock in a circle of any radius you like, until you run out of string.

A crude analogy may drive the point home. Let us liken our individual energy quanta to individual molecules of water. Suppose you had a drinking glass so small that it could hold only ten molecules of water. Clearly, any amount of water in the glass would be "quantized" in whole multiples of a water molecule. The microscopic glass could be one-tenth full (with one molecule) or two-tenths full (with two), and so on, because you could not add or remove a fractional part of a molecule. Similarly, the contents of an ordinary drinking glass are also quantized, but this glass can hold billions of billions of molecules. Addition of one or two makes no measurable difference. The quantization is not perceptible. The water pours in a steady stream, not in discontinuous jumps, one molecule at a time. (See Figure 23-4.)

Figure 23-4. The contents of both glasses are quantized. (a) Microscopic drinking glass, holding only a few molecules. (b) Ordinary drinking glass, holding billions of billions of molecules.

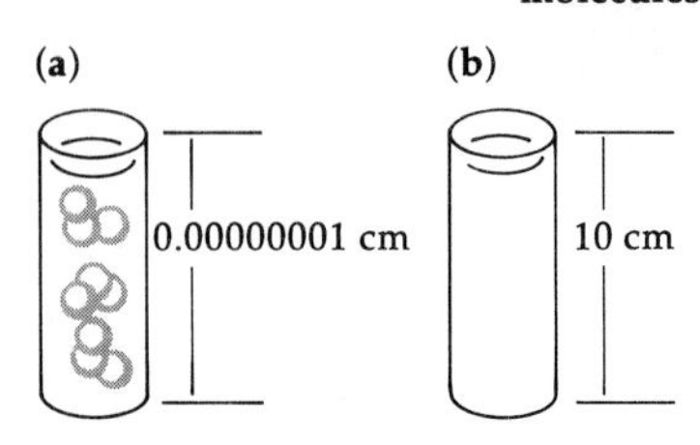

As you can see, we are talking about a size effect. We observe quite different behavior depending on whether the size of Planck's constant is significant with respect to the dimensions of the system involved. The energies and angular momenta of the electron are of magnitudes comparable to h, so the effects of h cannot be ignored. But Planck's constant is of no consequence for something the size of even a tiny grain of sand, much less for a rock or a person. It is essentially zero on this scale of things, and we need not worry about it.

However, it would be a mistake to suppose that the quantum theory involves only quantitative differences with classical mechanics. That would be like saying that the only distinction between a pauper and a king is the degree of wealth: The very fact of the *quantitative* difference implies important *qualitative* differences—in life-style, outlook, opportunities, quality of life. Similarly, quantum mechanics, by virtue of its success in interpreting nature at the submicroscopic level, has radically changed our picture of the world, our understanding of the role of the observer in experimentation, indeed our whole philosophy of science.

The Zone of Middle Dimensions

The everyday world in which we live might be called the "zone of middle dimensions." Objects we can see and touch and experience with our senses are intermediate in size, between atoms and molecules on the one hand, and stars and galaxies on the other. Classical physics is accurate and satisfying in accounting for our physical world in this middle zone. But it must give way to quantum mechanics in the realm of the very small and to general relativity for the very large. Likewise, the speeds at which objects in our middle zone travel are small compared with the velocity of light and, as we have already seen, special relativity takes over for motion at these higher speeds.

Because the zone of middle dimensions is the one to which we are accustomed, our thinking and hence our language are geared in a particular way. We have grown up in this zone. It has shaped the way we view the world. The words we use to describe our surroundings have become part of our vocabulary as a result of our experience. So, even though you might find the details of classical physics somewhat mystifying, the concepts are comfortable ones, easy to identify with.

Our description of what nature does in this zone lends itself to pictures. We can easily visualize what happens when we heat a pan of water. We draw pictures of falling objects or watch them in slow-motion movies. Such pictures, whether mental images or actual drawings, are highly useful in helping us understand some rules of the game as well as predicting new ones. We can readily see what classical physics is trying to teach us. If we don't understand an orbiting satellite, it takes only a little more effort to make a table-top model and watch what happens.

When it comes to electrons and photons, though, it is a whole new ball game. They behave in a way governed by the rules of quantum mechanics, which is like nothing you have ever seen before. Quantum mechanics seems to contradict common sense. But this is only because common sense has been fashioned by incomplete experience, limited to the zone of middle dimensions. In an earlier day, the notion that the earth was round violated common sense. People knew the earth was flat because their own backyards and the land and ocean looked flat as far as they could see. The realm of electrons and photons is equally puzzling to modern people. Is a photon both a particle and a wave? How can it be that way? It simply doesn't "feel" right, just like a round globe didn't to the flat-earthers.

> *I remember discussions with Bohr which went through many hours till very late at night and ended almost in despair; and when at the end of the discussion I went alone for a walk in the neighboring park I repeated to myself again and again the question: Can nature possibly be as absurd as it seemed to us in these atomic experiments?*
>
> *Werner Heisenberg*
> Physics and Philosophy

You have probably already experienced this uneasiness about light, or in trying to imagine why an orbiting electron can have only particular radii and cannot exist in between. Physicists have the same trouble, because their everyday experiences are pretty much the same as yours. We live on the same flat earth. The language really does fail us when we try to describe the quantum-mechanical world with the comfortable old classical ideas. We can make helpful analogies, but they always fall short. We can draw pictures, but something is always missing. The planetary model for the hydrogen atom is a good example. People were comfortable with it because it brought to mind something familiar: the sun and its planets. Bohr's energy-level picture, which torpedoed the solar-system model, was both good news and bad news—good because it worked, accounting for the observed line spectra, and bad because it introduced quantization into matter, thereby saddling us with an uncomfortable new idea.

Quantum theory has altered the philosophy of science by bringing into the spotlight a serious problem in science, that of describing the universe at *all* levels with a language and mode of thought developed and nurtured at only one level. Ours is the world of classical mechanics. It is quite natural that our ideas, concepts, and vocabulary should have emerged appropriate to this zone. We have little trouble visualizing sizes or distances ranging from a few millimeters to thousands of kilometers, or events that span a few hundredths of a second or several centuries. The space age has familiarized us with velocities of many thousands of kilometers per hour. In the zone governed by classical mechanics, we are very much at home. Our mental pictures and blackboard sketches are helpful representations of the objects and events with which we are somewhat familiar.

When we step through the gateway of the atom, we are in a world that our senses cannot experience. There is a new architecture, a way things are put together that we cannot know: we try to picture it by analogy, a new act of imagination. The architectural images come from the concrete world of our senses, because that is the only world that words can describe. But all our ways of picturing the invisible are metaphors, likenesses that we steal from the larger world of eye and ear and touch.

Twentieth-century physics has taught us to venture gingerly outside our middle-dimension realm. Our minds have difficulty producing an image of the nucleus, so small that a billion nuclei laid end to end would extend less than a centimeter. It is hard to visualize something traveling at the speed of light, seven times around the earth in less than a second. It is hard to imagine the size of our Milky Way galaxy, so large that light takes hundreds of thousands of years to make the trip from one edge to the other. So it isn't surprising that nature behaves differently at the exceedingly small dimensions and the exceedingly great speeds. And our difficulty in understanding parallels our difficulty in describing these strange zones.

> ***While one philosopher affirms***
> ***That by our senses we're deceived,***
> ***Another swears, in plainest terms***
> ***The senses are to be believed.***
>
> *Fables of La Fontaine*

Learning Checks

1. What is meant by the "zone of middle dimensions"?
2. How is quantum theory analogous to relativity as a description of nature?
3. Is the energy of the pendulum quantized? How about the angular momentum of a wagon wheel? Explain.
4. Explain the importance of the size of Planck's constant.

THE MATTER WAVES OF LOUIS DE BROGLIE

Neils Bohr did not suggest any fundamental rationale for his postulates that quantized the orbits of the electron in hydrogen. This was left to a graduate student at the École Polytechnique in Paris who, thirteen years after Bohr's work, introduced an idea that was to have far-reaching implications. Prince Louis Victor, duc de Broglie (pronounced BROH-lee), a son of French nobility, came to the study of physics by an unusual route. The de Broglie family had a long tradition of scholarship, and Louis began his studies in ancient history at the University of Paris. His move toward physics undoubtedly was influenced by the work of his older brother, Maurice, who was doing experimental work with X rays and radioactivity. The younger de Broglie attacked the problem of atomic theory. In his doctoral dissertation in 1926 he presented his work, which was the result of a brilliant bit of physical insight.

De Broglie had familiarized himself with the dual situation for light. As we have already discussed, light behaves like a wave in those circumstances, such as diffraction, where its nature as a continuous phenomenon is most evident. However, it behaves like a particle in those instances, such as the photoelectric effect, where quantization is the dominating feature.

Because he was a lover of classical music, de Broglie understood much about the physics of musical instruments. It occurred to him that music provides an example, within the zone of middle dimensions, of a phenomenon that also has this dual nature, showing wave properties and quantization effects. He was able to solve the mystery of Bohr's hydrogen atom by considering the electron in terms of standing waves analogous to those produced by musical instruments.

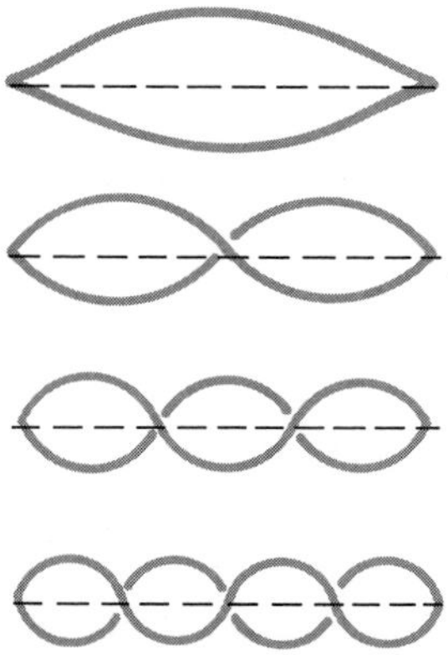

To understand this, we need to recall what we learned about standing waves. Figure 23–5 shows the possible standing wave patterns for a vibrating string clamped at both ends. We saw this picture in Chapter 17. The frequencies f_n at which the string can vibrate are restricted to whole number multiples of the lowest, or fundamental, frequency f_0. Corresponding to each frequency is a wavelength λ_n, given in terms of v, the speed of the waves traveling up and down the string:

$$\lambda_n = \frac{v}{f_n}$$

Figure 23-5. Quantization of guitar string frequencies.

So the wavelength is also restricted. For a string of length L, λ can only be equal to $2L$, L, $(2/3)L$, $(1/2)L$, $(2/5)L$, and so on. A frequency of 0.75 f_0 or a wavelength of $0.1372L$ simply cannot exist.

For standing waves, then, the frequency and the wavelengths are *quantized.* If the fundamental frequency is, say, 300 hertz, then the possible vibrations of the string are at frequencies of 300, 600, 900, 1200, . . .—that is, values of 1, 2, 3, 4, and so on, times the fundamental frequency. For a string 2 meters long, the wavelength could only be 4 meters, 2 meters, 4/3 meters, 1 meter, 4/5 meter, and so on—that is, values of the fundamental wavelength divided by 1, 2, 3, 4, ... This is a common example of quantization, but for a system much larger than a light wave or an electron.

Frequencies of sound waves in organ pipes are also quantized. Figure 23–6 shows an organ pipe open at one end. Drawn alongside are some of the various possible standing waves. For each wave that the pipe can contain, the closed end must be a *node* and the open end an *antinode.* This means that for a given pipe length L, only those waves are possible that have their wavelength λ so that $L = (1/4)\lambda$, $(3/4)\lambda$, $(5/4)\lambda$, and so on—that is, any odd number of quarter wavelengths. The allowed values of λ are then $4L$, $(4/3)L$, $(4/5)L$, $(4/7)L$, and so on, each of which corresponds to a particular frequency according to the equation given above.

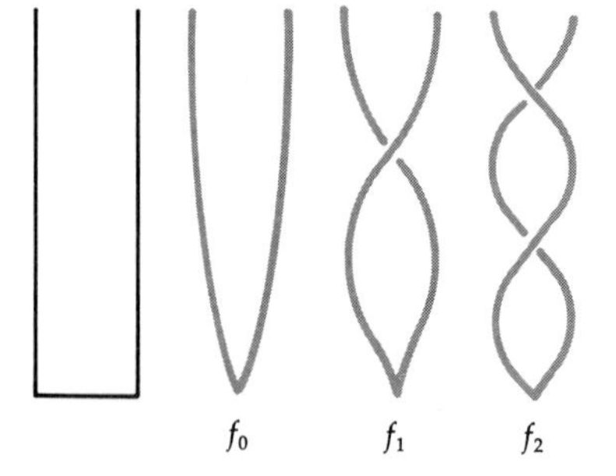

Figure 23-6. Quantization in an organ pipe.

Although the details of these examples of large-scale frequency quantization differ, they have a very important feature in common. Both are *bound* systems in the sense that the waves are confined to the length of the string or the interior of the pipe. The waves traveling up and down the string are restricted because the string is clamped at both ends, and any wave moving in one direction is reflected back the other way when it strikes the boundary. There are also boundaries in the organ pipe. Waves are reflected—from the closed end just like on the string, and from the open end because of differences between the air confined to the pipe and that on the outside. In both cases the boundaries or limitations of the systems lead to quantization of the frequencies and the wavelengths of standing waves. If the string were not clamped, then waves of any frequency at all could travel along it. But the presence of the boundary at the fixed ends, causing the "wrong" waves to be canceled, results in quantized frequencies and wavelengths.

De Broglie combined these two ideas—the wave-particle duality for light and quantization of waves in a bound system. His speculation was something like this: Since light, which we usually observe as *waves*, can also have *particle* features, is it possible that matter, which we observe as *particles*, can also have *wave* properties? Light shows its wave properties when it travels through space and through various materials such as glass and other optical devices—for instance, in refraction and diffraction. Its particle properties show up during emission and absorption, when it exchanges radiant energy with matter, such as the photoelectric effect and in atomic spectra. These observations led physicists to envision light as photons guided by characteristic waves.

De Broglie proposed a similar situation for matter, imagining it to be particles in some sense guided by associated waves. Playing around with these ideas, he established a mathematical way of describing his "matter waves." He took Einstein's famous equation for the equivalence of energy and mass,

$$E = mc^2$$

and the equation for the energy of a photon of frequency f (the one we have used several times recently),

$$E = hf$$

Let us set the righthand sides equal, since both are equations for energy:

$$mc^2 = hf$$

Going a little further, we can write the frequency f in terms of the speed c and the wavelength λ for light, $f = c/\lambda$. Then our equation becomes

$$mc^2 = \frac{hc}{\lambda}$$

This looks a little simpler if we divide both sides by c, and write

$$mc = \frac{h}{\lambda}$$

The same equation has another form if we just exchange λ and mc:

$$\lambda = \frac{h}{mc}$$

This equation is fine for light, which always travels at speed c. So de Broglie thought, why not have the same equation for a *particle* traveling with some arbitrary speed v? Simply putting v in place of c gives

$$\lambda = \frac{h}{mv}$$

Here we have an important equation, sometimes called the *de Broglie matter-wave relation.*

Look carefully at this equation. The lefthand side, λ, is the wavelength, a distinctly wave-associated quantity. Since the product mv, mass times velocity, is the momentum of a moving particle, then the righthand side can be said to carry a particle identification. The equation, then, uses Planck's constant h to connect the wavelike and particlelike properties of matter. In other words, de Broglie's equation says this: If Planck's constant is divided by the momentum of a particle, the resulting number is the wavelength of the wave associated with that particle.

That "waves" should be affiliated with "particles" at all is a bizarre notion, at least from the view of the mechanics of Galileo and Newton. But let us hold off criticizing that for a little and see how de Broglie used his unorthodox notion to account for Bohr's hydrogen. The key is that the electron is a *bound* system, much like the guitar string or the air in the organ pipe. As the gravity-bound earth orbits the sun, so the electron stays in its orbit around the proton because of the electrical attraction of the two opposite charges. De Broglie said that the electron is a standing wave. It can survive only if its wavelength "fits" around the circumference of the orbit in the same way that the wavelengths of the possible standing waves must fit the length of the vibrating string.

Imagine that the string is unclamped and its ends tied together (as in Figure 23-7) so that the string makes a circle whose circumference is the same as the length around the electron's orbit. If the standing waves on this circle are to guide the electron's motion, then the only allowed orbits are the ones with circumferences of just the right size to accommodate a whole number of wavelengths. The orbit distance might equal one wavelength or two, or three, and so on, but not, say, 1.17 or 0.26.

Restricting the circumference of the orbit in turn restricts the allowed values of the radius. The circumference is equal to 2π times the radius. Suppose the radius for a general orbit, the *n*th orbit, is called r_n. Then the circumference of this orbit will be $2\pi r_n$. De Broglie's idea was that this distance must be equal to some whole number of wavelengths; that is,

$$2\pi r_n = n\lambda$$

where n can take on the values 1, 2, 3, 4, and so on. The value of n, which is called the *principal quantum number,* is a label for allowed orbits—the first, second, third, and so on—that the electron can occupy. It can't be anywhere in between, just as the guitar string can vibrate in one, two, three segments, but not some fractional number.

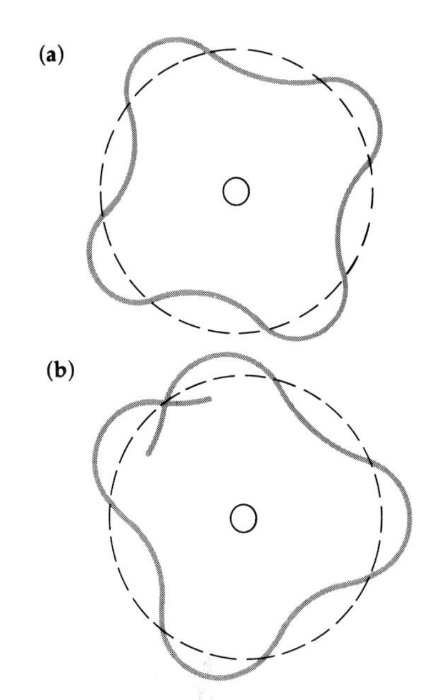

Figure 23-7. Standing waves for the circle. (a) A possible orbit: the wavelengths fit. (b) Not possible: the wavelength is the wrong size.

By carrying the mathematics just a little further, we can see how de Broglie's work linked up with Bohr's ideas. In the previous equation, let us substitute λ from the de Broglie matter-wave relation:

$$2\pi r_n = \frac{nh}{mv}$$

where m is now the mass of the electron and v is its orbital speed. Rearranging things slightly gives us

$$mvr_n = \frac{nh}{2\pi}$$

Now remember from Chapter 7 that when an object whose mass is m goes with speed v around a circle of radius r, its angular momentum L is mvr. And this is exactly what we have in this equation. The lefthand side is just the angular momentum for the orbiting electron. We will label this L_n because it is the angular momentum connected with the *n*th orbit. So we have

$$L_n = \frac{nh}{2\pi}$$

This equation says that the orbital angular momentum has to be some whole number multiplied by Planck's constant divided by 2π.

But this is just the first assumption that Bohr made about the electron in hydrogen. De Broglie, by extending his knowledge about quantization in a bound system, had achieved a mathematical expression of Bohr's ad hoc notion.

This was an immensely important result. Bohr made his assumption about quantized angular momentum out of necessity. Furthermore, it worked. But de Broglie, by looking at the electron in a way that no one ever had before, was able to arrive at the quantization automatically. He supposed the electron to be a standing wave and then simply let mathematics follow its nose. Both men's contributions are important as links in a chain put together to interpret the line spectrum of hydrogen. De Broglie's analysis is the more fundamental, because it serves to deepen our understanding of the very nature of matter at this submicroscopic level. Both men, by the way, were awarded the scientific community's highest honor, the Nobel prize, for their work. Bohr became a Nobel Laureate in physics in 1922, and de Broglie in 1929.

Learning Checks

1. For what kind of systems do we find quantized frequencies in the macroscopic world?
2. What were the two ideas that de Broglie combined in the concept of matter waves?
3. Is the electron orbiting the nucleus a bound system? Explain.
4. Explain how the electron's standing waves lead to quantization of angular momentum.

PARTICLES AS WAVES: A CONTRADICTION?

De Broglie's suggestion of associating a "wavelength" with matter seems strange and wild. We don't see anything about baseballs or coins or rocks that even remotely resembles waves. If particles have wave properties, why don't we notice them?

We can get an answer to this question by looking at de Broglie's matter-wave relation and the numbers it gives. Remember that diffraction of light waves and water waves takes place only when the diffracting object is about the same size as the wavelength. This is a general property of waves. So, for matter waves, we need to see what wavelength would be associated with a given particle.

Take a typical example. An object having a mass of 1 kilogram moving at 1 m/s has a momentum

$$mv = (1 \text{ kg})(1 \text{ m/s}) \\ = 1 \text{ kg} \cdot \text{m/s}$$

The de Broglie relation gives the wavelength:

$$\lambda = \frac{6.62 \times 10^{-34} \text{ J} \cdot \text{s}}{1 \text{ kg} \cdot \text{m/s}}$$

$$\lambda = 6.64 \times 10^{-34} \text{ m}$$

When you realize that the distance across the nucleus of an atom is about 10^{-15} meter, you can see what an incredibly small number this wavelength is—a *billion-billion* times smaller than the nucleus. To observe diffraction for this everyday-sized particle we would need a slit whose width is about a billion-billion times narrower than the nucleus. Clearly this is impossible. As the momentum becomes *larger,* either because of a bigger mass or a higher speed, or both, the corresponding de Broglie wavelength gets even *smaller.* This means that diffraction effects would never be observed for a grain of sand, much less for a bicycle or a television set.

But what about the de Broglie wavelength for an electron? Here the mass is very much smaller, about 10^{-30} kilogram. If we assume a velocity of, say, 1000 m/s, we have

$$mv = (10^{-30} \text{ kg})(1000 \text{ m/s}) = 10^{-27} \text{ kg} \cdot \text{m/s}$$

Putting this into the de Broglie relation gives a wavelength of

$$\lambda = \frac{6.62 \times 10^{-34} \text{ J} \cdot \text{s}}{10^{-27} \text{ kg} \cdot \text{m/s}} \cong 10^{-7} \text{ m}$$

A wavelength of 10^{-7} (0.0000001) meter is still very small compared with ordinary household objects. But the distance between atoms in a crystal, for example, is about 10^{-9} meter, so the electron's wavelength is at least on the order of a physically realizable dimension. This being the case, we would expect electrons to show diffraction effects, much like light waves in the Young double-slit experiment.

In fact, two American scientists at the Bell Telephone Laboratories, Clinton Davisson and Lester Germer, observed just such electron diffraction in 1928, 2 years after de Broglie's prediction. They fired beams of electrons at the surface of a nickel crystal. The electrons rebounded more strongly at *particular* angles, with hardly any rebound at intermediate angles. This "quantized rebound" produced a pattern of regions in which the electron density was alternately high and low, completely analogous to the bright-dark alternations in the two-slit diffraction of light. (See Figure 23-8.) If the electrons behaved like a stream of BB's, they would rebound randomly, not at preferred angles. The Davisson-Germer experiment was extremely important, for it dramatically demonstrated the wave behavior of particles.

The next year, in 1929, George Thomson (J. J.'s son), confirmed the wave aspects of electrons in similar studies involving electrons streaming through various crystalline materials. It is interesting that both Thomson men received Nobel prizes—J. J. in 1906 for showing the existence of electrons as *particles* and George, who shared the 1937 prize with Davisson, for proving the existence of electrons as *waves.*

Wave properties have also been established for protons, neutrons, and even atoms, objects thousands of times larger than the electron. The behavior of these "particles" under the conditions used in these experiments is not at all consistent with Newton's laws. The only feasible interpretation involves de Broglie waves.

It may appear that quantum mechanics has created something of a mess. After all, what are we to make of this business of light being waves on the one hand and particles on the other? For matter it seems even worse. To swallow the idea that material objects can be considered as waves requires a suspension of disbelief. And yet the experimental evidence is there: light *does* behave like streams of particles in the photoelectric effect; and electrons, protons, and even atoms *do* show wavelike diffraction effects. Everything begins to look the same—particles as waves and waves as particles. It all seems a bit screwy. Maybe it is some comfort that electrons and photons behave in the *same* screwy way!

So what are we to say about the electron? Or the photon? Is it a particle or is it a wave? To answer the question in either way—"It is a particle" or "It is a wave"—is to be misleading and restrictive. Whichever side we come down on, we leave out

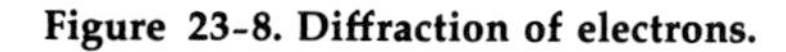
Figure 23-8. Diffraction of electrons.

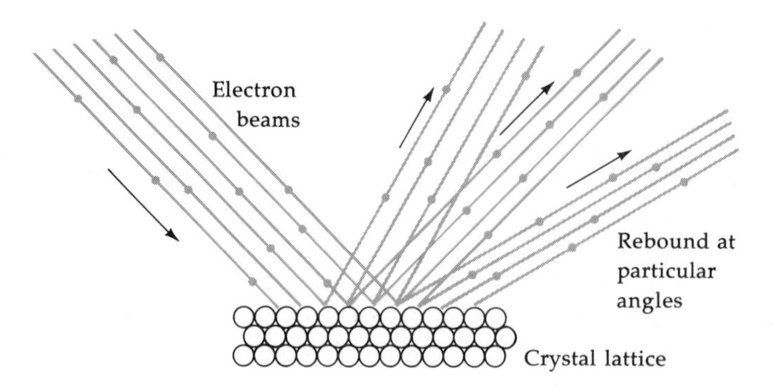

some very important parts of its behavior. It also gets us into the trouble of trying to describe nature at the submicroscopic level with thoughts and words developed in a severely limited environment and based on incomplete experiences.

The nature of matter (electrons) on the one hand and light (photons) on the other seems to have duality when the concepts of "waves" and "particles" are limited to their classical connotations. It is not much help, either, to say that a photon or an electron is somehow "both a wave and a particle" (someone has coined the term "wavicle" for such a beast). Given the constraints that our middle-zone language places on us, the best we can do is something like this: A photon is neither strictly a wave nor strictly a particle. It is simply a photon. It may reveal its wave nature or particle nature, depending on the conditions in which one measures its properties. For example, the beautiful nineteenth-century experiments that characterized light as a wave were those that happened to catch light in circumstances when its wave aspects were dominant. On the other hand, the particle aspects are more prominent in the photoelectric effect.

A similar set of statements can be made about matter. An electron is neither a particle nor a wave. It is an electron. It is like a particle in the sense that the effects of its energy and momentum are localized within a small region of space; and it is like a wave in that its motion is affected by conditions throughout a wider region. The experiments of Davisson and Germer and of George Thomson happened to "see" the electron wearing its wave costume. Ordinarily we see the particle aspects of matter. In fact, it is this aspect of the material world that gives the ordinary meaning to the word "particle." A baseball, for example, is so extremely unbalanced in favor of its particle features that we are simply never aware of its wave properties at all. Indeed, it sounds ridiculous to "diffract" a baseball by throwing it at a grating.

Our dilemma is rather like that of the three blind men touching the elephant. One said the elephant was a skinny rope, another a fat snake, the third a rough hunk of leather. All were right and yet none was right. The elephant is simply an elephant. It has the feel of all three of these things, depending on what part of the elephant you touch, but does not fit uniquely into the mold of any one of the three descriptions.

You can see how important the size of Planck's constant is in our entire discussion. The nearer an object's dimensions are to those for which the size of h is significant, the more nearly balanced are the wave and particle aspects. It is interesting to speculate on what the world would seem like if Planck's constant were, say, 10^{-5} or so, rather than 10^{-34}. Things would appear decidedly different, and, as we have pointed out, our language would likely be radically altered. The words "wave" and "particle," if they even existed, would have vastly different meanings than they have in our world.

Learning Checks

1. Is there such a thing as the "wave-particle duality"? If so, how does it arise?
2. We might say that what an electron *is* depends on what you use to observe it. Discuss this idea.

THE HEISENBERG UNCERTAINTY PRINCIPLE

Werner Heisenberg, an Austrian physicist and contemporary of Bohr and de Broglie, established the notion that the so-called duality of waves and particles is really not a duality at all, but a necessity of nature at its fundamental level. The result of his work is called the *Heisenberg uncertainty principle*, and it has provided a powerful tool for answering many of the important questions of modern physics.

Heisenberg's theory begins with the problem we have been discussing—that of applying methods and ideas of the zone of middle dimensions to nature on the atomic scale. When we say that a paper clip is a particle, how do we know that? Well, we can *see* it, and we can toss it up in the air, and we can watch what happens to it in a variety of situations. That is, we make *measurements* or *observations.* What Heisenberg showed was that, at the atomic level, the very process of making an observation on a system causes large changes in the system itself. It shifts under our gaze; the very act of our looking alters that at which we look. Observer and observed are not isolated. They are inextricably linked by interaction, which introduces an inherent uncertainty, or indeterminacy, in the result of the observation.

Interaction between the experimenter and the system being measured is not unique to the atomic level, of course. This happens in our everyday world as well. But here, the impact on the thing being observed is so small that it remains the same—or at least apparently the same—before, during, and after the measurement. For example, when you use a mercury thermometer to measure the temperature of a pan of water, some of the energy in the water is used to drive the mercury up the capillary of the thermometer. So you have actually lowered the temperature of the water by measuring its temperature. In order to determine how much current flows in a circuit, the electrician hooks up a meter that requires a very small amount of current to give a reading. As an even more common example, think of what happens when you "look" at something. In order for you to "see" the words on this page, some photons from a light source have to bounce off the page and strike your eye.

In all of these examples, the observer's intrusion causes only a negligible change in the object. The amount of energy transferred from the water to the thermometer is trivial, as is the current stolen from the circuit to be passed through the current-reading meter. The book doesn't change after being bombarded by the photons; it will look exactly the same tomorrow after sitting all night in the dark. All our experiences in the middle zone are like this. So we are led to believe that the measuring process and the phenomenon being observed are quite independent. Science has grown up with this notion—again born of what we know about our "normal" world—that the experimenter is separated from the experiment, the measurer from the measured. But is this really true? Again, let us step out of our familiar surroundings and pass again into the atomic level.

Let us take a simple example. Suppose we try to "look" at an electron as it flies through an evacuated box. We know it is in the box, and our job is to find it and track its motion. From our understanding of trajectories, we expect the electron to follow a parabolic path, like the flight of a golf ball off the tee. To confirm this prediction we will watch the electron, just as we would watch anything else—that is, by shining light on it. So we need at least one photon of light to hit the electron. But since the photon's energy and momentum are about the same size as the energy and momentum of the electron, the colliding photon and electron will behave much like two billiard balls. The photon will recoil. The electron will bounce away with a new speed and direction of motion. (See Figure 23-9.)

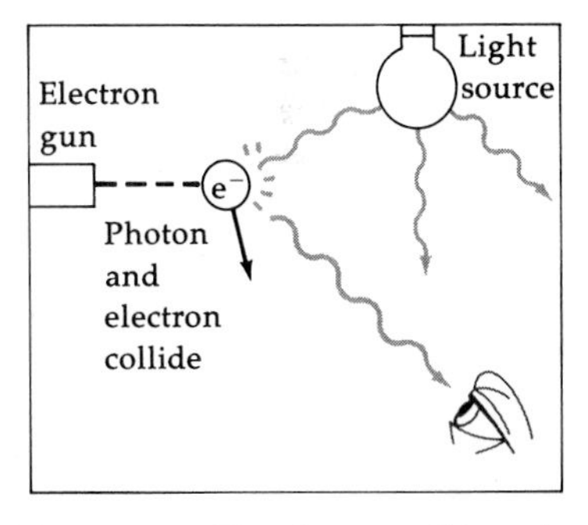

Figure 23-9. The photon striking the eye has collided with the electron, changing its speed and direction.

Now we have created a problem. Our looking at the electron has kicked it off track. The act of our observing its motion has changed the very motion we were trying to observe. There is a considerable uncertainty in the momentum we measure for the electron.

Can we do anything about this? Well, yes. We can reduce the impact of the photon by using light of a lower frequency, and thus lower energy, so as not to disturb the electron's motion as much, leaving it closer to its original path. But now we have a new problem. The lower the frequency of the light, the longer its wavelength and the less able we are to locate the electron, because of diffraction. So we can't determine the electron's exact position at any given time. Suppose we go the other way and use a shorter wavelength to cut down on the diffraction. This will give us a more accurate *position* measurement. But not without a price. Now, the photon's energy is greater, magnifying the effect of its collision with the electron and preventing a good determination of the electron's *momentum*.

Is There Any Certainty About Uncertainty?

The uncertainty principle was not embraced at once wholeheartedly, but was accepted into physics gradually. Here are two men of science remarking on this principle:

I count myself among those who are unable to believe, as Einstein could not and most philosophers and physicists today cannot, that the universe, or a stupendous event like the explosion of a supernova, is dependent for its existence on being observed by such paltry creatures as you and me.

Martin Gardner, "Quantum Weirdness," Discover, *October 1982, p. 76.*

People think the uncertainty principle actually applies to everyday phenomena. Nothing could be farther from the truth. Careless paraphrases have eroded and obscured the true meaning of the principle in the popular mind. . . . The pseudo uncertainty principle . . . an imitation version . . . states: The observer always interferes with the phenomenon under investigation.

Douglas R. Hofstadter, "Metamagical Themas," Scientific American, *July 1981, p. 18.*

What do you think about these views? Are they contradictory?

So our dilemma in trying to watch the electron move across the box is this: short wavelengths of light give good *position* measurements, but the photon's strong collision makes the *momentum* measurement uncertain. Long wavelengths give a softer collision and hence a better result for the electron's momentum, but its location is now uncertain because of the diffraction effects for the long-wavelength photon.

The genius of Heisenberg was his realization that this coupling between observation and event is at the heart of the way the universe is put together. It seems crazy only because we have been misled by our experience, all of which has taken place in the zone of middle dimensions. Classical mechanics has us a little hoodwinked. The precise trajectory of a rocket is a *consequence* of the happy accident of our living in the everyday world, but we have turned things around, unwittingly, until such a trajectory becomes a *necessity*. Given this necessity, the electron-photon idea seems weird.

Heisenberg taught us something different. Because of the particle-wave aspects of both the photon and the electron, he said, there is an intrinsic limit on the precision with which we can simultaneously measure the position and momentum. At best, we can only determine each of these quantities within a certain range, a particular tolerance. The very fact of making the measurement introduces this uncertainty, so that the observer and the event being observed are inseparable, linked together as a part of the phenomenon under investigation. Even in principle, there is no such thing as a physical phenomenon, per se. There is always an unavoidable connection between observer and experiment. They are never independent.

Heisenberg expressed this uncertainty principle in a formula. Suppose we let Δx be the uncertainty in the location of the electron. That is, we know it is somewhere in the interval Δx, but we don't know exactly where. (See Figure 23–10.) Similarly, we know its momentum to be in the range Δp, which is the uncertainty in the momentum. Then the Heisenberg uncertainty principle is

$$\Delta x \Delta p \geq \frac{\hbar}{2}$$

where $\hbar$ is Planck's constant divided by 2π, and the symbol $\geq$ means "greater than or equal to." In words: The product of the uncertainties in position and momentum must be larger than one-half of $\hbar$. We can never have zero uncertainty, or perfect certainty for both. Notice that this represents our imagined experiment with the electron in the box. As the measurement of position becomes more and more exact, Δx decreases but Δp must get larger in order that the product remain greater than $\hbar/2$, meaning that a simultaneous measurement of the momentum has a larger and larger uncertainty. On the other hand, if we arrange things so that the momentum is closely determined, then we are uncertain where the electron is located. This is much like squeezing a balloon. You might make it skinny in one place, but it bulges out somewhere else.

We emphasize that this has nothing to do with the quality of the instruments used to make the measurements. The limitations are intrinsic, inherent in nature. No amount of sophisticated technology can get around them.

Also, it is important to understand that the uncertainty principle applies to measuring *simultaneously both* position *and* momentum. Individually they can be determined with arbitrary exactness, but not together.

These measurement uncertainties don't show up in classical mechanics. If you drop a rock out the window, you can use Newton's laws to calculate the position and momentum at any time. Your simultaneous measurements of these quantities will be as accurate as your measuring instruments allow. But does the uncertainty principle apply to large objects? Let us look at the numbers.

Imagine that we have a 1 gram object; its mass is then 0.001 kilogram. The uncertainty principle gives

$$\Delta p \Delta x = m \Delta v \Delta x > 10^{-34}\ \text{J} \cdot \text{s}$$

We have written Δp as $m\Delta v$, the mass times the uncertainty in the speed. Dividing by the mass, we have

$$\Delta v \Delta x > 10^{-31}\ \text{m}^2/\text{s}$$

We can satisfy this by taking, say, $\Delta v = 10^{-16}$ m/s and $\Delta x = 10^{-15}$ meter. This says that the speed is uncertain by a fraction of a millimeter per *century,* and the position is uncertain by the diameter of an atomic nucleus. Such "uncertainties" hardly

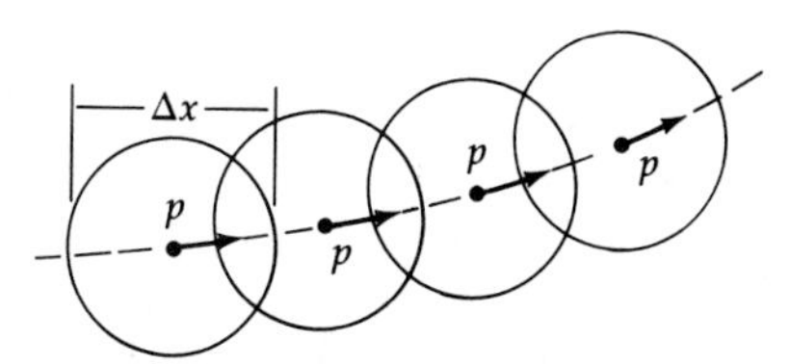

(a) Long-wavelength photons: the position is uncertain, but the momentum is sharply defined.

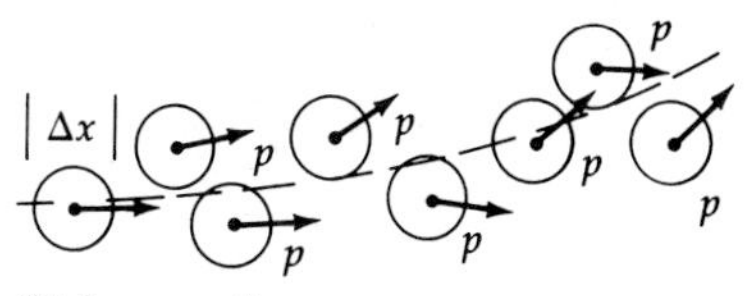

(b) Intermediate-wavelength photons: the position is more precise, but now the momentum is less certain.

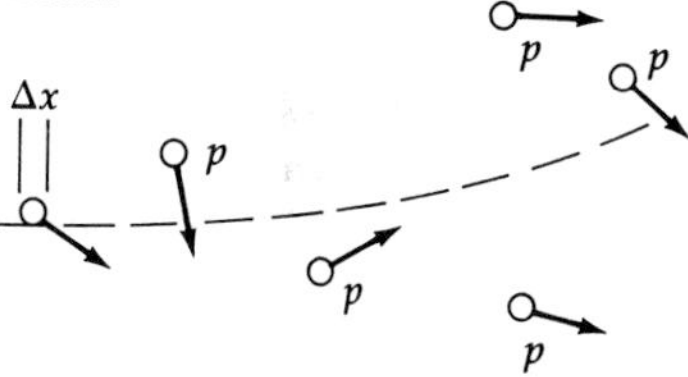

(c) Short-wavelength photons: the position is quite precise, but the momentum is very uncertain.

Figure 23-10. Position-momentum measurements for three levels of photon energy. The hypothetical undisturbed trajectory for the particles (circles) is shown as a dashed line.

make our measurements uncertain! The range within which the measured values lie is unimaginably tiny, so small we can take the uncertainties to be zero. Small wonder, then, that precise trajectories are a fact of life in our usual experience. Heisenberg's principle can be ignored for large objects because the indeterminacies are so fantastically small. We are safe in trusting the exactness of the classical laws for the zone of nature where they apply.

For an electron, whose mass is 10^{-30} kilogram, a similar calculation gives

$$\Delta v \Delta x = \frac{10^{-34}\ \mathrm{J \cdot s}}{10^{-30}\ \mathrm{kg}} = 10^{-4}\ \mathrm{m^2/s}$$

If the electron is in an atom, Δx is about 10^{-10} meter (roughly the diameter of the hydrogen atom), which gives

$$\Delta v = 10^{6}\ \mathrm{m/s}$$

This is a tremendous uncertainty in the electron's speed. It means that the *range* of speeds is almost as large as the speed of light.

The uncertainty principle tells us once again that it is not a question of objects being either particles or waves, or even particles *and* waves. The wave and particle aspects are mutually complementary ways of describing nature. They aren't opponents in a battle; they are partners in our attempts to understand the structure of the universe.

Learning Checks

1. What does it mean to say that "there is no such thing as a physical phenomenon, per se"?
2. Explain the symbols in $\Delta p \Delta x > \hbar/2$.
3. In tracking the electron across the box, why would a higher-frequency light give less diffraction?
4. Discuss the importance of the size of Planck's constant for the uncertainty principle.

Waves and Probabilities

The wave-particle description of light and matter, and the uncertainty principle that accompanies it, make necessary a completely different view of the world from the one classical physics and our ordinary experience give us. With Heisenberg, we must conclude that at the atomic level there is no such thing as the trajectory of a particle—at least not in the sense of the orbit a planet follows around the sun or the path of a baseball when an outfielder fires it home. For such large things we can measure location and speed, or calculate them from Newton's laws, almost as if a moving object is held in its path by some kind of track. The motion seems determined beforehand. By knowing what the situation is at some given time, we can correctly predict what sort of motion the object will have at some later time.

As we have seen, things are not so rosy in the realm of quantum mechanics. The determinism we are so used to isn't there any more. Since the electron and all other bits of matter have wavelike aspects, we should no longer think of them as localized in space. We can't say the electron is "here" or "there" in the way we are used to locating a stick or a house. This is in the nature of waves. A wave on a string has no particular location but is spread out all over the string. So it is with the electron. Rather than moving along a planet-type orbit, it is guided by the waves through a range of positions. And here is the key point: This guidance takes place in a *probabilistic* way. Instead of precisely locating the electron in such-and-such a place, we can only say that it has a certain *probability* of being there. Here is where the "uncertainty" that Heisenberg talked about comes into play. Rather than allowing us to state definitely that the electron has a particular location and momentum, all the information we have about the electron enables us only to state the probability that we will find it at a given place with a certain speed. But it *might* be somewhere else with some other speed—we cannot be absolutely sure.

Let us be careful here. The probabilistic feature is *not* due to a lack of information. For example, when we say that there is a certain probability of drawing four aces from a poker deck, we mean that we can't be sure about this draw because we don't know how the cards are arranged. If we knew how the deck was stacked, we could easily forecast the outcome of a particular draw. So probabilities for poker hands result from our ignorance. But in quantum mechanics, we *cannot* know how the deck is stacked. We can't, even in principle, have all the information necessary to predict events at some later time. What the uncertainty principle tells us is that perfect and complete knowledge about the electron still leaves us short of precisely determining its position and momentum.

As you might imagine, many physicists were not willing to accept this picture of nature. Einstein, who started all the trouble with his light quanta, simply could not buy the probabilistic interpretation of quantum mechanics. "I can't believe that God is throwing dice with the universe," he is supposed to have said. There are those who hold to what is called the "hidden variable" theory, which says that there is some as-yet-undiscovered, hidden information about any given quantum mechanical system that we just haven't been clever enough to find. When we do, determinism will be back on the throne where it belongs. But faced with the stunning success of quantum mechanical probabilities in describing the submicroscopic world, most scientists today have no trouble accepting this version of the theory.

Learning Checks

1. How do probabilities in quantum mechanics differ from those in poker?
2. What is the connection between probabilities and Heisenberg's uncertainty principle?

SUMMARY

The new theory of *quantum mechanics* was developed to aid in understanding nature at the level of very small sizes. A new feature that doesn't show up in the macroscopic world is *quantization* of such important physical quantities as angular momentum and energy. For a swinging pendulum, for example, the smallness of Planck's constant squeezes the energy levels so close together that they appear to be continuous. Because we are accustomed to experiencing objects in the zone of middle dimensions, our thought processes and language are not equipped to deal effectively with quantization.

In order to account for Neils Bohr's model of hydrogen, Prince Louis de Broglie introduced the idea of *matter waves.* He combined the wave-particle duality for light and the quantization of frequencies, as in standing waves on a string. The orbits of hydrogen become quantized when we assume that the electron has a wavelength that fits the circumference of the orbit. By extension, all objects have wave aspects that diminish with increasing size.

De Broglie's prediction of particles as waves was soon confirmed by Davisson and Germer, who observed *electron diffraction,* much like Thomas Young's two-slit diffraction of light. These results and the successful quantum theory teach us that radiation and matter both behave like particles and waves, depending on the way we observe them.

In the *Heisenberg uncertainty principle,* we see the wave-particle situation not as a dualism, but as a necessity of nature. Whenever we make measurements, we interfere with the process being measured. This has negligible effects for large objects, but at the size level of photons and electrons the measuring process can disturb the system greatly. In particular, we cannot simultaneously determine the position and speed of an electron with perfect accuracy.

According to quantum mechanics, we can only speak of the *probability* of finding the electron in a particular place or determining its speed to be a particular value.

Exercises

1. Comment on the size of Planck's constant as affecting the balance between the wave and particle aspects of an object.

2. For submicroscopic objects such as the electron, we have seen that the act of measuring its position changes the position. Comment on this problem for measuring the length of a bookshelf.

3. Discuss the uncertainty principle as a necessary consequence of the wave-particle aspects of nature.

4. Suppose we have one electron confined to a room and another confined to a cigar box. Which has the larger uncertainty in momentum?

5. As the speed of an object increases, what happens to the *frequency* of its de Broglie wave—does it increase, decrease, or remain the same? Explain.

6. Suppose you are a science historian, called upon to explain why, prior to the twentieth century, no one had imagined wave properties for particles. What would you say?

7. Suppose that an object has a de Broglie wavelength of 200 angstroms (that is, 2×10^{-8} centimeter). For a second particle of only one-tenth the mass of the first, and traveling at the same speed, what is the de Broglie wavelength?

8. Suppose that Planck's constant had the value 5×10^{-5} J · s instead of its true value. What would be the separation between the allowed energy levels for a pendulum whose frequency is 1 hertz?

9. Using the same hypothetical value of Planck's constant as in Exercise 8, calculate the number of equally spaced energy levels spanning a total energy of 20 joules.

10. Imagine again that Planck's constant has the value given in Exercise 8. What would the de Broglie wavelength be for a 1 kilogram object moving at 30 m/s? (The answer will be in meters.) Would it be difficult to diffract this object and thus observe its wave properties? Explain.

The applications of laser technology are nearly limitless, from delicate eye surgery to widespread use in the computer industry.

24

ATOMS AND LASERS: QUANTUM MECHANICS IN ACTION

So far in our study, we have learned that the quantum theory accounts for the wave properties of the electron, the particle properties of light, and the structure of the hydrogen atom. Now we are going to branch out and look at other situations where quantum mechanics explains the observed phenomena. In this chapter we will examine the structure of atoms other than hydrogen, those having more than one electron. Then we will see how an understanding of the way atoms are put together led to the development of a device that has revolutionized modern science and technology—the laser.

PAULI'S PRINCIPLE: THE PERIODIC TABLE OF THE ELEMENTS

For many years before the advent of quantum mechanics, scientists had gradually come to realize that every material object in the world consists of a relatively few different kinds of fundamental elements in a rich variety of combinations. Chemists had studied the properties of these elements by examining how they react with each other. They discovered that elements that are physically very different often have quite similar chemical properties. For example, fluorine is a gas and iodine is a dark brown liquid at ordinary room temperatures. Yet they both undergo similar chemical reactions with, say, hydrogen gas. Some other gases—helium, neon, and argon, to name a few—are likewise quite different in mass and appearance. Yet they have in common the property of being very unreactive. That is, they are so stable they rarely combine with other atoms to form molecules. Chemists began to think of these elements

Figure 24-1. The periodic table of elements.

Group	I	II											III	IV	V	VI	VII	VIII
Period 1	1.008 H 1																	4.00 He 2
2	6.94 Li 3	9.01 Be 4											10.82 B 5	12.01 C 6	14.01 N 7	16.00 O 8	19.00 F 9	20.18 Ne 10
3	22.99 Na 11	24.31 Mg 12											26.98 Al 13	28.09 Si 14	30.98 P 15	32.06 S 16	35.46 Cl 17	39.95 Ar 18
4	39.10 K 19	40.08 Ca 20	44.96 Sc 21	47.90 Ti 22	50.94 V 23	52.00 Cr 24	54.94 Mn 25	55.85 Fe 26	58.93 Co 27	58.71 Ni 28	63.54 Cu 29	65.37 Zn 30	69.72 Ga 31	72.59 Ge 32	74.92 As 33	78.96 Se 34	79.91 Br 35	83.8 Kr 36
5	85.47 Rb 37	87.62 Sr 38	88.91 Y 39	91.22 Zr 40	92.91 Nb 41	95.94 Mo 42	(99) Tc 43	101.1 Ru 44	102.91 Rh 45	106.4 Pd 46	107.87 Ag 47	112.40 Cd 48	114.82 In 49	118.69 Sn 50	121.76 Sb 51	127.61 Te 52	126.90 I 53	131.30 Xe 54
6	132.91 Cs 55	137.34 Ba 56	* 57–71	178.49 Hf 72	180.95 Ta 73	183.85 W 74	186.2 Re 75	190.2 Os 76	192.2 Ir 77	195.09 Pt 78	196.97 Au 79	200.59 Hg 80	204.37 Tl 81	207.19 Pb 82	208.98 Bi 83	(209) Po 84	(210) At 85	(222) Rn 86
7	(223) Fr 87	226.05 Ra 88	† 89–103															
			*Rare earths	138.91 La 57	140.12 Ce 58	140.91 Pr 59	144.24 Nd 60	(147) Pm 61	150.35 Sm 62	151.96 Eu 63	157.25 Gd 64	158.92 Tb 65	162.50 Dy 66	164.93 Ho 67	167.26 Er 68	168.93 Tm 69	173.04 Yb 70	174.97 Lu 71
			†Actinides	(227) Ac 89	232.04 Th 90	(231) Pa 91	238.03 U 92	(237.04) Np 93	(244) Pu 94	(243) Am 95	(247) Cm 96	(247) Bk 97	(251) Cf 98	(254) Es 99	(257) Fm 100	(256) Md 101	(254) No 102	(257) Lr 103

The number above the symbol of each element is its atomic mass, and the number below the symbol is its atomic number. The elements whose atomic masses are given in parentheses do not occur in nature, but have been prepared artificially in nuclear reactions. The atomic mass in such a case is the mass number of the most long-lived radioactive isotope of the element.

as a family, which they called the "inert gases," because of this lack of reactivity. Other elements also seemed to fall naturally into families.

In 1869, a Russian chemist named Dmitri Mendeleev took an important step toward understanding such behavior. He showed that by arranging the chemical elements in a particular way he could construct a chart that displayed the elements grouped according to their chemical similarities. In Mendeleev's day there were sixty-five known elements. Today there are just over a hundred. Figure 24–1 is a modern version of Mendeleev's table. This arrangement of the elements is called a *periodic table* because it shows that the chemical properties of the elements depend on their masses in a periodic way. The elements are arranged so that the masses increase as you read along the rows from the upper left to the lower right. Elements in a given column have similar chemical properties and undergo analogous chemical reactions. Notice, for instance, that fluorine and iodine appear in the same column. The inert gases are together in another column.

Remember, Mendeleev's table was constructed on an experimental basis, 50 years or so before quantum mechanics. So the question arises, can the quantum theory account for the form of the periodic table? We have seen how the new theory explains the hydrogen atom. Now we must see how it applies to these other, more complicated atoms.

For hydrogen, you will recall, the single negative electron

is held near the positive nucleus by electrical attraction. It can occupy any of certain allowed orbits, the Bohr orbits, whose radii correspond to the de Broglie standing waves. As we go along the periodic table to heavier atoms, the positive charge on the nucleus becomes larger, corresponding to a greater atomic number and a larger number of protons. This means that more and more electrons must be present to keep each atom electrically neutral. For example, the helium nucleus with two protons has a charge of +2 and thus requires two electrons. Lithium has three protons and three electrons, beryllium four of each, and so on.

Where do these additional electrons go? As a first guess, we might suppose that they would all drop into the first Bohr orbit and chase each other around the nucleus. (See Figure 24–2.) This is a reasonable suggestion, for such an arrangement would have the lowest energy. It would be much like rolling a bunch of golf balls down the stairs—they would all bounce to the lowest step, as classical mechanics would predict. But by now we are used to surprises in the domain where quantum mechanics reigns, and here we are in for another one. The truth is that all the electrons do *not* tumble down to the lowest orbit. If they did, every atom would be essentially like hydrogen. All the elements would have about the same properties, and the world would really be quite dull.

The fact that all atoms are *not* hydrogenlike tells us that there is some law, some rule that prevents the electrons from all dropping into the first orbit. A German physicist named Wolfgang Pauli, who was born the same year that Planck proposed the quantum, discovered the rule in 1924. It has come to be known as the *Pauli exclusion principle,* and we can express it like this: *No two electrons can occupy the same quantum state.* In order to see how this law sets quotas for the various electron orbits, we need to understand what we mean by a "quantum state."

Recall that the energy of the nth orbit in hydrogen is given by the equation

$$E_n = \frac{-R}{n^2} \qquad n = 1, 2, 3, 4, \ldots$$

Here the integer n is called the *principal quantum number.* For the single electron in its lowest circular orbit, it is the only label we need. But the higher orbits can have elliptical as well as circular shapes, so we need a *second* quantum number to distinguish the ellipses from the circles. This one is also an integer; we will call it ℓ. It can be as small as zero and as large as one less than n. Furthermore, the elliptical orbits can have different orientations in space, so we need a *third* quantum number (m) to specify whether the ellipse is oriented along the X, Y, or Z direction. Figure 24–3 summarizes these three quantum numbers. This description is only approximately correct—remem-

Figure 24-2. Atom with all electrons in the lowest Bohr orbit.

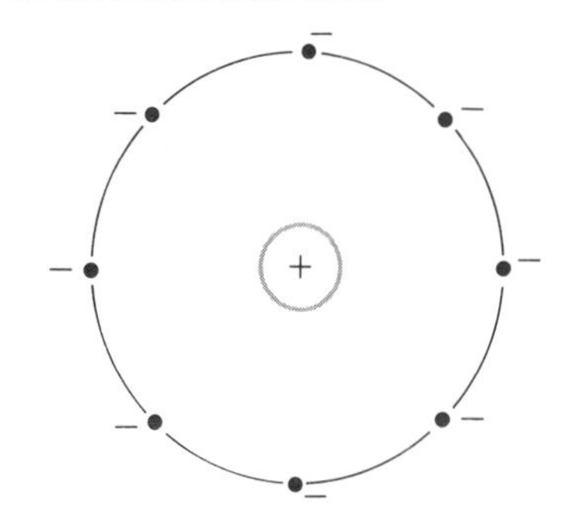

Figure 24-3. Representation of the n, ℓ, m quantum numbers.

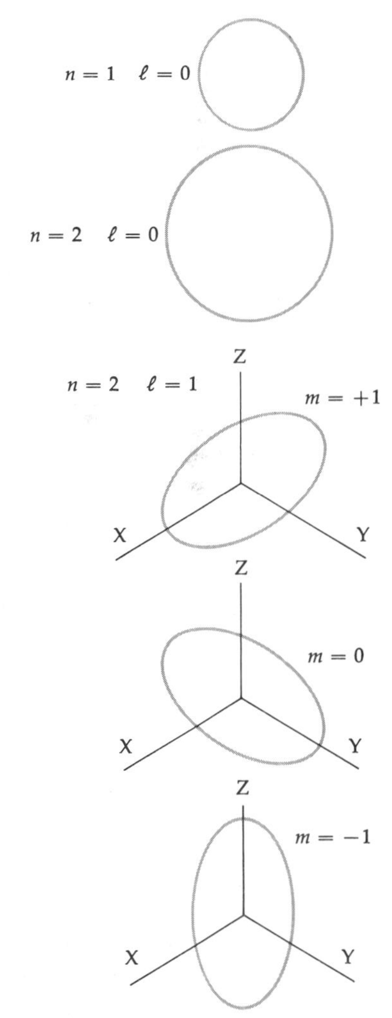

ber that any picture we draw about the atom is only a crude representation of what is actually going on.

Finally, there is a fourth quantum number, called s, the *spin* quantum number. (Recall that in Chapter 16 we attributed the magnetism of iron and other materials to the electron's spinning motion.) This quantum number arises because the electron is found to have an intrinsic angular momentum. Although strictly speaking there is nothing in classical physics with which to compare the electron spin, it is roughly analogous to the daily rotation of the earth about its axis. Using this analogy, we can see that an electron in its orbit can have its spin either in the same direction as its orbital motion or in the opposite direction. This is shown in Figure 24-4. These two possible spin directions mean that s can take on two values, which we arbitrarily will call $+1$ and -1. These four quantum numbers—n, ℓ, m, s—completely describe the quantum state of an electron in the atom. So we can express the Pauli principle this way: *No two electrons can have the same set of four quantum numbers.* Let us see how this works in practice.

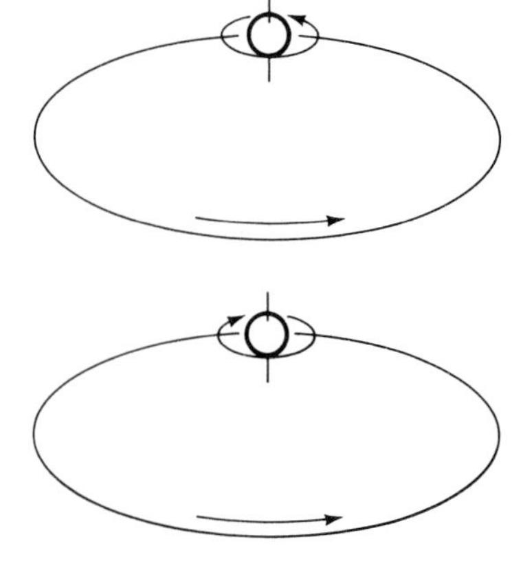

Figure 24-4. The electron can spin in either of two ways relative to its orbital motion.

For hydrogen, there is only one electron. It goes into the lowest ($n = 1$) orbit, which is circular or spherical in three dimensions. The spherical orbits all have $\ell = 0$. They also have $m = 0$ because we can't distinguish among the possible orientations of a sphere in space. This leaves only the spin quantum number s, and it can always be either $+1$ or -1.

When we come to helium, there are now two electrons. They both go into the lowest spherical orbit and hence have the same values of n, ℓ, and m. So, according to the Pauli principle, they must differ in their spin quantum numbers. That is, the two electrons must spin in opposite directions. They are said to be *paired.* Figure 24-5 shows this situation for helium.

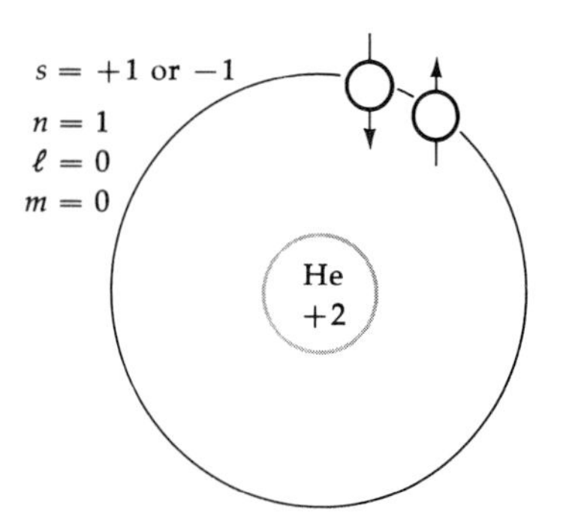

Figure 24-5. The two electrons in helium have the same values of n, ℓ, and m but are spin-paired.

The next atom is lithium, with three protons and three electrons. The first two electrons will be paired in the $n = 1$, $\ell = 0$, $m = 0$ orbit, one with $s = +1$ and the other, $s = -1$, just like helium. But the third cannot go into this orbit because it will have its spin in the same direction as one of the other two electrons. This means that the quota for the first shell is two electrons. So the third must go into the $n = 2$ shell. It turns out that this second shell can have one spherical orbit ($\ell = 0$) and three elliptical orbits, all with $\ell = 1$. Each of these in turn has $m = 1$, 0, or -1, corresponding to the X, Y, and Z directions. Since each of these four orbits can carry two spin-paired electrons, the $n = 2$ shell can accommodate a total of eight electrons as its quota. If you look at the periodic table in Figure 24-1, you can see that the second row contains eight elements. These correspond to the eight electrons that can go into the $n = 2$ shell. When we get to neon, we see that its ten electrons completely fill the two lowest shells—two into the $n = 1$ shell and eight into the $n = 2$ shell.

For the next element, sodium, the additional electron goes into the spherical orbit of the $n = 3$ shell. The rest of the elements in the third row are formed by filling in the $\ell = 0$ and $\ell = 1$ *subshells.* (A subshell is a group of orbits having the same n and ℓ values.) The eight electrons in these two subshells take us to the next inert gas, argon. It has a total of eight electrons.

After argon, we might suppose that the electrons would begin filling the subshell for $\ell = 2$. However, this doesn't happen. It turns out that the smaller ℓ values correspond to orbits of smaller energy, so the spherical shell for $n = 4$ and $\ell = 0$ has a lower energy than $n = 3$, $\ell = 2$. The outermost electrons for potassium and calcium, which begin the table's fourth row, fill in this low energy subshell. Then the next ten electrons, one each for the elements scandium through zinc, go back to fill the vacant $n = 3$, $\ell = 2$ subshell. The fourth row is completed by a return to the $n = 4$, $\ell = 1$ subshell, which can house the six electrons. This takes us to krypton at the end of the row.

From here on the situation becomes more complex. The shape of the orbits and the motion of the electrons are more difficult to describe by means of our crude pictures. However, the principle of only two spin-paired electrons in each orbit still holds throughout the periodic table.

This quantum mechanical scheme for feeding electrons into the elements one by one accounts for the similarity in chemical properties within columns of the periodic table. Let us return to sodium, for example. (See Figure 24–6.) It has a total of eleven electrons. The ten lowest ones are all spin-paired and fill up the shells for $n = 1$ and $n = 2$. The eleventh electron is by itself in the spherical orbit of the third shell. It therefore plays the role of the third electron of lithium, which is in the spherical orbit for $n = 2$, and the lone electron of hydrogen, in the $n = 1$ shell. That is, all these atoms have a single electron in an outer shell. This lone outer-shell electron accounts for the similarity in the chemical reactions of the elements in the first column of the table. Although *physically* these elements are quite different—hydrogen is a gas and sodium is a metal—they are very similar *chemically* because of this lone outer-shell electron in a spherical orbit.

Figure 24-6. All three atoms have an unpaired electron in an outer spherical orbit.

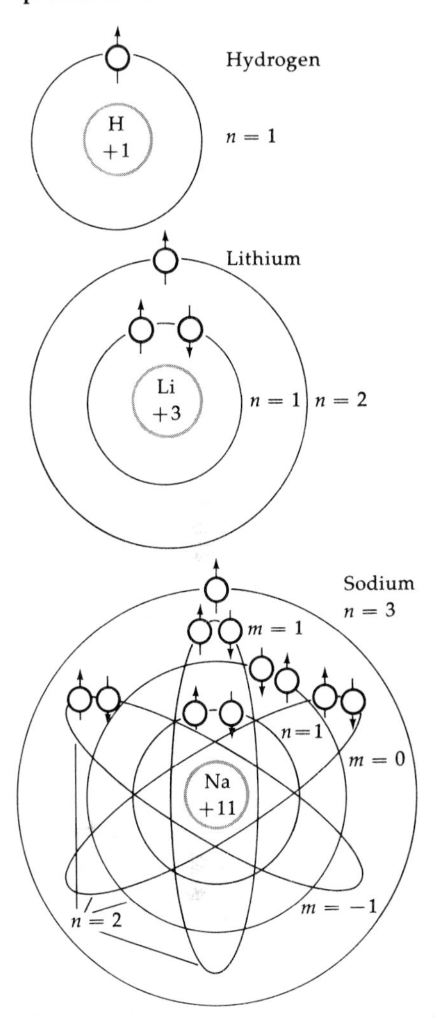

Notice that the inert gases in the far righthand column have all electrons paired and all shells or subshells filled. It turns out that this particular arrangement is an especially stable one. Atoms seem to "like" it very much. The fact that the inert gases do not readily combine with other atoms is good evidence of this preference for filled shells. (See Figure 24–7.)

We can account for many chemical reactions on the basis of this tendency of the elements to be in their most stable, filled-shell form. A good example is common table salt, sodium chloride. It results from the reaction of sodium (column 1) and chlorine (column 7). As we see in Figure 24–8, the sodium atom

Figure 24-7. The inert gases. (Distinction between spherical and elliptical orbits is not shown.)

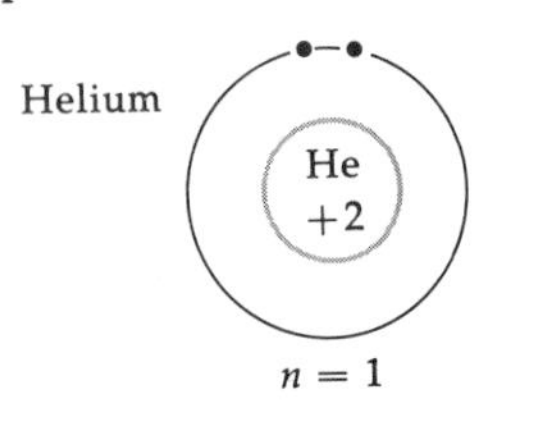

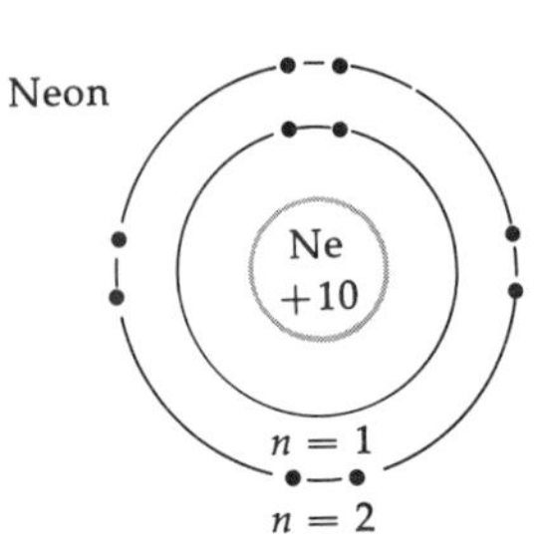

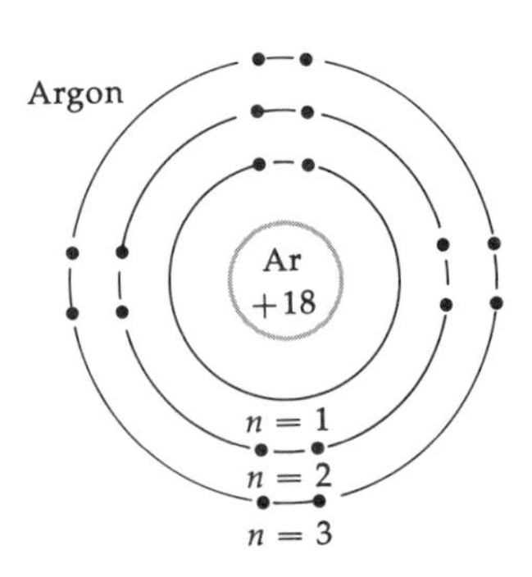

has one electron more than the filled-shell configuration of neon, and chlorine is one electron shy of the electron arrangement in the next inert gas, argon. So they get together in a sort of cooperative effort: sodium gives up its outer electron to the chlorine, and both have their appetites for filled shells satisfied. But notice something: Because the sodium *lost* an electron, it now carries a net charge of +1; and similarly the chlorine, having *gained* an electron, has a net charge of −1. These sodium and chloride *ions*, as they are called, now attract each other because of their opposite charges. When large numbers of them go through this same process, the result is the white crystalline salt you sprinkle on food. By this rather simple scheme, nature combines a dangerous metal and a deadly gas to form a necessary part of your diet.

By now, you can appreciate that the Pauli principle is quite important, not just for the esoteric domain of quantum mechanics but also in your everyday world. The Pauli principle, by establishing quotas for the various electron orbits, prevents the different elements from all behaving alike. In this sense, it is responsible for the wide variety of material things in the universe.

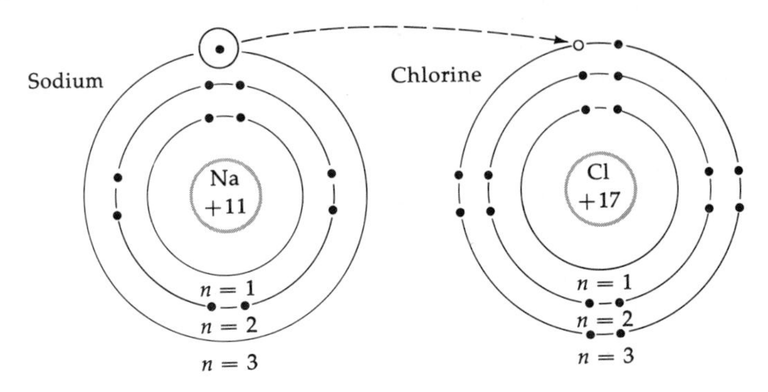

Figure 24-8. Reaction of sodium and chlorine.

Learning Checks

1. Explain the name *periodic table* for Mendeleev's arrangement of the elements.
2. For the atoms in each column in the periodic table, how many electrons are in the outermost shell?
3. What is meant by the "quantum state" of an electron?
4. Which quantum number corresponds to a given row of the periodic table?
5. Why would all atoms have the same chemical behavior if the Pauli principle did not operate?
6. How do we account for the relative chemical inactivity of such atoms as helium and argon?

TRANSITIONS IN ATOMS: THE LASER

We saw in Chapter 22 that the spectrum of hydrogen is due to photons given up when the electron makes transitions from high orbits down to lower ones. Spectra of other atoms in the periodic table are more complicated because of the larger number of electrons and allowed orbits, but we can interpret them in just the same way. Each line in an atomic spectrum corresponds to the transition of an electron between allowed Bohr orbits.

An atom is said to be in its *ground state* if all its electrons are in the lowest orbits allowed them by the Pauli principle. It attains any of several *excited states* whenever one or more electrons occupy higher orbits, that is, higher energy levels. Under normal conditions, it is the outermost electrons that become involved in transitions. Being farther away from the nucleus, they are not bound so tightly as the inner-shell electrons, and hence are more readily moved around among the higher orbits.

We can transfer an outer electron from the ground state to an excited state by supplying the atom with energy in a variety of ways. Collisions with other atoms, the spark from an electric discharge, bombardment with light—all of these will kick an electron from its ground state into a higher orbit. Once it is in the excited state, it doesn't hang around for long. After a billionth of a second or so, it makes a transition back to its original level. It can do this in a single jump, or in a series of jumps if there are intermediate levels along the way.

You don't need any fancy equipment to watch the electrons fall back to ground state. In many atoms, the gaps between energy levels correspond to photon frequencies in the visible region, and you can see the emitted light with the naked eye. For instance if you take a pair of tweezers and hold some crystals of table salt in a flame, you can see a bright yellow light characteristic of the sodium ion. The energy of the fire boosts some sodium electrons into an excited state. When they cascade back down to ground level, they give off the photons of yellow light that you see. Salts of copper give a similar effect. In this case, the light is green.

You can buy mixtures of salt crystals that give different colors when sprinkled among the flames of a fireplace. The various atoms making up the crystals have their own patterns of energy levels. So the variety of colors that you see is due to the different frequencies of photons emitted when the electrons fall back home after having been kicked out of their ground state by the flames.

Each atom has its own characteristic set of energy levels and hence its own particular spectrum. This serves as a sort of fingerprint, a means of identifying an atom. The atomic make-up of stars, for example, can be determined by analyzing the spectra of their light.

How the Laser Works

Laser beams are in the news quite often these days. They are being used in an increasing number of such widely diverse areas as medicine, weaponry, and communications. The laser is characterized by an intense beam of single-frequency light able to focus its energy on a very small region. It is probably the most dramatic example of a phenomenon which, although discovered purely within the context of attempts to understand the fundamental workings of nature, nevertheless has important practical applications. With our knowledge of wave motion and atomic transitions, we are in a good position to understand how the laser works.

The word "laser" itself describes the process. It is an acronym of the phrase *l*ight *a*mplification by *s*timulated *e*mission of *r*adiation. First of all, we need to understand what we mean by "stimulated emission."

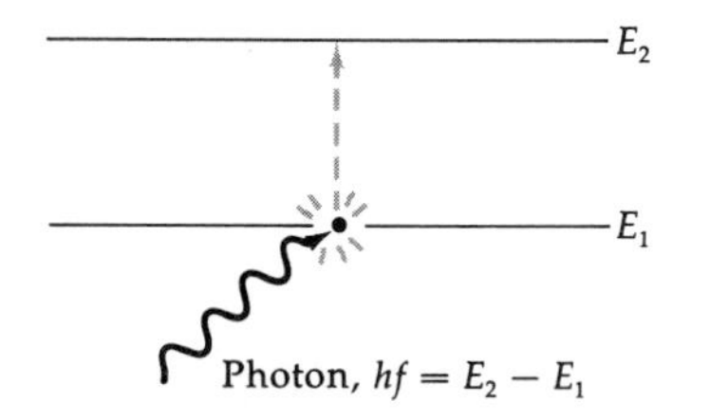

Figure 24-9. For the electron to jump to level 2, the photon energy, *hf*, must be $E_2 - E_1$.

Consider some hypothetical atom having the energy level diagram shown in Figure 24-9. Of course, the atom will have many more levels than these two, but they are all we need to illustrate the point. Suppose that an electron is in the lower level, E_1. As you already know, one way of getting the electron to jump up to E_2 is to bathe it with light whose frequency corresponds to the difference in energy between the two levels. That is, the frequency f is such that it satisfies

$$hf = E_2 - E_1$$

The atom will absorb a photon of that frequency, and the electron can make its transition.

Now it turns out that light having this energy-gap frequency has another effect. Suppose that an atom for some reason already has an excited electron in E_2. We know that after some short period of time it will spontaneously hop back down to E_1, giving off a photon in the process. However, in addition to this spontaneous transition, the electron can be *stimulated* to jump down to E_1 if a photon is present whose frequency matches the $E_2 - E_1$ gap. Again, of course, the atom will emit a photon of the same frequency during this process. Furthermore, the stimulated transition takes place faster than does the spontaneous one.

This is what is meant by "stimulated emission of radiation." The cumulative effect for many atoms produces the beam of laser light. Normally, the atom's electrons will all be in the lowest available energy levels permitted by the Pauli principle. And if an electron is boosted out of its ground state, it will

An Out-of-This-World Laser

Technicians started developing laser devices on earth about a quarter of a century ago. It was only in 1980, however, that a team of NASA astronomers found one on another planet: The upper reaches of the atmosphere of Mars form a natural laser, the first ever detected.

It turns out that the thready streaks of nearly pure carbon dioxide that surround Mars form perfect lasing conditions. This planet emits in the infrared range the very same wavelength as some commercial lasers on earth. The only difference between the earth and Martian versions is that the terrestrial models use mirrors to build up the high intensity.

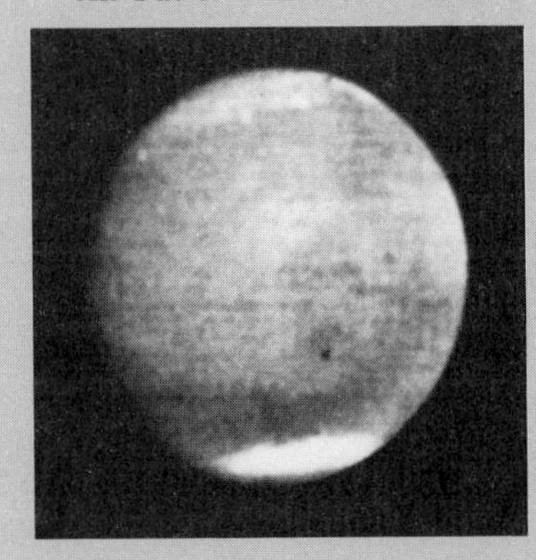

A natural laser was discovered in the Martian atmosphere.

NASA researchers maintain that Martian lasing yields over a billion kilowatts of continuous power (which is about four times the total annual electric output of the United States). They speculate that earth-orbiting satellites with atmospheres mimicking that of Mars could generate power in the same manner. The search is under way for other planets that lase, though none have been found to date.

Source: "Martian Laser," *Science Digest*, December 1981, p. 20.

usually hop right back down again almost immediately. But the atoms will have some excited levels that are special in the sense that once an electron arrives there it gets stuck and hangs around for a relatively long time—perhaps a thousandth of a second instead of a billionth. If this happens for many electrons, then we have what the laser physicists call a *population inversion.* That is, the population of the energy levels is upside down, because more electrons are in higher energy levels than in lower ones. Such a condition leaves the system ripe for stimulated emission.

Suppose we have a few identical atoms all with an electron in the level E_2. (See Figure 24–10.) An incident photon of just the right frequency strikes the first atom, stimulating it to emit a photon. These two photons, the original one and the emitted one, can in turn stimulate the emission of photons from two more atoms, so now there are four. These can generate four more, and so on. You can see that as this process continues, very shortly we will have many, many photons. Thus, the light is *amplified.* A key feature of this emitted radiation is this: Not only do the photons all *have the same frequency, they also travel in the same direction and are in phase with each other.* This is quite different from a spontaneously emitted photon—it can move in any direction and is not necessarily in phase with any other photon.

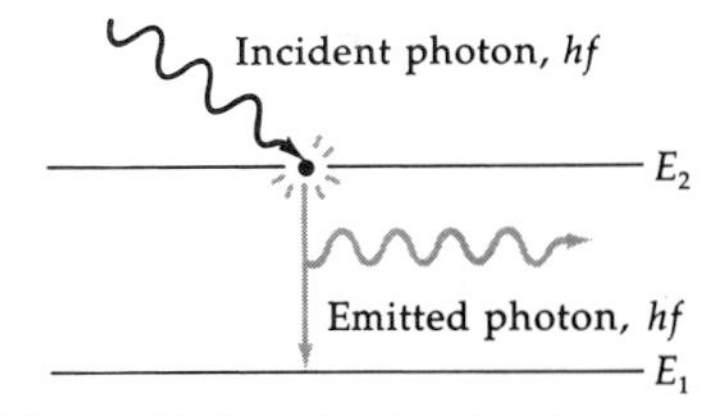

Figure 24-10. Stimulated emission. The emitted photon is identical to the incident photon. It can stimulate further emissions.

Figure 24–11 reminds us of what we mean by two light waves being "in phase." That is, they have their crests and troughs lined up together. It is this feature that gives the laser some of its most important properties.

Figure 24-11. Light waves in phase.

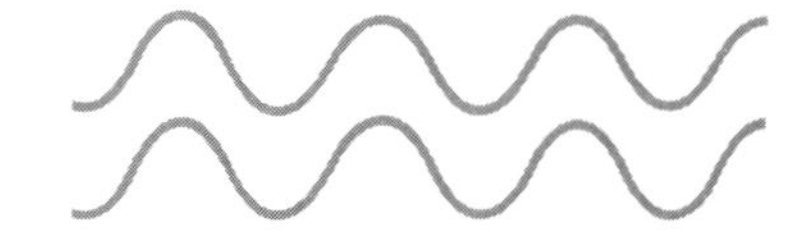

In practice, the chances are quite slim that a given photon traveling through a swarm of atoms will pass close enough to

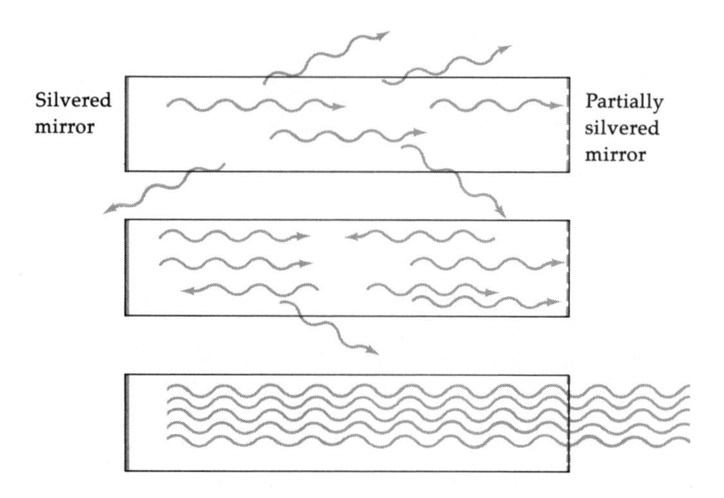

Figure 24-12. Laser.

any one of them to stimulate emission. So the laser must be designed to increase these odds. Figure 24–12 illustrates one way of boosting the chances of stimulated emission. The laser atoms are confined to a tube closed at both ends. At one end is a silvered mirror that reflects all the light incident upon it. The other end contains a mirror that is partially silvered. It reflects most incident light, but it is designed to let a little light pass through if the intensity is high enough.

Now consider a photon bouncing back and forth between the mirrors, along the axis of the tube. Obviously, its chances of striking an atom and hence stimulating some emission are much greater than if it merely made one pass through the system. Furthermore, any photon traveling in some direction other than along the length of the tube will strike one of the mirrors a glancing blow, rather than head-on, and will eventually bounce out of the tube altogether. So ultimately the only photons left will be the original ones running parallel to the axis of the tube plus those emitted through the stimulation process, which travel in the same direction. When there are enough of these photons, some will "leak" out through the partially silvered mirror. This is the laser beam.

Laser light is said to be *coherent,* which means the photons all are in phase and move in the same direction. Such a beam will spread very little, in contrast to the incoherent beam from a flashlight. If you aim a flashlight across the room, the spot it illuminates on the wall gets larger as you back away. This spreads out the energy of the light, also. But with a laser beam, which spreads very little even over many thousands of miles, the energy is concentrated in a small point. This is one of the characteristics that makes lasers so important commercially: the energy of the narrow beam is focused into an extremely small region.

Uses of the Laser

Lasers have a wide variety of applications in industry, the marketplace, the medical field, and scientific research. The list seems to grow every day. Here we will mention only a few of the laser's many uses.

The laser has the important capacity for being focused with extreme precision. This makes it quite useful in construction and in measurement of distances. Lasers were used, for example, in laying out the tunnels in San Francisco's subway system, the Bay Area Rapid Transit (BART). Several of the Apollo missions to the moon carried mirrors to bounce laser light back to earth, enabling scientists to redetermine the earth-moon distance with greatly improved accuracy.

This focusing feature has important applications in surgery. Because its energy can be concentrated in such a small beam spot, the laser can be used by eye surgeons to repair a detached retina. The laser welds the retina back to the inner surface of the eye, eliminating the need to cut into the eye.

You have run across a laser use in shops and stores. The Universal Product Code, shown in Figure 24-13, carries pricing and inventory information for a product. The clerk uses a laser to scan the lines and spaces, whose darkness and arrangement are coded and stored in a computer. This reduces the possibility of mistakes in punching the cash register and makes possible an up-to-date inventory.

The laser as a cutting tool has made possible the microelectronics industry. Incredibly small circuits are "drawn" on microprocessor chips by lasers.

Researchers use lasers to study processes in nature that take place on time scales of a billionth of a billionth of a second. Also, chemists can induce specific reactions in molecules that were not possible before the laser.

Figure 24-13. Universal Product Code.

Some of the countless applications of the laser. (a) Silicon wafers are annealed to a platform using a laser. The angled glass directly behind the wafer is a mirror that is precisely positioned to direct the laser beam to the area to be annealed. (b) Lasers are also used to treat conditions of the eye.

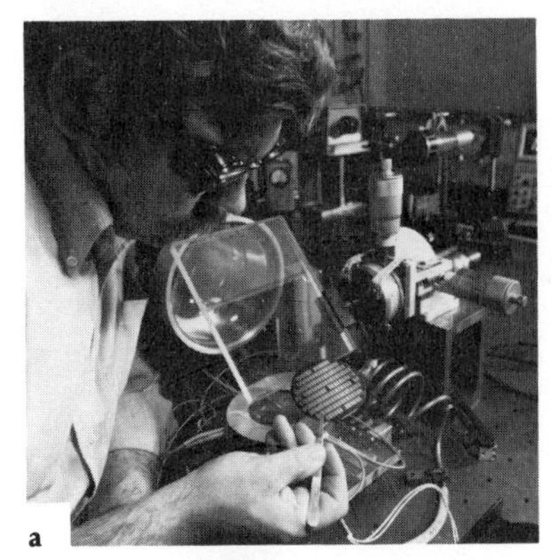

a

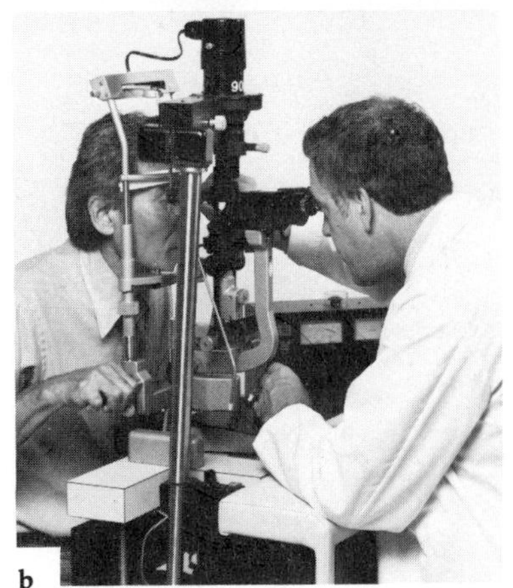

b

Learning Checks

1. What is meant by the *ground state* of an atom? What is the *excited state?*
2. How do we account for the fact that it is the outermost electrons that participate in chemical reactions?
3. What is the origin of the word *laser?*
4. Originally, the process of stimulated emission was first observed for microwave radiation, instead of light. Can you guess what such a device would be called?
5. What is meant by *coherent light?*

QUANTUM MECHANICS IN YOUR EVERYDAY WORLD

It may seem to you that quantum mechanics is a strange and mystifying theory that only has importance in keeping physicists amused as we study the submicroscopic world of atoms and molecules. Because Planck's constant is so tiny, we don't see energy quantization or matter waves in our ordinary world. But the fact is that the "building blocks" of our world, the atoms and molecules of which all things around you are made, obey the rules of quantum mechanics; this by itself makes the theory important in our lives.

We have already noticed that we wouldn't even exist if it weren't for the quantization of energy levels in atoms. The rules of quantum mechanics prevent the orbiting electrons from falling into the nucleus, as they would have to do if the classical theory of electromagnetism governed the atomic world. Atoms would disappear in a puff of energy, and there would be no life, at least not in any form that resembles what we know.

Also, we have seen that another feature of quantum mechanics, the Pauli principle, is responsible for the staggering variety of nature. Without it, the ground states of atoms would all be similar, because all the electrons of each atom would assemble in the lowest and simplest quantum orbit. Chemical reactions probably would not take place. The Pauli principle, by establishing quotas for the various electron shells, insures that there are many different kinds of atoms and allows for the possibility of reactions among them.

But it turns out that quantum mechanics relates directly to other circumstances in our large-scale world as well. We will take only a single commonplace example. Have you ever wondered why a rock is as hard as it is? Although rocks and electrons are a long way apart in relative size, it turns out that the rock's hardness is related to the Heisenberg uncertainty principle.

Nothing is infinitely hard. Any solid can be compressed if we apply enough pressure to it. The resistance of an object to this compression is what we mean by hardness. So let us see how the uncertainty principle applies.

You will recall that one of the consequences of this principle is that the position and momentum of an electron cannot simultaneously be determined with 100 percent certainty. We can put this another way: An electron cannot be completely at rest when it is confined to a given region of space, such as in an atom, for instance. Otherwise, we would know precisely its momentum (zero, when it is at rest) *and* we could locate its position to within the dimensions of the atom. But the principle doesn't allow this combination. The smaller the confinement of the electron—that is, the more we box it in—the greater its motion becomes. So there is a sort of pressure that the particle exerts on its confinement due to the kinetic energy of its motion. This is sometimes called the *Schroedinger pressure,* after the Austrian physicist Erwin Schroedinger, one of the pioneers of quantum mechanics. The Schroedinger pressure tends to expand the region in which the electron is trapped, because this would lead to a decrease in its kinetic energy. We can imagine a hydrogen atom as a "happy medium" of two opposing effects: the electrical attraction serving to pull the electron and proton together, balanced by the Schroedinger pressure pushing them apart.

If we have many electrons, such as in metal or a rock, it turns out that the Schroedinger pressure becomes larger as more electrons are present in a given volume. It can be a very large number, equivalent to thousands of atmospheres, enough to explode the object were it not for the attraction between the electrons and the nuclei. Again we have a balance. Compressing the object means acting against this pressure. So the hardness of a rock—its resistance to being compressed—depends on the Schroedinger pressure.

SUMMARY

The variety of materials in the physical world is ultimately due to the different chemical behavior of atoms, which in turn depends on the number and arrangement of their electrons. The Russian chemist Mendeleev developed the *periodic table,* an arrangement of the atoms according to increasing mass and similarity of chemical behavior.

The chemistry of the atoms results from the *Pauli exclusion principle,* a rule forbidding any two electrons in the same atom to occupy the same quantum state. This rule establishes quotas for the various allowed electron orbits and accounts for the structure of the periodic table.

The *laser* is an important technological advance that has its roots in *quantum mechanics.* Lasing results when electrons in a higher energy level are stimulated to emit photons by falling to lower levels. Light emitted in this way is *coherent,* enabling the laser's energy to be focused on a very small region.

Exercises

1. The elements aluminum (Al) and silicon (Si) have similar atomic weights. Would you expect them to have similar chemical properties? Explain.
2. Oxygen (O) is a gas and sulfur (S) is a yellow solid. Would you expect them to have similar chemical properties? Explain.
3. The water molecule is H_2O, two atoms of hydrogen for each oxygen atom. This compound is *not* formed in the way described in the text for sodium chloride. Speculate on how it is formed.
4. How many electrons are there in each of the first five shells of beryllium (Be) and of barium (Ba)? Comment on any possible similarity in their chemical properties.
5. One important property of laser light is that it is *monochromatic,* meaning "of one color." Explain what this means and why it is so.

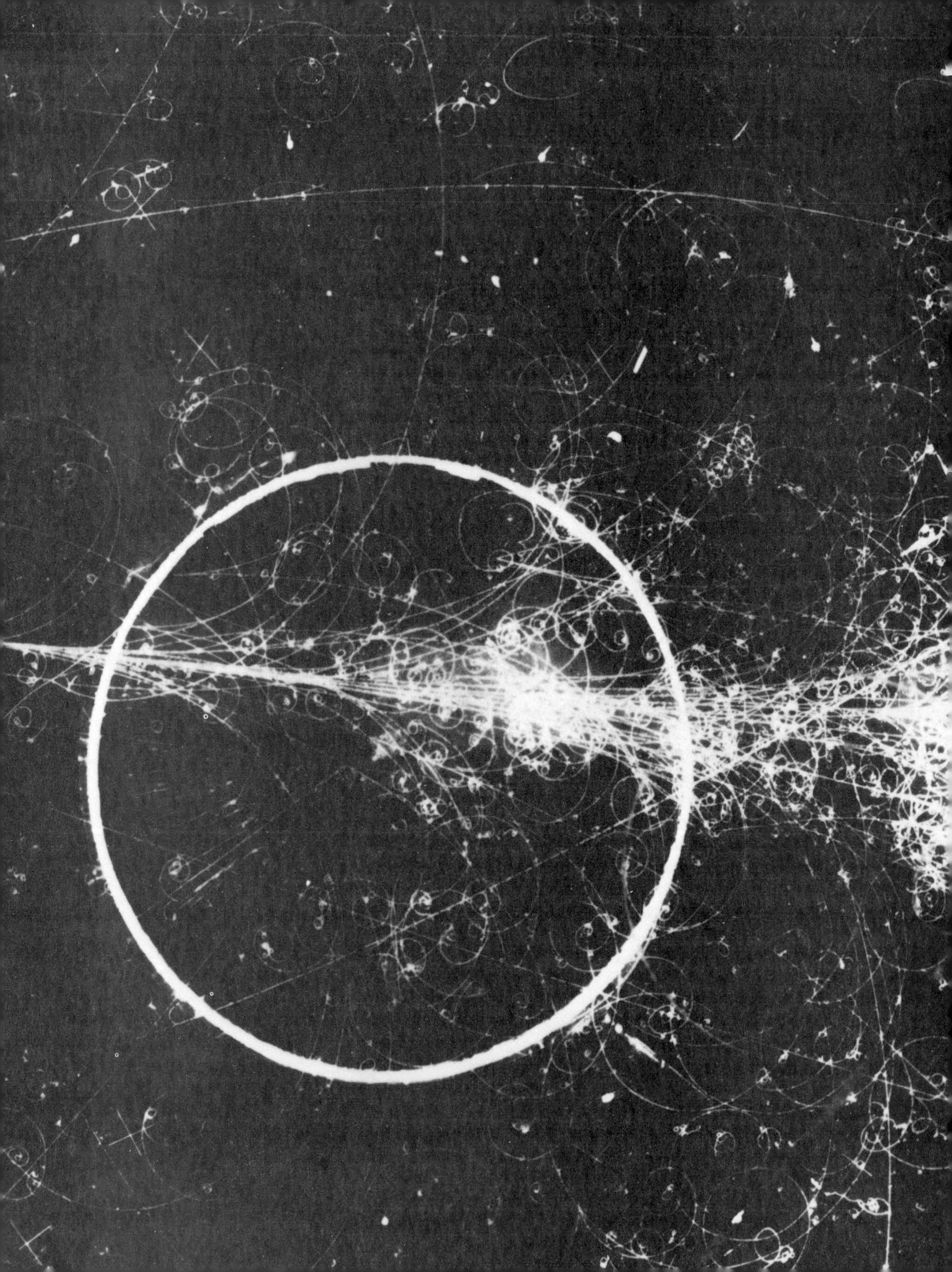

25

INSIDE THE ATOM: THE NUCLEAR WORLD

The ghostly image of the neutrino reveals itself by its trail. One sees only where it has been and not where it actually is.

So far in our study of atoms we have concentrated on the arrangement and behavior of the electrons. Except for mentioning the nucleus as the seat of the atom's positive charge, we have essentially ignored it. After all, the nucleus takes up an unbelievably small fraction of the atomic volume, having a diameter of about 10^{-15} meter compared with 10^{-10} meter for the atom as a whole. In fact, if the atom were expanded to the size of the earth, the nucleus would be a sphere only about 200 meters across.

However, more than 99.9 percent of the mass of the atom is due to the nucleus. It also accounts for a staggering amount of energy—so much that the transitions involving the electrons seem puny in comparison. Furthermore, the nucleus and what goes on inside it are responsible for a great deal of our ignorance about nature. Much is known about the objects that make up the nucleus, but there is much more that we don't understand. As a result, a great portion of the research effort today in physics is directed toward unraveling those questions about the nucleus.

RADIOACTIVITY

Our story begins with the discovery that a French physicist named Henri Becquerel made, quite by accident, in 1896. Becquerel was studying a phenomenon called *fluorescence,* in which certain materials absorb some incident ultraviolet radiation and emit visible light. He was working with a material called *potassium uranyl sulfate,* a salt containing uranium atoms. During one of his experiments, he exposed a crystal of this material to several hours of sunlight while it was sitting on a photographic plate. The plate was wrapped in black paper to shield it from the sun. When Becquerel developed the plate, he found a dark spot where the uranyl crystal had been sitting. This was something of a surprise because the rest of the plate, covered by the black paper, had not been darkened. Becquerel decided to do more experimenting with this material.

The next several days were cloudy in Paris, preventing Becquerel from continuing his experiments—or so he thought, for he believed that the sunlight was responsible for the exposure of the plates. So he put the uranyl crystal in a desk drawer and turned his attention to other matters until the weather had cleared. As it happened, in the same desk drawer he had also stored some fresh photographic plates that he intended to use for some other experiments. But when he developed them, he found them darkened, as if they had been previously exposed to light. The same was true of the rest of the plates in the same drawer. Because the plates had been wrapped and in the drawer, Becquerel was convinced that the culprit was not stray sunlight. There had to be something else, some other source of radiation. He traced the cause to a chunk of uranium ore, known as pitchblende.

The surprising point about Becquerel's discovery is this: the new radiation required *no external source* of energy. Contrast this with the line spectra of atoms, which we discussed earlier. These spectra arise because the atom becomes excited by some outside provider of energy. But the radiation that Becquerel discovered is generated inside the material itself.

Becquerel also found that there wasn't anything he could do to the uranium ore to make it stop radiating. He dissolved it in acids. He crushed it to a powder. He heated it, chilled it, did all the things he could think of. But the material kept churning away, pouring out radiation with always the same intensity, no matter what form it was in. Thus, Becquerel knew that this property—*radioactivity,* as it came to be called—had nothing to do with the outer electrons of the atoms and molecules. It is the electrons that determine the chemical structure and behavior of materials. But the radioactivity was insensitive to chemical or physical changes. This new radiation must come from deep within the atom.

Many chemists and physicists began to study radioactivity during the years following Becquerel's work. Among them was Marie Sklodowska Curie. She was a Polish-born graduate student in chemistry at the Sorbonne, who had married the physicist Pierre Curie. Madame Curie found that some uranium ores were more highly radioactive than their uranium content would warrant, indicating the presence of other radioactive atoms. When they separated the ores' constituents by a tedious chemical procedure, the Curies isolated two new radioactive elements: *polonium,* named in honor of her birthplace, and *radium,* an element two million times more radioactive than uranium. Madame Curie, her husband Pierre, and Becquerel shared the 1903 Nobel Prize in physics for their pioneering work in radioactivity.

Three Kinds of Radioactivity

As physicists and chemists continued to study radioactivity, they soon established that the stream of radiation coming from a radioactive substance has three components. For lack of better

Health Physics: Alpha, Beta, Gamma — Radiation Protection

Today, many people may risk exposure to nonnatural radiation where they work—at nuclear facilities, in any vehicle that transports radioactive material, or at large hospitals (which often hold a great number of radioactive sources, both sealed and unsealed). How harmful are alpha, beta, and gamma rays to humans, and what do we use to shield ourselves from these kinds of radiation?

Of the three, the penetrating power of alpha rays is the smallest, with beta rays being about 100 times, and gamma rays about 10,000 times, more penetrating. Alpha rays do not penetrate matter to any appreciable depth. They do not, in fact, even travel very far. (If an alpha source is more than 5 centimeters from you, the alpha particles will be stopped before they reach your body.) Being feeble (of short range), alpha particles cannot penetrate the dead layer of skin surrounding the body. Most alpha particles can be completely stopped by a thin piece of paper, most normal clothing, or a thin sheet of aluminum foil. Because they are so easily stopped, they never present shielding difficulties and ordinarily do not pose any external radiation danger. If, however, an alpha-emitting source gets inside the body (by being inhaled or swallowed), a great deal of local tissue damage can occur as a result of intense ionization and excitation of cell molecules.

Externally, beta radiation tends to be much more hazardous than alpha. Beta particles may not be stopped by the dead skin layer around the body or by normal clothing. (Some very heavy clothing may do the trick.) They can penetrate up to 3 meters of air and can damage your skin tissue and your eyes. Hence, shielding must be used. Beta particles can pass through a sheet of paper but can be stopped by materials such as wood, glass, and plastic. Customarily, plastic or aluminum at least 1 centimeter thick is used for this purpose; if necessary, these materials are lined with thin lead. Beta particles ingested into the body (that is, breathed or eaten) also pose a serious radiation hazard—they can cause extremely severe internal injury.

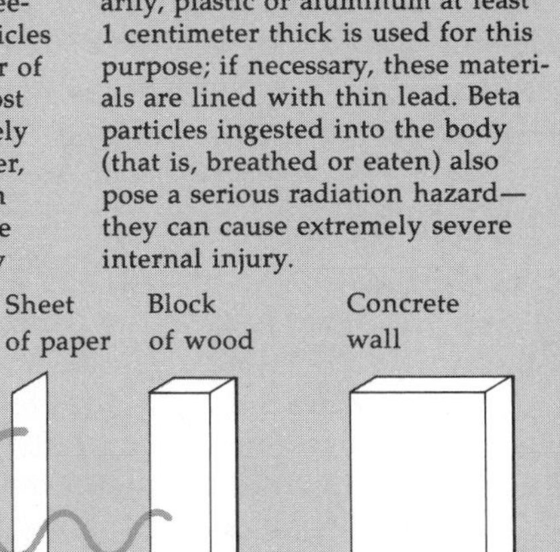

Alpha, beta, and gamma rays differ in their ability to penetrate various materials.

The international symbol for radioactivity.

With gamma rays, we cannot rely on just the air between us and the source, nor on our own garments, to shield us. Gamma rays are a form of high-energy electromagnetic radiation; they are not particles. They travel at speeds near the speed of light and are so penetrating that they can be halted only by thick layers of concrete or lead (as much as 20 centimeters of lead may be required for highly active sources). These rays pass through the skin with ease and can effect intense harm to internal organs. The degree of penetration depends on the amount of energy involved; thus, penetration increases as energy increases.

Since many activities result in radioactive materials becoming airborne or finding their way into discharge water despite contamination control, all citizens should know something about this subject regardless of where they work or live.

Sources: J. C. Robertson, *A Guide to Radiation Protection.* New York: Halsted Press, 1976.

R. V. Scheele and J. Wakley, *Elements of Radiation Protection.* Springfield, Illinois: Charles C Thomas, 1975.

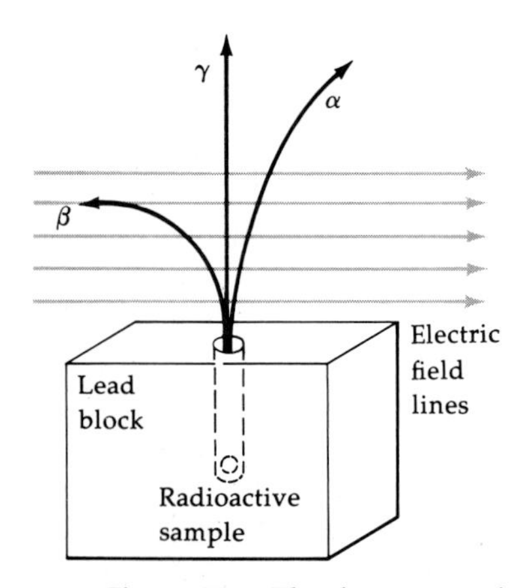

Figure 25-1. The three types of radioactivity separated by an electric field.

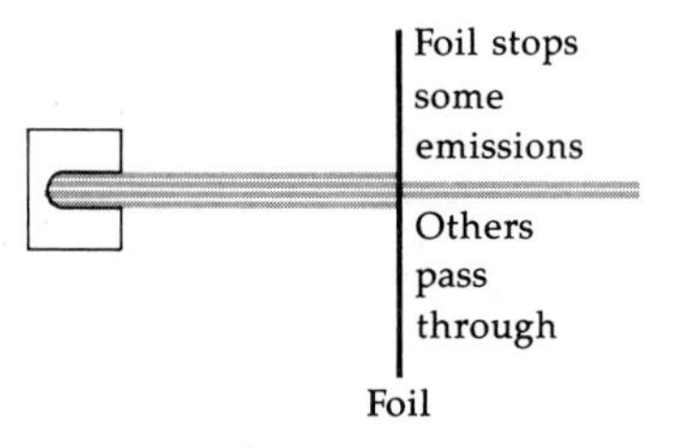

Figure 25-2. Rutherford's experiment.

labels, these earlier investigators simply dubbed them *alpha* (α), *beta* (β), and *gamma* (γ). These names have stuck until the present time.

One of the ways in which the three differ is in their behavior with respect to an electric field, as shown in Figure 25-1. The lead block (a good absorber of all kinds of radiation) contains a sample of radioactive material that provides a beam of radiation. When the beam passes through the electric field, the part called *alpha* bends slightly toward the negative pole, the *beta* component breaks sharply toward the positive pole, and the *gamma* radiation continues undeflected as if the field were not there. These experiments indicate that the alpha component is made up of particles carrying positive charges, the beta particles are negatively charged, and the gamma rays have no electrical charge at all.

Ernest Rutherford, who at this time was still at McGill University in Montreal, made some further important studies on the nature of this radiation, during the years between about 1900 and 1907. (Remember, it was not until 1911 that Rutherford did the scattering experiments that gave birth to the idea of the nuclear atom. At this time J. J. Thomson's "plum pudding" model was still in vogue.) Rutherford placed different thicknesses of foil between a radioactive source and the detector. (See Figure 25-2.) He found that the foil stopped the alpha particles but not the beta particles.

Rutherford also discovered from additional experiments that the alpha particles each had a charge of +2 and a mass four times that of the proton. These particles were later identified as helium nuclei made up of two protons and two neutrons, but at the time neither the proton nor the neutron, much less the nucleus, was known. The beta particles were found simply to be high-energy electrons, while gamma rays are pure electromagnetic radiation of extraordinarily high energy.

Learning Checks

1. What is meant by radioactivity? How do you suppose the name originated?
2. What led the Curies to search for other radioactive sources besides uranium?
3. How did Becquerel conclude that radioactivity has nothing to do with atomic electrons?
4. Describe the behavior of alpha, beta, and gamma rays moving through an electric field. How does this observation lead to a description of the radiation?

Composition of the Nucleus

Suppose we summarize what we know about the radiation emitted by radioactive elements:

1. Alpha particles are high-energy helium nuclei, each carrying a charge of $+2$ and weighing four times as much as the proton.
2. Beta particles are high-energy electrons.
3. Gamma radiation is made up of high-energy photons and can accompany emission of either alpha or beta particles.

In order to understand what this emission of radiation means, let us get ahead of the game a little bit and look at the make-up of the atomic nucleus.

We have already learned that Rutherford in 1911 established that the atom's positive charge is concentrated in its central core. This positive charge is due to the aggregate of protons residing in the nucleus. But the protons alone are not enough; there must be some other kind of particle. For example, oxygen, with an atomic weight of 16, would have sixteen protons in the nucleus if it were made up only of protons. But oxygen has only eight electrons and thus, being electrically neutral, it would also need eight "nuclear electrons" to cancel the charge of eight of the sixteen protons. However, electrons cannot exist in the nucleus without having thousands of times more energy than is observed in nuclear phenomena. So Bohr and Rutherford and others reasoned that there must be some neutral particle whose mass is about equal to that of the proton that also has its home in the nucleus. This particle is called the *neutron.* James Chadwick, a student of Rutherford's at Cambridge, discovered the neutron in 1932.

Here are the masses of the three particles:

proton: 1.6725×10^{-27} kg

neutron: 1.6748×10^{-27} kg

electron: 9.1091×10^{-31} kg

As you can see, the neutron is only slightly larger than the proton, and both are about 2000 times more massive than the electron. Because the neutron and proton inhabit the nucleus, they are called *nucleons.*

We can identify each element according to the number of protons in the nucleus—which, of course, is equal to the number of electrons in the atomic orbits. This is called the *atomic number.* The periodic table (see Figure 24–1) shows the atoms arranged according to their atomic number. The *mass number* of each element is the number of nucleons—that is, the sum of all the protons and neutrons in the nucleus. It turns out that a given element can have varying numbers of neutrons accompanying the fixed number of protons. This gives rise to *isotopes,* which we are going to study later in this chapter.

Let us take carbon as an example. We see from the periodic table that it has atomic number 6, so it has six protons. One form of carbon has a mass number of 12, meaning that its nucleus also has six neutrons. A standard way of symbolizing this particular element is this:

$$^{12}_{6}\mathrm{C}$$

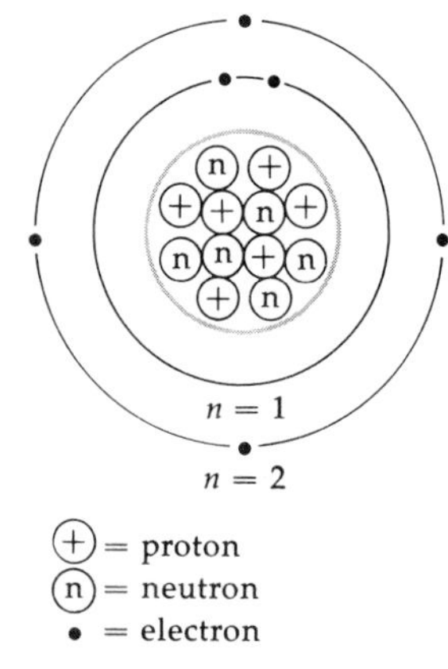

Figure 25-3. The carbon-12 atom.

The chemical symbol for carbon is the letter C. Its atomic number (6) is written as a *sub*script to the left of the symbol, and its mass number (12) is a *super*script to the left. Notice that it is redundant to have both the atomic number *and* the chemical symbol, since if it is the element carbon, it must be number 6. However, this notation is standard and we shall use it. Besides, it is easier to be redundant than to go flipping to the periodic table all the time.

With what we now know about the nucleus, combined with what we learned about atomic electrons earlier, we can sketch a picture of the carbon atom as shown in Figure 25-3. The six protons and six neutrons are shown in the nucleus. The two paired electrons having principal quantum number $n = 1$ occupy the inner shell, and the four unpaired $n = 2$ electrons make up the outer orbit.

Learning Checks

1. A certain atom contains ten electrons and twelve neutrons. What is its nuclear symbol? (See the periodic table, Figure 24-1.)

2. What evidence is there for the existence of the neutron?

TRANSMUTATION: NATURE'S MAGIC SHOW

In this section we will try to understand what happens when a radioactive element gives off alpha, beta, and gamma radiation. Let us look first at the emission of alpha particles. We will take as our example a process that actually takes place: alpha emission by a particular isotope of uranium, called uranium-238. Its symbol is

$$^{238}_{92}\mathrm{U}$$

You can see that uranium has 92 protons, a mass number of 238, and $238 - 92 = 146$ neutrons. When it emits an alpha particle, it gives up two of these protons and two neutrons, because the alpha particle is $^{4}_{2}He$. This means that the resulting nucleus will have only 90 protons and 144 neutrons. But this is no longer uranium; it is the element with atomic number 90 and mass number 234. By looking in the periodic table, you will find that this is an isotope of a *different* element, called *thorium* (Th).

Let us be clear about what has happened here. The uranium has spontaneously ejected two protons and two neutrons and in the process has become a *new element*, thorium. It is convenient to write such nuclear reactions in symbols like this:

$$^{238}_{92}U \rightarrow ^{4}_{2}He + ^{234}_{90}Th$$

Note that the sum of the subscripts on the right side equals that on the left; the same is true of the superscripts. That is, we have conserved the charge and the number of nucleons.

This particular type of break-up of the nucleus is called *alpha decay*. It occurs for several different nuclei. In every case, the nucleus that results is of an element two steps to the left in the periodic table and has a mass number lower by four units. We use the general term *transmutation* for the production of one nucleus due to the decay of another.

Now what about emission of beta particles? It turns out that this also causes transmutation, this time by *beta decay*. You have learned that the beta particle is a high-energy electron and that an electron cannot exist in the nucleus. So how can the nucleus emit an electron when it has no electrons in the first place?

The answer to this question involves another of the four interactions in nature, called simply the *weak force*. We will study more about this interaction in the next chapter. For now, be content to know that the neutron undergoes a reaction to create a proton and the emitted beta particle, as well as some other particles that we will talk about later. (See Figure 25-4.) Here we can see why the neutron has a larger mass than the proton: it has to include the mass of the electron and some other things. The proton created in this process remains in the nucleus. The electron appears as the observed beta particle.

$$\text{neutron} \longrightarrow \text{proton} + {}^{0}_{-1}e + \text{others}$$

Figure 25-4. Beta decay produces a proton and an electron from a neutron.

It is not quite correct to consider the neutron as simply a proton and an electron clamped together. Rather you should think of this as a process whereby, for whatever reasons, a proton and an electron are produced, with the simultaneous disappearance of a neutron.

In terms of our symbols, we can write for the beta particle:

$$\text{beta particle} = {}^{0}_{-1}e$$

Here, e stands for electron. Compared with the proton, its mass can be considered zero, and hence its mass number is zero. It makes little sense to talk of an atomic number of -1. However, when we realize that the atomic number of a nucleus is also its positive charge, we can see that it makes sense to present the beta particle in this way, showing a charge of -1.

The thorium-234 isotope produced in the alpha decay of $^{238}_{92}U$ undergoes beta decay, changing to the element protactinium. We can write this transmutation as

$$^{234}_{90}Th \rightarrow {}^{0}_{-1}e + {}^{234}_{91}Pa$$

Notice that this time the mass number remains the same but the atomic number increases by one unit. Thus, beta decay moves us one step to the *right* in the periodic table with no change in mass number. This is consistent since we have *lost* a neutron and *gained* a proton. Notice that again on both sides of this expression the sums of the subscripts are the same, as are the sums of the superscripts.

We can summarize these transmutations in this way:

α decay: *reduces* the *atomic number* by 2 and the *mass number* by 4

β decay: *increases* the *atomic number* by 1 and does not change the *mass number*

The emission of gamma radiation does not involve transmutation because gamma rays do not contain protons or neutrons. We can think of gamma radiation as the emission of nuclear energy accompanying alpha and beta decay.

Isotopes

When thorium decays by emitting a beta particle, the resulting protactinium nucleus also undergoes beta decay to produce another element. Since beta decay must increase the atomic number of protactinium (91) by one unit, the resulting nucleus has an atomic number of 92—so we are back to uranium! However, it must be somehow different from the uranium we started out with, $^{238}_{92}U$, since it has a mass number of 234:

$$^{234}_{91}Pa \rightarrow {}^{0}_{-1}e + {}^{234}_{92}U$$

The atoms $^{234}_{92}U$ and $^{238}_{92}U$ are called *isotopes* of uranium. The name is taken from the Greek words meaning "in the same place," indicating they are located in the same position in the periodic table. Both have ninety-two protons and ninety-two electrons. They differ in the number of *neutrons* and hence in atomic weight. Uranium-238 has $238 - 92 = 146$ neutrons, while U-234 has 142.

Many other elements besides uranium have isotopes. In fact, it is more the rule than the exception for an element to have two or more isotopes. The various isotopes of a given element appear in nature with varying relative abundances. For example, carbon has two isotopes, $^{12}_{6}C$ and $^{13}_{6}C$, commonly called carbon-12 and carbon-13. An ordinary sample of carbon contains 98.9 percent carbon-12 and 1.1 percent carbon-13. Chlorine also has two isotopes, $^{35}_{17}Cl$ and $^{37}_{17}Cl$, which have relative abundances of 75.4 percent and 24.6 percent respectively. Only a few elements have only a single isotope. Gold and iodine are examples.

The isotopes of hydrogen are particularly important and so have been given special names. The normal isotope is $^{1}_{1}H$, our old friend with one proton and one electron. However, this proton can hook up with a neutron to give $^{2}_{1}H$, called *deuterium*, and with two neutrons to give $^{3}_{1}H$, *tritium*. Deuterium, given the symbol D, has a mass number of 2, meaning that it is twice as massive as ordinary hydrogen. Tritium (T) is three times as massive as H, since it has a mass number of 3.

Notice that since isotopes of a given element have the same number of *protons* they also have the same number of *electrons*. Hence, you would expect them to have the same chemical properties and, indeed, this turns out to be true. As a general rule, isotopes undergo the same chemical reactions. Occasionally chemists find exceptions to this, especially among the very light isotopes such as hydrogen. The variances in chemical reactivity come about because the differences in mass are more pronounced for these elements. Deuterium, for example, is twice as heavy as hydrogen, while uranium-238 is only about 2 percent heavier than uranium-234.

Learning Checks

1. Explain the term *transmutation*.
2. What kind of radioactive emission changes both the *mass* number and the *atomic* number? What kind changes only the atomic number? Is it possible to change only the mass number?
3. What are isotopes?
4. Why do isotopes of the same element have the same chemical properties?

Decay Times for Nuclei: Half-Lives

It is probably obvious to you that in a given amount of a radioactive element, transmutation cannot go on forever. Material is being lost all the time. After all, if you start out with a kilogram of $^{238}_{92}U$, it continuously and gradually disappears as its nuclei emit alpha particles and change to $^{234}_{90}Th$. It turns out that the rate at which transmutation takes place—the "death rate," we might say—is different for different elements. It is also different for isotopes of the same element. These death rates, or decay times, vary all over the map. Some nuclei disintegrate extremely rapidly, while others take a long time.

So for radioactive materials, there is a certain "life expectancy," just as there is for any collection of people or cats or houseflies. If you begin with a group of 20-year olds, you can be sure that not all of them will die on the same day. But if the group is large enough, you can predict with some accuracy the percentage that will survive after some length of time. This is the way insurance companies make their money. Using such statistics, they are betting that enough people will live long enough to pay more than enough money in premiums to cover the death benefits the companies must pay out. Of course, they can't predict which individuals will die when. What they are interested in is the group as a whole.

Similar things can be said for radioactive nuclei. There is one very important difference, however. The death rate for people increases with advancing age. A person 80 years of age is much more likely to die than a 25-year-old. But for radioactive nuclei, the death rate remains *fixed* no matter how long a given nucleus has been around. This means that a nucleus of thorium-234 formed 5 minutes ago by the alpha decay of a uranium-238 nucleus has *exactly* the same life expectancy as one that has been hanging around for centuries. Since the rate of decay is fixed, the fraction of nuclei that disintegrates in some unit of time is also constant. So we can conclude that the number of nuclei that will decay in a given length of time is proportional to the number originally available but does not depend on how old they are.

Figure 25-5. The death rate for radioactive atoms.

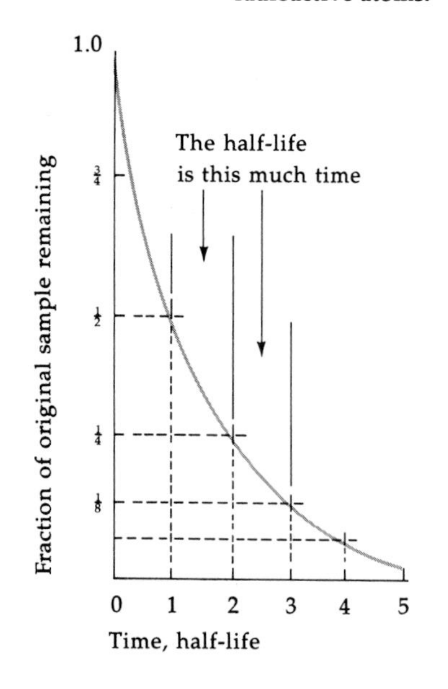

We find it convenient to talk about a time period during which one-half of the nuclei decay. This time period is called the *half-life* of the isotope. Suppose the half-life of some isotope is 5 minutes. Then, if you start out with 16 grams, 5 minutes later there will be 8 grams, 5 minutes after that there remain 4 grams, then 2 grams, and so on. At the end of any 5 minute interval there remains one-half the material that was on hand at the beginning of that interval. The graph in Figure 25–5 illustrates this behavior and shows the half-life.

A curve like that in Figure 25–5 is called an exponential curve. Any quantity that is depleted at a fixed rate is said to *decay exponentially.* Exponential decay is just the reverse of exponential growth, which we studied in Chapter 11. You may want to refer there to our discussion of doubling time, the time required for an exponentially growing quantity to double. The half-life is the reverse—that is, it is the time required for an exponentially decaying quantity to be decreased by one-half.

Remember, the radioactive half-life has nothing to do with the life expectancy of any *individual* nucleus. We can't predict which ones will do the disintegrating. All we know, after careful experimentation, is the rate for *large numbers* of nuclei. Just like the insurance companies, we find that our statistical predictions are accurate when we apply them to a sample containing many nuclei.

As we said, half-lives can vary tremendously from isotope to isotope. The protactinium-234, which undergoes beta decay, has a half-life of 1.2 minutes; the uranium-234 it produces has a half-life of about 250,000 years. Two isotopes of polonium have widely different half-lives: 3 minutes for $^{218}_{84}Po$ and 138 days for $^{210}_{84}Po$.

Learning Checks

1. What is meant by *half-life?* How is it related to *doubling time?*
2. If the half-life of an element is 1 minute, what fraction of the original sample remains after 3 minutes?
3. How does the death rate for radioactive nuclei differ from that for humans?
4. Can we predict which particular nucleus will disintegrate in any given time span? Explain.

Radioactive Families

By going through a series of alpha and beta decays, a given radioactive element can change into a variety of others, one by one, until it produces a stable element—one whose nucleus does not decay. One "parent" nucleus can give rise to an offspring, and so on, as long as the process continues to produce new radioactive species. Figures 25-6 and 25-7 show the family trees of uranium and thorium, both of which ultimately become stable isotopes of lead, element number 82.

THE NUCLEAR FORCE

We have seen that the nucleus is made up of protons and neutrons crowded together into an incredibly small region at the core of the atom. This crowding requires the positively charged protons to be extremely close to one another, so immediately the question arises, "How can the nucleus exist at all?" You learned in your study of electricity and magnetism that objects having the same charge repel each other. The closer they are, the stronger they tend to push apart. So, simply on the basis of electrostatic considerations, we would expect that protons could never live together quite happily, in a stable situation, in such close quarters. (See Figure 25-8.)

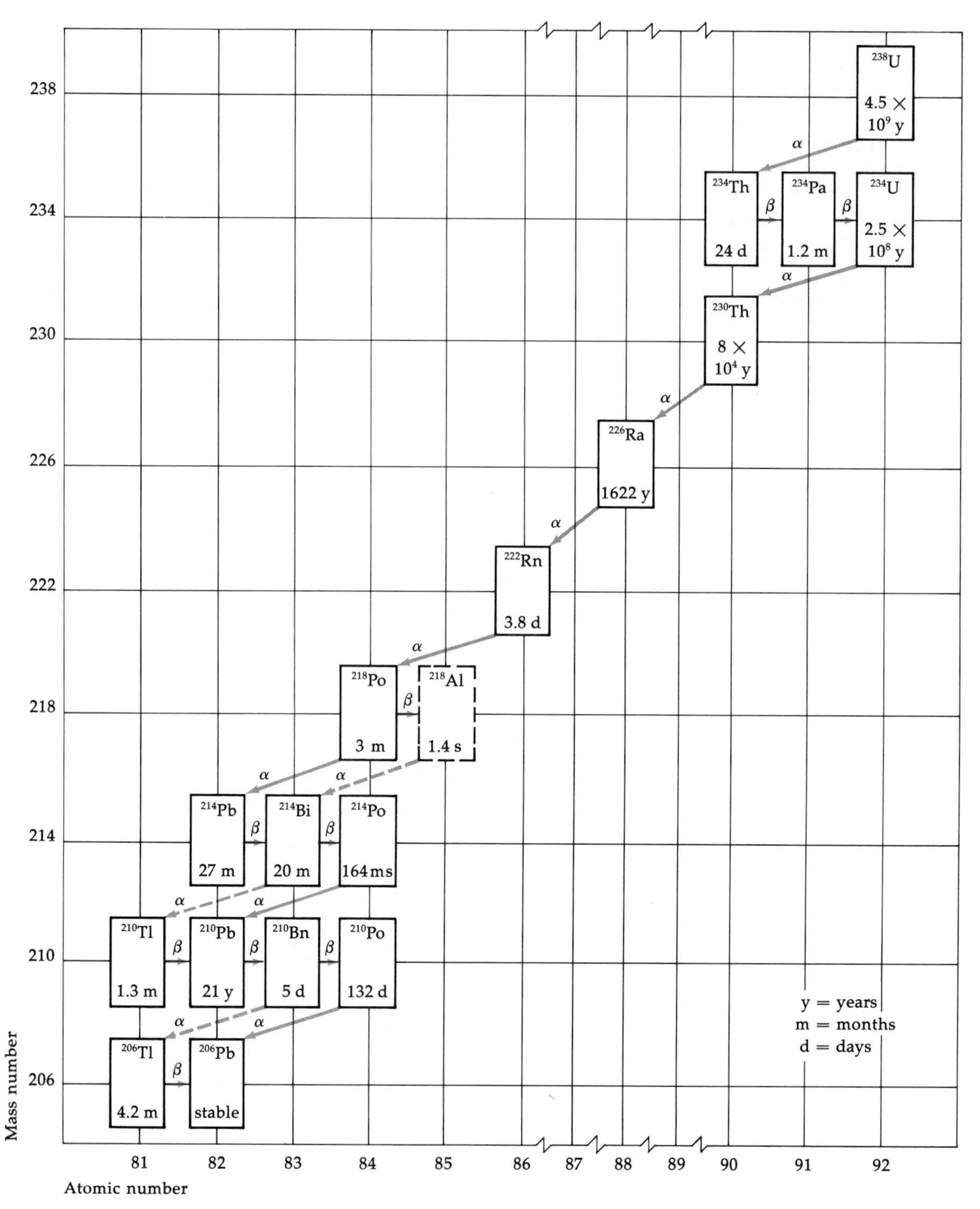

Figure 25-6. The uranium family tree. Boxes show half-lives.

Furthermore, there is the matter of the neutrons. Because they have no electrical charge, they are not attracted or repelled by the protons. We would suppose them to have no effect on each other either, except for the insignificant gravitational attraction. So again, just based on the electrostatics, there is no particular reason why a neutron should team up with protons and other neutrons in forming the nucleus.

Nevertheless, although there are solid electromagnetic arguments against such a thing as a stable nucleus, there it is. It exists, in spite of the reasons that say it shouldn't. The only way out of this dilemma is to suppose that there is *another kind of force* holding the nucleus together. This new force is the fourth and last of the fundamental forces that nature provides between particles. It must be an exceptionally powerful force, mighty enough to overwhelm the electrostatic repulsion the protons have toward each other. For this reason, this new nuclear force is called simply the *strong force*, or, more properly, the *strong interaction*.

The strong interaction is not nearly so well characterized as are the electrostatic and gravitational forces. For example, we don't know exactly how it varies with the distance between particles. We know that in the region where it is strongest, which must be over a range of around 10^{-15} to 10^{-14} meter, it is considerably stronger than the Coulomb force. (See Figure 25-9.) Otherwise, it would not overcome the repelling force between protons and they would fly apart. But it must also be extremely weak for distances greater than these, since the electromagnetic interaction completely swamps it outside the nucleus. So it must fall off much faster than the square of the distance—perhaps as the fifth or sixth power instead. Experiments tell us this. For distances greater than the dimensions of the nucleus, the electrostatic repulsion between the protons takes over again, and they no longer feel the strong attraction that exists inside the nucleus.

Electrical charge plays no part in the strong force. We know this because neutrons attract protons as well as other neutrons. It is also interesting to note that neutrons never repel each other. They only attract by virtue of the strong force and are not affected by the electrostatic interaction.

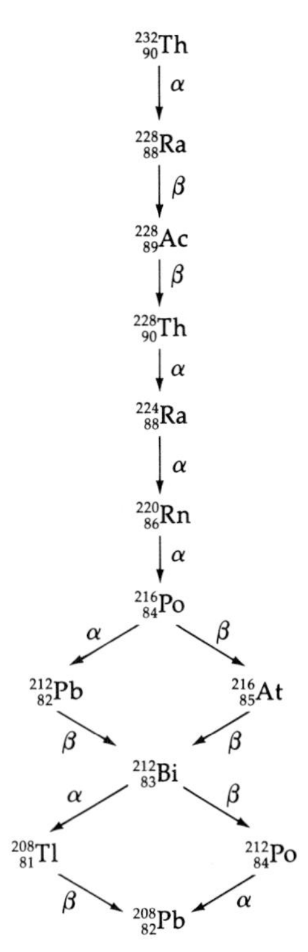

Figure 25-7. Decay scheme for thorium.

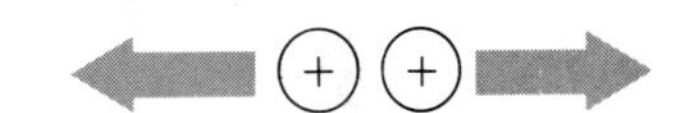

Figure 25-8. If the electrical force operated alone, the nucleus could not exist.

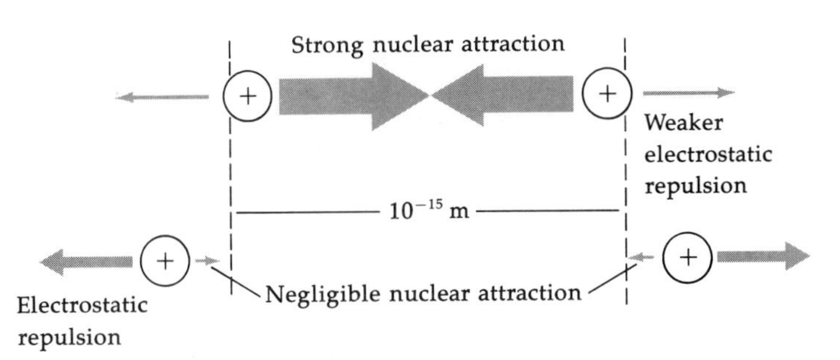

Figure 25-9. Comparison of the two forces at different proton separations.

Mysteries remain about this force that holds the nucleus together. Its dependence on distance must be sorted out. Also, physicists are still trying to understand what it is about protons and neutrons that make them attract each other in the range over which the strong force operates.

Learning Checks

1. What observations hint at the existence of the strong force?
2. The nuclear force must fall off faster than $1/r^2$, r being the distance between particles. Explain.
3. Does electric charge play any role in the nuclear force?

Numbers of Nucleons

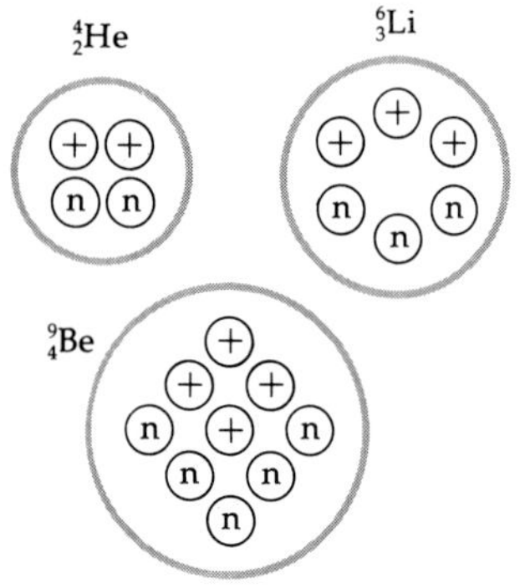

$^4_2$He $^6_3$Li $^9_4$Be

Figure 25-10. Nuclei contain equal (or nearly equal) numbers of protons and neutrons.

The neutron apparently plays a significant role in the structure of the nucleus. Neutrons seem to serve as a sort of "nuclear glue," holding the protons together in stable nuclei. As we already know, a single proton can be a nucleus: the nucleus of the normal hydrogen atom. But it turns out that we can't have more than one proton in a nucleus without at least one neutron. In fact, among the lighter elements, there is a tendency for the stable nuclei to contain equal numbers, or nearly equal numbers, of neutrons and protons. (See Figure 25–10.) For example, helium-4 has two of each, lithium-6 has three of each. Beryllium has four protons and five neutrons. Boron, carbon, and nitrogen each have two stable isotopes, in which the number of neutrons either equals or is one greater than the number of protons.

This trend continues through calcium, having atomic number 20. Of its six stable isotopes the lightest, calcium-40, has twenty protons and neutrons. From here on, as we proceed to more massive nuclei, the number of neutrons is always greater than the number of protons. And the imbalance becomes more pronounced as the mass number increases.

We can follow this tendency by looking at the ratio of neutrons to protons for a given nucleus. Let us call this ratio n/p. We list a few values in Table 25–1.

The last nucleus listed, uranium-238, with a half-life of more than a billion years, is the most massive of the naturally occurring isotopes. You can see that, with a neutron-to-proton ratio of 1.59, it has more than half again as many neutrons as protons.

We can account for the need for more and more neutrons in stable nuclei by recalling some things about the nuclear

Table 25-1
Ratio of Neutrons (n) to Protons (p) for a Given Nucleus

Element name	Symbol	n/p
Iron	$^{56}_{26}Fe$	1.15
Copper	$^{63}_{29}Cu$	1.17
Arsenic	$^{75}_{33}As$	1.27
Silver	$^{107}_{47}Ag$	1.28
Iodine	$^{127}_{53}I$	1.40
Tungsten	$^{180}_{74}W$	1.43
Gold	$^{197}_{79}Au$	1.49
Uranium	$^{238}_{92}U$	1.59

strong force. For each pair of protons, there is competition between the attractive nuclear force and the repelling influence of the electrical force. For the smaller nuclei all the protons are very close together, close enough for the strong nuclear force to dominate. But as more and more protons get into the act, those on opposite edges of the nucleus become farther apart. (See Figure 25-11.) For these separated protons, the strong force is less likely to dominate. They are too far apart, and the Coulomb repulsion begins to take over. So there is a greater need for more neutrons to hold the nucleus together. Remember, the neutrons are only affected by the strong force and not at all by the electrostatic one. Their attraction for the protons counterbalances the repulsion between protons a nucleus-width apart.

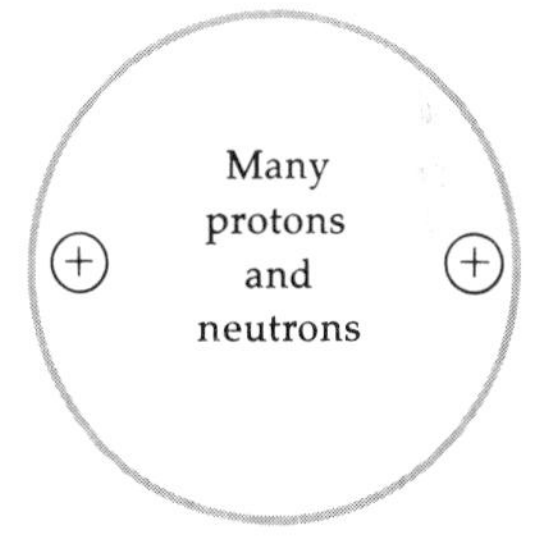

Figure 25-11. The two protons are almost too far apart for the strong force to hold the nucleus together.

It is interesting that nuclei containing an even number of protons are not too likely to decay. Apparently there is something especially stabilizing about *pairs* of protons. Notice that the six most common elements on earth have even atomic numbers: iron, oxygen, magnesium, silicon, sulfur, and nickel. The numbers of protons in these elements are 26, 8, 12, 14, 16, and 28, respectively. These elements make up about 98 percent of the material of the earth.

Proton-pair stability is quite evident in our old friend, the alpha particle. It contains two protons and two neutrons and is particularly stable, as evidenced by the alpha decay process. When radioactive atoms give off nucleons they never do so in units of less than an alpha particle. Furthermore, the nucleus made up of a pair of alpha particles is extremely fragile. This is beryllium-8, $^{8}_{4}Be$, having a half-life of about 10^{-16} second. The fact that it breaks up almost as soon as it forms implies that the two alpha particles are so stable by themselves they have essentially no tendency to join together.

Learning Checks

1. What does the ratio n/p indicate?
2. Why does n/p increase with mass number?
3. Is it likely that a nucleus containing forty-five protons and forty-five neutrons will be stable? Explain.

Artificial Transmutation

It became clear to Rutherford and others that the transmutation of radioactive elements is due to their instability. The protons are involved in a tug-of-war between the strong-force attraction on the one hand and the electrostatic repulsion on the other. Neutrons lined up on the strong-force team tip the battle in favor of stability for the lighter nuclei. But instability, in the form of alpha and beta decay, wins out for the heavier ones whose protons can be farther apart.

Rutherford felt that it should be possible to produce the break-up of nuclei by artificial means. He speculated that a non-radioactive nucleus subjected to powerful forces might be induced to become unstable and change to a different element. It was obvious, beginning with the work of Becquerel, that the magnitude of energies involved in chemical reactions would not do the trick. Forces on this scale affect only the electrons, which serve as a sort of shield protecting the nucleus. Rutherford's idea was to blast a stable nucleus with a stream of high-energy particles emitted during the spontaneous decay of larger, radioactive nuclei. The energies here are some *billions* of times larger than those producing chemical changes.

He put his notion to work in 1919 by bombarding nitrogen gas with alpha particles streaming from the element radium. This is the nuclear reaction that resulted:

$$ {}^{4}_{2}\mathrm{He} + {}^{14}_{7}\mathrm{N} \rightarrow {}^{17}_{8}\mathrm{O} + {}^{1}_{1}\mathrm{H} $$

Here we see that the two protons and two neutrons making up the helium-4 nucleus collide with nitrogen-14. This collision kicks out a proton (${}^{1}_{1}\mathrm{H}$) and leaves behind an isotope of oxygen, ${}^{17}_{8}\mathrm{O}$. This was the first instance of *artificial transmutation,* in which Rutherford deliberately changed nitrogen into oxygen.

Figure 25-12. The alpha particle's kinetic energy must overcome repulsion by the nucleus.

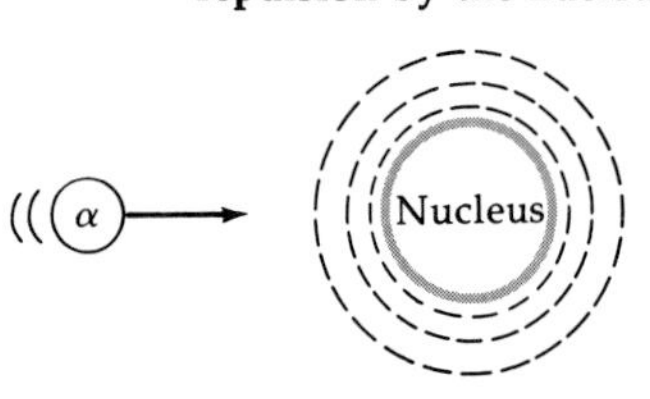

As you can imagine, the alpha particle must be traveling extremely fast. Once it gets past the electrons and heads for the nucleus, it will feel an enormous electrical repulsion, because both the alpha particle and the nucleus have positive charges. (See Figure 25-12.) Its kinetic energy must be large enough for it to push on in the face of this strong repulsion. Obviously, this problem becomes more severe as we try to do this with heavier nuclei, since they carry an even bigger positive charge. In practice, Rutherford could make light target nuclei, such as aluminum, transmutate. But for anything more massive than argon, with eighteen protons, the speeding alpha particles are simply deflected, scattered away. The electrical repulsion is so great that they never have a chance to collide with the nucleus.

Radioisotopes

The change of nitrogen-14 into oxygen-17 represents a transmutation in which both the initial isotope and the product are stable nuclei, and hence found in nature. However it is also possible to transmutate artificially a stable nucleus into a radioactive one. Irene and Frederic Joliot-Curie, the daughter and son-in-law of Pierre and Marie Curie, were the first to demonstrate such artificial radioactivity, in 1934. These artificially produced radioactive nuclei are called *radioisotopes.* The Joliot-Curies found that when they bombarded aluminum with alpha particles, protons and neutrons were emitted. These emissions stopped when the alpha bombardment was interrupted.

Let us see what happens when an alpha particle collides with the most abundant isotope of aluminum to produce a proton:

$$^{4}_{2}\text{He} + ^{27}_{13}\text{Al} \rightarrow ^{30}_{14}\text{Si} + ^{1}_{1}\text{H}$$

This reaction produces the stable isotope silicon-30, which occurs in nature with about 3 percent abundance. It is completely analogous to the nitrogen-to-oxygen transmutation that Rutherford saw 15 years earlier. But when the helium-aluminum collision produces a neutron, we get this reaction:

$$^{4}_{2}\text{He} + ^{27}_{13}\text{Al} \rightarrow ^{30}_{15}\text{P} + ^{1}_{0}\text{n}$$

Here we have written the neutron as $^{1}_{0}\text{n}$, since it has no charge and a mass number of 1. The resulting species must have fifteen protons and fifteen neutrons, and hence must be phosphorous-30.

Now it turns out that phosphorous is one of those elements having a single stable isotope, $^{31}_{15}\text{P}$, with sixteen neutrons rather than fifteen as in the isotope the Joliot-Curies produced. Apparently the odd proton in excess of the fourteen tied up in pairs requires two neutrons to stabilize the nucleus. In any event, the $^{30}_{15}\text{P}$ nucleus is unstable with a half-life of about 3 minutes.

Radioisotopes are characterized by half-lives that are short compared with the age of the earth. This means that any nuclei that may have been formed at the time of creation have all had plenty of time to decay away. Since the Joliot-Curies' pioneering production of phosphorous-30, hundreds of radioisotopes have been produced. Their half-lives range from several seconds to hundreds of thousands of years. Although a half-life of, say, a million years is quite long by human lifetime standards, it is short on the scale of the 5-billion-year-old earth, and so plenty of time has elapsed for complete disappearance of even the hardiest of the radioisotopes.

Applications of the Radioactive Isotopes

Radioactive materials, including both those occurring naturally and those artificially produced, find many uses in such widely diverse areas as medicine, agriculture, archaeology, and industry. All of these applications take advantage of the fact that radioactive isotopes behave chemically just like their stable siblings but are detectable because of the emissions of radioactivity. We are going to give a few examples.

Carbon is far and away the most important element in living tissue of both plants and animals. It has two stable isotopes, carbon-12 and carbon-13, which occur in natural abundances of 99 percent and 1 percent, respectively. An important radioactive nucleus is carbon-14, with a half-life of about 5730 years. Carbon-14 can be produced by bombardment of carbon-13 with the nucleus of deuterium ("heavy hydrogen"):

$$^{13}_{6}C + ^{2}_{1}H \rightarrow ^{14}_{6}C + ^{1}_{1}H$$

or by the neutron bombardment of the common stable isotope, nitrogen-14:

$$^{14}_{7}N + ^{1}_{0}n \rightarrow ^{14}_{6}C + ^{1}_{1}H$$

Carbon-14 in turn is a beta emitter, decaying back into nitrogen-14:

$$^{14}_{6}C \rightarrow ^{14}_{7}N + ^{0}_{-1}\beta$$

In 1946, an American chemist named Willard Libby proposed the idea that since cosmic rays showering down on the earth from outer space produce high-energy neutrons by collisions with atoms in the upper atmosphere, these neutrons should react with atmospheric $^{14}_{7}N$ to furnish a continuous supply of carbon-14. Since this isotope is chemically identical to the stable and abundant carbon-12, it should also be taken up by plant usage of carbon dioxide and subsequently should show up in animals that eat the plants. (See Figure 25-13.) Libby suggested that since carbon-14 is produced by cosmic rays at an approximately constant rate and decays at a constant rate, the atmosphere should contain a fairly constant but low-level amount.

Figure 25-13. A way of getting carbon-14 into living substances.

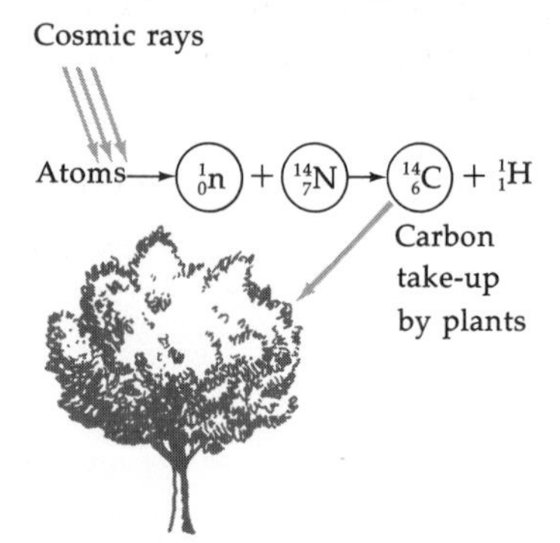

In a living tissue, then, the ratio of carbon-12 to carbon-14 should be fixed. However, as soon as the tissue dies, it no longer takes in carbon dioxide, so its concentration of carbon-

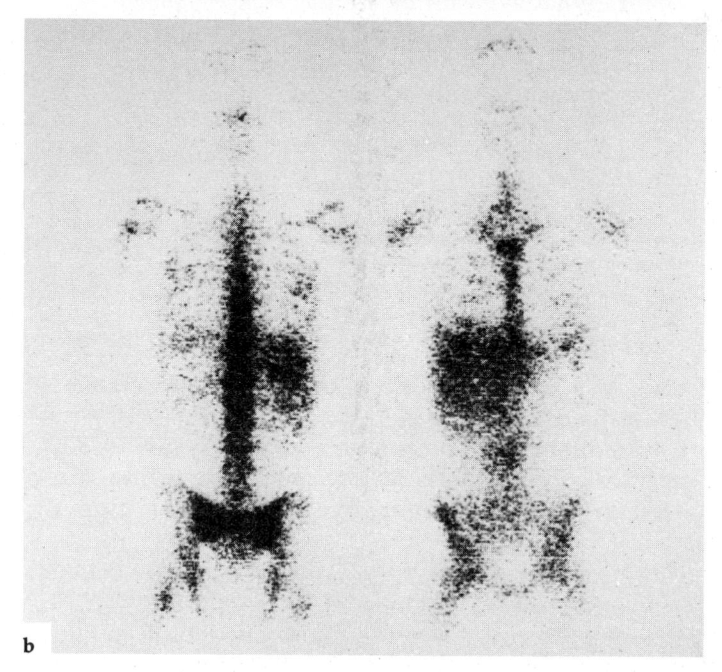

Some uses of radioactive isotopes. (a) Archeologists can fix the ages of remains through carbon-14 measurements of charcoal and other organic matter. (b) Physicians can determine the distribution of bone marrow by injecting indium-111 chloride. The figure on the left is a back view (posterior); the figure on the right is a front view (anterior).

14 will diminish by radioactive decay. Libby began studying this as a means of establishing the time of death of an old, formerly living object. If you compare the relative amounts of carbon-12 and carbon-14 in a sample of freshly cut wood with those in, say, trees felled by ice-age glaciers, you can use the half-life of carbon-14 to find out when the old trees died. In this way, scientists have recently determined that it was not until about 6000 B.C. that the glaciers retreated from the Great Lakes region.

An important medical application of radioactive isotopes is in their use as "tracers." Chemical compounds can be prepared in which a radioactive atom replaces its normal twin. Sensitive detectors can then locate the chemical by the radiation emitted. For example, there is a larger concentration of iodine in the thyroid gland than in any other tissue. The isotope $^{131}_{53}I$, a beta emitter with a half-life of about 8 days, can be used to locate the gland and determine its size and functioning. This isotope is also important in locating and identifying brain tumors.

There are also several industrial applications of radioactivity, which again take advantage of the "tagging" features of radiation. Chemists have made great strides in determining the structure of large complex molecules by selectively replacing normal atoms with their radioactive twins.

Learning Checks

1. What is meant by *artificial transmutation?*
2. Explain Rutherford's failure to achieve artificial transmutation among the heavier elements by alpha bombardment.
3. What are radioisotopes?
4. What are the distinctions and similarities of radioactive versus "normal" nuclei?

FISSION AND FUSION: THE NUCLEUS AS AN ENERGY SOURCE

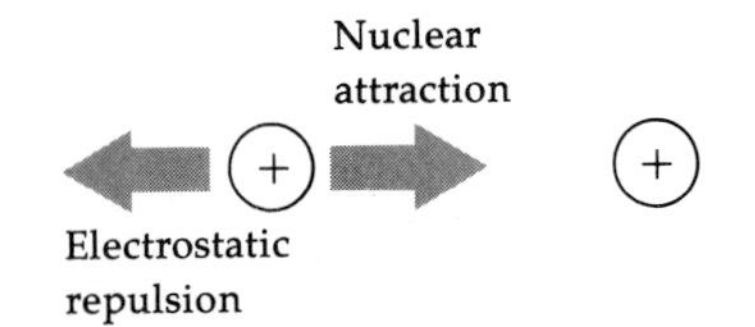

Figure 25-14. The strong nuclear force.

We have seen that the neutrons and protons are crammed together inside the nucleus and held there by the strong nuclear force. It is obvious that quite a lot of energy is involved in this arrangement, for it enables the protons to live together in very crowded quarters. The strong force prevents them from flying apart. (See Figure 25–14.)

Because of the attraction that binds the nucleons together, we would have to expend a certain amount of energy to pull them apart. (See Figure 25–15.) For this reason, it is called the *binding energy.* This situation is exactly analogous to lifting a rock off the ground. The energy you exert overcomes the gravitational attraction that the earth and rock have for each other. A similar thing occurs in moving an electron from an orbit near the nucleus to one further out. Energy is required to counteract the electrical attraction between the positive nucleus and the negative electron.

Figure 25-15. Much energy is needed to pull apart the nucleons.

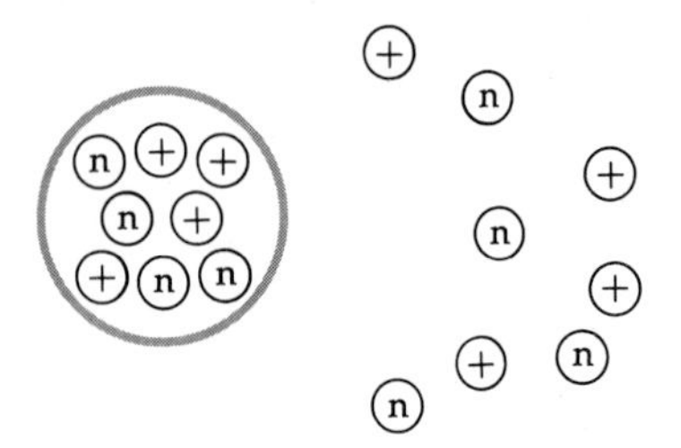

For the nucleus, let us see where the binding energy comes from. One of the consequences of Einstein's theory of relativity is that mass and energy are equivalent, two sides of the same coin. This idea is expressed in the famous equation, $E = mc^2$.

(We encountered this equation in Chapter 23 when we discussed de Broglie waves.) Here m is the mass, E the energy stored in matter in the form of mass, and c is the speed of light. Because c is such a large number, 3×10^8 m/s, a small amount of matter is equivalent to a tremendous quantity of energy. This mass-energy equivalence is the reason the nucleus is stable.

Suppose we look at some numbers. On page 449 we listed the masses of the proton, neutron, and electron. We can use these to predict the mass of any given atom or nucleus. Rather than use the masses in kilograms and have to deal with powers of ten, physicists find it convenient to switch to the *atomic mass unit* (abbreviated amu), defined as 1/12 the mass of the carbon-12 isotope, or 1.66055×10^{-27} kilogram. On this basis here are the masses of these particles:

proton: 1.00728 amu

neutron: 1.00867 amu

electron: 0.00055 amu

As an example, suppose we take the nucleus of, say, nitrogen-14, ${}^{14}_{7}N$. It has seven protons and seven neutrons. So we simply add up their masses:

$$\begin{aligned} 7 \text{ protons} \times 1.00728 \text{ amu} &= 7.05096 \text{ amu} \\ 7 \text{ neutrons} \times 1.00867 \text{ amu} &= \underline{7.06069 \text{ amu}} \\ & 14.11165 \text{ amu} \end{aligned}$$

But the *actual* mass of the nitrogen-14 nucleus is only 13.99922 amu. That is, there is a discrepancy between the mass of the nucleus itself and the total masses of its individual nucleons. The difference is

$$\begin{array}{r} 14.11165 \text{ amu} \\ \underline{-13.99922 \text{ amu}} \\ 0.11243 \text{ amu} \end{array}$$

This discrepancy is sometimes called the *mass defect*. It tells us that the combined mass of the seven protons and seven neutrons is *smaller* by 0.11243 amu (or 1.8669×10^{-28} kilogram) when they are tied up in the nitrogen-14 nucleus than when they are separated. This much mass has been converted into energy in the process of combining the individual photons and neutrons to form the nucleus. Or, turning the situation inside out, this mass is equivalent to the amount of energy we would have to supply in order to pull the individual nucleons apart. So the mass defect is the binding energy.

Let us use Einstein's equation to convert this mass defect to energy units. We have

$$\begin{aligned} E &= mc^2 \\ &= (1.8669 \times 10^{-28}\ \text{kg}) \times (2.996925 \times 10^8\ \text{m/s})^2 \\ &= 1.6768 \times 10^{-11}\ \text{J} \end{aligned}$$

A common energy unit that nuclear physicists use is the *electron volt* (eV), the energy an electron gains when it accelerates through an electric field with a potential difference of 1 volt. The conversion factor is

$$1\ \text{eV} = 1.602 \times 10^{-19}\ \text{J}$$

It is also convenient to designate an energy of a million electron volts (MeV):

$$10^6\ \text{eV} = 1\ \text{MeV} = 1.602 \times 10^{-13}\ \text{J}$$

This means that the binding energy of the nitrogen-14 nucleus is

$$1.6768 \times 10^{-11}\ \text{J} \times \frac{1\ \text{MeV}}{1.602 \times 10^{-13}\ \text{J}} = 104.67\ \text{MeV}$$

To get a feeling for how much energy this is, let us look at some comparisons. It takes about 10^{16} MeV of energy to boil a gram of water. This would require the binding energy of only 1.5×10^{14} nitrogen-14 nuclei. That may seem like a lot of nuclei, but consider this: a single gram of nitrogen is made of 4×10^{22} nuclei. So we would need the binding energy of only about 10^{-9} gram of nitrogen-14—a billionth of a milligram—to boil the water.

Contrast this with the chemical energy released by burning natural gas, which is mostly methane. Each molecule of methane furnishes only about 0.00001 MeV. So we would need to burn around 10^{21} molecules of methane—roughly 0.05 gram—to get the 10^{16} MeV that only 10^{-9} gram of the nitrogen nuclei could supply by this binding energy.

The difference, of course, is due not to the nitrogen versus the methane but to the amount of energy involved with the nucleons versus that of the electrons. For any atom or molecule, its *nuclear* energy is many millions times greater than its *chemical* energy.

The Curve of Binding Energy

Now let us look at these numbers in a slightly different way. The nitrogen-14 nucleus has a binding energy of 104.67 MeV and a total of fourteen nucleons. Dividing the binding energy by the number of nucleons gives

$$\frac{104.67 \text{ MeV}}{14 \text{ nucleons}} = 7.48 \text{ MeV per nucleon}$$

So the binding energy per nucleon for $^{14}_{7}N$ is 7.48 MeV, which tells how much, on the average, each proton and neutron contributes to the mass-to-energy conversion holding this nucleus together.

For comparison, we will do a similar calculation for a very heavy nucleus, uranium-235, which we will be discussing shortly. This isotope has the symbol $^{235}_{92}U$, so it contains 92 protons and 143 neutrons. Their combined mass is

$$\begin{aligned} 92 \times 1.00728 \text{ amu} &= 92.66976 \text{ amu} \\ 143 \times 1.00867 \text{ amu} &= \underline{144.23981 \text{ amu}} \\ \text{Total} &= 236.90957 \text{ amu} \end{aligned}$$

The mass of the $^{235}_{92}U$ nucleus is 234.99422 amu, giving a mass defect of 236.90957 − 234.99422 = 1.91535 amu. This is equivalent to 1783.38 MeV, so the binding energy per nucleon is

$$\frac{1783.38 \text{ MeV}}{235 \text{ nucleons}} = 7.59 \text{ MeV per nucleon}$$

which is somewhat different from that for nitrogen-14. On the other hand, an identical calculation for a much lighter nucleus, $^{7}_{3}Li$ (lithium-7), gives a smaller binding energy per nucleon, 5.64 MeV.

Figure 25–16 shows a plot of the binding energy per nucleon for the various stable and long-lived nuclei. As you can see, the binding energy per nucleon varies slowly for the heavier nuclei, but can be vastly different for nuclei whose mass number is less than about 40. Notice also that the curve has a maximum that occurs at a mass number of around 56.

This curve tells us how the contribution of each proton and neutron to the stability of the nucleus *differs* depending on *which* nucleus we are considering. The relationship demonstrated by the curve of binding energy is what allows us to get energy from the nucleus by two processes—fission and fusion. Now we are ready to see how this happens.

Fission

Fission is the process by which a big nucleus splits and forms two smaller nuclei, with a large amount of energy released. Two German radiochemists, Otto Hahn and Fritz Strassman,

Figure 25-16. The curve of binding energy per nucleon.

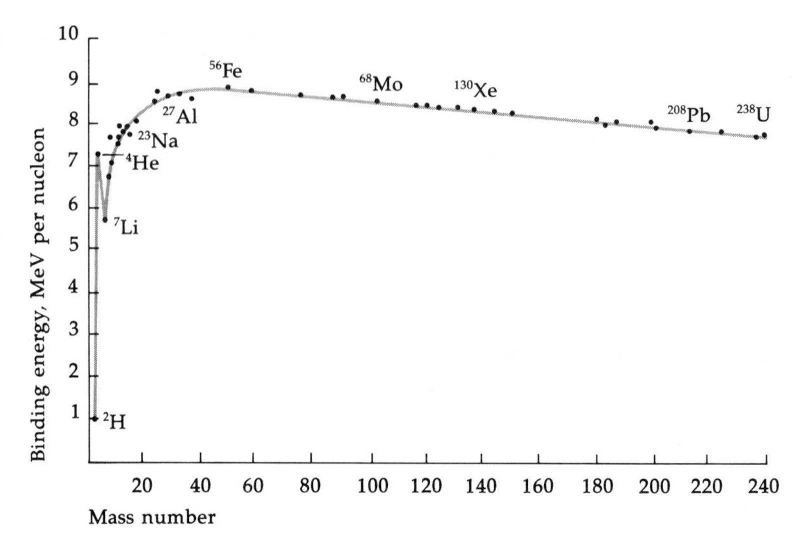

discovered this process early in 1939. They were bombarding uranium with neutrons, expecting to produce elements with larger mass numbers. But their results indicated that instead they were producing elements with *smaller* mass numbers, in the middle range of the periodic table. Hahn's longtime associate, Lise Meitner, and her nephew, Otto Frisch, confirmed that the uranium nucleus split into two medium-sized fragments accompanied by the release of about 200 MeV of energy.

Meitner and Frisch, who were Jewish, had fled Germany in 1938 to escape the threat of Hitler's rising power and joined Niels Bohr in Copenhagen. Bohr soon reported their experiments on uranium fission in America. There followed a flurry of activity as physicists sought to learn more about this new discovery.

Uranium-235 was soon identified as the isotope responsible for most of the fissions. This isotope has only about 0.7 percent natural abundance compared with 99 percent for uranium-238, which, as we have seen, is the parent of a radioactive family. One reaction that splits the uranium-235 nucleus is

$$^{1}_{0}n + ^{235}_{92}U \rightarrow ^{142}_{56}Ba + ^{91}_{36}Kr + 3(^{1}_{0}n)$$

We can schematically illustrate this equation, as shown in Figure 25-17. Uranium-235 absorbs a neutron and then breaks up, yielding a barium-142 nucleus and a krypton-91 nucleus. Notice also that three neutrons are kicked out. We will return to this important point later.

As we said, this approximately 60-40 split of the large nucleus produces about 200 MeV of energy. To see where this energy comes from, we need to return to the curve of binding

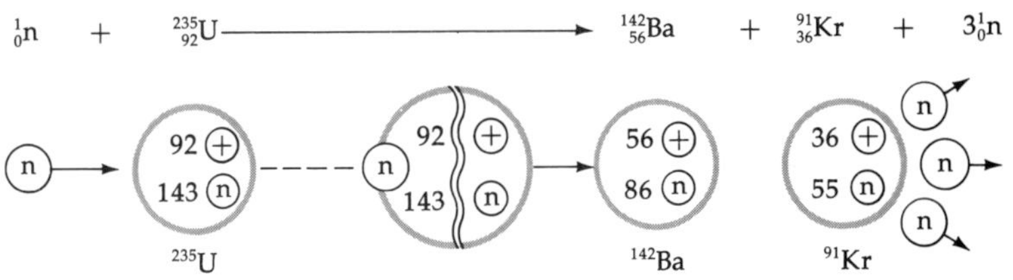

$$^{1}_{0}n + {}^{235}_{92}U \longrightarrow {}^{142}_{56}Ba + {}^{91}_{36}Kr + 3\,^{1}_{0}n$$

Figure 25-17. The fission of uranium-235.

energy, Figure 25–16. The uranium nucleus, located at mass number 235, contains around 7.6 MeV per nucleon, as we saw earlier. With 235 nucleons, this gives a total binding energy of about 1786 MeV. The particular isotopes of barium and krypton give these numbers:

Isotope	Binding energy per nucleon		Binding energy
$^{142}_{56}Ba$	8.3 MeV		1178.6 MeV
$^{91}_{36}Kr$	8.7 MeV		791.7 MeV
		Total	1970.3 MeV

The total binding energy of 1970.3 MeV is about 184 MeV *greater* than that for $^{235}_{92}U$ alone. This much energy is released when the U-235 fissions. To drive home the point about the magnitude of this energy: exploding a single molecule of TNT gives only 30 eV. The energy from a single U-235 fission is over *6 million times greater.*

Remember what the binding energy means. It is the energy that we must supply to separate the nucleons completely. We can recover this same energy by allowing the neutrons and protons to re-form the nucleus, very much like recovering gravitational potential energy by allowing an object to fall to earth. In the present case, we would have to provide 1786 MeV to drag the nucleons of U-235 apart. But by allowing them to "fall back together," so to speak, in forming the barium and krypton nuclei, we recover 1970 MeV. Figure 25–18 illustrates this idea.

We can get an idea about why fission takes place by tapping some of our knowledge of the strong force. Uranium-235 does not normally fission spontaneously but requires neutron bombardment. We can imagine that when the nucleus absorbs the speeding neutron, the crash makes the nucleus vibrate somewhat. The protons' delicate balance between electrostatic repulsion and nuclear attraction tips slightly in favor of repulsion since the alien neutron makes things more crowded. (See Figure 25–19.) Two protons on opposite edges of the nucleus are now slightly farther away than before, too far away for the

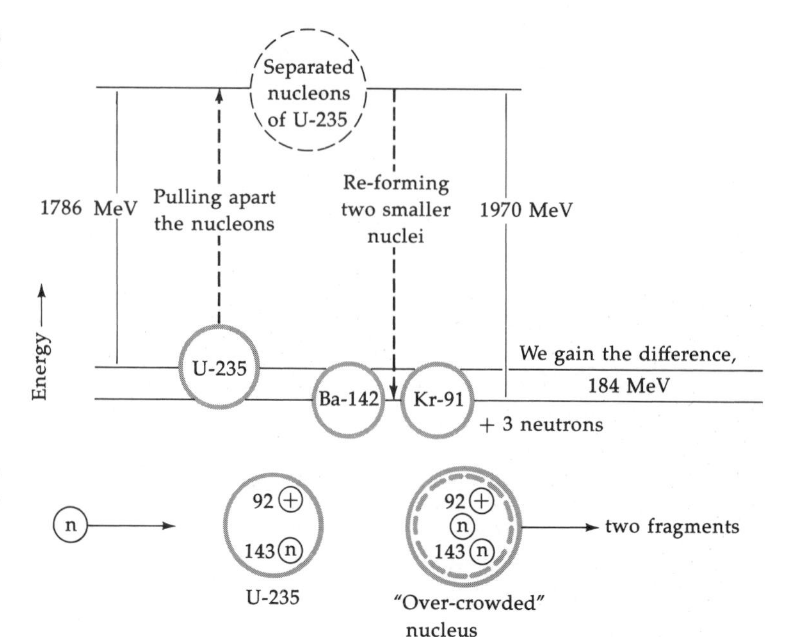

Figure 25-18. How fission provides energy.

Figure 25-19. The extra neutron pushes the protons apart.

strong force to dominate. Hence, they are more likely to push each other apart. Neighboring protons and neutrons "choose up sides," and the nucleus ruptures.

Practical Aspects: Nuclear Reactors

In order for us to learn how nuclear fission may be put to use—whether as a weapon or as a source of energy for peaceful purposes—there are some nuts-and-bolts things we should understand. First, there is the matter of resource supply. In naturally occurring uranium, only seven atoms in 1000 are fissionable U-235, by far the greater number being U-238, which does not fission. This introduces the problem of how to get the two isotopes separated from each other. Primarily this is done by *gaseous diffusion.* A gaseous compound of uranium, called uranium hexafluoride, can be separated on the basis of the mass differences into portions that contain a single isotope of the uranium.

In discussing the fission reaction for U-235, we pointed out that when the nucleus ruptures into the two smaller fragments it also kicks out three neutrons. These neutrons could possibly lead to three more fissions, each of which would also release three neutrons, and so on. This gives the possibility of a *chain reaction,* in which each fission generates others, and a tremendous quantity of energy is released in a very short time. This, of course, was the idea behind the *fission bomb,* more popularly called the *atomic bomb,* that the United States exploded over the Japanese cities of Hiroshima and Nagasaki in the summer of 1945.

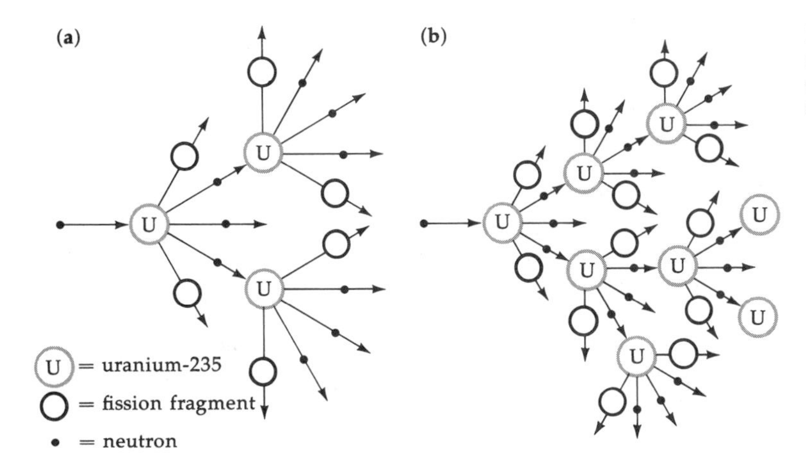

Figure 25-20. Critical mass for fission. (a) Subcritical mass (too many neutrons escape). (b) Critical mass (chain reaction sustained).

A chain reaction cannot be sustained, even with pure uranium-235, unless a large enough volume of material is available. If too many neutrons simply escape without a fission, the process will fizzle out. The minimum amount of material needed to sustain the chain reaction is called the *critical mass.* (See Figure 25–20.) For an explosive device the critical mass should be as small as possible, making the device easier to move from place to place.

One way of reducing the critical mass is by slowing down the fission-producing neutrons. It turns out that slow neutrons are more efficient for causing fission than fast ones. Graphite is a good material for slowing the neutrons down. When some graphite is included in the core of U-235, a larger fraction of neutrons will be suitable for generating fission. These are called *thermal* neutrons, since they have a kinetic energy corresponding to ordinary room temperature.

The control room of a nuclear plant. In establishing a nuclear facility such as this, planners take into account such factors as the safety aspects, the types of supporting services required, and the construction and operating costs.

In a nuclear reactor, as opposed to a bomb, the idea is not to produce an explosion, but to release energy at a slower, more controlled rate. Again, this can be done by controlling the speed and supply of neutrons. The first nuclear reactor used cadmium rods for this purpose. Cadmium is a good neutron absorber. By pushing cadmium rods in and out of the fissionable material it is possible to maintain control over the rate at which the chain reaction proceeds. This first reactor was built during 1942, on a squash court under the stands of Stagg Field, the University of Chicago football stadium. The project, directed by the Italian physicist Enrico Fermi, was conducted in great secrecy by the Manhattan Project and ultimately led to production of the atomic bomb.

Radioactivity and Health Hazards

As the use of nuclear reactors becomes more and more important in the scheme of the nation's energy production, greater attention is being drawn to the potential hazards of radioactive materials to human health. There is no question that the alpha, beta, and gamma radiation emitted by radioactive nuclei can do damage to the cells of living organisms. The concerns relate to the extent of the damage and the degree that nuclear power plants intensify the problem.

You should understand that exposure to radiation is not something invented by nuclear reactors. There are many natural sources that have been around since the beginning, undoubtedly affecting the evolution of the species. The cosmic rays that continuously shower down upon us are responsible for a large fraction of this natural radiation. In addition, there is uranium in the soil. There are radioactive elements in our food and water, in the materials from which our houses are built. All these sources provide an amount of radiation that, on the average, in a year's time is only about one-fourth of the amount that would be fatal if taken in a single dose.

Emissions of radiation come from the various radioactive products that result when uranium and other fissionable isotopes rupture. The fission of uranium-235 that we used as an example is not unique; it is only one of several possibilities. One dangerous fission product is strontium-90, a beta-emitting isotope of strontium, element number 38. Strontium-90 has a half-life of about 23 years, ejecting a beta particle with an energy of 0.5 MeV. As you can see from the periodic table, strontium is in the same column as calcium, so its chemical properties are quite similar. For this reason, it can replace calcium in milk and subsequently in bone structure. Bone marrow suffers damage by the radiation from strontium-90, and hence the body's red-blood cell production is affected.

Genetic damage by radiation is perhaps more dangerous than the pathological effects caused by elements such as strontium-90. Radiation damage can cause gene mutations that not only are harmful but also are passed from generation to generation. Information about genetic damage is sketchy, at best, since little time has passed since large radiation doses were unleashed at Nagasaki and Hiroshima. Gene mutation is a fertile field for science fiction writers to cultivate, but it will take many generations of descendants of the bombing survivors to provide information about mutation in humans that is caused by radiation.

Learning Checks

1. Explain what is meant by *binding energy.*
2. What role does the mass-energy equivalence play in holding the nucleus together?
3. What does the curve of binding energy represent?
4. Explain *mass defect.*
5. Is a proton more massive or less massive when tied up in a nucleus? Explain.
6. Compare the energy of a proton bound to the nucleus with that of an electron bound to an atom.
7. Why does fission of a large nucleus into two smaller ones release energy?

Fusion

For heavy nuclei, fission corresponds to climbing the curve of binding energy toward higher energy values. But we can also climb the other side of the curve, where the mass numbers are less than about 56, and gain energy by putting together two lighter nuclei to form a heavier one. This is called *fusion:* we say that two less massive nuclei *fuse* to form a more massive one.

As a hypothetical example, we might consider fusing two nuclei of mass number 10 to form one with mass number 20. For mass number 10, the binding energy is about 6.5 MeV per nucleon, giving the two nuclei a total binding energy of 130 MeV. The nucleus with mass number 20 has 8.0 MeV per nucleon, or 160 MeV altogether. We gain the difference, $160 - 130 = 30$ MeV.

Fusion provides tremendous quantities of energy in the sun and other stars. As such, it is the source of the energy and light necessary for life on the earth. The simplest possible fusion reaction, conversion of four hydrogen nuclei to form one helium nucleus, is responsible for nearly all the sun's energy. We can again use our binding energy curve to find out how much energy is involved. The single proton of $^{1}_{1}H$ has no binding energy since, of course, it is not bound to any other nucleon. Deuterium ($^{2}_{1}H$) and tritium ($^{3}_{1}H$) have, respectively, 1.1 and 2.6 MeV per nucleon. The next point on the curve is

The Fusion Race

"We really think it's almost in our hands," one fusion expert avers. "We can almost feel it." How close, exactly, are we to this glamorous source of power?

The teams on two gigantic research projects—one at Princeton University and one at the University of Rochester—feel they are within "spitting distance" of success. Each school is on a different path to this goal: Princeton with the magnetic fusion method, Rochester with laser fusion.

The fusion process circumvents a lot of problems posed by existing fission plants: fusing atoms gives no "hot" waste and there is no danger of meltdown or accidental runaway reactions. Because fusion seems to provide the answer to many of the world's problems, the United States is funding its development with billions of dollars. Despite the optimism of its backers, however, formidable engineering challenges must be met before we can see this power running our nation.

Sources: Philip Bofley, "An Earthly Furnace Fueled by Fusion Nears a Crucial Test," *Smithsonian*, 1980, pp. 129–136.

Hans Fantel, "Fusion—We're Harnessing the Power of the H-Bomb," *Popular Mechanics*, pp. 86–90.

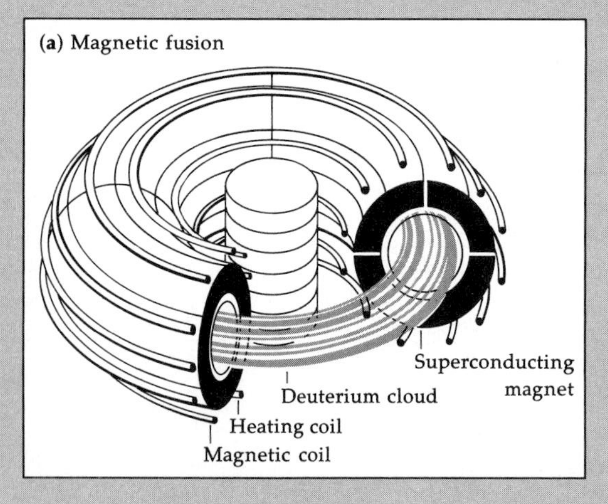

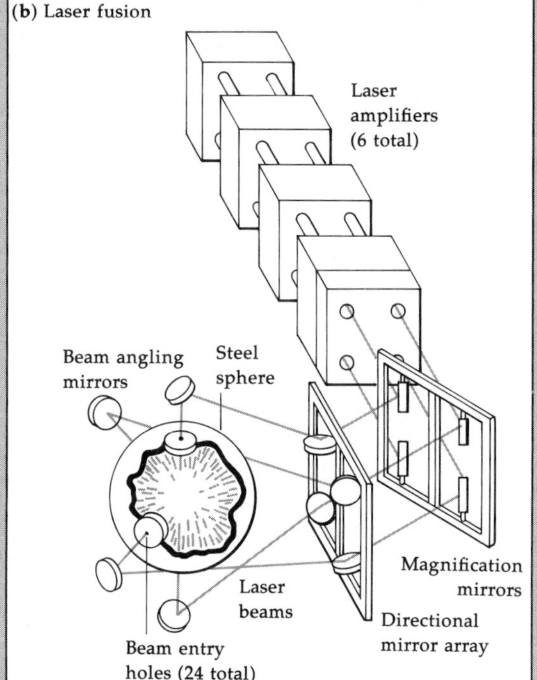

(a) In the Tokamak magnetic fusion reactor at Princeton University, a deuterium cloud is heated and jolted with a neutral beam of electricity. This forces the atoms to fuse. A superconducting coil prevents the cloud from breaking up and losing its energy too quickly. The deuterium from a single gallon of water would produce energy the equivalent of approximately 300 gallons of gasoline. (b) A section of the Laser Energetic Laboratory project at the University of Rochester. The light of 24 giant laser beams is shot at a tiny pellet of deuterium and tritium smaller than a pinhead. The temperature inside the pellet is raised to the point required for atomic fusion. Thus ignited, each pellet becomes a miniature hydrogen bomb whose energy is captured to generate electricity. Exploding pellets one after the other at half-hour intervals produce enormous energy.

for helium-4, ${}^{4}_{2}He$, at about 7.1 MeV per nucleon. So, fusing four hydrogens to make a helium releases around 7 MeV per nucleon.

We can see how very energetic this reaction is by comparing it with the uranium fission. There we have a net gain of about 200 MeV for 235 nucleons, or less than 1 MeV per nucleon. The energy gain of 7 MeV per nucleon for hydrogen-to-helium fusion represents a much more efficient reaction.

Practical Aspects of Fusion

Because the fusion process joins two small nuclei to form a larger one, it requires mashing two positively charged particles close enough for the nuclear strong force to dominate. This means they must be rammed together at very high speeds, high enough to overcome the electrostatic repulsion that tends to push them apart. A large amount of "starter" energy, then, is required for the fusion process. And, of course, for fusion to be a useful energy source, we must be able to recover more energy than we put in.

The likeliest candidate for fusion fuel is deuterium, ${}^{2}_{1}H$. This is one feature that makes fusion attractive, since deuterium is readily available in the oceans. It is easily separated from normal hydrogen. The reaction involves five nuclei of deuterium:

$$5\,{}^{2}_{1}H \rightarrow {}^{3}_{2}He + {}^{4}_{2}He + {}^{1}_{1}H + 2\,{}^{1}_{0}n$$

It releases about 25 MeV. To put this in perspective, a cubic meter of water provides enough deuterium to fuse and release the energy equivalent to burning 200 barrels of oil.

The starter conditions on the sun, 20,000,000° Celsius and extremely high pressures, are not likely to be duplicated on earth. So alternative means of obtaining fusion are currently being sought. One attractive possibility is to focus the beam of a powerful laser on tiny frozen pellets of deuterium. The resulting vaporization should provide the temperature and pressure conditions to achieve fusion. Research into this and other ways of getting fusion is underway in many parts of the world.

Learning Checks

1. How is fusion an energy-releasing process?
2. Is there a loss of mass during fusion? Explain.
3. What are some practical considerations in obtaining energy from fission? From fusion?

SUMMARY

Becquerel discovered that some atoms are *radioactive,* emitting radiation of three kinds: *alpha particles, beta particles,* and *gamma rays.* Alpha particles are nuclei of helium containing two protons and two neutrons. Beta particles are high energy electrons. Gamma rays consist of high energy photons. This radioactivity comes from the nucleus, composed of protons and neutrons, collectively called *nucleons.*

Nuclei are labeled according to their *atomic number* (number of protons) and *mass number* (number of nucleons).

Radioactive nuclei can spontaneously *transmutate,* or change to another element, by emitting alpha or beta radiation. *Alpha decay* decreases the atomic number by two units and the mass number by four units. *Beta decay* increases the atomic number without changing the mass number.

Isotopes are nuclei of the same element (same atomic number) having different numbers of neutrons. Examples are carbon-12 ($^{12}_{6}C$) and carbon-13 ($^{13}_{6}C$). Isotopes of the same element have the same chemical properties but differ in their nuclear structure. They may also have different radioactive properties.

The decay time for a given nucleus is expressed in terms of its *half-life:* the time required for one-half of a given sample to transmutate. The half-life is fixed for a particular element, no matter what the age of a sample. Half-lives may vary from small fractions of a second to billions of years.

The force holding the nucleus together is called the *nuclear strong interaction.* It only attracts acts over a very short range (of the order of 10^{-14} meter), and does not depend on the electric charge. Only the neutrons and protons are affected by this force. It is much stronger than the electrostatic force.

The neutrons play a major role in stable nuclei. For the lighter elements, the nuclei have approximately equal numbers of neutrons and protons. But for the more massive elements the number of neutrons may exceed the number of protons by as much as 50 percent, indicating that the neutrons act as a "nuclear glue."

Rutherford discovered that stable nuclei could undergo *artificial transmutation.* One result is the production of *radioisotopes,* artificially-produced radioactive nuclei.

Applications of radioactivity take advantage of the chemical similarity of isotopes. *Carbon dating* is one example, in which ages of formerly living tissue can be determined by comparing the ratio of carbon-12 to the radioactive carbon-14 with that in live tissue. Isotopes are also used as "tracers" in medicine and industry.

Energy from the nucleus comes from its *binding energy,* which is the energy that would be necessary to separate all the nucleons completely. It arises from *mass defect*—that is, the mass of a nucleon is less when it is in the nucleus than when separate. This missing mass is converted to binding energy according to Einstein's mass-energy equivalence, $E = mc^2$.

From the curve of binding energy per nucleon, we see that *fission* of heavier nuclei into lighter ones releases energy, as does *fusion* of very light nuclei into heavier ones. Fission of such species as U-235 releases tremendous quantities of energy, much more than the chemical energy obtained from the burning of fossil fuel. Fission has been used in the destructive power of the atomic bomb and currently provides energy for peaceful purposes in nuclear reactors.

Fusion is a theoretically promising energy source, but technological problems remain.

Exercises

1. Is it possible for a hydrogen nucleus to emit an alpha particle? Explain.

2. Explain how we know that the nuclear force is stronger than the electrostatic force.

3. The beryllium-8 nucleus is very unstable, with a half-life of about 10^{-16} second. Account for this instability.

4. Which is most like an X ray: alpha, beta, or gamma radiation?

5. Where does the energy come from that is produced in the fission process? In the fusion process?

6. Does the electric force play a role in either radioactivity or fission? Explain.

7. The following list shows the number of neutrons in the most abundant isotope for each element. Give the standard-notation symbol for each nucleus.

Element name	Number of neutrons
Helium	2
Boron	6
Phosphorus	16
Chromium	28
Nickel	30
Tin	70

8. Give the number of electrons, protons, and neutrons for each of the following:
$^{10}_{5}B$, $^{31}_{15}P$, $^{35}_{17}Cl$, $^{106}_{46}Pd$, $^{239}_{94}Pu$

9. Write nuclear equations for the following reactions:
a. Chlorine = 36 decays by beta emission.
b. Potassium = 40 decays by beta emission.
c. $^{216}_{84}P$ emits an alpha particle.

10. Neptunium-237 (atomic number 93) undergoes eight alpha and four beta decays. Give the label for the stable isotope that finally results.

11. An aluminum-27 nucleus is bombarded with an alpha particle to give a new element plus a free neutron. Write the nuclear equation for this process.

12. Explain the comment that the neutrons serve as a "nuclear glue." Could the protons fulfill that role? Explain.

13. Some radioisotopes have half-lives of several thousands of years, and yet they are not found in nature. Explain.

14. Suppose, as a nuclear physicist, you had to reduce critical mass by slowing down neutrons. How would you do this?

15. How does the process of radioactive decay differ from processes that follow Newton's laws?

16. Why is it easier for a neutron than for a proton to penetrate a nucleus?

17. Uranium-235 releases an average of 2.5 neutrons per fission, compared to 2.7 for plutonium-239. Which of these two nuclei would you think has the smaller critical mass?

18. Calculate n/p for these nuclei: $^{3}_{2}He$, $^{9}_{4}Be$, $^{14}_{7}N$, $^{44}_{20}Ca$, $^{103}_{45}Rh$, $^{121}_{51}Sb$, $^{81}_{35}Br$, $^{40}_{20}Ca$.

19. An archeologist finds a piece of charcoal from an ancient buried campfire. Its ratio of C-14 to C-12 is about one-eighth that in a sample of freshly cut wood. About how long ago did the campfire burn?

20. The mass of $^{7}_{3}Li$ nuclei is 7.01436 amu. What is its binding energy? What is its binding energy per nucleon? (Note: 1 amu = 931 MeV)

21. Suppose that two $^{7}_{3}Li$ nuclei are fused to form a new nucleus. What isotope is formed? If the new isotope has a binding energy per nucleon of 7.7 MeV, how much energy will this fusion provide?

22. A particular isotope of cobalt has a half-life of 5 years. If you started with a 1 kilogram sample in 1975, how much would remain by the year 2000?

23. Suppose that the half-life of an element is 5 minutes. You obtain a sample at 12 noon but decide to go to lunch before measuring its radioactivity. How much sample must you originally obtain in order that 5 grams remain when you return from lunch at 1 o'clock?

24. If only one-sixteenth of a radioactive sample remains after 1 year, what is the half-life of the material?

25. A nucleus of radium ($^{224}_{88}Ra$) decays in several steps to lead ($^{208}_{82}Pb$). How many alpha particles and how many beta particles are emitted in the process?

A representation of how quarks may piece together like a puzzle to form a proton.

26

THE ATOM WITHIN THE ATOM

A practical joker's way of giving Christmas presents is to use a device called Chinese boxes. Under the Christmas tree you find a large, beautifully wrapped box. When you open it, you find another box, which in turn contains a smaller box, inside of which is a still smaller box, and on, and on. If you're lucky, there might be a gift inside the last box.

For the last few decades or so, physicists trying to unravel the nature of matter at its most fundamental level have felt much like the child opening Chinese boxes. Macroscopic matter was found to be made up of invisible atoms. Upon cracking open the atoms, the physicists found electrons and a nucleus. The nuclear box, as we have just seen, contains neutrons and protons. For a while it seemed as though that might be the end of it. Ah, but don't go away, folks—we're not through yet. With new machines capable of higher and higher energies, physicists have unearthed what amounts to a zoo of other particles.

In this chapter, we want to talk about these particles, what their discovery means, and how they fit into the scheme of the universe. We shall also discuss a particle, as yet undiscovered, called the *quark*, which many physicists believe could possibly be the very building block of nature and the last of the Chinese boxes. Whether or not this is the case, of course, only the future will tell.

The world of these elementary particles is a fast-changing one, in which new results appear almost monthly. This means that there is something of a risk: what is written here may be out of date before the ink is dry. But the picture presented is that nature, on a scale several million times smaller than the atom, has a rather remarkable symmetry and simplicity about it. So, in a spirit of "No guts, no glory," let's forge ahead and see what we can learn.

> *It is interesting that there is not one iota of direct experimental evidence for supersymmetry [a current particle theory], yet we study it because it looks so much like the sort of theory we would like to believe in. This is symptomatic of the terrible state we are in . . . The salvation of elementary particle physics is, at least for the moment, in the hands of the experimentalists.*
>
> Steven Weinberg

THROUGH THE LOOKING GLASS: ANTIMATTER

Let us begin our discussion of elementary particles by recalling the two great twentieth-century revolutions in physics—relativity and the quantum theory. By 1929 both of these theories were well established, having succeeded in explaining a variety of observations. The two had flourished more or less independently of each other. But both are applicable in many situations, for example, an electron (described by quantum mechanics) traveling at close to light speed (which is the realm of relativity).

A British physicist, Paul A. M. Dirac, set about to combine the two theories. His desire was not merely to tack one onto the other as an afterthought but to create a new theory that would apply in circumstances where the two overlap. One of the most interesting results to come out of Dirac's work was the prediction of a new form of matter that had not been detected up to that time. Dirac, in 1929, showed that there should exist a particle having the same mass as the electron but carrying a *positive* charge rather than a negative one. In a sense, this new particle would be sort of a "mirror image" of the electron.

(a) The Wilson cloud chamber. (b) Particle tracks revealed in a typical cloud-chamber photograph.

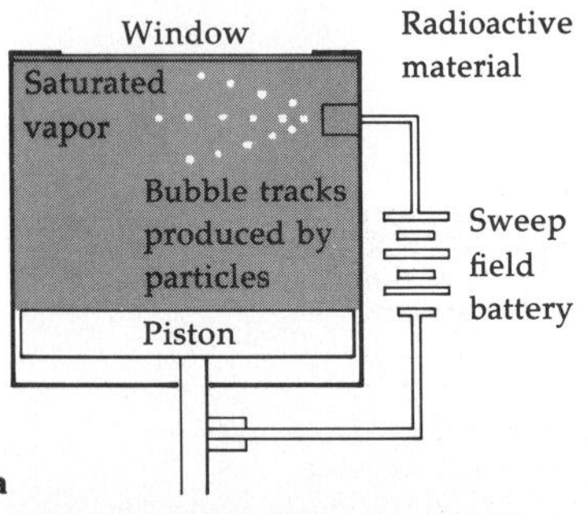

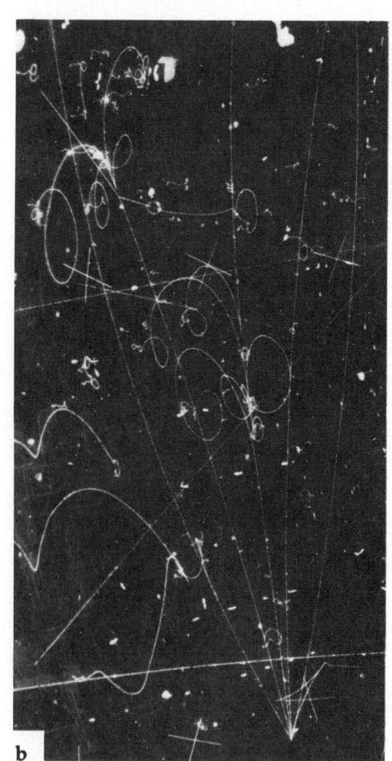

Not many people took Dirac's proposal seriously. But then, three years later, in 1932, an American physicist named Carl Anderson, at the California Institute of Technology, discovered the new particle in cloud-chamber photographs. A *cloud chamber* is a device, invented by C. T. R. Wilson at the Cavendish Laboratory, that allows the detection of charged particles. You can't see the particle itself, but you can follow the trail it leaves behind. When an electron, say, plows through air, it ionizes some of the atoms. The air in a cloud chamber is moist, and the water droplets that condense on the ionized atoms show the path the electron has taken. In a strong magnetic field, positive and negative charges follow paths curving in opposite directions. So the electron and its predicted mirror image should show curved paths that are also mirror images. And this, in fact, is what Anderson saw.

The new particle came to be called the *positron*, a nickname for "positive electron." It is the first member in the family of alternate forms of matter called *antimatter*. The idea of antimatter goes something like this: Since Dirac's positron is the antiparticle of the electron, then presumably there should also be an antiproton—that is, a negatively charged particle identical in every other respect to the proton. Carrying this argument further, we would expect antihydrogen, made up of a positron orbiting an antiproton. In fact, the antiproton was discovered at Berkeley in 1955. However, antihydrogen has yet to be observed.

There is also an antineutron, which may seem surprising because the neutron has no electrical charge. However, besides charge, there are other properties of particles that are reversed for the corresponding antiparticles. Antineutrons have been detected in various high-energy experiments.

Because ordinary matter is made up of various combinations of protons, electrons, and neutrons, there is no reason to doubt the existence of "antiordinary antimatter," made from antiprotons, positrons (antielectrons), and antineutrons. We can let our imaginations run wild with this and conjure up antipeople living in antihouses eating antifood and studying antiphysics.

What happens when matter and antimatter collide? The term physicists use is *annihilation*. The particles no longer exist as such but become photons whose energy is equal to the rest mass energy plus the kinetic energy of the colliding objects. So the collision of an electron and a positron produces a flash of light, pure energy, consistent with the mass-energy equivalence, $E = mc^2$, of special relativity. In this particular case, since

Antimatter: Fact and Fancy

Twenty-eight out of one million is not many, but that was the number of antiprotons found when a balloon-borne instrument package was sent up to scan for antimatter in 1979. Why the imbalance? Recent theory holds that equal amounts of matter and antimatter existed in the first instant after the big bang, but that violent collisions among particles led to an asymmetry that has been locked into the universe ever since.

Though rare in the universe at large, antimatter is commonly seen in the high-energy physics laboratory, and we have found a way to put some of it—positrons—to work in our behalf with the PET (for *p*ositron *e*mission *t*omography) scanner. A high point in the cooperative efforts of physicists and medical scientists, the PET scanner can, among other capabilities, distinguish different patterns of brain activity associated with such abnormal mental states as schizophrenia. Technical speculation about antimatter outside medicine has included suggestions for using it as fuel and bombs.

Antimatter has stimulated science fiction as well as technology. Many science fiction writers have used the concept thematically to envision varieties of antiplanets and antimatter counterparts to the inhabitants of earth. The annihilative aspects have also been exploited. A. E. Van Vogt's story "The Storm" concerns a storm in space that occurs when an ordinary gas cloud encounters a cloud of antimatter gas. In the story "Beep," James Blish used the electron-positron interaction as the basis of a technique for instantaneous signaling.

Thus, antimatter plays a prominent part in modern physics, both the real and imagined varieties.

Sources: Ted Agres, "Particles of Antimatter Detected," *Industrial Research and Development*, December 1979, pp. 56–57.

A. Oberg and D. Woodward, "Antimatter Mind Probes," *Science Digest*, April 1982, pp. 54–59.

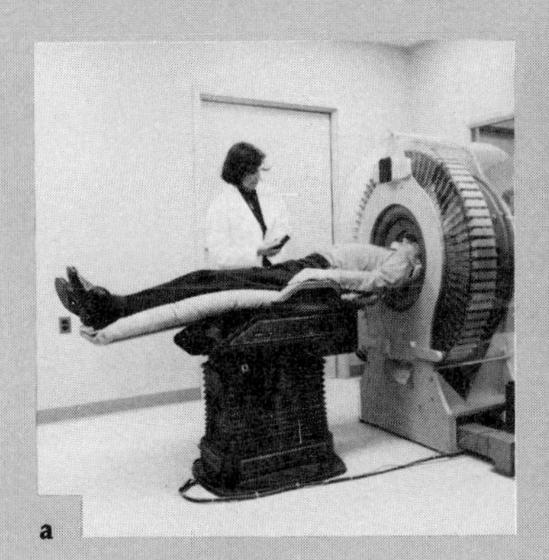

a

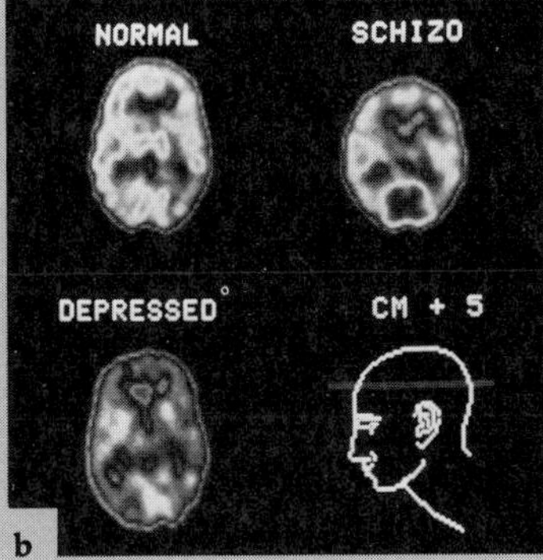

b

(a) The positron-emission tomography (PET) scanner.
(b) PET scans of brains of normal, schizophrenic, and depressed patients. Scans can reveal an organ's metabolic activity: A nutrient with a positron-emitting isotope is injected into the patient. A positron collides with an electron, its antiparticle. They annihilate each other, emitting two gamma rays. A computer, calculating the source of many rays over time, determines the metabolism involved.

the electron and positron each have a rest mass equivalent to slightly more than 0.5 MeV of energy, the resulting photons would have an energy of about 1.02 MeV, at least. This ensures that the energy is conserved.

Note that since the momentum must also be conserved, this annihilation must produce at least two photons. If, for example, the electron and positron travel at the same speed from opposite directions, their total momentum is zero *after* the collision. However, a single photon cannot have zero momentum, so there must be created a second photon carrying an equivalent amount of momentum in the opposite direction.

Particle-antiparticle annihilation is much like a movie that can be run backward. Photons passing through matter can change into a particle and its antiparticle twin, a process called *pair production.* Figure 26–1 is a cloud-chamber photograph of electron-positron pair production. You can see that the tracks bend in opposite directions because of the applied magnetic field. The paths are identical in all other respects. They are mirror images.

An "Anti" Exchange

In 1956 Dr. Edward Teller, the famous nuclear physicist, had given a popularized lecture in which he talked about matter and antimatter and the fact that they would explode into photons on contact. This inspired Harold Furth, a physicist at Berkeley, to write this poem, which later appeared in the November 10, 1956, issue of *The New Yorker:*

Well up beyond the tropostrata
There is a region stark and stellar
Where, on a streak of anti-matter
Lived Dr. Edward Anti-Teller.

Remote from Fusion's origin,
He lived unguessed and unawares
With all his anti-kith and kin,
And kept macassars on his chairs.

One morning, idling by the sea,
He spied a tin of monstrous girth
*That bore three letters: A.E.C.,**
Out stepped a visitor from Earth.

Then, shouting gladly o'er the sands,
Met two who in their alien ways
Were like lentils. Their right hands
Clasped, and the rest was gamma rays.

Dr. Teller wrote an answering letter and closed by pointing out that, since only Dr. Anti-Teller had been mentioned in the poem, perhaps somewhere there existed *The Anti-New Yorker* writing about him.

Source: Harold Furth, "Perils of Modern Living," *The New Yorker,* November 10, 1956.

*Atomic Energy Commission, a government agency that gave way to the Energy Research and Development Administration, and more recently, the Department of Energy.

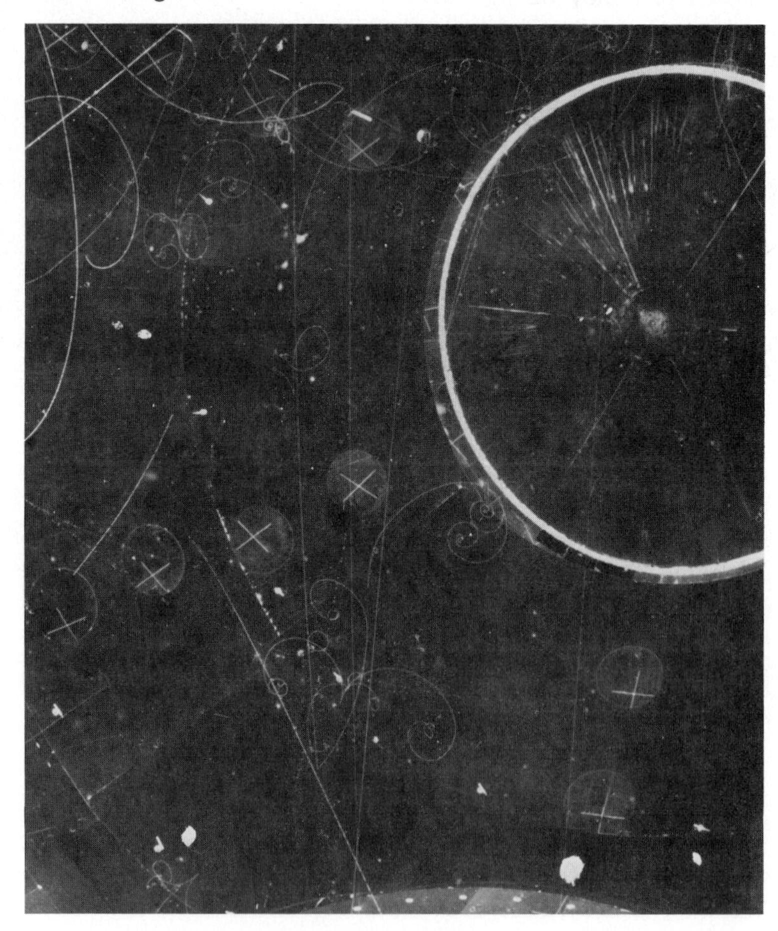

Figure 26-1. Oppositely bending tracks of electrons and positrons.

Cosmic Gall

Neutrinos, they are very small.
They have no charge and have no mass
And do not interact at all.
The earth is just a silly ball
To them, through which they simply pass,
Like dustmaids down a drafty hall
Or photons through a sheet of glass.
They snub the most exquisite gas,
Ignore the most substantial wall,
Cold-shoulder steel and sounding brass,
Insult the stallion in his stall,
And, scorning barriers of class,
Infiltrate you and me! Like tall
And painless guillotines, they fall
Down through our heads into the grass.
At night, they enter at Nepal
And pierce the lover and his lass
From underneath the bed—you call
It wonderful; I call it crass.

John Updike

The combination of antimatter and its annihilation upon contacting matter can produce all kinds of interesting speculation. There may be pockets of antimatter somewhere in the universe—antistars, antigalaxies, an antiearth. Antimatter is now a well-established fact of modern physics. All of the particles we shall discuss in this chapter have their antiparticle twins, which have also been observed. In the case of the photon, it is its own antiparticle.

Learning Checks

1. What is *antimatter?*
2. What difference is there between a positron and an electron? Are their masses different?
3. Explain the different path curvatures of a charged particle and its antiparticle twin in a magnetic field.

WEAK INTERACTIONS: THE NEUTRINO

We mentioned in Chapter 25 that beta decay, in which a nucleus gains a proton at the expense of a neutron, is caused by the weak interaction. Of the four basic interactions in nature, this one is understood the least. Like the strong interaction, it acts over an extremely short range, much less than 10^{-16} meter. The weak interaction, as its name implies, is much weaker than the strong, by a factor of about 10^{-5}.

Early in the observation of beta decay, physicists noticed that there was a problem with energy conservation. For instance, the free neutron—that is, the one outside a nucleus—is not stable but decays after several minutes into a proton and an electron. But these products have less energy than the original

"What's most depressing is the realization that everything we believe will be disproved in a few years."

neutron, meaning that, unless something else is involved in this decay, energy disappears somewhere along the way. Furthermore, there are decay reactions in which the spin angular momentum doesn't come out right, either. So there is also a problem with angular momentum conservation.

Physicists were naturally quite puzzled by all this. Of course, one solution would be to say that the laws of conservation of energy and of angular momentum do not hold in these situations. But this is not an easy idea to accept. Energy and angular momentum are known to be conserved in all other situations, encompassing classical mechanics, relativity, and quantum theory. The people concerned about beta decay believed strongly in these conservation laws. So firm was Pauli's belief that he mentally constructed a new particle that was very much like a ghost but that could provide the necessary energy and angular momentum to resolve the dilemma.

Thus was born the idea of the *neutrino,* named by Enrico Fermi, who worked out the theory for this new particle in 1936. Its name means "little neutral" in Italian. The neutrino has no charge and no mass, and so travels at the speed of light, like a photon. However, unlike a photon, which interacts strongly with matter, the neutrino hardly knows the matter is there. This is the reason for describing the neutrino as "ghost-like." A beam of neutrinos is as indifferent to a thick chunk of matter as gnats are to a fence of chicken wire. It would take a pail of water hundreds of light-years thick to cut the neutrino beam intensity in half. For all practical purposes, neutrinos go right through any material without even slowing down.

At this extremely low level of interaction with matter, it is not surprising that it was a long time before anyone detected a neutrino. Although physicists had no doubt about its existence, it was not until 1955 that somebody caught one. Fred Reines and Clyde Cowan at the Los Alamos Scientific Laboratory finally snared some of these ghosts, detecting their collision with a proton to produce a neutron and a positron:

$$^{1}_{1}\mathrm{H} + \nu_e \rightarrow {}^{1}_{0}\mathrm{n} + {}^{0}_{1}\mathrm{e}$$

Here we have labeled the positron as $^{0}_{1}\mathrm{e}$ (to distinguish it from its antiparticle twin, the electron, $^{0}_{-1}\mathrm{e}$) and the neutrino as ν_e. The subscript e implies that this is an "electron neutrino." We label it like this in anticipation of another kind of neutrino, called the "Muon neutrino" (ν_μ) that we shall meet later on. We can write the scheme of the free neutron decay as

$$^{1}_{0}\mathrm{n} \rightarrow {}^{1}_{1}\mathrm{H} + {}^{0}_{-1}\mathrm{e} + \nu_e$$

That is, the neutron produces a proton, an electron, and an electron neutrino.

Physics and Astronomy: The "Ghost" Catcher

Deep in mines under hundreds of tons of earth, detectors have been searching for elusive neutrinos. This has been the traditional way to "catch" low-energy neutrinos from the sun—all those tons of earth screen out the interference from cosmic rays.

Now, an international team of astrophysicists are exchanging miner's helmets for diving helmets as project DUMAND gets under way. (The acronym stands for *d*eep *u*nderwater *m*uon and *n*eutrino *d*etector.) Seawater will have the same screening effect as earth.

Deep under the coastal waters of Hawaii, 2000 of the most sensitive light detectors in the world will be housed in a pattern of glass spheres spread over a square kilometer of ocean floor. The sensors will detect the characteristic flashes of light (called Cerenkov radiation) that are created when neutrinos collide with other particles. These flashes will be registered by onshore computers. The astrophysicists are looking for neutrinos originating from certain galactic sources and quasars to try to answer some basic questions about the universe.

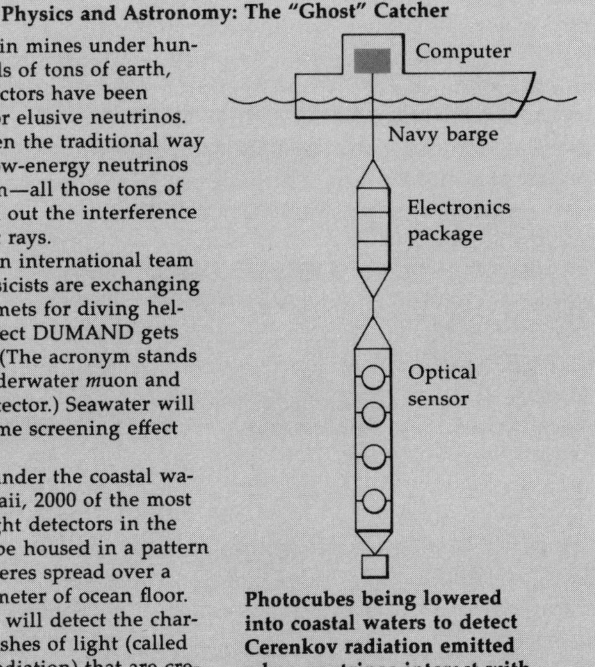

Photocubes being lowered into coastal waters to detect Cerenkov radiation emitted when neutrinos interact with other particles. Astrophysicists seek information about the galaxies that produced the neutrinos.

Source: "Neutrinos: Have Energy Will Travel," *Technology Review*, October 1981, p. 80.

We can use the neutrino to gauge the relative strengths of the weak and electromagnetic interaction. When a neutrino collides with an electron, the two interact by the weak interaction. For a collision energy of 1 MeV, the weak force is about one ten-millionth (10^{-7}) of the electromagnetic force between two electrons crashing together at the same energy.

Learning Checks

1. Why did Pauli propose the neutrino as a new particle?
2. What is the origin of the name "neutrino"?
3. What are similarities and distinctions between the neutrino and the photon?
4. Compare the strength of weak and electromagnetic interaction.

THE FOUR BASIC INTERACTIONS REVISITED

By 1933, the number of known particles had grown quite a bit from the photon, electron, and proton known about 1913. We can illustrate the collection of particles schematically, as shown in Figure 26–2, whereby we give the masses of the particles in units of energy, million electron volts (MeV), because of the relativistic mass-energy equivalence. The masses are not shown to scale in this figure. Remember, each particle also has its twin antiparticle.

Before we discuss the blizzard of particles that were observed following the discovery of the positron and the prediction of the neutrino, let us take another look at the four basic interactions of nature. Our examination of elementary particles rests on an interpretation arising out of the marriage of relativity and quantum theory that gives a different slant to these interactions and how they take place.

Exchange Forces

Consider first the gravitational and electromagnetic forces. These are the most familiar to us and the ones physicists understand the best. Recall their similarities—both follow the in-

Figure 26-2. The particles in 1933.

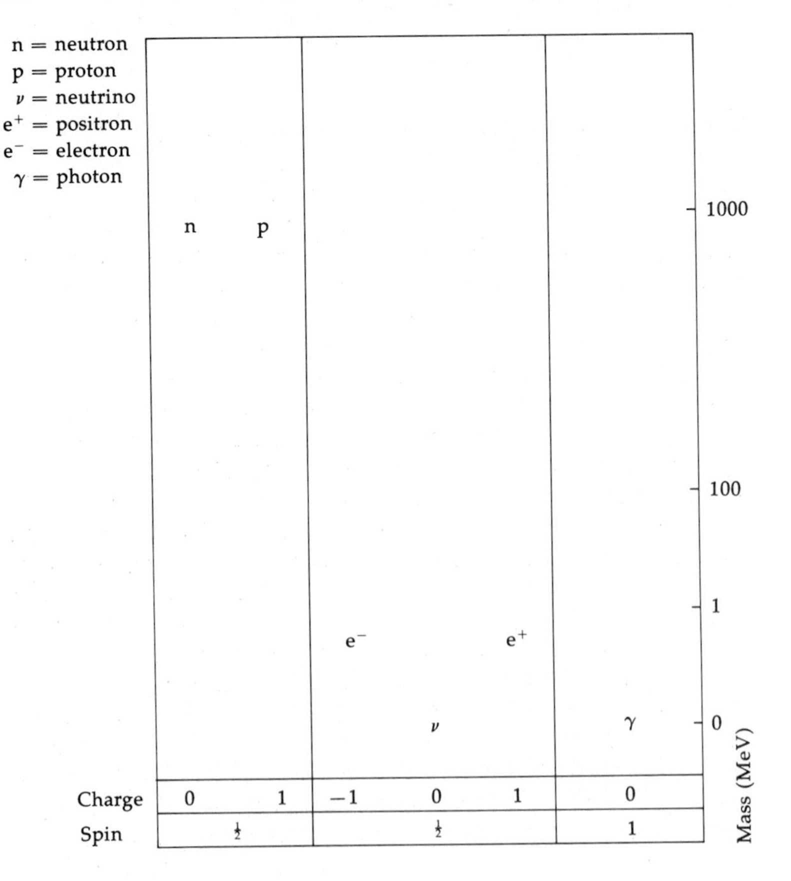

verse-square laws, both have an infinite range. To get around the old "action-at-a-distance" problem, you remember, we associate with each force a field, produced by a mass (gravitational) or a charge (electromagnetic), that influences the behavior of any other mass or charge in the vicinity. Here is where the combination of relativity and quantum mechanics enters the picture, in the form of what is called the *quantum field theory.*

Relativity leads us quite naturally to the field idea. If we have a charged particle—an electron, say—and we shake it up and down, a second charge some distance away will not know instantaneously that the electron has moved. According to relativity, no signal can travel from one place to another faster than light can make the same trip. So there is a time delay between the instant the electron moves and the instant the second charge "feels" the new force caused by the electron's having a different position. In order to satisfy the conservation laws all along the way, we say that the field carries energy and momentum from the electron through the surrounding space and delivers some to the second charge. It takes a finite amount of time for the charge to get the message.

Now quantum mechanics gets into the act. The second charge responds to the force by altering its momentum and energy. These changes, says quantum mechanics, must take place in whole chunks, or quanta. Put another way, the field has to deliver individual packages of energy and momentum. Its delivery boy is the photon that travels from the electron to the second charge. So the point of view that the quantum field theory takes is this:

> The electromagnetic interaction between charged particles is due to an exchange of photons between them.

Extending this idea to the other forces (gravitational, strong, and weak), we can say that in each case the force arises because the two objects participating in the interaction exchange a particle. This gives rise to the name *exchange forces* to describe the general process.

We can go through the same song and dance for the gravitational interaction. If some sadistic monster were to grab the earth and wiggle it back and forth, the news wouldn't get to the moon for several minutes, the time it takes for light to make the trip. In this case it is not a photon that transfers the energy and momentum. The field quantum here goes by the name *gravitational radiation,* and the corresponding particle is sometimes called the *graviton.* Such a particle has never been detected, but there is no serious doubt among physicists that it exists. Remember, the gravitational force is much, much weaker than the electromagnetic force, by about thirty-nine powers of ten. So the graviton is correspondingly more diffi-

Figure 26-3. The baseball represents the exchange particle for the "force" pushing the skaters apart.

cult to detect than the photon. Nevertheless, we can confidently believe that the gravitational attraction between two masses is due to the exchange of a quantum of gravitational radiation, the graviton, between them.

This idea of exchange forces is a new one and may seem a bit strange. You have become accustomed by now to the difficulty of describing quantum mechanical effects with ordinary language, but perhaps here we can give a crude analogy. Suppose that you and a friend are wearing ice skates on a frozen pond and you begin tossing a baseball back and forth between you. Because of the energy and momentum that the baseball transfers, the two of you would gradually glide farther and farther apart. The effect would be exactly the same as if there were some repulsive force pushing you apart. In fact, someone hovering above you in a helicopter, watching all this take place, would naturally conclude that the exchanged baseball was the cause of your being pushed apart. (See Figure 26-3.)

Particle exchange that produces an attractive force might be a little harder to imagine. Perhaps we could make an analogy with two children fighting over who gets to hold a toy. As each grabs it away from the other, they get closer and closer until they finally lock together.

Continuing in this vein, we can reasonably assume that the strong and weak nuclear forces also take place by the exchange of elementary particles. In the case of the strong force, a Japanese physicist named Hidekei Yukawa did some calculations in 1935 to predict what sort of particle the protons and neutrons must exchange to hold the nucleus together. (George Gamow tells us that the graduate students having to deal with Yukawa's complicated mathematics dubbed him "Headache" Yukawa.) An important consideration is to determine the mass of the exchange particle. Heisenberg's uncertainty principle can help us here.

Recall that we earlier expressed the uncertainty principle in terms of the position and momentum as

$$\Delta x \, \Delta p \geq h/2\pi$$

It turns out that another way of stating the same thing is in terms of the energy and the time; that is

$$\Delta E \, \Delta t \geq h/2\pi$$

In this form, the uncertainty principle says that we can make very accurate energy measurements only if there is a great deal of time available to make them. This is completely analogous to the notion that we can make accurate momentum measurements only if the system can roam around in a large region of space.

Now this says something about the law of energy conservation. In classical mechanics there is no problem with measuring the energy very accurately. But in the realm of quantum mechanics, the inherent uncertainty in measuring the energy means that we can't be sure exactly what the energy of a particle is at any given time, and so it is possible for energy conservation to be violated. The uncertainty principle places some restrictions on this, however: it is all right to violate the principle, as long as the violation doesn't last too long. The books must eventually balance. The time (Δt) during which energy conservation breaks down, and the extent to which the energy is not conserved (ΔE), must satisfy $\Delta E \, \Delta t \geq h/2\pi$.

In Supplement 10, we show a calculation that illustrates how one can use the energy-time uncertainty principle in relating the mass of the exchanged particle to the range over which the corresponding force operates. The upshot can be stated as a generalization: *The rest mass of the exchanged particle is inversely proportional to the range of the force.* If the force is effective over a small distance, then the exchange particle has a large mass. The gravitational and electromagnetic interactions operate over an infinite range; hence the exchanged particles should be massless. And this is what we find, at least for the photon: it has zero rest mass. Presumably the same is true of the graviton, although this particle has yet to be observed.

For the strong interaction, where the range of the force is on the order of 10^{-15} meter, Yukawa's calculations predicted a particle of from 200 to 300 times the mass of the electron. Carl Anderson discovered such a particle in 1936, the *muon* (symbol μ), which in all respects is identical to the electron except that it is 207 times more massive, having a mass of about 105 MeV. However, further studies showed that the muon belongs to the lepton family and has essentially no interaction with the atomic nucleus. So it can't be the source of the strong interaction. The particle that fits Yukawa's prediction actually turns out to be the *pi-meson,* or pion. There are three kinds. Two carry an electric charge, π^+ and π^-, and are the antiparticles of each other. The neutral pi-meson (π^0) is its own antiparticle. C. F. Powell, a British physicist, is credited with discovering the pi-meson in 1947.

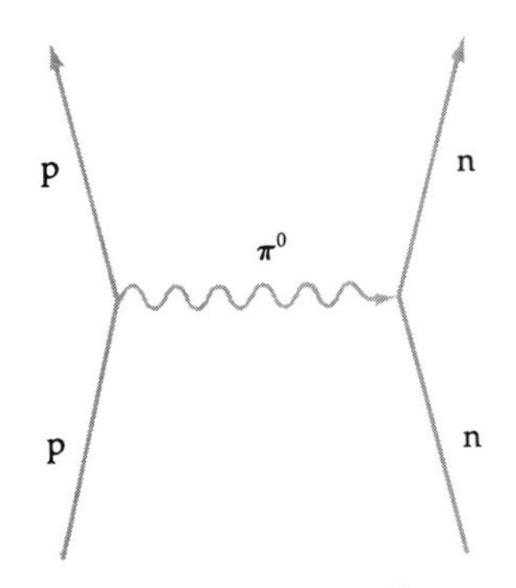

Figure 26-4. Feynman diagram for proton-neutron interaction.

Let us look more closely at the strong force between, say, a proton and a neutron. We can illustrate this by what is known as a *Feynman diagram* (see Figure 26–4), invented by Richard Feynman, an American Nobel laureate in physics. The proton emits a neutral pion. This violates the energy conservation law, of course, since initially we had only the proton, then subsequently both the proton and the pion. But, as we have seen, this is all right as long as the pion is absorbed by the neutron fast enough to be consistent quickly with the uncertainty principle. One way of looking at the range of the strong force is to see how far the exchanged particle can travel in the time during which the energy is temporarily not conserved. For the pion, this distance is of the order of 10^{-15} meter. If the neutron is farther away than that, there is no force, since the uncertainty principle does not let the pion exist long enough to make the trip.

Because the weak force has a much shorter range than the strong, we would predict that the exchange particle for the weak force would have a much larger mass than the pi-meson. These particles have not yet been observed. Their properties have been predicted, however, and it turns out that the weak

Figure 26-5. The particles in 1947.

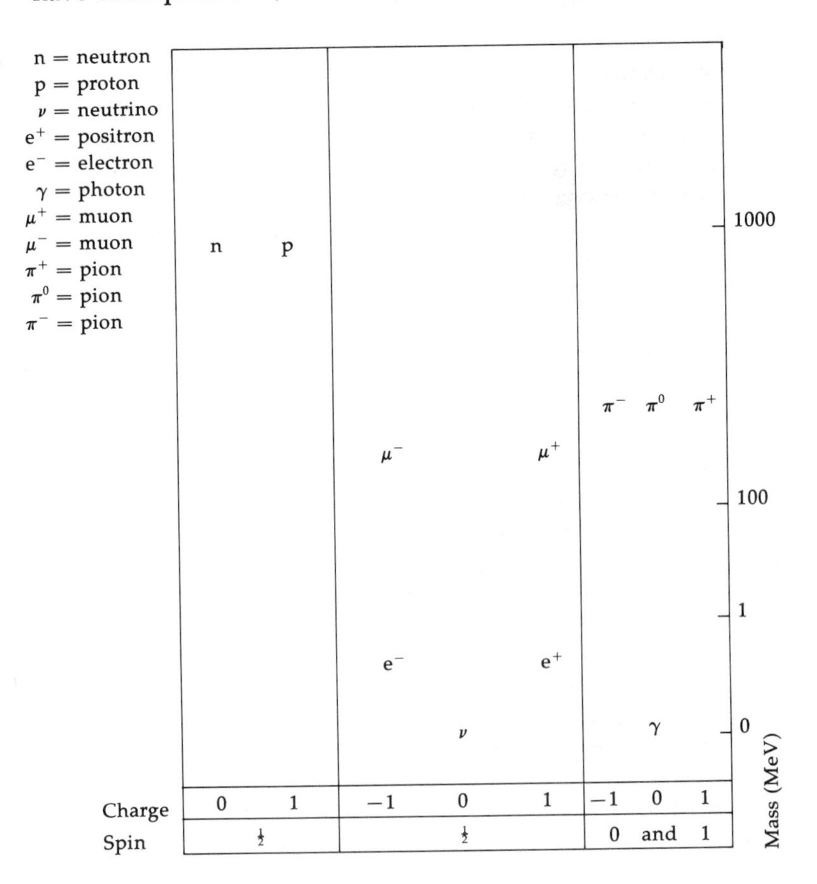

force may have some deep relation with the electromagnetic interaction. We are going to return to this point later.

By 1947, the list of elementary particles had grown again, as we have just described. Figure 26–5 illustrates the situation as it existed in that year.

Learning Checks

1. What particles were known in 1933? In 1947?
2. What is the picture of the four basic forces presented by the quantum field theory? What is the origin of the name of this theory?
3. What are *exchange forces?*
4. Describe how the uncertainty principle allows momentary violation of the law of conservation of energy.
5. What is the role of the pi-meson in the nuclear force?

Particles and More Particles

The situation that existed in 1947 (illustrated in Figure 26–5) was not particularly complicated, and each particle fit neatly into the then-existing picture of material structure. The photon of electromagnetic radiation was a familiar friend. So were the electron, proton, and neutron composing the atom. The pi-meson, the carrier of energy and momentum between nucleons, accounted for the strong interaction. The ghostly neutrino provided the means for conserving energy and momentum in the weak-force phenomenon of beta decay. Only the muon remained a puzzle. It is hard to see why nature would choose to manufacture a particle whose only distinction is that it is a sort of heavy electron. But leaving that one small riddle aside, one might have hoped that the structure of nature at its most fundamental level had been discovered.

> ***I don't think physics will ever have an end. I think that the novelty of nature is such that its variety will be infinite—not just in changing forms but in the profundity of insight and the newness of ideas . . .***
>
> *Isidor Rabi*

But this happy state of affairs didn't last very long. By probing the nucleus with collision experiments of higher and higher energy, physicists uncovered many new particles. These particles have a variety of physical properties, such as mass, charge, and spin, as well as some others that are a little more obscure. They are mostly unstable, decaying to particles of lower mass and greater stability by either the strong, weak, or electromagnetic interaction. The situation appeared to be even more complex because what turned out to be excited states of existing particles were originally taken as distinct particles in themselves.

Confronted with such a bewildering array of particles, physicists sought some classification scheme that would reveal similarities among them, analogous to the periodic table of the chemical elements, as a means of understanding their properties. The first division of the particles is made on the basis of whether or not they are involved in the strong interaction. Those involved are called *hadrons* and those not involved are labeled *leptons.*

In Table 26-1 we show the leptons and some of the many hadrons. The leptons consist of the electron, the muon, and both neutrinos (and, of course, the corresponding antiparticles of all of them). Note that the leptons are all charged particles and hence participate in the electromagnetic force. In general, the hadrons are heavier than the leptons. But the distinguishing feature, again, is that the hadrons are affected by the strong force while the leptons are not.

Note also that we have included the photon as a separate entity in this table. It is neither a lepton nor a hadron but stands in a class by itself, serving solely as the mediator of the electromagnetic interaction.

Table 26-1 also shows a further division of the hadrons into *baryons* and *mesons.* We have already met one of the mesons,

Table 26-1
Leptons and Some Hadrons

		Particle	Symbol	Charge	Spin	Mass (MeV)	Lifetime (s)	Strangeness	Charm	Quark composition
		Photon	γ	0	1	0	∞			
Leptons		Electron	$e(e^+)$*	$-1(+1)$	$\frac{1}{2}$	0.511	∞			
		Muon	$\mu(\mu^+)$	$-1(+1)$	$\frac{1}{2}$	105.66	10^{-6}			
		e-Neutrino	$\nu_e(\bar{\nu}_e)$	0	$\frac{1}{2}$	0	∞			
		μ-Neutrino	$\nu_\mu(\bar{\nu}_\mu)$	0	$\frac{1}{2}$	0	∞			
Hadrons	Mesons	Pion	$\pi^+(\pi^-)$	$+1(-1)$	0	139.57	10^{-8}	0	0	$u\bar{d}(\bar{u}d)$
			π^0	0	0	134.97	10^{-16}	0	0	$u\bar{u} + d\bar{d}$
		Kaon	$K^+(K^-)$	$+1(-1)$	0	493.71	10^{-8}	$+1(-1)$	0	$u\bar{s}(\bar{u}s)$
			K^0	0	0	497.71	10^{-10}	1	0	$u\bar{u} + s\bar{s}$
		Phi	ϕ	0	1	1020	10^{-22}	0	0	$s\bar{s}$
		J/Psi	ψ	0	1	3095	10^{-20}	0	0	$c\bar{c}$
	Baryons	Proton	$p(\bar{p})$	$+1(-1)$	$\frac{1}{2}$	938.26	∞	0	0	$uud(\overline{uud})$
		Neutron	$n(\bar{n})$	0	$\frac{1}{2}$	939.55	1000	0	0	$udd(\overline{udd})$
		Lambda	$\Lambda(\bar{\Lambda})$	0	$\frac{1}{2}$	1115.59	10^{-10}	$-1(+1)$	0	$uds(\overline{uds})$
		Charmed lambda	$\Lambda_c(\bar{\Lambda}_c)$	0	$\frac{1}{2}$	2260	?	0	$+1(-1)$	$udc(\overline{udc})$
		Sigma	$\Sigma^+(\bar{\Sigma}^+)$	$+1(-1)$	$\frac{1}{2}$	1189.42	10^{-10}	$-1(+1)$	0	$uus(\overline{uus})$
			$\Sigma^0(\bar{\Sigma}^0)$	0	$\frac{1}{2}$	1192.48	10^{-16}	$-1(+1)$	0	$uds(\overline{uds})$
			$\Sigma^-(\bar{\Sigma}^-)$	$-1(+1)$	$\frac{1}{2}$	1197.34	10^{-10}	$-1(+1)$	0	$dds(\overline{dds})$

*Antiparticle information is shown in parentheses.

the pion, responsible for the strong interaction. The baryons all have half-integral spin and the mesons all have integral spin, either 1 or 0. Some of the hadrons have an electric charge and some do not. Of course, the ones that do, such as the proton, are affected by the electromagnetic force, while the others, such as the neutron, are not.

There are two other properties of particles that we have not mentioned, indicated by the columns labeled *strangeness* and *charm*. We will wait until our discussion of quarks to talk about charm. As for strangeness, a Caltech physicist named Murray Gell-Mann introduced the concept as a way of accounting for the unusually slow decay of certain hadrons discovered back in the 1950s. Massive hadrons are created by the strong interaction; they generally decay quickly by way of the same interaction. They usually exist for only about 10^{-23} second before breaking up; in a sense the strong interaction is said to be strong because it does its job so quickly. But certain hadrons are thought to be "strange" because they hang around for much longer times, decaying after about 10^{-8} second. This is still an extremely short time on the scale of ordinary human activities, but it is a million billion times longer than expected. Their long lifetimes indicate that these particles decay by the weak interaction. Gell-Mann discovered that he could assign to each known particle a strangeness quantum number in such a way that for the strongly decaying hadrons the numbers add up to be the same before and after the reaction, while for the weakly decaying hadrons the numbers are different. Put another way, strangeness is conserved in the rapid decays, and it is not conserved in the slow ones.

With what we now know about the new particles and the basic interactions in nature, we are in a position to summarize this information. Table 26-2 gives in condensed form properties of the interactions and how the various particles relate to them.

Table 26-2
The Four Basic Interactions

	Gravitational	Electromagnetic	Strong	Weak
Range	infinite	infinite	$10^{-15}-10^{-16}$ m	$\ll 10^{-16}$ m
Examples	astronomical forces	atomic forces	nuclear forces	beta decay
Relative strength	10^{-39}	0.007	1	0.00001
Particles acted upon	everything	charged particles	hadrons	hadrons leptons
Particles exchanged	gravitons	photons	hadrons	?

Learning Checks

1. Distinguish between hadrons and leptons.
2. What are the two subclasses of hadrons and how is this classification made?
3. With the strong force assigned a strength value of 1, what are the relative strengths of the other three interactions?
4. How did the concept "strangeness" arise?

QUARKS: THE LAST OF THE CHINESE BOXES?

We have listed in Table 26–1 only a few of the known hadrons, but there are many others. They have varying degrees of stability and include such completely unstable species as the rho-meson, which lives about as long as it takes light to get across a nucleus. If one counts excited states in which the hadrons have properties different from their ground-state parents, then there are literally hundreds of such particles.

In recent years it has become very difficult for physicists to consider all these particles "elementary" in the same sense in which the atoms are elementary from the chemical standpoint. It would be like ignoring the atoms and considering the millions of different molecules as basic and fundamental and elementary. So in 1963, Gell-Mann and George Zweig, also at the California Institute of Technology, independently put forward a new elementary particle theory widely accepted today.

According to this line of thought, the leptons and the photon are truly elementary. There is no indication that they are composed of any smaller building blocks but seem to be absolutely fundamental in their own right. Apparently they exist as point particles, having no structure. On the other hand, *none* of the hadrons is considered elementary. The idea is that the hadrons are themselves composed of the ultimate fundamental particles, which Gell-Mann dubbed "quarks." In its original version, the theory included three quarks, which accounts for the name. It is taken from *Finnegan's Wake,* a novel by the Irish writer James Joyce, in which a bartender named Earwicker says on occasion, "Three quarks for Muster Mark!" The three kinds, or "flavors," of quarks are named *up* (symbol u), *down* (d), and strange (s). There are also the three corresponding antiquarks, labeled $\bar{u}$, $\bar{d}$, and $\bar{s}$.

Recently, a fourth flavor has been added to the list, bearing the whimsical name *charm* (symbol c—$\bar{c}$ for the antiquark). Although the fourth quark is not in keeping with the literary origin of the name, it does provide an appealing symmetry. The four quarks match up nicely with the four leptons. Table 26–3

Physics and Language

At one point, physicists, feeling whimsical, referred to the three quarks as "chocolate," "vanilla," and "strawberry," which is why the word "flavor" came into play. The vocabulary of the ice-cream shop eventually went the way of the words "wavicle" and "aces" (another term for quarks used by an American physicist in Europe who thought of this substructure independently but was not able to get his article into print). But the use of "flavor" as a generic label had staying power. It remained to attain the official standing it enjoys today.

Source: H. R. Pagels, *The Cosmic Code.* New York: Simon and Schuster, 1982.

Table 26-3
Quarks

Symbol	Mass (MeV)	Charge	Spin	Strangeness	Charm
d($\bar{d}$)	338	$-\frac{1}{3}\left(+\frac{1}{3}\right)$	$\frac{1}{2}\left(-\frac{1}{2}\right)$	0	0
u($\bar{u}$)	336	$+\frac{2}{3}\left(-\frac{2}{3}\right)$	$\frac{1}{2}\left(-\frac{1}{2}\right)$	0	0
s($\bar{s}$)	540	$-\frac{1}{3}\left(+\frac{1}{3}\right)$	$\frac{1}{2}\left(-\frac{1}{2}\right)$	$-1(+1)$	0
c($\bar{c}$)	1500	$+\frac{2}{3}\left(-\frac{2}{3}\right)$	$\frac{1}{2}\left(-\frac{1}{2}\right)$	0	$+1(-1)$

gives the properties of the four flavors of quarks. The really different thing about the quarks is that they carry *fractional electric charges.* All the "ordinary" particles are neutral or have a charge $+1$ or -1 times the charge on the electron.

To form the hadrons, we need only play a simple game. Merely combine the quarks and the antiquarks in the proper combinations to give the desired particle. By "proper combinations" we mean putting the quarks and antiquarks together so that by adding their quantum numbers we get the quantum numbers of the desired hadron. One possibility is to tie a quark and an antiquark. By taking all possible combinations (sixteen of them), we can construct the *mesons.* For example, the positively charged pi-meson (π^+) consists of a u quark and a $\bar{d}$ antiquark. As you can see from Table 26-3, the charge that results is $2/3 + 1/3 = +1$ and the spin is $1/2 + (-1/2) = 0$. Both of these quantum numbers are correct for the positive pi-meson.

Similarly, the baryons consist of three quarks (and the antibaryons of three antiquarks). The two most familiar, and indeed the only two that play any role in ordinary matter, are the proton (quark combination uud) and the neutron (udd). (See Figure 26-6.) You can satisfy yourself that these combinations do indeed give the correct properties for these hadrons. It turns out that there are twenty possible three-quark combinations to give baryons. All the other 300 or so hadrons are excited states of these twenty combinations.

Figure 26-6. The quark make-up of the proton and neutron.

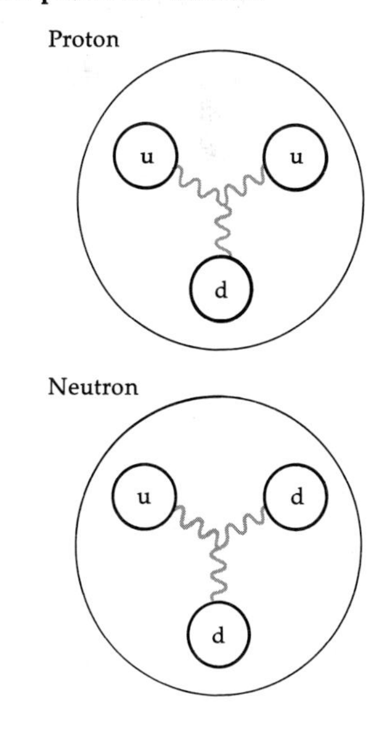

The last column of Table 26-1 shows the quark composition for the hadrons. Notice that the "strange particles" are characterized by the presence of an s quark or $\bar{s}$ antiquark, which respectively have values of -1 and $+1$ for the strangeness quantum number. For example, the positive K-meson, a strange particle, has the composition (u$\bar{s}$), giving the correct strangeness quantum number of $+1$.

November 1974 saw the best evidence for the necessity of the fourth, or *charmed* quark. Two groups of researchers, one at Brookhaven National Laboratory and the other at the Stanford Linear Accelerator, discovered a new, quite massive meson, which one group called "psi" and the other "J" (let us compromise and call it "J/psi"). Its unusual mass (3095 MeV) and lifetime (10^{-20} second, long for a strongly decaying particle) could not be accounted for on the basis of the original three u, d, s quarks. The postulated c quark presumably has the necessary properties to allow the J/psi particle to exist.

WHAT ABOUT THE FUTURE?

As we have seen, the search for the most fundamental building blocks of nature has led physicists through a bewildering array of particles. Greater energies made accessible by larger accelerators have enabled us to peer deeper and deeper into the structure of matter, with discoveries at each level revealing particles more "fundamental" than before. This search has seemed endless, like the one described by Jonathan Swift:

> So, naturalists observe, a flea
> Hath smaller fleas that on him prey;
> And these have smaller still to bite 'em;
> And so proceed ad infinitum.

The hope of the quark model was that it would be the last of the Chinese boxes, bringing an end to this succession of fleas-upon-fleas.

But the model has begun to show symptoms of the very disease it was meant to cure. In addition to the quarks and leptons we have already mentioned, others have been uncovered in more recent research. A new lepton, called the *tau particle*, has been discovered. It has an electric charge, like the electron and the muon, but its mass is much greater. Presumably it has a neutrino associated with it as well. This brings to six the membership in the lepton family.

The four quarks have expanded to six with the establishment of one called *top* and the almost-certain existence of its partner, given the flavor *bottom.* (These are less frequently referred to as *truth* and *beauty,* but the "t" and "b" labels are consistent in either nomenclature.) A characteristic feature called *color* is also attributed to all six quarks. Any of three colors is possible for each. Hence, the complement of seemingly "elementary" particles consists of quarks in six flavors and three colors, for a total of eighteen quarks, plus the six leptons. Each of these twenty-four particles has an antiparticle. So any theory that embraces all of them must have room for at least forty-eight different forms of matter.

You can see the problem. Again it begs the question, can these forty-eight particles be truly "fundamental"? Or must the search go on? Do the quarks and leptons have structure as well, themselves being made up of fundamental particles?

Several theories incorporating structure for the quarks, called substructure models, have been proposed. The one we shall look at is the work of Haim Harari, an Israeli physicist. Harari proposes just two fundamental entities, which he calls *rishons,* from the Hebrew for "first" or "primary." In various combinations the two rishons can account for all the quarks and leptons, and hence all subatomic particles.

The two fundamental species are labeled "T" and "V" (from a phrase in the Book of Genesis, "Tohn va-Vohm," translated as "without form, and void"). Both have mass. Each has a spin angular momentum of 1/2. The T rishon has an electric charge of 1/3; the V rishon is neutral.

In this model, a quark or a lepton is formed by putting together three rishons or three antirishons. For example, TTT is the positron, having an electric charge of +1 (or 1/3 + 1/3 + 1/3). The three combinations TTV, TVT, and VTT all give a net charge of 2/3 and correspond to the three possible colors of the *up* quark. Similarly, the three colors of the *down* antiquark are obtained from TVV, VTV, and VVT, each of which gives a charge of 1/3. The combination VVV is electrically neutral and represents the electron-type neutrino.

Harari's model is appealing because of its economy, involving only two types of particles (and their antiparticles). It also naturally accounts for such seemingly arbitrary properties as the three colors of quarks. But we must remember that it is a speculative model only, in that it cannot be verified experimentally. The same holds true of the other substructure models.

The quark picture itself suffers from the same flaw. Although the quark model accounts for the zoo of hadrons in a straightforward and simple way, there is a hooker: no quark has ever been observed. Physicists using the highest-energy equipment known have searched for a long time, with no luck. This is one of the most important problems facing theoretical physics at the present time.

Several rationales have been offered. One is that unlike the electromagnetic or gravitational forces, which diminish with distance, the force—whatever its name—that holds the quarks together *increases* as they get farther apart. According to this idea, quarks making up a proton, say, are so close together as to be essentially free particles, but when you try to separate them the force holding them together gets larger. As you supply more and more energy to try and separate them, you can create a quark-antiquark pair—much like an electron-positron pair can appear from a photon. One physicist has described this as like trying to isolate one end of a piece of string—when you pull very hard, what you eventually wind up with is two pieces, each having two ends.

The jury is still out on the quark model and the substructure theories. Whether or not we are on the threshold of opening the last Chinses box is unknown at this time.

ELEMENTARY PARTICLES AND COSMOLOGY

We have seen that physicists classify all the relationships among various parts of the universe in terms of the four basic interactions. Given the tremendous variety of form and substance that we see about us, this is a highly simplifying assumption. And yet, among the speculations growing out of the study of fundamental particles is the idea that things are even simpler. The notion has been around for some time that perhaps there is some deep relationship among these four interactions, that they are in fact special cases of a single interaction governing everything in the universe. We have seen, for example, the tantalizing similarity in form between the electromagnetic and gravitational forces, both being inverse-square laws.

Lately, theorists have been developing a connection between the weak and electromagnetic interactions. As we discussed earlier, the mass of the exchange particle that generates an interaction must be inversely proportional to the range of the interaction. This means that the weak interaction, having an extremely short range of around 10^{-17} meter, must have an exceptionally large exchange particle. (See Supplement 10.) According to the theories, this is the only fundamental difference between the weak and electromagnetic forces. Further evidence comes from the fact that the angular momentum exchanged in weak processes such as beta decay is exactly the same as that exchanged in the electromagnetic interaction—that of a photon, numerically equal to Planck's constant. Although it has yet to be observed, the weak-force exchange particle has already been given a name, the *intermediate vector boson.* The origin of this term is a little too technical for us to go into; all you need to understand is that it plays the same role as the pi-meson in strong interactions and the photon in electromagnetic ones.

We can see the analogy between the weak and electromagnetic interactions by comparing their Feynman diagrams. Figure 26–7 shows such a comparison for the scattering of a neutrino by a neutron. Note that the weak interaction in part (b) can be obtained from part (a) by simply changing some of the particles into others of different charge.

Our ideas about fundamental particles have profound implications for modern theories of cosmology. If this quark theory is correct, during the first one-hundredth of a second

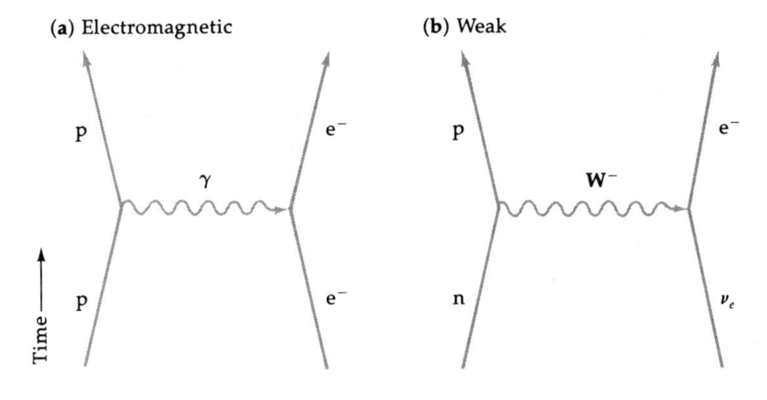

Figure 26-7. Feynman diagrams for two interactions. W^- is the intermediate vector boson.

following the "big bang" the universe consisted of an undifferentiated soup of leptons, antileptons, quarks, antiquarks, and photons, all in thermal equilibrium at a temperature of several million degrees Kelvin. At such high temperatures, they were all moving as essentially free particles. As the universe expanded and cooled down, the free quarks were either annihilated with antiquarks or separated far enough from each other for the force between them to come into play, tying them up in neutrons and protons.

In the case of the weak interactions, there is associated with them a sort of "phase transition," much like the one water passes through as it is cooled down on its way to becoming ice. The theory is that above the critical temperature of about 3×10^{15} K (3,000 million-million degrees) the weak and strong interactions were essentially identical. They both had about the same strength and both obeyed an inverse-square law. As the universe cooled below this critical temperature, it experienced a kind of "freezing" in which the symmetry between the weak and strong interactions was broken. One possibility is that we now live in a sort of "domain" of the universe corresponding to the particular way in which this symmetry was broken. Perhaps there are other domains caused by different patterns of broken symmetry.

Of course, this is all speculation, based on mathematical theories. We can't really know in an experimental sense what the universe was like in the beginning. But the possibility of a theory that unifies the basic interactions and also requires only leptons, photons, and quarks presents a beautifully simple picture of how the universe might have come into existence.

SUMMARY

About the time that quantum mechanics was being developed, several new particles were discovered in addition to the electron, proton, and neutron. Dirac predicted the existence of the *antielectron,* which Anderson discovered in 1932. It is now called the *positron*. There is also an *antiproton*, discovered in 1955. In fact, every particle has its corresponding antiparticle, implying the existence of *antimatter* in complete analogy with ordinary matter.

When matter and antimatter collide, they *annihilate*, exploding into pure energy (photons). Conversely, photons can produce a particle and its antiparticle in a process called *pair production*.

The *neutrino* is a massless particle that participates in the weak interaction. It has very little interaction with matter but provides for energy conservation in such processes as beta decay.

The modern way of looking at the *four basic interactions* is to consider them as being due to the *exchange of particles* between interacting objects. The mass of the exchange particle varies inversely with the range of the force. In the electromagnetic interaction, the exchange particle is the photon. For the strong force, it is the *pi-meson*, discovered by Powell in 1947.

Many more particles are known. We classify them as *leptons* and *hadrons*, the latter participating in the nuclear strong force. The hadrons are further divided into *baryons* that have half-integral spin and *mesons* that have integral spin.

Physicists do not believe that all these many particles can be elementary. The current theory, invented by Gell-Mann and Zweig, is that hadrons are made of truly elementary particles called *quarks*, which so far have not been observed. There are six flavors of quarks, with various masses and fractional electric charges. A quark and an antiquark make up the mesons, while the baryons are composed of three quarks. The quantum numbers for the hadrons are obtained by simply adding the appropriate quark quantum numbers.

Recent theoretical work has been aimed at trying to unite the four basic interactions into a single superbasic force. This has some implications about the formation of the universe.

Exercises

1. Explain why the annihilation of a particle and its antiparticle must produce at least two photons.
2. Suppose that an electron and a positron annihilate. Can the resulting photons have energy less than 1.0 MeV? Explain.
3. Using the four flavors of quarks in Table 26–3, list all the mesons that are possible. How many should there be?
4. Match the force in the lefthand column with the appropriate exchange particle in the righthand column:

Force	Exchange particle
Gravitational	Intermediate vector boson
Strong nuclear	Photon
Electromagnetic	Graviton
Weak nuclear	Pi-meson

5. When we create a new particle in the accelerator, where does the energy come from?
6. How can conservation laws be used in determining properties of new particles, even when we don't directly observe the particles?
7. What are the reasons for believing that the tau neutrino exists?

8. Classify the following particles as leptons or hadrons (or neither): (a) neutrino, (b) neutron, (c) baryon, (d) proton, (e) electron, (f) pi-meson, (g) muon.

9. A certain hadron has the quark composition cds. Give its charge, spin, strangeness, and charm.

10. The J/psi particle is a meson with the quark composition $c\bar{c}$. Does J/psi have charm? Explain.

11. Give one reason why a neutron cannot decay into a proton and a neutrino.

12. From the quarks u, d, s, and their antiparticles, work out all possible combinations of three quarks with zero net strangeness. Determine the net charge for each such combination.

13. Of the six postulated quarks, which are necessary to explain ordinary matter?

14. Suppose that the lifetime of a certain particle is 1 second. What will be the uncertainty in the measurement of its energy?

15. Beginning with the rishons T and V and their antiparticles, list the combinations that give the following: (a) electron, (b) the up antiquark, (c) the down quark, (d) the electron-type antineutrino.

SUPPLEMENT 1

VECTORS

Chapter 2 stated that velocity and acceleration are examples of vectors—quantities that have direction as well as numbers associated with them. There are other such directional quantities—force, for example—so we will discuss some things that are common to all vectors.

DEFINITIONS

As discussed earlier, we distinguish between *scalars*, which have only numerical value, or *magnitude*, and *vectors*, which have both *magnitude* and *direction*. Speed and distance are examples of scalar quantities. Velocity, displacement, and acceleration are vector quantities. We may represent a vector by an arrow whose length is proportional to the magnitude of the vector and whose direction is determined by the direction of the vector. Thus, a vector **A** may be drawn as:

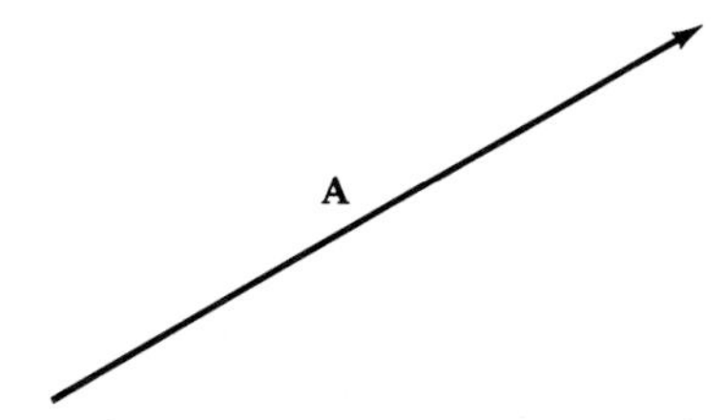

We will use bold-face type (**A**) to refer to vectors. For example, suppose that we have two velocity vectors, $\mathbf{V}_1$, which is 10 km/h north, and $\mathbf{V}_2$, which is 20 km/h east. We represent these velocities by the following arrows, using road map directions:

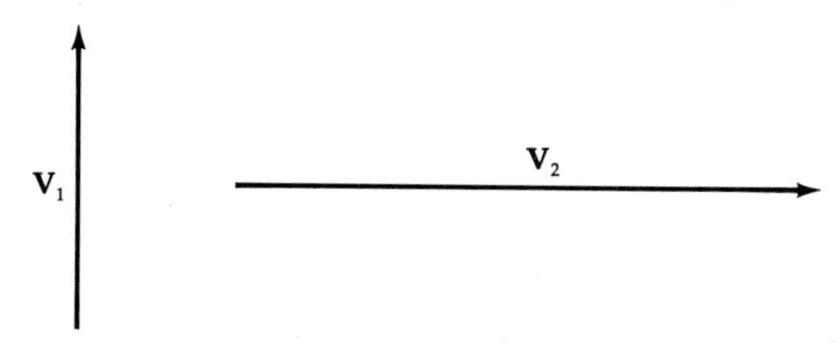

The magnitude of $\mathbf{V}_2$ is 20 km/h and that of $\mathbf{V}_1$ is 10 km/h. Thus, the length of the arrow representing $\mathbf{V}_2$ is twice the length of arrow $\mathbf{V}_1$. We specify the direction of the vector quantity by the direction of the arrow: $\mathbf{V}_1$ points north and $\mathbf{V}_2$ points east.

A vector is completely specified by its length and direction. It has no fixed location in space. Therefore, two vectors are equal if they have the same length and direction, even though the arrows representing them are drawn in different places.

Two vectors pointing in the same direction are said to be *parallel,* like the vectors **B** and **C** below:

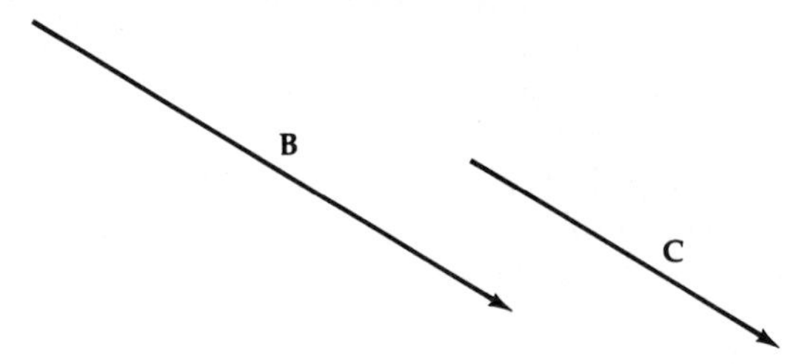

The vector −**A** is a vector whose magnitude is the same as that of **A** but whose direction is exactly opposite to that of **A**:

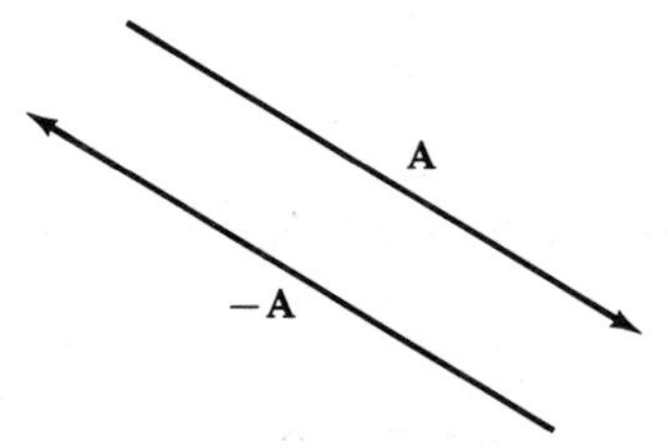

Two vectors, such as these, that point in opposite directions are said to be *antiparallel.*

VECTOR ADDITION
Head-to-tail Method

We may combine two vectors to form a third vector, called the *resultant,* by two equivalent methods. The first of these is the *head-to-tail method.* To illustrate, suppose we want to add the two vectors **A** and **B**.

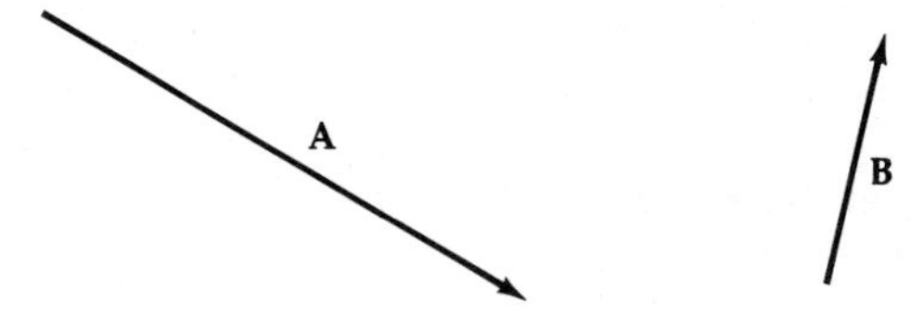

We begin with either vector, in this case **A**, and place **B** so that its tail is at the head (arrow end) of **A**. However, we must preserve the direction of **B**:

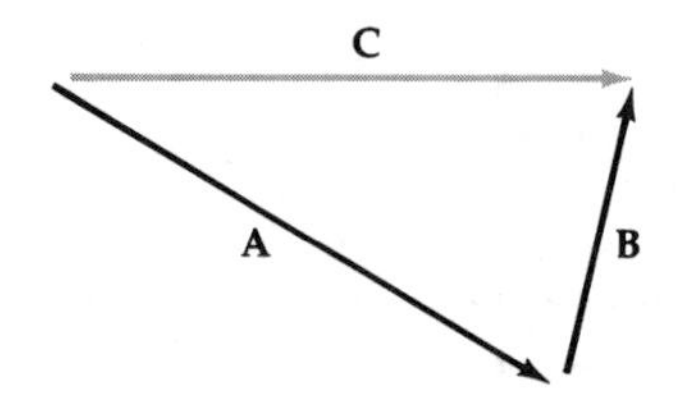

Next, we draw an arrow from the tail of **A** to the head of **B**. This new vector, **C**, is the resultant. Thus we may say

$$\mathbf{C} = \mathbf{A} + \mathbf{B}$$

if we are careful to note that this is *vector addition*, which means paying attention to the directions. For example, if **A** is 10 feet long and **B** is 2 feet long, **C** will in general *not* be 12 feet long. Only in the special case in which **A** and **B** are parallel will the length of **C** be equal to the sum of the lengths of **A** and **B**. In general, the length of **C** will be less than this sum.

Let us look at examples of addition of parallel vectors. Suppose the vectors **F** and **K** are parallel; **F** is 3 inches long and **K** is 2 inches long:

Following the head-to-tail method, we place the tail of **K** at the head of **F**:

The resultant **R** is drawn from the tail of **F** to the head of **K**, **R** = **F** + **K**, and its length is 5 inches. In this *special case* of parallel vectors, *vector* addition of **F** and **K** gives the same length as *ordinary* addition of the quantities 3 inches and 2 inches: 5 inches. However, if **F** and **K** are *not* parallel, then the length of the resultant **R** is less than 5 inches. It is still true that **R** = **F** + **K**, as this denotes vector addition, but we do *not* obtain the length of **R** by ordinary addition of the lengths of **F** and **K**.

Another simple case of vector addition involves antiparallel vectors. Let the antiparallel vectors **V** and **W** be 4 inches and 2 inches long, respectively.

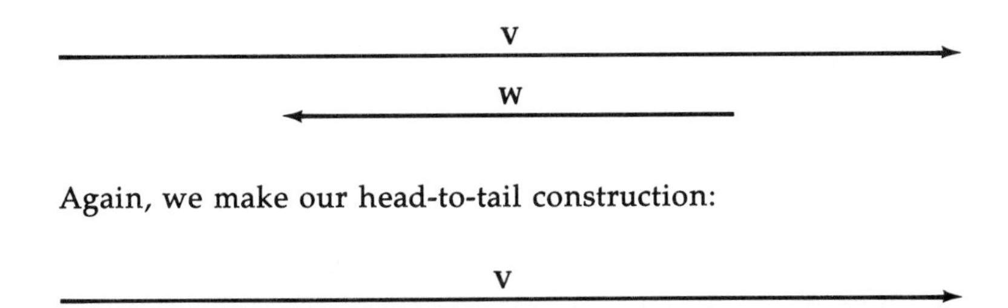

Again, we make our head-to-tail construction:

The resultant is $\mathbf{T} = \mathbf{V} + \mathbf{W}$, and measurement shows that **T** is 2 inches long. For antiparallel vectors, the length of the resultant is obtained by *subtracting* the lengths of **V** and **W**: $4 - 2 = 2$ = length of **T**.

We return now to our original example of the vectors **A** and **B**, whose resultant is **C**. Note that, because $\mathbf{C} = \mathbf{A} + \mathbf{B}$, we can replace the two vectors **A** and **B** by a single vector, the resultant **C**. This is what the *vector equation* means. Thus, if **A** represents walking from my home to the gas station, and **B** represents walking from the gas station to the drugstore, then **C** represents walking directly from my home to the drugstore, without ever visiting the gas station. So the result of combining **A** and **B** is the single vector **C**.

Clearly this method of head-to-tail addition can be applied to any number of vectors:

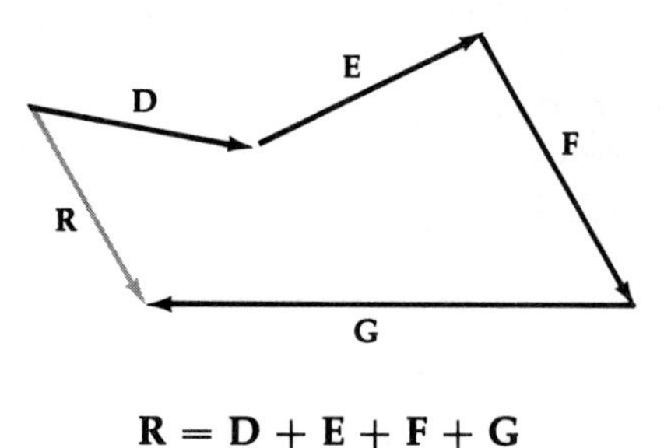

$$\mathbf{R} = \mathbf{D} + \mathbf{E} + \mathbf{F} + \mathbf{G}$$

where **R** is the resultant of vectorially adding **D**, **E**, **F**, and **G**.

Parallelogram Method

A second method of vector addition is called the *parallelogram method*, which we can illustrate by returning to our example vectors **A** and **B**. This time we draw **A** and **B** as originating from the same point:

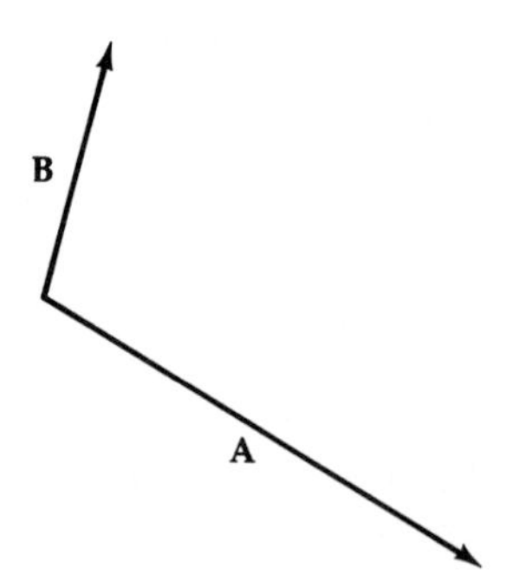

We then draw two dashed lines, one from the head of **A** parallel to **B** and the other from the head of **B** parallel to **A**:

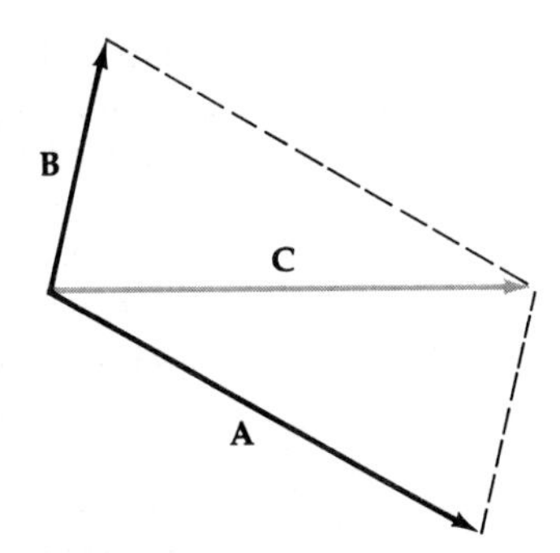

These two dashed lines cross and form with **A** and **B** a four-sided figure called a *parallelogram.* We get the resultant **C** by drawing along the diagonal of the parallelogram a vector that begins at the origin **A** and **B** and ends with its arrow at the intersection of the constructed dashed lines. You can see, of course, that the resultant **C** is the same as that obtained in the head-to-tail method, as it must be.

Let us use velocity and acceleration in an example of vector addition. Suppose that an airplane flies in a straight line with a velocity of 400 km/h east. We represent this by a vector labeled $\mathbf{v}_1$:

$$\mathbf{v}_1 = 400 \text{ km/h east}$$

Over a 10-second time interval, the velocity is increased to 430 km/h east, which we show as the vector $\mathbf{v}_2$:

$$\mathbf{v}_2 = 430 \text{ km/h east}$$

From Chapter 2, we know that the acceleration is

$$\mathbf{a} = \frac{\mathbf{v}_2 - \mathbf{v}_1}{t}$$

We can write this as

$$\mathbf{v}_2 - \mathbf{v}_1 = \mathbf{v}_2 + (-\mathbf{v}_1)$$

in which $-\mathbf{v}_1$ has the same length as $\mathbf{v}_1$ and is antiparallel to it. Our head-to-tail method will give us the resultant, which we will call **V**. The vector diagram looks like this:

$\mathbf{v}_2$

V $\qquad -\mathbf{v}_1$

Since $\mathbf{v}_2$ and $-\mathbf{v}_1$ are antiparallel, we can get the length of **V**: it is 430 km/h − 400 km/h = 30 km/h. The diagram shows that its direction is east, so **V** = 30 km/h east.

We can now calculate the acceleration. Our time interval is $t = 10$ s, so

$$\mathbf{a} = \frac{30 \text{ km/h east}}{10 \text{ s}}$$

$$\mathbf{a} = 3 \text{ km/h/s east}$$

PERPENDICULAR VECTORS

In the general case of adding two arbitrary vectors, as with **A** and **B**, we can find the magnitude and direction of the resultant **C** in two ways. One is a graphical method: we measure with a ruler the length and with a protractor the angle of the arrow representing **C**. The other is an analytical method too complicated for our discussion. However, in the special case in which **A** and **B** are *perpendicular* to each other—that is, when the angle between the arrows **A** and **B** is equal to 90°—then the resultant **C** forms the hypotenuse of a right triangle. We can calculate the length of **C** by a famous theorem credited to the Greek philosopher and mathematician Pythagoras (sixth century B.C.). The Pythagorean theorem says that the square of the hypotenuse of a right triangle is equal to the sum of the squares of the other two sides. We may illustrate this graphically:

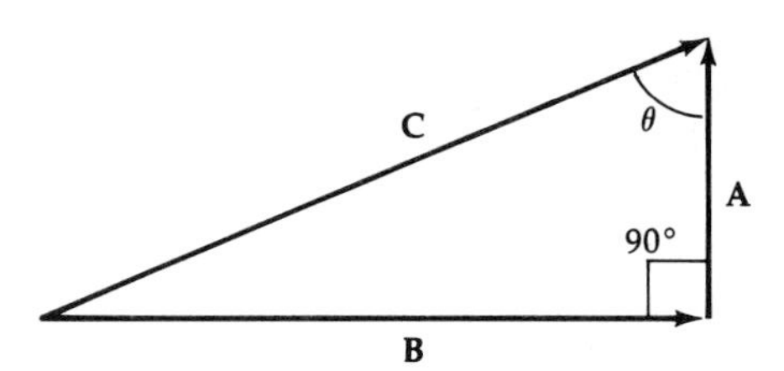

Suppose we let A represent the length of the vector **A**, and B and C the lengths of the vectors **B** and **C**, respectively. Then, the theorem tells us that

$$C^2 = A^2 + B^2$$

For example, if $A = 3$ and $B = 4$, then $C = 5$ because

$$3^2 + 4^2 = 9 + 16 = 25$$

and $5^2 = 25$. This is a particularly simple example because C^2 comes out to be a perfect square. However, the principle is the same for any right triangle.

RESOLUTION OF VECTORS INTO COMPONENTS

Recall that we can replace two vectors added together by their resultant. The ease of working with perpendicular vectors often makes it convenient to turn this process inside out. That is, we replace one vector by two perpendicular vectors that have the original vector as their resultant. This seems as if it would complicate matters—why use two when one will do? The answer is that, by doing so, we can greatly simplify our analysis of a physical situation involving vectors.

We begin by constructing a two-dimensional *coordinate system*, or axis system, by drawing two perpendicular dashed lines:

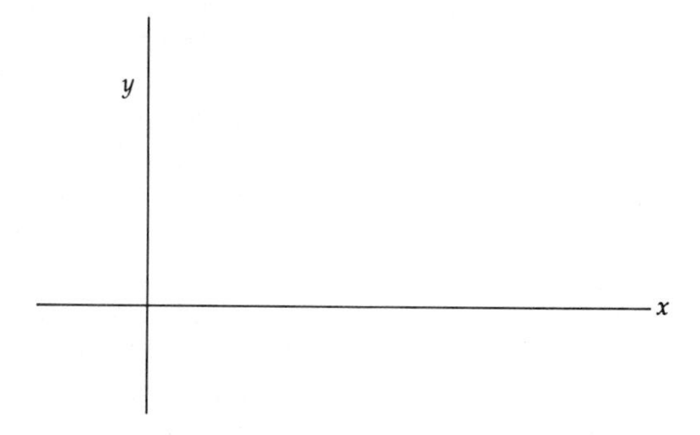

We call the horizontal line the *x-axis* and the vertical line the *y-axis*. The intersection of the lines is called the *origin* of the axis system.

Now consider the arbitrary vector **A**. Because we are free to move **A** about as long as we maintain its correct length and direction, we may position **A** such that its tail is at the origin of the system.

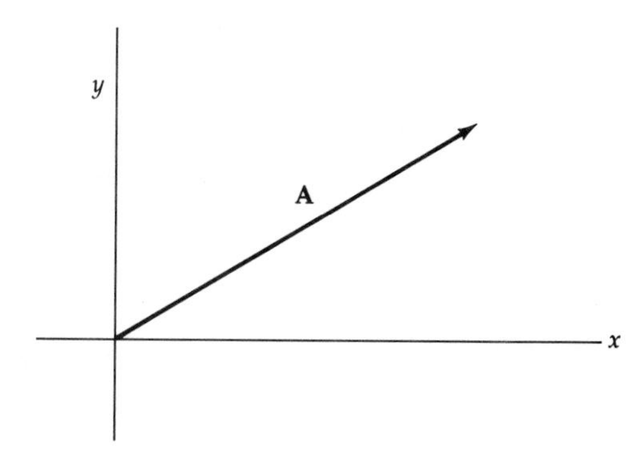

By resolving **A** into components, we mean drawing two vectors—one parallel to the x-axis and the other parallel to the y-axis—such that **A** is the resultant.

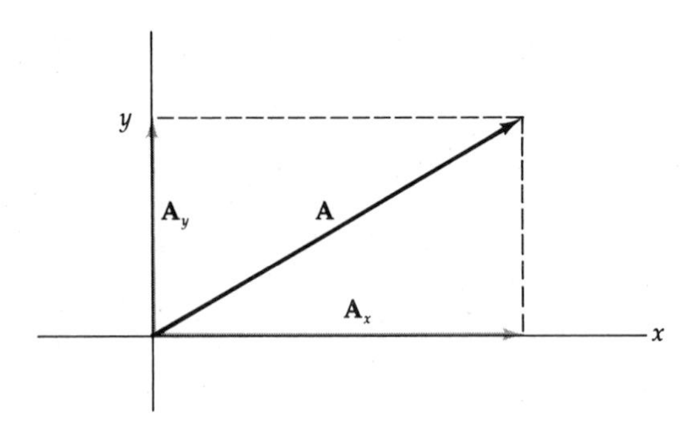

We label these new vectors $\mathbf{A}_x$ and $\mathbf{A}_y$, where $\mathbf{A}_x$ is called the *x-component* of $\mathbf{A}$, and $\mathbf{A}_y$ is the *y-component* of $\mathbf{A}$. We see that, by the parallelogram method,

$$\mathbf{A} = \mathbf{A}_x + \mathbf{A}_y$$

Since we have fixed things so that this is a right triangle, we also know that

$$A^2 = A_x^2 + A_y^2$$

Exercises

For some of the following exercises, you will need a ruler (in inches or centimeters) and a protractor. Use the cheapest ones you can find.

1. What are parallel vectors? What are antiparallel vectors?

2. Consider two vectors $\mathbf{X}$ and $\mathbf{Y}$ that add together to give the resultant $\mathbf{Z}$. The lengths of these vectors are X, Y, and Z, respectively. Is it always true that $\mathbf{X} + \mathbf{Y} = \mathbf{Z}$? Is it always true that $X + Y = Z$? Why or why not? Is it possible that $X - Y = Z$? If so, under what circumstances?

3. You leave your house and walk to the drugstore, which is 60 meters to the east, and then go to the ice cream store located 80 meters north of the drugstore. Draw a vector diagram, and from it find how far you are from home.

4. In problem 3, find the answer without using your ruler.

5. You leave home and drive 3.5 kilometers north, then 6.1 kilometers west, and then 5 kilometers southwest (that is, 45° south of west). What is your final position relative to home?

6. In problem 5, suppose you drive the three segments of the trip in reverse order; that is 5 kilometers southwest, then 6.1 kilometers west, and finally 3.5 kilometers north. What is your final position relative to home in this case?

7. Consider two vectors 3 meters and 4 meters long. Determine the length of the resultant if:
   a. the vectors are parallel;
   b. the vectors are antiparallel;
   c. the vectors are perpendicular.

8. Construct an x,y axis system and draw a vector from the origin 8 units (either centimeters or inches) long at a 30° angle from the x-axis. About how long are the x- and y-components of this vector? Check the accuracy of your measurement with the Pythagorean theorem.

9. Suppose you are in a boat at the south bank of a river that is flowing with a velocity of 8 km/h west. You start rowing due north across the river with a speed of 6 km/h relative to the river. Draw a vector diagram to scale representing this situation, and determine the magnitude of your resultant velocity relative to the ground. Also calculate the resultant by the Pythagorean theorem.

SUPPLEMENT 2

GRAPHS

Chapter 2 explained that, when describing motion, one must determine the relationships among different quantities. *Speed* is characterized by the dependence of *distance* on *time. Velocity* and *time* go together in defining *acceleration.* Finding relationships among various quantities in nature is an extremely important activity for physicists in their continuing task of explaining the universe. Merely collecting data is not enough. The scientist is interested in reducing the results of measurements to relationships and mathematical forms, so that patterns emerge that aid understanding.

Such relationships are also important in other areas of our lives. Market analysts want to know how stock prices fluctuate with respect to such factors as unemployment, inflation, and so on. The performance of a baseball player is partially measured by his batting average—the ratio of the number of hits to total times at bat. The consumer price index, a team's winning percentage, the Dow-Jones average—we are flooded daily with statistical relationships describing some part of our world.

A convenient device for studying how two quantities depend on each other is the *schematic diagram,* or *graph.* In Chapter 2 we used graphs of velocity and time, and distance and time, in examining the motion of falling bodies. Graphs give a clearer picture of what is happening than tabular columns of numbers do. In this supplement we discuss some properties and uses of graphs.

Figure S2-1. Speed versus time for freely falling body.

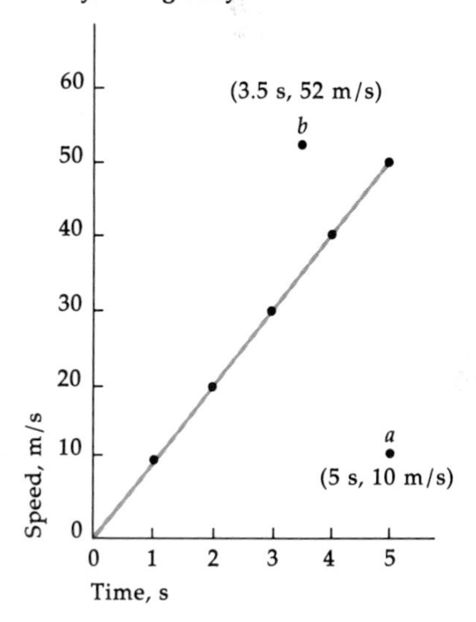

PLOTTING THE GRAPH

The graphs we will investigate are *line graphs.* We will use one for speed and time as an example (Figure S2–1). First we construct two intersecting perpendicular lines, or *axes*—one horizontal and the other vertical. In mathematical jargon, the horizontal axis is called the *abscissa* and the vertical axis is the *ordinate.* Each axis is labeled by one of the quantities being studied, called *variables,* and is marked off in some convenient unit. In our illustration, the abscissa is labeled *time,* with one division of the axis corresponding to 1 second, and the ordinate is labeled *speed,* with one division equivalent to 10 m/s. In this

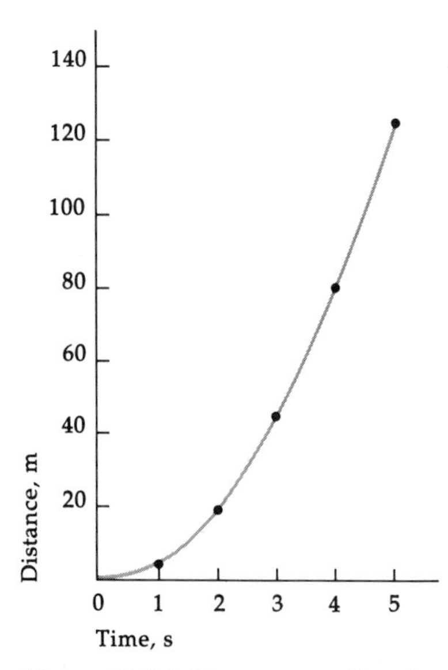

Figure S2-2. Distance versus time for freely falling body.

way, each point on the page represents one possible pair of values of time and speed. For example, we might choose two points at random, designated *a* and *b* in the figure. The point *a* corresponds to a time of 5 seconds and a speed of 10 m/s. To reach this position we go along the time axis to the reading "5 s," and then parallel to the speed axis a distance corresponding to "10 m/s." It is common to label this (5 s, 10 m/s), listing the horizontal member of the pair first. Point *b* is at 3.5 s and 52 m/s, denoted (3.5 s, 52 m/s).

Next we use the axis system to plot the data of speed versus time for the freely falling body, contained in the table on page 19. We show these as the unlabeled points in Figure S2–1. We can gain a better picture of our results by connecting these points by a line, or *curve*. The word "curve" is used in a general sense to mean any line, including a straight line. In Figure S2–1, we can see by laying a straightedge along the points that they can be connected by a straight line. In contrast, the data in the table for distance and time (p. 20) for the same falling body do not lie along a straight line. So we connect them by a smooth curve that fits the points (Figure S2–2).

A word of caution: in connecting our data points by a curve, we are exercising a little faith. We have measurements only for the times indicated by the points on the graph. We don't really know what the speed was after, for example, 2.3 seconds because we didn't take a measurement at that time. So our *interpolation*, as it is called, assumes that the speed-time relationship for falling bodies is the same for *all* the times included in our range of measurement, not just those particular instants when we recorded data. In this case, our experience with falling bodies assures us that nothing unusual is happening between measurements, so we can make the interpolation with confidence. Indeed, one of the conveniences of the graphical approach is that we get a picture of the relationship between two variables based on a relatively small number of observations. However, in situations where we are not so certain about what is going on between measurements, interpolation could give us an incorrect picture of the results.

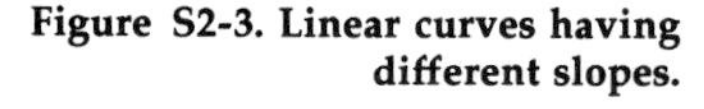

Figure S2-3. Linear curves having different slopes.

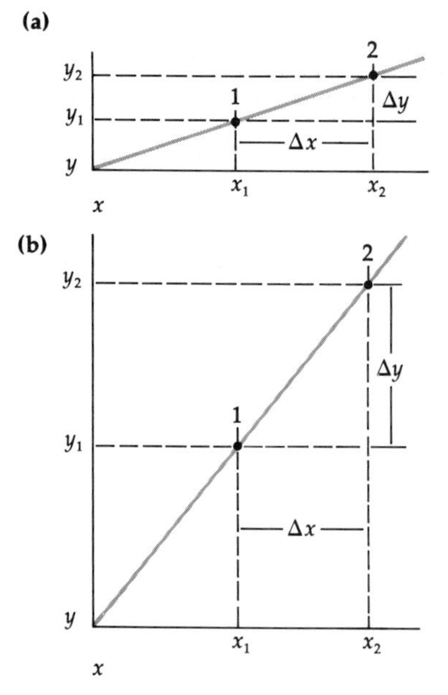

Another hazard of using graphs is *extrapolation*, which means extending the curve beyond the last data point. This is even riskier than interpolation. For example, extrapolating the speed-time curve gives the prediction that the body is moving 100 m/s after 10 seconds, when in fact it may have reached the ground by that time. You have to be extremely careful in projecting results outside the range of the observations.

SLOPE
Linear Graphs

Figure S2–3 shows two graphs of straight-line curves, drawn to the same scale. We have labeled the axes x and y to represent arbitrary variables. How do these curves differ? You would probably say that the line in **(b)** is steeper than the line in **(a)**. Mathematically we express the same idea by saying that the *slope* for **(b)** is greater than that for **(a)**. By definition,

$$\text{slope} = \frac{\text{change in vertical quantity}}{\text{change in horizontal quantity}} = \frac{\Delta y}{\Delta x} = \frac{y_2 - y_1}{x_2 - x_1}$$

The Greek letter delta, Δ, means "change in," or "difference in." Here, the pair of numbers x_1 and y_1 give the x and y readings for point 1 on each line, and x_2 and y_2 are for point 2. We have chosen Δx to have the same numerical value for both graphs. You can easily see that, because Δy is larger for graph **(b)**, the slope for **(b)** is larger than that for **(a)**.

To do a sample calculation of the slope, let us return to our falling-body graph of speed versus time (Figure S2–4). Suppose we take the (x_1, y_1) pair to be $x_1 = 1$ s, $y_1 = 10$ m/s, and the (x_2, y_2) pair to be $x_2 = 4$ s, $y_2 = 40$ m/s. Then we have

$$\Delta y = y_2 - y_1 \qquad \Delta x = x_2 - x_1$$
$$= 40\ \text{m/s} - 10\ \text{m/s} \qquad = 4\ \text{s} - 1\ \text{s}$$
$$\Delta y = 30\ \text{m/s} \qquad \Delta x = 3\ \text{s}$$

$$\text{slope} = \frac{\Delta y}{\Delta x} = \frac{30\ \text{m/s}}{3\ \text{s}}$$

$$\text{slope} = 10\ \text{m/s}^2$$

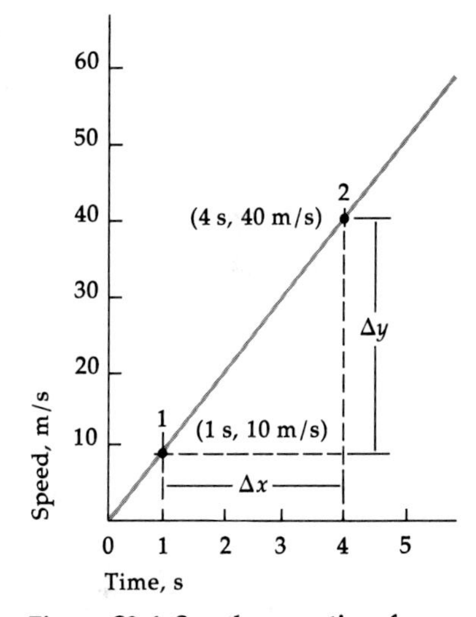

Figure S2-4. Speed versus time for freely falling bodies.

The slope is 10 m/s^2. Note that the curve itself is the hypotenuse of a right triangle whose legs are Δx and Δy.

To see that the slope of a straight-line curve is constant, suppose we take $x_1 = 2$ s, and $x_2 = 3$ s. The corresponding values are y and $y_1 = 20$ m/s and $y_2 = 30$ m/s. Then we have

$$\Delta y = y_2 - y_1 \qquad \Delta x = x_2 - x_1$$
$$= 30\ \text{m/s} - 20\ \text{m/s} \qquad = 3\ \text{s} - 2\ \text{s}$$
$$\Delta y = 10\ \text{m/s} \qquad \Delta x = 1\ \text{s}$$

$$\text{slope} = \frac{\Delta y}{\Delta x} = \frac{10\ \text{m/s}}{1\ \text{s}}$$

$$\text{slope} = 10\ \text{m/s}^2$$

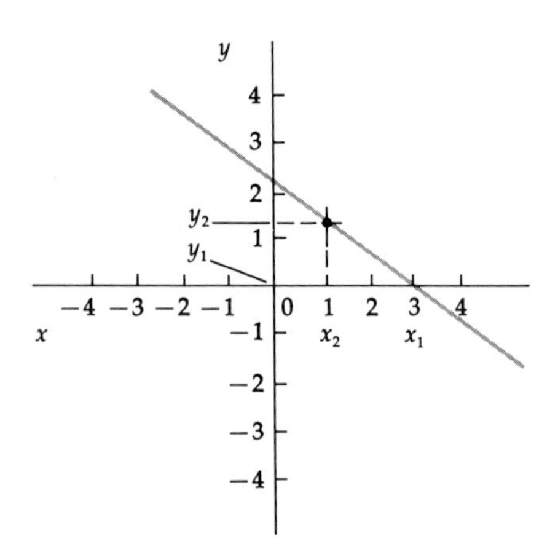

Figure S2-5. Linear curve with negative slope.

Again, the slope is 10 m/s^2. We may conclude that the *slope of a straight-line graph is constant.*

The slope of our speed-time curve has an important interpretation. Note that the slope is given in m/s^2, which is the unit for the magnitude of acceleration. This is consistent with our earlier definitions. The quantity Δy in the equation for slope is simply the change in speed, and Δx is the time interval over which the speed change takes place; that is, for this particular case,

$$\text{slope} = \frac{\text{change in speed}}{\text{time interval}}$$

But this is just the definition of the magnitude of acceleration. So, we may state as a general result that *the slope of a straight-line curve for speed versus time is equal to the magnitude of acceleration.*

The slope of a curve may also be a negative number. Consider the graph in Figure S2–5:
The curve represents the relationship between two arbitrary variables, x and y. Because we know that the slope of this straight line is constant, we are free to choose any two (x,y) pairs to make the calculation. Let $x_1 = 3$ and $x_2 = 1$. Then, the curve gives $y_1 = 0$ and $y_2 = 1.5$. We have

$$\begin{aligned} \Delta y &= y_2 - y_1 & \Delta x &= x_2 - x_1 \\ &= 1.5 - 0 & &= 1 - 3 \\ \Delta y &= 1.5 & \Delta x &= -2 \end{aligned}$$

$$\begin{aligned} \text{slope} &= \frac{\Delta y}{\Delta x} \\ &= \frac{1.5}{-2} \\ \text{slope} &= -0.75 \end{aligned}$$

Figure S2-6.

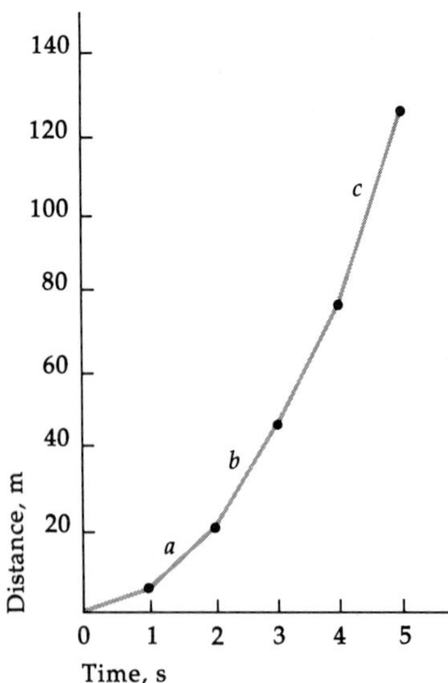

This is a negative number, which reflects the fact that *y decreases as x increases.* In our previous example, the positive slope indicated that *y increases as x increases.* More simply, this line

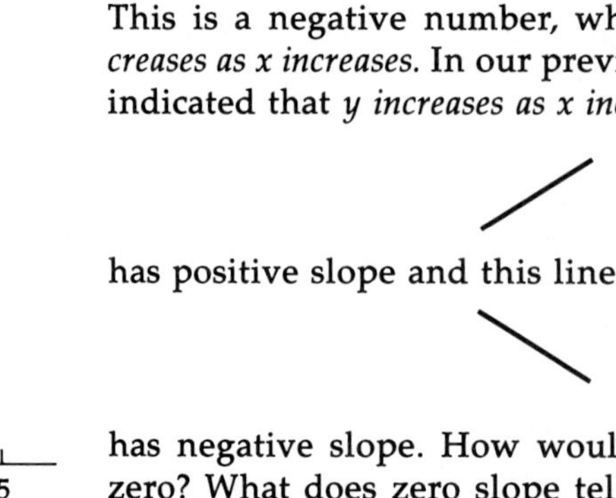

has positive slope and this line

has negative slope. How would a curve look whose slope is zero? What does zero slope tell you about the way y changes with x?

NONLINEAR GRAPHS

Physicists like straight-line or *linear* graphs because they yield the simplest picture of the relation between two variables. However, not all curves are linear. For example, we saw earlier that the graph of distance versus time for the freely falling body is a parabola. How can we determine the slope of such a nonlinear curve, and what does it mean?

Look again at the distance-time graph. Rather than connecting the lines by a single smooth curve, we can obtain a crude approximation to the true result by connecting successive points by straight lines, as shown in Figure S2–6. Clearly, each of the segments has a different slope. You should satisfy yourself that the line segments labeled *a*, *b*, and *c* in Figure S2–6 have slopes of 15 m/s, 25 m/s, and 45 m/s, respectively.* Therefore, we cannot speak of "the" slope for a nonlinear graph.

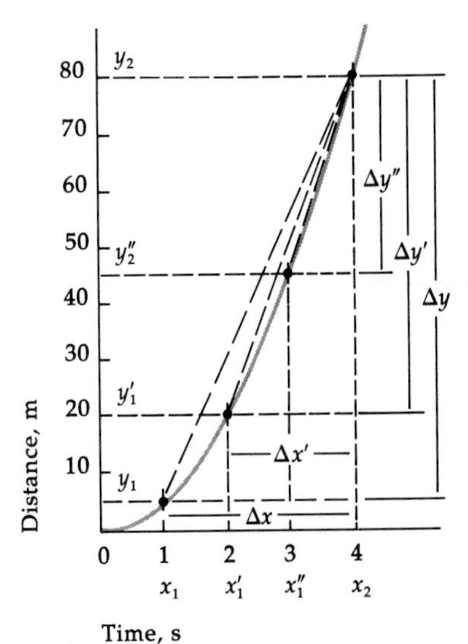

Figure S2-7.

We can, however, define the slope at each particular point of the curve. Figure S2–7 is a blowup of a portion of our distance-time graph. Here Δx and Δy are shown for $x_1 = 1$ s, $y_1 = 5$ m, and $x_2 = 4$ s, $y_2 = 80$ m. The line connecting the two points does not follow the curve as it does in the linear case; we have no triangle because the "hypotenuse" is bent. By selecting values of x_1 and x_2 that are closer together—that is, by letting Δx become smaller—we make the hypotenuse of the dashed-line triangle approach coincidence with the curve. In this example, we have kept $x_2 = 4$ s and let x_1 take on the values 2 s and then 3 s. If we let Δx become infinitely small, the hypotenuse of the triangle becomes a line *tangent* to the curve at the point $x = 4$. A tangent is a straight line that touches the curve only at one point. It is natural, then, to define the slope of a curve at a point as the slope of the line tangent to the curve at that point. By drawing tangent lines at several points along the curve, you can see that the slope of the distance-time curve becomes greater as the time increases. The tangent lines become steeper as you move to the right.

The top half of Figure S2–8 shows the distance versus time graph with tangent lines drawn for each second. The slopes of these lines are indicated in red and plotted in the bottom half. This lower curve should look very familiar; it is the speed-time curve we have seen several times. We have thus established an important relationship between the two graphs describing motion: *the slope of the free-fall distance-time curve at any time is equal to the speed at that same time.*

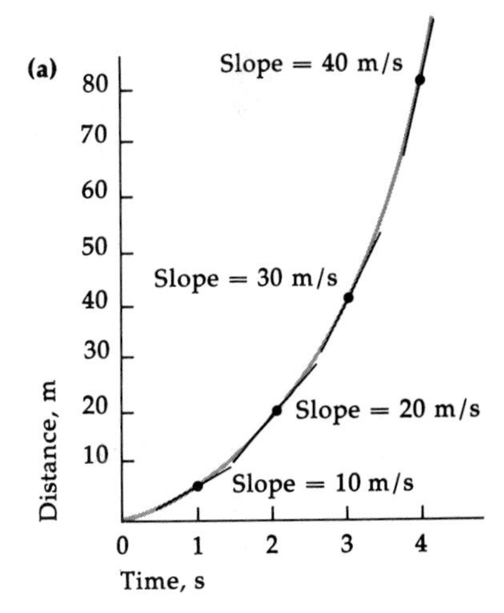

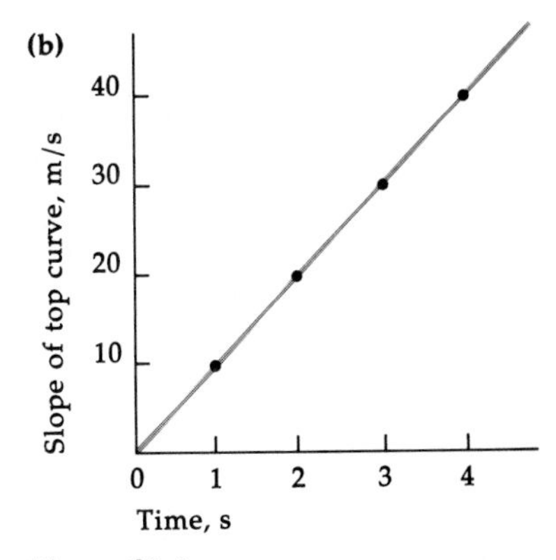

Figure S2-8.

In determining the slope of the curve by taking smaller and smaller values of Δx, we have used what is called in calcu-

*In calculating the slope of a given line segment, we are free to extend the line for convenience in either direction, even though there may be no data in the extended region. This is because the slope is a property of the curve and remains constant no matter how far we extend the line.

lus the *limit*. The slope of the curve of y versus x is given by

$$\text{slope} = \lim_{\Delta x \to 0} \frac{\Delta y}{\Delta x}$$

where the right side is the limit of $\Delta y/\Delta x$ as Δx approaches zero. This limit is called the *derivative of y with respect to x*, and is given the symbol dy/dx.

$$\frac{dy}{dx} = \lim_{\Delta x \to 0} \frac{\Delta y}{\Delta x}$$

So, *the derivative is the slope of the curve.* For the curve of distance versus time, the derivative of the distance with respect to the time equals the instantaneous speed.

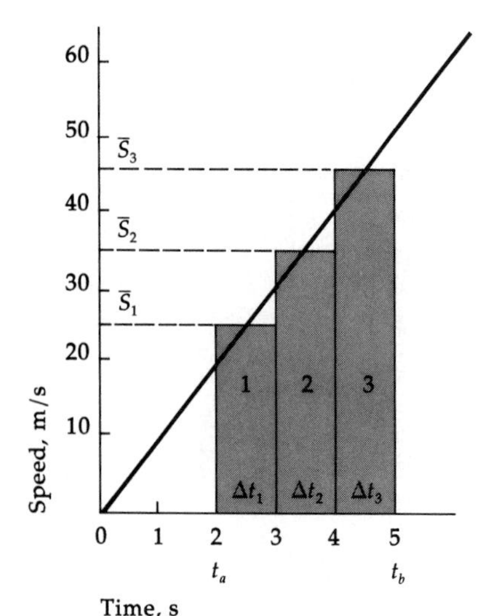

Figure S2-9.

THE AREA UNDER THE CURVE

Figure S2–9 shows the graph of speed and time for the freely falling body. Rectangles have been drawn for time intervals Δt_1, Δt_2, and Δt_3. For rectangle 1, $\bar{S}_1$ is the average speed. It is the instantaneous speed at the midpoint of time interval Δt_1. Similarly, $\bar{S}_2$ is the average speed for Δt_2, and so on for Δt_3. Note also that $\bar{S}_1$ is the height of the first rectangle, and Δt_1 is its width. Thus, the product $\bar{S}_1 \, \Delta t_1$ is the *area* of the rectangle. Similarly, $\bar{S}_2 \, \Delta t_2$ and $\bar{S}_3 \, \Delta t_3$ are the areas of rectangles 2 and 3.

For a given rectangle the shaded portion above the curve is equal in area to the unshaded region below the curve. This means that the area of the rectangle is equal to the area under the curve.

The definition of average speed was given on page 12:

$$\bar{S} = \frac{\Delta d}{\Delta t}$$

Multiplying both sides by Δt gives the distance traveled during the time interval

$$\Delta d = \bar{S} \, \Delta t$$

Therefore, since $\Delta d_1 = \bar{S}_1 \, \Delta t_1$ is the distance traveled during time interval Δt_1, we can say that the area under the portion of the curve corresponding to Δt_1 is equal to the distance traveled during this time. Obviously, similar statements apply to the time intervals Δt_2 and Δt_3. If we add up the areas for the three time intervals, we get the total distance traveled between times t_a and t_b.

In calculus, such an area under the curve is called the *integral*, given the symbol $\int$:

$$\text{area under curve between } t_a \text{ and } t_b = \int_{t_a}^{t_b} S\,dt$$

The integral symbol is an elongated "S," denoting that the integral is a *sum* of areas. So, the distance is the integral of the speed over the time.

We have illustrated the integral for a straight-line graph, but the principle is the same for any curve: *the integral is equal to the area under the curve.*

We may summarize what we learned about freely falling bodies from our study of graphs:

1. A graph of speed versus time is linear. Its slope is constant and is equal to the numerical value of acceleration. The area under the curve is equal to the distance traveled.
2. A graph of distance versus time is a parabola. Its slope for any time is equal to the speed at that time.

Exercises

1. Plot the following data on an x-y axis system and answer the questions.

x	y
0	−3
1	0
2	3
3	6
4	9
5	12

a. Is the curve a straight line?
b. What is the slope of the curve?
c. Where does it intersect the y-axis?

2. Figure S2-10 shows a graph of the distance of a car from the owner's house in relation to the time.

a. How far away from the house was the car at the end of 3 seconds?
b. What is the slope of this curve?
c. Show that the slope is equal to the car's speed.
d. Was the speed constant?
e. What was the speed at the end of 4 seconds?
f. If the motion continued, how far would the car have gone at the end of 6 seconds?

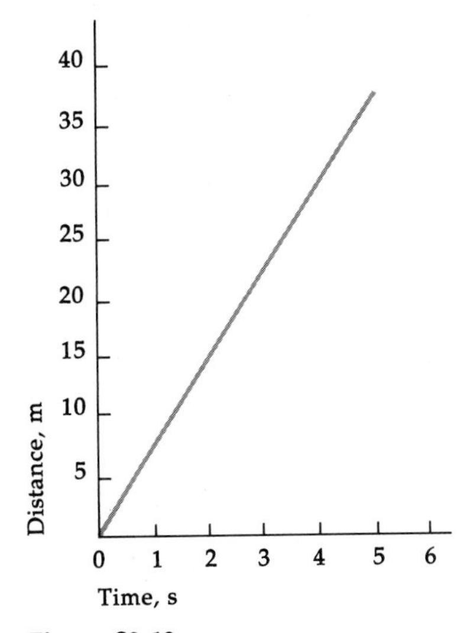

Figure S2-10.

3. Plot the following data of speed and time, the vertical axis being for speed and the horizontal axis for time.

Speed (m/s)	Time (s)
50	0
50	1
50	2
45	3
40	4
35	5
30	6
25	7
20	8
15	9
10	10
5	11
0	12

a. How long was the speed constant?
b. What is the slope of the curve between 0 and 2 seconds? What was the magnitude of the acceleration during this time?
c. What is the slope for the rest of the curve?
d. What was the speed at $t =$ 8 seconds?

4. Figure S2–11 shows the results of an experiment in which a ball was thrown straight up. Recall from problem 2 that the slope at any time is equal to the speed at that time.

a. Determine as accurately as you can the slope for $t = 0, 1, 2, 3, 4, 5$, and 6 seconds.
b. When was the speed zero?
c. For the first 3 seconds, was the speed increasing or decreasing?
d. When was the object 25 meters from the ground?
e. From your determination in part a, plot a curve of speed versus time for the ball. What is the numerical value of the acceleration?
f. How large was the acceleration when the speed was zero?

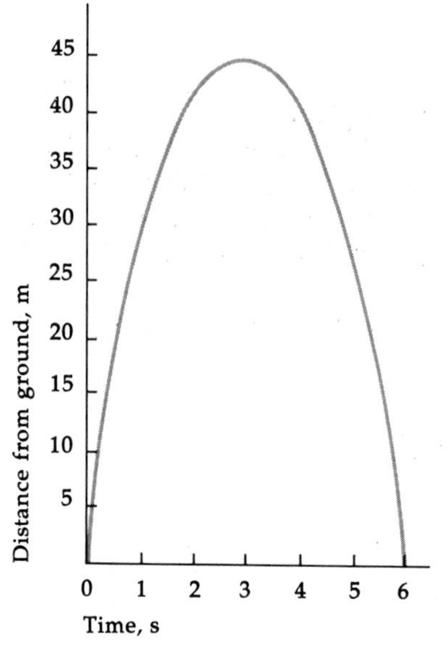

Figure S2-11.

5. A lady drives a car at a constant speed from Gravel Switch to Rabbit Hash, a distance of 60 kilometers, in 1.5 hours. Make plots of position versus time and speed versus time for her trip. A man also makes the same trip in 1.5 hours, but he reaches Possum Trot, a city 40 km from Gravel Switch, in 45 minutes, and then drives on to Rabbit Hash at a slower speed. Make plots of distance versus time and speed versus time for his trip.

THE VECTOR NATURE OF FORCE

FORCES ACTING AT ANGLES

In our discussion in Chapter 3 of Newton's second law of motion, we saw that the acceleration of an object results from the *net force.* In the language of vectors developed in Supplement 1, net force means the *resultant* of the vectors representing all the forces acting on the object. We looked at some examples of forces applied along the same straight line. Now we are going to consider some more complicated situations.

Again take as a simple example a box sitting on the floor, and we assume the box is frictionless. This time, instead of applying forces antiparallel to each other, we will pull on the box in perpendicular directions. The diagrams in Figure S3–1 illustrate this situation for two forces, each being 5 newtons. One is directed north, the other east. Part b of this figure shows how to get the resultant by the head-to-tail method. The resultant $\mathbf{F}_{net}$ is a vector pointing in a northeasterly direction. We can calculate its magnitude by the Pythagorean theorem:

$$\mathbf{F}_{net}^2 = \mathbf{F}_1^2 + \mathbf{F}_2^2$$
$$= (5^2 + 5^2)\ \mathrm{N}^2$$
$$= 50\ \mathrm{N}^2$$
$$\mathbf{F}_{net} = \sqrt{50}\ \mathrm{N}$$
$$\mathbf{F}_{net} \approx 7.1\ \mathrm{N}$$

If we measure the angle between $\mathbf{F}_{net}$ and $\mathbf{F}_2$, we find that $\mathbf{F}_{net}$ is directed 45° north of east.

We can extend this treatment to any number of forces. If you and four of your friends pull on the box with forces of various magnitudes and directions, you would have the situation shown in Figure S3–2. The box would accelerate in the direction of the resultant of all these forces. We cannot calcu-

(a)

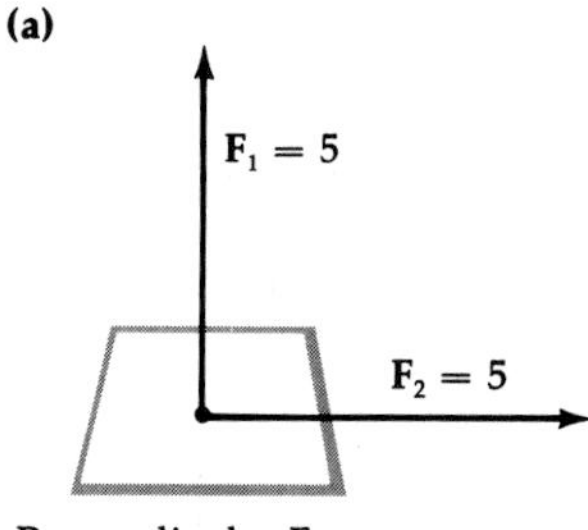

Perpendicular Forces

(b)

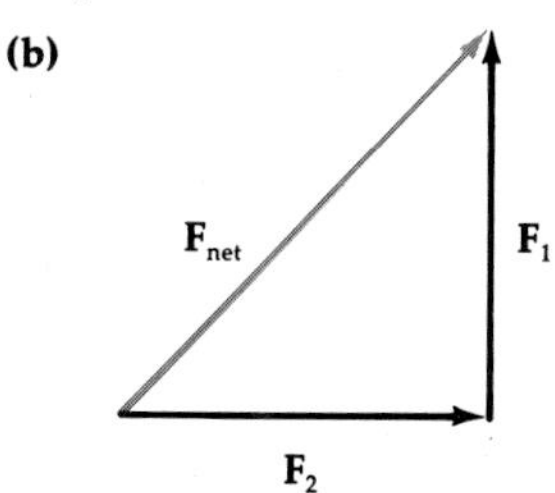

Head-to-tail Addition

Figure S3-1. Force vectors: perpendicular directions.

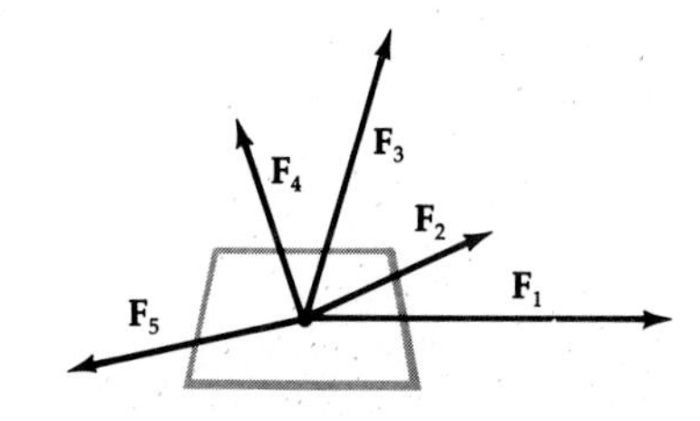

Figure S3-2. Several force vectors: various directions.

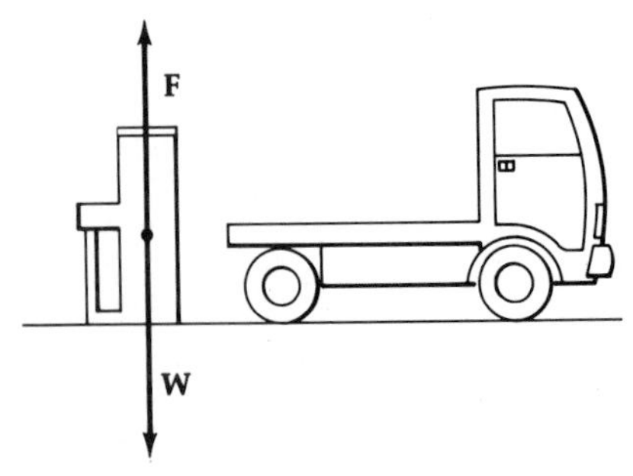

Figure S3-3. F is the force, equal in size to the weight W, which must be applied to lift the piano straight up.

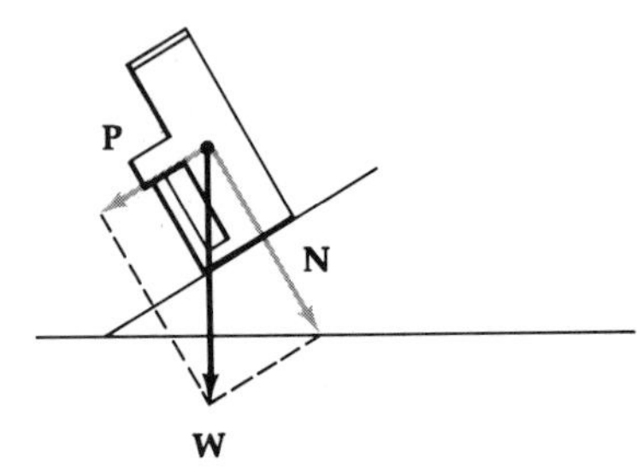

Figure S3-4. The ramp resolves W into N and P.

late the magnitude of the net force by the Pythagorean theorem because the forces are in several different directions, not all perpendicular. However, we could draw the various force vectors to scale and measure the length and direction of the resultant.

We can use the idea of force vectors to illustrate the convenience of resolving vectors into components, discussed in Supplement 1. Suppose a friend asks you to help move a piano by truck: what is the easiest way of getting the piano onto the truck? Well, one way of doing it is to lift the piano straight up onto the truck. As shown in Figure S3–3, the force you have to apply must be enough to overcome the force of gravity, so you would have to furnish a force at least equal to the weight of the piano. That might be several hundred pounds, so this method is not a particularly good one. What should you do to reduce the force you have to apply (other than, say, breaking the piano up into little pieces)?

Figure S3–4 shows how a ramp can help: the weight of the piano is still **W**, a vector pointing toward the center of the earth. But the piano, if released, will roll down the ramp, not drop straight through it; that is, the ramp has effectively resolved the **W** vector into two components: **N**, which is *perpendicular* to the ramp (sometimes this is called the *normal* force), and **P**, a vector *parallel* to the ramp. Both **P** and **N** are smaller than **W**, as you can see by the lengths of the corresponding vectors. It is the vector **P** that must be overcome in order to roll the piano up the ramp. So, the ramp helps you by allowing you to use a smaller force.

We can improve things even more by making the ramp longer. This is illustrated by the diagrams in Figure S3–5. As you can see, the longer the ramp is, the shorter the vector **P**, and hence the easier it is to move the piano. In each case, note that **W** is always the resultant of **P** and **N**.

CENTRIPETAL FORCE

If you tie a stone on the end of a string and whirl it around, the stone follows a circular path. The moon's orbit around the earth is nearly circular, as is the earth's orbit around the sun. Because these objects do not move in straight lines, they are

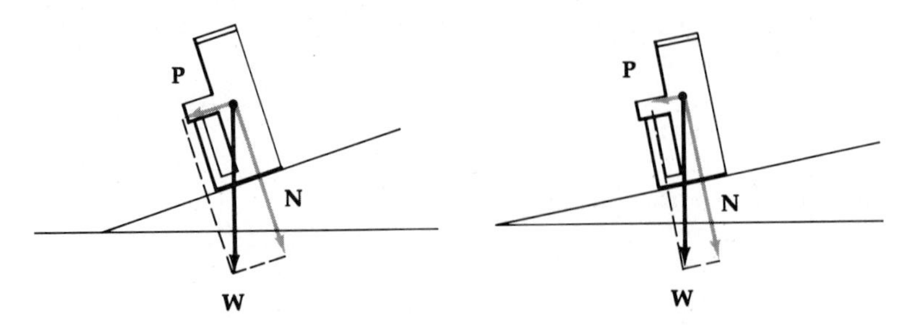

Figure S3-5. A longer ramp reduces the length of the vector P, easing the task of moving the piano. Always, W = P + N.

accelerating, meaning that a net force acts on them. We can use what we know about vectors to study this force.

Figure S3–6 shows a body moving at constant speed v in a circle of radius r. When it is at the point P_1 its velocity is $\mathbf{v}_1$, and at P_2 the velocity is $\mathbf{v}_2$. The velocity vectors are drawn tangent to the circle at these points. Their directions are different but, because the speed is constant, the lengths of the vectors are both equal to v.

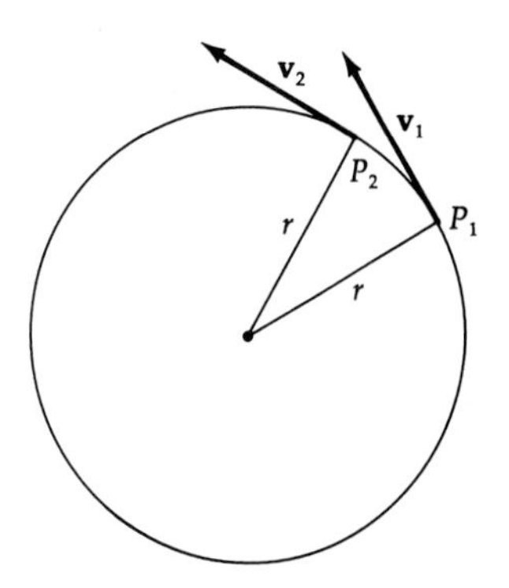

Figure S3-6. The object travels in a circle of radius r_1 at constant speed v.

We can use the head-to-tail method of vector addition to see how the velocity changes as the object moves from P_1 to P_2. The change in velocity is $\Delta\mathbf{v}$:

$$\Delta\mathbf{v} = \mathbf{v}_2 - \mathbf{v}_1$$

From the head-to-tail diagram in Figure S3–7 you can see that the direction of $\Delta\mathbf{v}$ is toward the center of the circle. Therefore, this is also the direction of the acceleration $\mathbf{a}$, since

$$\mathbf{a} = \frac{\Delta\mathbf{v}}{\Delta t}$$

This is called the *centripetal acceleration,* which means acceleration "toward the center."

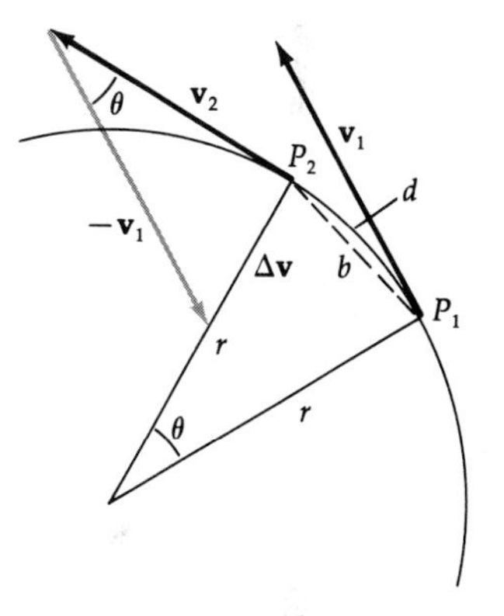

Figure S3-7. Determination of $\Delta\mathbf{v}$.

What about the size of the acceleration $\mathbf{a}$? Well, we can also get this from Figure S3–7. You can see that $\mathbf{v}_1$, $\mathbf{v}_2$, and $\Delta\mathbf{v}$ form a triangle where the angle between $\mathbf{v}_1$ and $\mathbf{v}_2$ is called θ. There is another triangle in the picture made up of the radii r drawn from the center to P_1 and P_2, and the line connecting these points, whose length is b. Because these triangles are similar, ratios of corresponding sides are equal. That is,

$$\frac{\Delta v}{v} = \frac{b}{r}$$

Now, suppose that Δt is the time it takes the object to go from P_1 to P_2 along the circle. Call this distance d; then

$$d = v\,\Delta t$$

For small values of Δt, this distance *along* the circle will be approximately equal to the *straight-line* distance between P_1 and P_2; that is, we can write

$$b \approx v\,\Delta t$$

Substituting into the ratio equation gives

$$\frac{\Delta v}{v} \approx \frac{v\,\Delta t}{r}$$

Rearranging this expression, we get

$$\frac{\Delta v}{\Delta t} \approx \frac{v^2}{r}$$

In the limit as Δt approaches zero, this result is exact. The left side is the magnitude of the acceleration, **a**. So

$$\mathbf{a} = \frac{v^2}{r}$$

We can conclude, then, that the acceleration for an object traveling in a circle of radius r and constant speed v is a vector that points toward the center of the circle and has a length v^2/r.

Since Newton's second law says that $\mathbf{F} = m\mathbf{a}$, we see that the net force on our object is also directed toward the center of the circle and that its magnitude is

$$\mathbf{F} = \frac{mv^2}{r}$$

This is called the *centripetal force.*

DEVELOPMENT OF THE GRAVITATIONAL FORCE EQUATION

In this supplement we are going to arrive at the equation that Newton discovered for the universal law of gravitation. For simplicity, we will assume that the planets travel in *circular* rather than *elliptical* orbits, which is not a very bad approximation. We will use what we already know about circular motion, as well as Newton's second law and Kepler's third planetary law.

As we have seen earlier (Supplement 3), the acceleration of an object traveling at constant speed v in a circle of radius r is

$$a = \frac{v^2}{r}$$

Because the speed is constant, it is the same as the average speed, or the total distance around the circle divided by the time required for one revolution. That is, if T is the time for one trip around the circle, then

$$v = \frac{2\pi r}{T}$$

from which we get

$$v^2 = \frac{4\pi^2 r^2}{T^2}$$

Replacing v^2 in the equation for a by this formula gives

$$a = \frac{4\pi^2 r}{T^2}$$

According to Newton's second law of motion, the magnitude of the force holding the planet in its circular orbit is the mass of the planet times this acceleration:

$$F = ma = m\left(\frac{4\pi^2 r}{T^2}\right)$$

Here is where Kepler's third law enters the picture, because it tells us how the time T is related to the radius r. The law says

$$k = \frac{D^3}{T^2}$$

where D is the length of the major axis for an ellipse. For a circle, the major axis is just the diameter, which is twice the radius—that is, $D = 2r$. So the equation for k becomes

$$k = \frac{(2r)^3}{T^2} = \frac{8r^2}{T^2}$$

Solving for T^2 gives

$$T^2 = \frac{8r^3}{k}$$

With this expression for T^2, the force equation for F becomes

$$F = m\left(\frac{4\pi^2 rk}{8r^3}\right)$$

$$= m\left(\frac{\pi^2 k}{2r^2}\right)$$

To clarify the next step, we can rewrite this equation like this:

$$F = \frac{m}{r^2}\left(\frac{\pi^2 k}{2}\right)$$

Look at the factor in parentheses. Since k is a constant for all the planets, we may replace this entire factor by a single number. Newton felt that the force somehow depended on the sun. Since the mass m of the planet is involved, it seemed reasonable to him that the mass of the sun should also enter in. So, he defined

$$MG = \frac{\pi^2 k}{2}$$

in which M is the mass of the sun and the proportionality factor G is the universal gravitational constant. This gives us the final form of the law of gravitation:

$$F = \frac{GmM}{r^2}$$

MORE ABOUT MOMENTUM, TORQUE, AND ANGULAR MOMENTUM

In Chapters 6 and 7, we briefly mentioned that momentum, torque, and angular momentum are vectors. We will look more closely at the vector nature of these quantities.

MOMENTUM AS A VECTOR

Because momentum is a vector, we must consider its direction as well as its magnitude. This is especially true when we discuss the *conservation of momentum.* This expression means that the *vector sum* of the various momenta of a system does not change as long as the net external forces are zero.

An excellent example of applying conservation of momentum is found in billiards. When billiard balls run into one another, their rolling motion complicates the picture somewhat, but we can ignore this effect. We can suppose that the balls slide along a frictionless table. If you want the eight ball to move straight ahead, strike the cue ball so that it hits the eight ball dead center. However, if you want it to go off to one side, you make the cue ball strike a glancing blow. This situation is shown in Figure S5–1. The cue ball has mass m and moves with velocity $\mathbf{v}$ toward the eight ball at rest. So, the initial momentum is

$$\text{initial momentum} = m\mathbf{v}$$

Figure S5-1. Collision in two dimensions.

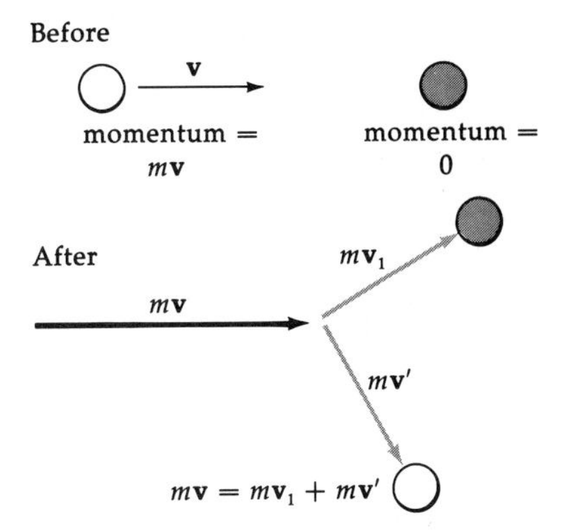

After the collision, the cue ball moves off at an angle with velocity $\mathbf{v}'$ (the apostrophe just indicates a new velocity for the cue ball), so the momentum is $m\mathbf{v}'$. The eight ball goes in the other direction with velocity $\mathbf{v}_1$, and its momentum is then $m\mathbf{v}_1$. There is no net external force, so the momentum is conserved; that is,

$$m\mathbf{v} = m\mathbf{v}' + m\mathbf{v}_1$$

where, remember, this means vector addition, not just the ordinary addition of numbers.

As an exercise, suppose that a stationary bomb in outer space explodes into three equal-mass fragments. Draw a vector diagram showing the paths of the three pieces.

TORQUE

When we discussed torque in Chapter 7, we looked at the situation in which the force is applied perpendicularly to the lever-arm line. We can generalize this for forces applied at an arbitrary angle, as shown in Figure S5-2. From our discussion of vectors in Supplement 1, we know that we can resolve F into components, one of which is *parallel* ($F_{\parallel}$) and one *perpendicular* ($F_{\perp}$) to the lever-arm line. This is shown in the lower half of the figure. The perpendicular force $\mathbf{F}_{\perp}$ is the one that causes the rotation, whereas $\mathbf{F}_{\parallel}$ has no effect as far as rotating the stick is concerned. So, the magnitude of the torque is

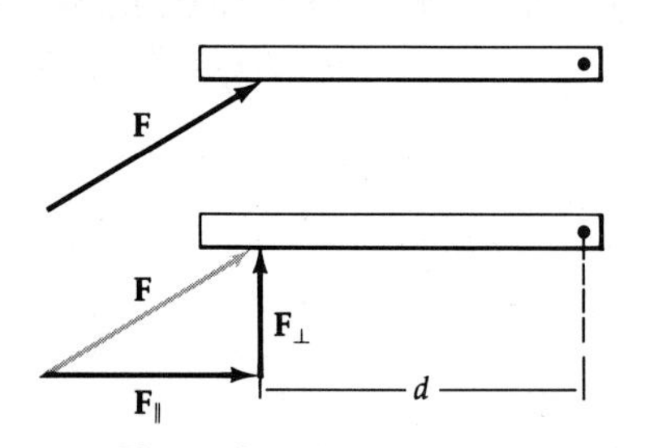

Figure S5-2. Torque is produced by $\mathbf{F}_{\perp}$.

$$\text{torque} = \mathbf{F}_{\perp} \cdot d$$

Because $\mathbf{F}_{\perp}$ is smaller than $\mathbf{F}$, the resulting torque is less than it would have been had the full $\mathbf{F}$ been applied perpendicularly to the stick. As $\mathbf{F}$ gradually becomes more nearly parallel to the lever-arm line, the component $\mathbf{F}_{\perp}$ becomes progressively smaller, and so does the torque. Finally, when $\mathbf{F}$ is exactly parallel there is *no* perpendicular component. There is thus no torque, and the stick won't rotate.

So we see that torque depends not only on force and distance but also on the relative direction of the two—it is the perpendicular component that counts. For example, when you whirl a rock around in a circle at the end of a string, the force on the rock is toward the center of the circle, as we have seen previously. Because this force is always parallel to the lever-arm line, it produces no torque. This is generally true of central force systems, such as satellites orbiting the earth or planets going around the sun. The gravitational forces are like $\mathbf{F}_{\parallel}$ in the preceding diagram: they produce no torque.

The *direction* of the torque vector obeys the right-hand rule given in Chapter 7: if you curl the fingers of your right hand in the rotation direction, the thumb indicates the torque direction.

We discussed the analogies between linear and rotational motion earlier, and now we can elaborate on them. Recall that our momentum-impulse equation (page 77) is

$$\mathbf{F}\,\Delta t = \Delta\mathbf{p}$$

If we replace **F** by its rotational counterpart torque (τ) and **p** by the angular momentum **L**, we have $\tau\,\Delta t = \Delta\mathbf{L}$, whence

$$\tau = \frac{\Delta\mathbf{L}}{\Delta t}$$

which shows that the torque is the rate at which angular momentum changes in time; torque and angular momentum are related in the same way as force and momentum. We also see that conservation of angular momentum comes from this equation: if $\tau = 0$, then $\Delta\mathbf{L} = 0$, which says that the angular momentum doesn't change.

ANGULAR MOMENTUM

We can now generalize our definition of angular momentum, rather than talking only about circular motion. Figure S5–3 shows an object of mass m moving with velocity **v** and located a distance r from a given point, O. Since its momentum is thus $\mathbf{p} = m\mathbf{v}$, this picture is just like Figure S5–2, with **p** replacing **F**. The magnitude of angular momentum of the object about O is given by the component of **p** perpendicular to r multiplied by r:

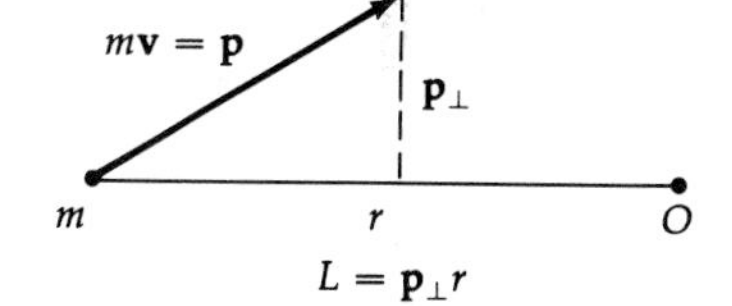

Figure S5-3. Angular momentum about the point O.

$$L = p_\perp r = mv_\perp r$$

Just as with force and torque, it is the *perpendicular* component of the momentum that is involved in the angular momentum.

Now let's see how Kepler's law about the equal areas is related to angular momentum conservation. Figure S5–4 shows a planet having momentum $m\mathbf{v}$ at some point of its orbit. The dashed line gives the motion perpendicular to the line to the sun, and its length is $v_\perp\,\Delta t$ for a time interval Δt. Thus we have a triangle whose base is r and whose area is $\frac{1}{2}rv_\perp\,\Delta t$ (the area of a triangle is one-half the base times the height). Since the force on the planet is a central force, the angular momentum of the planet does not change as it moves around the sun. This means that the number $rv_\perp$ will always remain the same (since the mass is constant), so for any other equal time interval Δt, we will get the same area, $\frac{1}{2}rv_\perp\,\Delta t$. And this is what Kepler found.

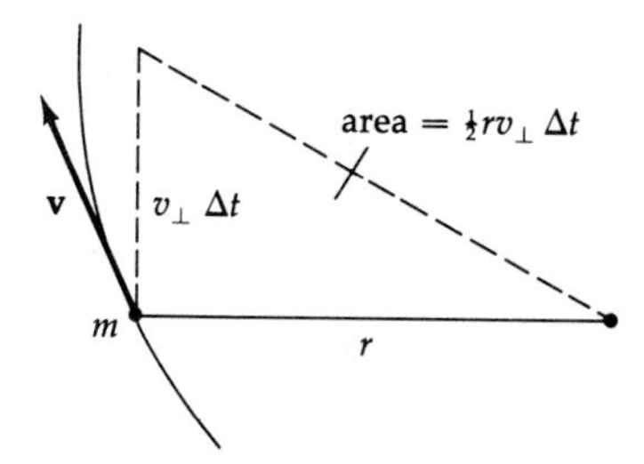

Figure S5-4. Angular momentum and Kepler's law.

SUPPLEMENT 6

THE MICHELSON-MORLEY EXPERIMENT

The experiment by A. A. Michelson and E. W. Morley that demolished the ether hypothesis is considered one of the definitive experiments in the history of physics, primarily because of its role in the special theory of relativity. In this supplement we look at this famous experiment in some detail.

A FAMILIAR ANALOGY

We will consider an ordinary situation as a way of introducing the principle behind the experiment. Let us suppose that you are about to row a boat in a river that flows eastward with speed v. We shall see that it takes longer to complete a round trip parallel to the riverbank than to complete one of equal distance across the river and back.

First look at the trip parallel to shore (Figure S6–1). Imagine that you start rowing west (upstream) with a speed u relative to the water. The boat-speed u must be greater than v if you are to make any progress. Since you are rowing against the current, your speed relative to the riverbank will be $(u - v)$. The time required to reach a point a distance L away is therefore

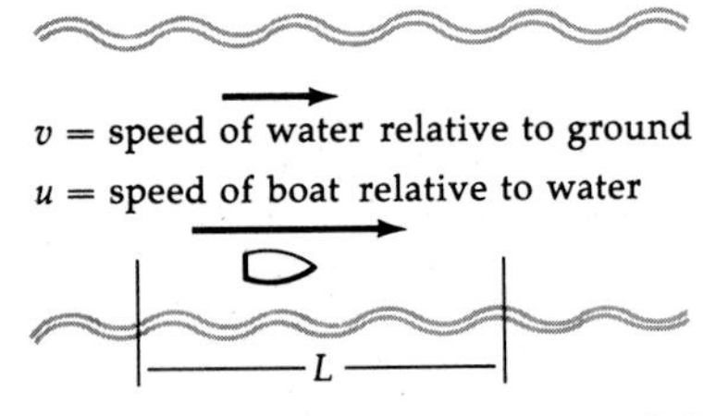

Figure S6-1. The trip parallel to shore.

$$t_{\text{up}} = \frac{L}{(u - v)}$$

On the return trip, you are rowing downstream *with* the current, so your speed relative to the riverbank is $u + v$, making the time required to return to the original starting point

$$t_{\text{down}} = \frac{L}{(u + v)}$$

Thus, the total time $t_{\parallel}$ required for the round trip parallel to the shore is the sum of t_{up} and t_{down}:

$$t_{\parallel} = \frac{L}{(u - v)} + \frac{L}{(u + v)}$$

$$= \frac{2Lu}{(u^2 - v^2)}$$

We write this in a more convenient form by dividing both numerator and denominator by u^2:

$$t_{\parallel} = \frac{\frac{2L}{u}}{1 - \left(\frac{v}{u}\right)^2}$$

Note that if the river is not flowing (that is, if $v = 0$), then the denominator collapses to 1, and $t_{\parallel}$ becomes $2L/u$. If the river has any current at all, $t_{\parallel}$ will be *greater* than $2L/u$ because v is smaller than u. This makes the denominator smaller than 1; any river speed lengthens the time required for a round trip.

Now let us look at the situation in which the boat is to land directly across the river. To reach a point a distance L away from the starting point, you must aim the boat upstream a little to compensate for the flowing water. This is illustrated in Figure S6-2. Let D represent the length of the line in the actual direction in which the boat is aimed. This is the distance the boat will cover with respect to the water. The distance R represents the amount of downstream drift caused by the current. The relative sizes of R and D depend on the relative magnitudes of the two velocities. It is not hard to see that

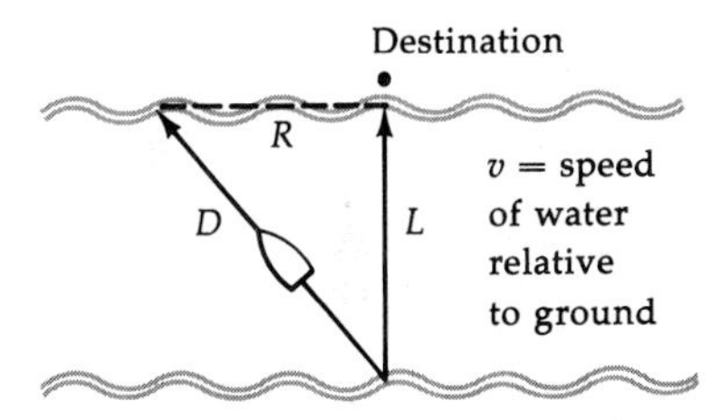

Figure S6-2. The trip across the river.

$$\frac{R}{D} = \frac{v}{u} = \frac{\text{water speed}}{\text{boat speed}}$$

or,

$$R = \frac{Dv}{u}$$

Now, the sides D, L, and R form a right triangle, so according to the Pythagorean theorem,

$$D^2 = R^2 + L^2$$

Substituting for R gives

$$D^2 = \left(\frac{Dv}{u}\right)^2 + L^2$$

from which we get

$$D = \frac{L}{\sqrt{1 - \left(\frac{v}{u}\right)^2}}$$

The time required for the perpendicular crossing involves the distance D and the boat's speed u. For the round trip, it is

$$t_\perp = \frac{2D}{u} = \frac{\frac{2L}{u}}{\sqrt{1 - \left(\frac{v}{u}\right)^2}}$$

This equation looks like the one for $t_\parallel$ except that here the denominator has the square root sign. The ratio of the times for the two round trips is found by dividing one time by the other:

$$\frac{t_\perp}{t_\parallel} = \frac{\dfrac{\frac{2L}{u}}{\sqrt{1 - \left(\frac{v}{u}\right)^2}}}{\dfrac{\frac{2L}{u}}{1 - \left(\frac{v}{u}\right)^2}} = \sqrt{1 - \left(\frac{v}{u}\right)^2}$$

Since u is larger than v, the right side is smaller than 1. This means that $t_\perp$ is smaller than $t_\parallel$; the cross-current trip requires a shorter time than the one made with and against the current.

Figure S6-3. The Michelson-Morley experiment.

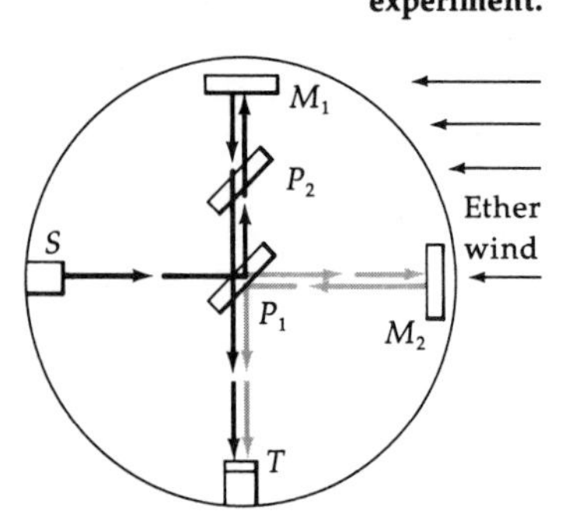

THE MICHELSON-MORLEY EXPERIMENT

The Michelson-Morley experiment is analogous to the situation we have just described, with the boat replaced by a beam of light and the river replaced by the ether wind. The experimental arrangement, shown in Figure S6-3, was mounted on a solid marble slab about 3 meters in diameter floating on a pool of mercury to allow easy rotation and eliminate any problems due to building vibrations, and so forth.

A beam of light from the source S is split into two perpendicular segments by the plate P_1, which is half-silvered so that part of the incident light is reflected and part is allowed to pass through. The reflected part travels to the mirror M_1 where it is reflected again back to plate P_1, which also splits this beam into two parts, one of which passes through to the telescope T (the reflected part is unimportant). The transmitted portion of the original beam strikes mirror M_2, equidistant with M_1 from P_1, and travels back to P_1, where part is reflected down to the telescope T. Thus the two light beams received at T will have traveled the same distance* but along paths that are perpendicular to each other for a part of the trip.

Suppose that the ether wind were from right to left, as indicated. Beginning at plate P_1, where the beam is originally split, and following the paths of the halves, we see that the beam reflected by mirror M_1 always travels perpendicularly to the ether wind, whereas the beam striking M_2 travels parallelly to the wind for awhile. These two beams of light create an *interference pattern*, similar to the way that water waves created by pebbles dropped in a pond interfere with one another (see Chapter 17). If there were no ether wind, the interference pattern would not change as the slab is rotated. The presence of an ether wind would show up as a change in the interference pattern for different positions of the slab, corresponding to variations in the time the light travels along the two paths.

When Michelson and Morley performed this experiment, they observed no difference whatsoever in the travel times of the two beams of light. They repeated the experiment with the slab rotated to various positions, again with null results. They did it in different locations and at different times of the year, always obtaining the same answer: there is no ether wind. The ether simply does not exist.

*Plate P_2 was introduced to make the portion of the paths going through glass equal for both beams. This makes the two paths as nearly identical as possible.

INTERVALS IN SPACETIME

INTRODUCTION

In our discussion of Einstein's special relativity in Chapter 6, we saw that observations of nature are not as absolute as they may seem. For example, clock readings keep track of "proper time" but are of little help in measuring time for other frames of reference. Length measurements are subject to the same limitations. And the question of the simultaneity of two events cannot have the same answer for all observers.

In essence, the concept established by special relativity that was not realized before is this: measurements of space and time depend on the relative motions of observers. Thus these measurements are subjective. Part of the business of science is to distinguish those quantities that are *objective* physical facts, independent of any observer, from those *subjective* measurements that depend on who is making the observation. We have completed part of this task by noting those things that *are* observer related, such as time and the sequence of events. Now the question arises: Does anything remain that *is* an objective measurement? Can we measure a "distance" or "time" or a combination of the two that is the same for all observers, regardless of their uniform motion relative to one another or to the events in question? Well, it turns out that there is such a quantity, called the *interval in spacetime*. Let's see what these words mean.

SPACETIME

The rationale for the change in phraseology from the old "space and time" to the hybrid term *spacetime* is a thread that will run through this entire discussion. The fundamental point about the special theory of relativity is that space and time are not isolated from each other, as our usual slow-velocity experience would lead us to believe. We must no longer speak of events separated in "space" in the usual three-dimensional sense, or separated in "time" in the ordinary cosmic meaning of the word, because two or more observers moving in relation to each other will disagree about these time and space separations. Rather, in the new terminology, three-dimensional "space" is joined with "time" as the fourth dimension to create "spacetime" as the continuum in which all events take place.

Suppose we take the meter as the unit for the three dimensions of ordinary space. We may convert time in seconds to this same system of units by multiplying it by a conversion factor whose units are meters per second; that is,

$$\text{seconds} \times (\text{meters/second}) = \text{meters}$$

This gives us time in the same units (meters) as distance. We choose this conversion factor to be the speed of light c, which is in meters per second. This is a convenient choice because the speed of light is the same for all observers and because it serves as a "speed limit" in the universe. In general, if the time is t, then the product ct gives the distance that light travels in this time, called the *light-travel time.* In this way, all four dimensions of spacetime are in the same units: meters.

Although we have now expressed time and space in the same units by way of the conversion factor c, we cannot say that time and distance are identical. They are not interchangeable in the way that the three dimensions of ordinary space are. For example, the height of a rectangular box becomes its length if we turn the box on its side. But no amount of "turning" will change a clock into a yardstick or vice versa, as would be the case if time and space were indistinguishable. This will become more evident as we proceed with a discussion of intervals in spacetime.

THE INTERVAL

We come now to the quantity used to describe an event in spacetime. This is called the *interval,* and it is determined in the following way. Suppose that you observe two events that take place a distance x meters apart and are separated in time by t seconds. The distance that light would travel in this time—that is, the light-travel time—is ct. Take x^2 and $(ct)^2$, subtract the smaller from the larger, and the result is the square of the interval. That is,

$$(ct)^2 - x^2 = (\text{interval})^2$$

if the time part dominates, or

$$x^2 - (ct)^2 = (\text{interval})^2$$

if the distance part dominates.

Now, suppose that your friend, who is moving relative to you, observes the same two events, for which he or she measures the space separation to be x' and the time separation t'. Then, if your friend makes the analogous calculations, he or she gets the *same answer* for the interval. For example, if the time part is larger,

$$(ct')^2 - (x')^2 = (\text{interval})^2$$

and the right-hand side will be the same number as in the first equation in this section. This is the important characteristic of the interval: *it is the same for all observers for a given pair of events.* It thus represents a genuine, objective and physical relationship between the two events, even though the time and the space measurements, separately, do not.

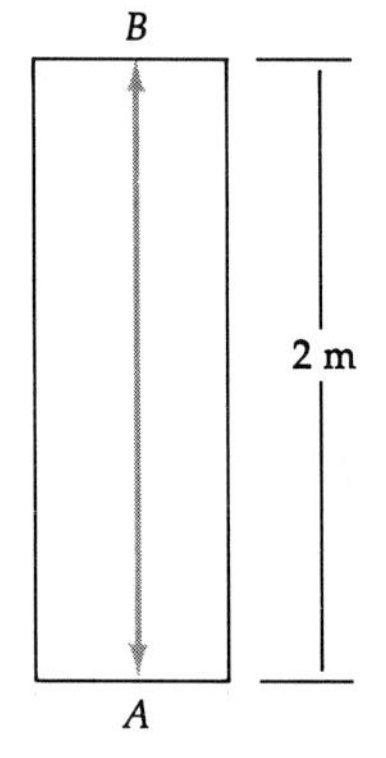

Figure S7-1. Stationary light pulse clock.

Perhaps a geometrical illustration will help you to visualize what the interval means. Let us return to the case of the two light-pulse clocks, one on board the rocket and one on the ground (Figure S7–1). Suppose for the sake of argument that the length of the earth clock (the distance AB) is 2 m. Let us define our two events as (1) the leaving of the light pulse from mirror A and (2) the reception of the pulse when it returns to mirror A. What will these events look like from the two frames of reference, the earth and the rocket?

First, put yourself in the position of someone stationary with respect to the earth. Since the light pulse leaves from and returns to mirror A, the two events occur at the same place. Their separation in space is zero meters. What about the time separation? Well, it is the time required for the light to travel from the bottom to the top and back to the bottom, a total distance of 4 meters. Therefore, the time separation in light-travel time is 4 meters. In the earth frame, then, we have

$$\text{distance separation: } x = 0 \text{ m}$$

$$\text{time separation: } ct = 4 \text{ m}$$

According to our formula for calculating the square of the interval, we need to square these two numbers and subtract the smaller from the larger:

$$\begin{aligned}(\text{interval})^2 &= (ct)^2 - x^2\\ &= (4 \text{ m})^2 - (0 \text{ m})^2\\ &= 16 \text{ (m)}^2\end{aligned}$$

Taking the square root gives the interval

$$(\text{interval}) = 4 \text{ m}$$

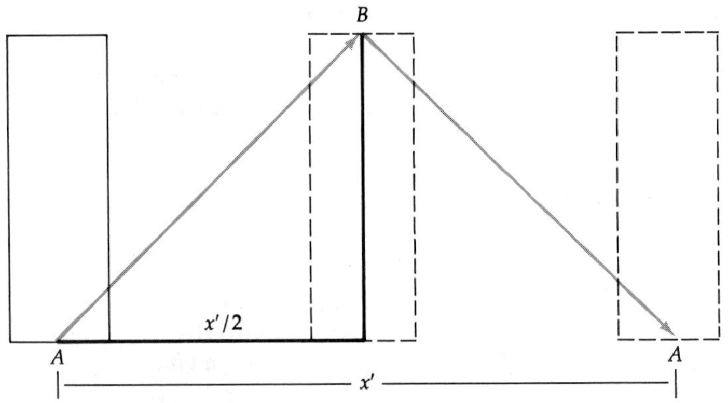

Figure S7-2. Moving light pulse clock.

Now, how does the astronaut see these events? He or she notices that, as the earth whizzes by below at speed v, the light travels along the zigzag path in Figure S7–2. For him, the mirror A moves to a new location while the light pulse is making its journey to the top mirror and back. The two events take place at different locations in space, separated by a distance we shall call x'. The time separation in light-travel time is the length of the path A to B to A. This is twice the hypotenuse of the right triangle shown in the figure. From the Pythagorean theorem,

$$AB = \sqrt{\left(\frac{x'}{2}\right)^2 + 2^2}\ \text{m}$$

So, we have

$$\text{distance separation} = x'\ \text{m}$$

$$\text{time separation} = ct' = 2\sqrt{\left(\frac{x'}{2}\right)^2 + 4}\ \text{m}$$

Calculating the interval for this frame,

$$\begin{aligned}(\text{interval})^2 &= (ct')^2 - (x')^2\\ &= 4\left[\left(\frac{x'}{2}\right)^2 + 4\right] - (x')^2\\ &= 4\left[\frac{(x')^2}{4}\right] + 16 - (x')^2\\ &= 16\ (\text{m})^2\end{aligned}$$

The interval is then 4 m.

This is the same answer we got for the earth frame of reference. So, even though the individual time and distance separations in the two frames are quite *different* for the events, the interval is the *same.*

SPACE-LIKE INTERVALS AND TIME-LIKE INTERVALS

We saw earlier that although space and time are related, they are not identical. The distinction can be pushed even further by looking more carefully at events. For any pair of events in the universe we can say one of two things. It is either possible for a body to be present at both events, or it is not, owing to the limit that the speed of light places on travel speeds for objects. Let's look at these two possibilities from the point of view of determining the interval.

Suppose that in calculating the squares of the space between events and the distance in light-travel time we find that the space-separation quantity (x) is larger than the time-separation quantity (ct). This means that the *spatial distance* between the events is *greater* than the distance that light could travel in the time between them. Therefore, it is physically impossible for one observer to be present at both events, because, in order to do so, he or she would have to travel faster than light. In this instance we have what is called a *space-like interval,* because it is always possible to move with an appropriate speed such that the time separation between the events is reduced to zero. The events then become simultaneous as viewed from that particular spacetime frame of reference.

For example, suppose that lunch is served on the earth at noon and dessert is served on the moon at exactly 1 second past noon. The moon is some 384,000 kilometers from earth, so a beam of light takes almost 1.3 seconds to travel between them. Thus, it is physically impossible to be present at the noon lunch on earth and also be on time for dessert on the moon 1 second later, because this would require traveling faster than light. But by moving with some particular speed we can reduce the time separation to zero; the interval is then space-like.

In contrast, suppose that the time-separation quantity is the larger of the two. This means that the spatial distance between the events is *smaller* than the distance in light-travel time, making it physically possible for one observer to be present at both events. In this case the interval is called *time-like,* because it is always possible to move in such a way as to make the separation in space between the events zero. This sort of interval is what we observe in the everyday slow-speed world. For example, it is a simple matter to attend a football game on Monday night in Dallas and another on Tuesday night in Houston simply by leaving from Dallas after the first game and driving to Houston. The distance between the cities is much less than the distance light would travel in 24 hours. By choosing our car as the frame of reference (and ignoring such trivia as walking to the stadium from the parking lot), we can take the view that both games occur in the same place separated by the 24-hour time span. Hence, the interval is time-like. In our light-pulse clock example given earlier, the interval of 4 meters is time-like, since it is possible to choose a reference frame (in this case, the earth) such that the separation in space between the two events is zero.

The notion of a time-like interval for two events is a comfortable one for us, because we are used to it. Time-like intervals are a part of our ordinary experience, and we have no trouble visualizing them. For example, if the stewardess on an airplane serves you a cocktail and gives you dinner a half-hour later, a person standing on the ground sees these events as separated in space by several hundred kilometers. But to you, riding on the plane, the two events (drinking the cocktail and eating dinner) occur at the same location (your seat). In your frame of reference, the separation in space is zero.

Space-like intervals, however, seem quite strange to us. This is because the speed of light is so great compared with "ordinary" speeds, we have a hard time visualizing the very small time differences that arise. Again, consider the situation in the airplane. Suppose that two stewardesses standing at opposite ends of the cabin light their cigarettes simultaneously from the point of view of a passenger. For our eagle-eyed observer on the ground, these events take place at different times. We can find this time separation quite simply. If the stewardesses are separated by 30 meters, then the time required for light to travel this distance is 30 meters divided by c, or 0.0000001 second. This small time interval is not readily apparent to our senses.

A GEOMETRICAL VIEW OF INTERVALS

We may generalize the notion of an interval by using a geometrical picture. In Figure S7–3 the length of the line XY represents the distance in light-travel time between two events as observed from a particular frame of reference. The length of the line XZ represents the distance in space between the same two events as measured from the same frame. We join the X-ends of the lines together and orient them so that a line drawn from Y to Z is perpendicular to XZ, shown in the lower part of the figure. Then, according to the Pythagorean theorem

$$(XY)^2 = (XZ)^2 + (YZ)^2$$

or

$$(YZ)^2 = (XY)^2 - (XZ)^2$$

Figure S7-3. The length of *XY* is light-travel time; the length of *XZ* is separation in space.

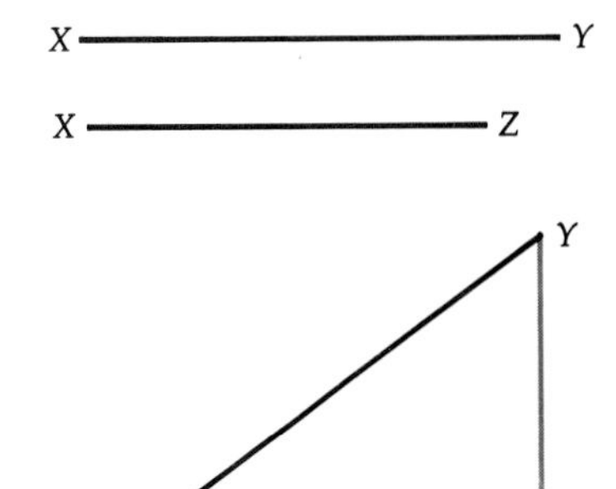

Figure S7-4. Different combinations of time (XY) and distance (XZ) give the same interval (YZ).

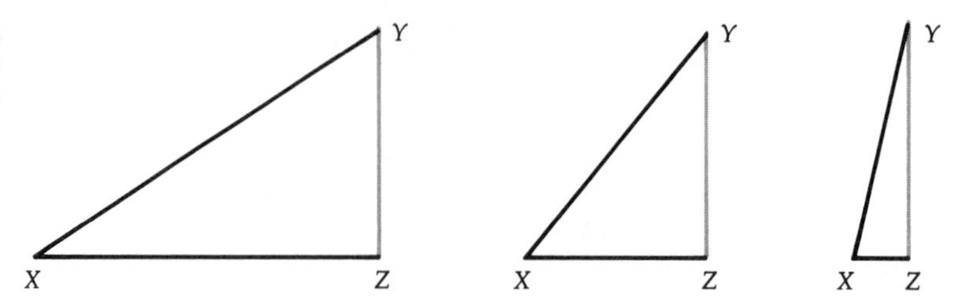

Here YZ fits our definition of an interval, since its square is the difference of the squares of the spatial distance and the distance in light-travel time.

Note that there are an infinite number of distance and time separations that are compatible with a given interval. Three such combinations are shown in Figure S7–4. Each of these represents the observation of the same two events seen from different frames of reference moving at different speeds. The separations in time (XY) and the separations in distance (XZ) are different as observed from each frame, but the interval (YZ) is the same for all.

Figure S7-5. A space-like interval.

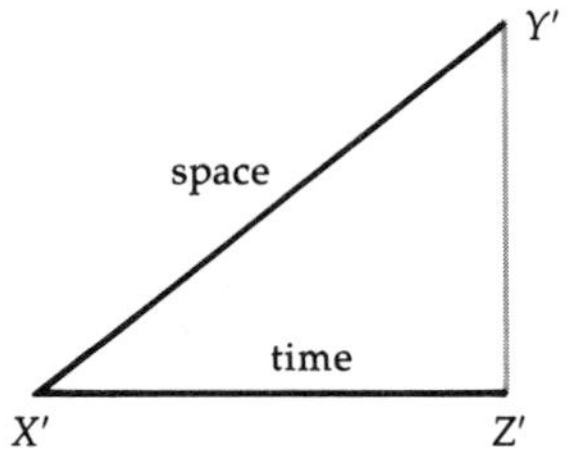

These illustrations all contain *time-like* intervals. It is physically possible to be present at both events because the spatial separation between them is smaller than the distance light could travel in the intervening time. For *space-like* intervals, the spatial distance is larger than the distance in light-travel time, making it impossible to be present at both events (see Figure S7–5). Here, the roles of the spatial and time separations are reversed. Now the hypotenuse ($X'Y'$) of the triangle represents the distance in space between the two events, larger than the distance light can travel in the time between ($X'Z'$).

SUPPLEMENT 8

DEVIATIONS FROM THE IDEAL GAS LAW

The equation for the ideal gas law given in Chapter 13 is

$$PV = nkT$$

where P, V, and T are the pressure, volume, and absolute temperature, respectively. The number of molecules of gas is given by n, and k is a proportionality constant. This equation is based on the assumptions that the molecules occupy no volume of their own (that is, they are point particles) and that they do not interact with one another.

How good is this equation? That is, how well does the ideal gas law represent the behavior of *real* gases? It turns out that, at normal temperatures and pressures, air, which is mostly nitrogen, obeys this equation very well, even though the distance a given nitrogen molecule travels under these conditions is only about 0.00001 centimeter. At lower temperatures, however, the behavior of nitrogen and other gases deviates quite a bit from the ideal. For example, we know that there are attractive forces between the molecules, because the gas will liquify if the temperature is low enough or if the pressure is high enough. Nitrogen becomes a liquid at about −195° C.

We can get a feeling for the validity of the ideal gas equation by plotting graphs of pressure against volume at several temperatures. Such plots are shown in Figure S8-1. At temperature T_1, the gas strictly obeys the ideal gas law. At the lower temperatures T_2 and T_3, there are humps in the curve, which indicate how far the behavior of the gas deviates from the ideal.

The question then arises: Can we make some adjustments to the ideal gas law so that it more nearly represents the real world? It turns out that this has been done, and we are going to take a brief look at the reasoning.

Figure S8-1. Deviations from the ideal gas law.

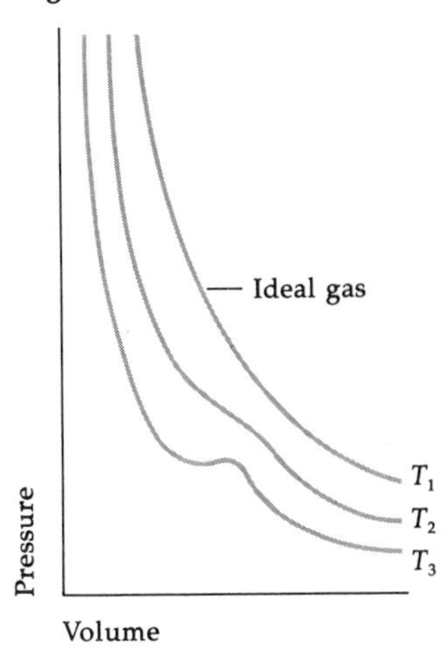

First, consider volume. Of course it is not strictly true that molecules have no volume. However, if the volume in which they can move around is quite large, then we are not far wrong in making this assumption. But as the temperature decreases, so does the volume available to the molecules. Thus, their own volume becomes an appreciable fraction of the volume of their container. Look at it this way. If you are alone in Grand Central Station, you can move around at will, and the volume of your body is not very significant. Compare this to being in a telephone booth. Here your freedom of motion is severely restricted because you are nearly as large as the phone booth. In the ideal gas equation, we might assign the letter b to represent the volume occupied by the n molecules. Then the volume factor would read $(V - b)$.

Now what about the pressure? Increasing the pressure on the molecules pushes them closer together, and the attractive forces between them become more significant. Put another way, the mutual attraction of the molecules actually *helps* the external pressure, so we need to add a term to the pressure factor. This term should become larger as the volume increases. It turns out that we can replace P by the factor $(P + a/V^2)$. The equation then reads

$$\left[P + \frac{a}{V^2}\right](V - b) = nkT$$

The numbers a and b are called *empirical constants.* This means that their values are not determined from the theory, but from fitting the equation to the experimental curves such as those in Figure S8–1. This equation is called *van der Waals equation,* after J. D. van der Waals, who developed it.

SUPPLEMENT 9

MORE ABOUT THE HYDROGEN ATOM

In this supplement, we are going to derive the equation (given on page 400 in Chapter 22) that specifies hydrogen atom energy levels. Let us imagine that an electron and a proton are infinitely far apart. We define the potential energy of the electron to be zero at this point. Because of the electrical attraction, the electron begins to "fall" toward the proton, losing potential energy along the way. For any separation r between the proton and the electron, the potential energy is

$$E_p = \frac{-ke^2}{r} \tag{1}$$

where e is the charge in coulombs (C) and k is the Coulomb law constant, 9×10^9 Nm^2/C^2. The electron also has kinetic energy because of its speed, and this is given by

$$E_k = \left(\frac{1}{2}\right) mv^2$$

So, its total energy is $E = E_p + E_k$, or

$$E = \left(\frac{1}{2}\right) mv^2 - \frac{ke^2}{r} \tag{2}$$

We can simplify Eq. (2) by getting rid of the term containing the speed v. Combining Newton's second law and Coulomb's law will do this for us. If the proton exerted no force, the electron would zoom right on past. But the electrostatic attraction pulls it into an orbit around the proton. This force is

$$F = ma$$
$$= \frac{mv^2}{r} \tag{3}$$

since, as you recall, v^2/r is the centripetal acceleration. From Coulomb's law, the force F is

$$F = \frac{ke^2}{r^2} \tag{4}$$

Equating the right sides of Eqs. (3) and (4) gives

$$\frac{ke^2}{r^2} = \frac{mv^2}{r}$$

We can solve this for mv^2 to give

$$mv^2 = \frac{ke^2}{r} \tag{5}$$

Now we substitute this into Eq. (2) to get an expression for the total energy,

$$E = \left(\frac{1}{2}\right)\left(\frac{ke^2}{r}\right) - \frac{ke^2}{r}$$
$$E = -\left(\frac{1}{2}\right)\frac{ke^2}{r} \tag{6}$$

It is at this point that Bohr's quantization enters the picture. He assumed (Chapter 22) that the angular momentum is a whole number multiple of $h/2\pi$:

$$mvr = n\frac{h}{2\pi} \qquad (n = 0, 1, 2, \ldots)$$

This gives us a relationship between r and v,

$$r = \frac{nh}{2\pi mv} \tag{7}$$

A second relationship between r and v comes from Eq. (5):

$$r = \frac{ke^2}{mv^2} \tag{8}$$

Setting the right sides of Eqs. (7) and (8) equal gives

$$v = \frac{2\pi ke^2}{nh} \tag{9}$$

Putting this into Eq. (8) yields

$$r = n^2 \times \frac{h^2}{4\pi^2 kme^2} \tag{10}$$

This allows us to write E in Eq. (6) in terms of the quantum number n:

$$E = \frac{1}{n^2} \times \frac{4\pi^2 k^2 e^4 m}{h^2}$$

or

$$E = \frac{-R}{n^2} \tag{11}$$

where R, called the *Rydberg constant,* is a composite of several constants,

$$R = \frac{4\pi^2 k^2 e^4 m}{h^2}$$

Putting in the numbers gives a numerical value of $R = 2.18 \times 10^{-18}$ J. Thus, $R/h = (2.18 \times 10^{-18}\ \text{J})/(6.62 \times 10^{-34}\ \text{J} \cdot \text{s})$, or

$$\frac{R}{h} = 3.289 \times 10^{15}\ \text{s}^{-1}$$

which is the factor quoted in Chapter 22 (page 401).

SUPPLEMENT 10

EXCHANGE FORCES AND PARTICLES

In Chapter 26 we pointed out that whenever the exchange of a particle generates a force between two objects, the mass of the exchange particle is inversely proportional to the range over which the force is effective. This conclusion arises from the Heisenberg uncertainty principle. Suppose we call the uncertainty in the particle's position Δx. Because the particle must be located within the range of the force during the time the force is "turned on," then Δx also corresponds to the range of the force.

The other quantity in the uncertainty principle, the change in momentum, is related to energy in the following way. The energy is given in terms of the momentum as

$$E = \frac{p^2}{2m_{ex}}$$

where m_{ex} is the mass of the exchange particle. Thus, the energy uncertainty is related to that of the momentum by

$$\Delta E = \frac{2p\,\Delta p}{2m_{ex}} = \frac{p\,\Delta p}{m_{ex}}$$

or, since $p = m_{ex}v$,

$$\Delta E = v\,\Delta p$$

Because v is the velocity of the exchange particle, it will be given by $\Delta x/\Delta t$—that is, the uncertainty in position divided by the time interval, Δt, during which the force operates. So we have

$$\Delta E = \frac{\Delta x}{\Delta t}\,\Delta p$$

or

$$\Delta E\,\Delta t = \Delta x\,\Delta p \geq \frac{h}{2\pi}$$

where the last inequality is, of course, the uncertainty principle. So, our conclusion is the uncertainty principle for energy and time:

$$\Delta E\,\Delta t \geq \frac{h}{2\pi}$$

We can use this expression to estimate the mass of the pi-meson, exchanged in the nuclear strong force. Recall that this force operates over a range of about 10^{-15} meters. If we suppose, as is reasonable, that the pi-meson travels at two-thirds the speed of light, its speed is

$$v = \left(\frac{2}{3}\right)c = 2 \times 10^{8}\ \text{m/s}$$

And, since $v = \Delta x/\Delta t$, we have that $\Delta t = \Delta x/v$, or

$$\begin{aligned}\Delta t &= \frac{10^{-15}\ \text{m}}{2 \times 10^{8}\ \text{m/s}} \\ &= 5 \times 10^{-24}\ \text{s}\end{aligned}$$

As discussed in Chapter 26, we can interpret this as the length of time nature will allow the law of conservation of energy to be violated. The uncertainty principle specifies what sort of energy discrepancy is allowed; that is,

$$\begin{aligned}\Delta E &= \frac{\dfrac{h}{2\pi}}{\Delta t} = \frac{6.6 \times 10^{-35}\ \text{J}\cdot\text{s}}{2\pi \times 5 \times 10^{-24}\ \text{s}} \\ &= 2 \times 10^{-11}\ \text{J}\end{aligned}$$

We have become accustomed to mass in electron-volt units; conversion gives a value of 131.09 MeV for the mass of the pi-meson. The observed value is 139.57 MeV for the neutral species.

INDEX

S

T

U

V

W

X

Y